Lecture Notes in Computer Science 12116

More information about this series at http://www.springer.com/series/7412

Xiang Bai · Dimosthenis Karatzas ·
Daniel Lopresti (Eds.)

Document Analysis Systems

14th IAPR International Workshop, DAS 2020
Wuhan, China, July 26–29, 2020
Proceedings

 Springer

Editors
Xiang Bai 🆔
Huazhong University of Science
and Technology
Wuhan, China

Dimosthenis Karatzas 🆔
Autonomous University of Barcelona
Barcelona, Spain

Daniel Lopresti 🆔
Lehigh University
Bethlehem, PA, USA

ISSN 0302-9743   ISSN 1611-3349 (electronic)
Lecture Notes in Computer Science
ISBN 978-3-030-57057-6   ISBN 978-3-030-57058-3 (eBook)
https://doi.org/10.1007/978-3-030-57058-3

LNCS Sublibrary: SL6 – Image Processing, Computer Vision, Pattern Recognition, and Graphics

This Springer imprint is published by the registered company Springer Nature Switzerland AG
The registered company address is: Gewerbestrasse 11, 6330 Cham, Switzerland

Preface

Welcome to the 14th IAPR International Workshop on Document Analysis Systems (DAS 2020). For the first time in our long history, DAS was held virtually. The workshop was originally set to take place in Wuhan, China, in May 2020. However, given the worldwide pandemic, we decided to host DAS during July 26–29, 2020, and converted the workshop into a fully virtual event.

As a result, instead of welcoming you to Wuhan, we brought Wuhan to you. While the organization of DAS 2020 was still based in Wuhan, the workshop was not confined to a specific location. In a sense, this was the first truly worldwide edition of DAS, taking place around the world in a coordinated fashion, employing a schedule we designed to support participation across a wide range of time zones. Of course, this comes with some challenges, but also with interesting opportunities that caused us to rethink how to foster social and scientific interaction in this new medium. It also allowed us to organize an environmentally friendly event, to extend the reach of the workshop, and to facilitate participation literally from anywhere to those with an interest in our field and an internet connection. We truly hope we managed to make the most out of a difficult situation.

DAS 2020 continued the long tradition of bringing together researchers, academics, and practitioners from all over the world in the research field of Document Analysis Systems. In doing so, we build on the previous workshops held over the years in Kaiserslautern, Germany (1994); Malvern, PA, USA (1996); Nagano, Japan (1998); Rio de Janeiro, Brazil (2000); Princeton, NJ, USA (2002); Florence, Italy (2004); Nelson, New Zealand (2006); Nara, Japan (2008); Boston, MA, USA (2010); Gold Coast, Australia (2012); Loire Valley Tours, France (2014); Santorini, Greece (2016); and Wien, Austria (2018).

As with previous editions, DAS 2020 was a rigorously peer reviewed and 100% participation single-track workshop focusing on system-level issues and approaches in document analysis and recognition. The workshop comprises presentations by invited speakers, oral and poster sessions, a pre-workshop tutorial, as well as the distinctive DAS discussion groups.

We received 64 submissions in total, 57 of which in the regular paper track and 7 in the short paper track. All regular paper submissions underwent a rigorous single-blind review process where the vast majority of papers received three reviews from the 62 members of the Program Committee, judging the originality of work, the relevance to document analysis systems, the quality of the research or analysis, and the presentation. Of the 57 regular submissions received, 40 were accepted for presentation at the workshop (70%). Of these, 24 papers were designated for oral presentation (42%) and 22 for poster presentation (38%). All short paper submissions were reviewed by at least two of the program co-chairs. Of the 7 short papers received, 6 were accepted for poster presentation at the workshop (85%). The accepted regular papers are published in this proceedings volume in the *Springer Lecture Notes in Computer Science* series. Short papers appear in PDF form on the DAS workshop website.

The final program includes six oral sessions, two poster sessions, and the discussion group sessions. There were also two awards announced at the conclusion of the workshop: the IAPR Best Student Paper Award and IAPR Nakano Best Paper Award. We offer our deepest thanks to all who contributed their time and effort to make DAS 2020 a first-rate event for the community.

In addition to the contributed papers, the program included three invited keynote presentations by distinguished members of the research community: Tong Sun, who leads the Document Intelligence Lab in Adobe, spoke about "The Future of Document: A New Frontier in the New Decade;" Lianwen Jin, from South China University of Technology, spoke on the topic of "Optical Character Recognition in the Deep Learning Era;" and C.V. Jawahar, from IIIT Hyderabad, shared his vision about "Document Understanding Beyond Text Recognition."

We furthermore would like to express our sincere thanks to the tutorial organizers, Zhibo Yang and Qi Zheng from Alibaba, for sharing their valuable scientific and technological insights. A special thanks is also due to our sponsors IAPR, Meituan, Hanvon Technology, Huawei Technologies, and TAL Education Group, whose support, especially during challenging times, was integral to the success of DAS 2020.

The workshop program represents the efforts of many people. We want to express our gratitude especially to the members of the Program Committee and the external reviewers for their hard work in reviewing submissions. The publicity chairs Koichi Kise (Japan), Simone Marinai (Italy), and Mohamed Cheriet (Canada) helped us in many ways, for which we are grateful. We also thank the discussion group chairs Alicia Fornés (Spain), Faisal Shafait (Germany), and Vincent Poulain d'Andecy (France) for organizing the discussion groups, and the tutorial chairs Jun Sun (China), Apostolos Antonacopoulos (UK), and Venu Govindaraju (USA) for organizing the tutorials. A special thank goes to the publication chair Yongchao Xu (China), who was responsible for the proceedings at hand. We are also grateful to the local arrangements chairs who made great efforts in arranging the program, maintaining the Webpage, and setting up the virtual meeting platform. The workshop would not have happened without the great support from the hosting organization, Huazhong University of Science and Technology.

Finally, the workshop would not be possible without the excellent papers contributed by authors. We thank all the authors for their contributions and their participation in DAS 2020! We hope that this program will further stimulate research and provide practitioners with better techniques, algorithms, and tools for the deployment. We feel honored and privileged to share the best recent developments in the field of Document Analysis Systems with you in these proceedings.

July 2020

Cheng-Lin Liu
Shijian Lu
Jean-Marc Ogier
Xiang Bai
Dimosthenis Karatzas
Daniel Lopresti

Organization

Organizing Committee

General Chairs

Cheng-Lin Liu	Institute of Automation of Chinese Academy of Sciences, China
Shijian Lu	Nanyang Technological University, Singapore
Jean-Marc Ogier	University of La Rochelle, France

Program Chairs

Xiang Bai	Huazhong University of Science and Technology, China
Dimosthenis Karatzas	Universitat Autònoma de Barcelona, Spain
Daniel Lopresti	Lehigh University, USA

Program Committee

Alireza Alaei	Southern Cross University, Australia
Adel Alimi	University of Sfax, Tunisia
Apostolos Antonacopoulos	University of Salford, UK
Xiang Bai	Huazhong University of Science and Technology, China
Abdel Belaid	Université de Lorraine, LORIA, France
Jean-Christophe Burie	University of La Rochelle, France
Vincent Christlein	University of Erlangen-Nuremberg, Germany
Andreas Dengel	German Research Center for Artificial Intelligence, Germany
Markus Diem	Vienna University of Technology, Austria
Antoine Doucet	University of La Rochelle, France
Véronique Eglin	LIRIS-INSA de Lyon, France
Jihad El-Sana	Ben-Guion University of the Negev, Israel
Gernot Fink	TU Dortmund University, Germany
Andreas Fischer	University of Fribourg, Switzerland
Alicia Fornés	Universitat Autònoma de Barcelona, Spain
Volkmar Frinken	University of California, Davis, USA
Utpal Garain	Indian Statistical Institute, Kolkata, India
Basilis Gatos	National Centre of Scientific Research Demokritos, Greece
Lluis Gomez	Universitat Autònoma de Barcelona, Spain
Dafang He	Pinterest, USA
Masakazu Iwamura	Osaka Prefecture University, Japan

Sponsors

Meituan, Hanvon, Huawei, TAL, and IAPR

Logos

Organization

Sponsors

Meituan, Samsung, Hanvon, TAL, and IAPR

Logos

Contents

Segmentation and Layout Analysis

Word Embedding and Spotting

Font Design and Classification

Character and Text Recognition

Maximum Entropy Regularization and Chinese Text Recognition

Changxu Cheng$^{(\boxtimes)}$, Wuheng Xu, Xiang Bai, Bin Feng, and Wenyu Liu

Huazhong University of Science and Technology, Wuhan, China
{cxcheng,xwheng,xbai,fengbin,liuwy}@hust.edu.cn

Abstract. Chinese text recognition is more challenging than Latin text due to the large amount of fine-grained Chinese characters and the great imbalance over classes, which causes a serious overfitting problem. We propose to apply Maximum Entropy Regularization to regularize the training process, which is to simply add a negative entropy term to the canonical cross-entropy loss without any additional parameters and modification of a model. We theoretically give the convergence probability distribution and analyze how the regularization influence the learning process. Experiments on Chinese character recognition, Chinese text line recognition and fine-grained image classification achieve consistent improvement, proving that the regularization is beneficial to generalization and robustness of a recognition model.

Keywords: Regularization · Entropy · Chinese text recognition

1 Introduction

Text recognition is a popular topic in the deep learning community. Most of the existing deep learning-based works [13,15,20,22–24] pay attention to Latin script and achieve good performance.

However, Chinese text differs much from Latin text. There are thousands of common Chinese characters appearing as various styles. Chinese text recognition can be regarded as a kind of fine-grained image classification due to the high inter-class similarity and the large intra-class variance, as shown in Fig. 1(a). Besides, there is usually a great data imbalance over character classes [32,33]. These features cause a large demand for training data. Thus character-based recognition model is prone to overfit. Radical-based methods [16,29,30,35] convert Chinese characters to radicals as the basic class to simplify the character structure and decrease the number of classes, thus reducing the demand for training data. They are mentioned here to prove the large demand for data if we use character-based recognition. Nevertheless, they are not flexible enough for various circumstances, e.g., handwritten text [37]. And current methods in that way cost much time in practice due to the use of RNN-based decoder.

A common practice to deal with the overfitting problem is to regularize the model during training. There are several techniques aiming at this, including

---

C. Cheng and W. Xu—Equal contribution.

X. Bai et al. (Eds.): DAS 2020, LNCS 12116, pp. 3–17, 2020.
https://doi.org/10.1007/978-3-030-57058-3_1

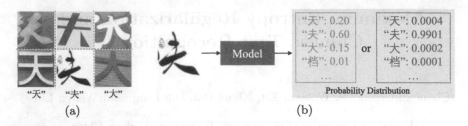

Fig. 1. Example of Chinese characters and 2 kinds of model prediction. (a) The 2 images in each column are a same character class and the 3 in each row are different. It shows the high inter-class similarity and the large intra-class variance which is a fine-grained attribute. (b) Although the 2 different probability distributions have a same prediction in training set, the left one holds a higher entropy that can describe the learned feature better. Obviously, "天" is far more similar to "夫" than "档", so the confidences on them is supposed to have a large distinction.

dropout [25], L2 regularization, batch normalization [10]. They act on model parameters or hidden units like a blackbox. We consider regularization from the perspective of entropy. The predicted probability distribution are an indication of how the network generalizes [19]. In Chinese character recognition, we hope that a similar negative class is assigned a larger probability than the dissimilar one given an input image. It requires the probability of positive class to be not that large to leave probability space for other negative classes, which causes a big entropy. This is okay because we can recognize correctly as long as the probability of the positive class is the largest. The maximum entropy principle [12] also points out that the model with the largest entropy of output distribution can represent features best. Figure 1(b) illustrates the idea by comparing the 2 probability distributions with high and low entropy respectively. Hence we adopt Maximum Entropy Regularization [19] to regularize the training process.

In this paper, we perform an in-depth analysis of Maximum Entropy Regularization theoretically. The cross-entropy loss and the negative entropy term behave like two 1-dimensional forces which function on the output probability distribution. The elegant gradient function illustrates the regularized behaviour from the perspective of backward propagation. Under an assumption similar to label smoothing, we formulate the convergence probability distribution in training set, which is exactly a relationship between the convergence probability of positive class and the coefficience.

We conduct experiments on Chinese character recognition, Chinese text line recognition and fine-grained image classification and gain consistent improvement. In addition, we find that model trained with MER can attend on more compact and discriminative regions and filter much noisy area. MER also makes model more robust when label corruption is exerted to our training data.

2 Related Works

Our work focuses on model regularization and its application to Chinese text recognition. Here we briefly review some recent works about these two aspects.

2.1 Model Regularization

Large deep neural network is prone to overfit in many cases. To relieve the problem, model regularization is commonly used during training. Dropout [25] is to randomly drop some neurons or connections with a certain probability in layers. L2 regularization is also called weight decay, which restricts the magnitude of model weights. Batch normalization [10] normalizes hidden units in a training batch to reduce internal covariate shift. These methods act on model parameters or hidden layers, which is hard to control and not intuitive. Mixup [34] simply uses linear operation on both input images and their labels with an assumption that linear interpolations of features should lead to linear interpolations of the associated targets.

Recently, output distribution of neural network has earned much attention. Knowledge distillation [8] is proposed to train a small-size model to have a similar output distribution to a large model since the "soft targets" can transfer the generalization ability, which indicates the effect of output distribution. Label smoothing [27] is proposed to encourage the prediction to be less confident by disturbing the one-hot ground truth label with a uniform distribution, which actually adds a KL-divergence (between the uniform and the output distribution) term to the cross-entropy loss in the view of loss function. Label smoothing is able to improve generalization and model calibration, and benefit distillation [17]. Softmax loss [1] prompts the summation of top-k output probabilities to be as great as possible to alleviate the extreme confidence caused by cross-entropy loss. But the hyperparameter k is hard to choose. Bootstrapping [21] aims at leveraging ground truth distribution and output distribution as expected distribution, thus the model can train well on dataset with noisy label [3]. But the output entropy is still as low as that with cross-entropy loss. According to maximum entropy principle [12], the model whose probability distribution represents the current state of knowledge best is the one with the largest entropy. Correspondingly, a maximum-entropy based method, called confidence penalty [19], includes a negative entropy term in loss function, which acts similarly to label smoothing but performs better. As a result, it has attracted researcher's interest in applying it to multiple tasks, e.g., sequence modeling [14], named entity recognition [31] and fine-grained image classification [2]. CTC [5]-based sequence model tries to penalizes peaky distributions by using maximum-entropy based regularization [14]. However, none of these works make a deep analysis on the regularization term or the whole loss function theoretically or experimentally.

2.2 Chinese Text Recognition

Chinese text recognition is more challenging than Latin script due to the more character categories and the more complicate layout. There are usually two tasks derived from Chinese text recognition: Chinese character recognition (CCR) and Chinese text line recognition (CLR).

As for Chinese character, recent methods can be divided into two streams: character-based CCR (CCCR) [39] and radical-based CCR (RCCR) [35]. Taken as a single class, every Chinese character is well classified with deep learning [36]. However, CCCR has no capability to handle unseen characters. By considering the structure of Chinese character, RCCR methods exploit radicals to represent a character [16,29,30,35]. Multi-label learning has been used to detect radicals [29]. Radical analysis network (RAN) [35] takes the spatial structure of a single Chinese character as a radical sequence and decodes with an attention-based RNN. JSRAN [30] improves RAN by jointly using STN [11] and RAN. However, RAN is very time-consuming during inference, and RCCR is hard to tackle some nonstandard handwritten text. In this paper, we choose CCCR and use SE-ResNet-50 [9] as backbone for CCR.

As for Chinese text line, currently there are also two streams: one is Convolutional Recurrent Neural Network (CRNN) [23] with CTC [5], the other is attention-based Encoder-Decoder [24]. The vanilla version of them can only process image one-dimensionally. 2D-attention [13,15] decodes an encoded text image from two-dimensional perspective, which is an extension of the latter. In our experiments, we simply use 1-d attention-based Encoder-Decoder for CLR.

3   Analysis on Maximum Entropy Regularization

Unlike the previous works which only analyze the entropy term, we study both the term and the complete loss function to discover the joint effect.

3.1   Review of Cross-Entropy Loss

We first review the common operation in a classification problem. Given an input sample x with label y, a classification model produces C scores $\{z_i\}_{i=1}^{C}$. Then we canonically get the output probability distribution $\mathbf{p}$ by softmax function:

$$p_i = \frac{e^{z_i}}{\sum_j e^{z_j}} \tag{1}$$

The derivative of softmax is:

$$\frac{\partial p_i}{\partial z_j} = \begin{cases} p_j(1 - p_j), & i = j \\ -p_i p_j, & i \neq j \end{cases} \tag{2}$$

The cross-entropy (CE) loss and its derivative are:

$$L_{CE} = -\log p_y \tag{3}$$

$$\frac{\partial L_{CE}}{\partial z_i} = \begin{cases} p_i - 1 < 0, & i = y \\ p_i > 0, & i \neq y \end{cases} \tag{4}$$

Optimized by using gradient descent, the model prompts the probability of the y-th class to be higher and higher and that of other classes to be lower and lower, which leads to a approximation of one-hot vector in training dataset. Consequently we get confident output with low entropy that is often a symptom of overfitting [27]. In image classification, a confident model tends to focus on many regions, even including noisy background, to have sufficient clues for assertive prediction.

3.2   Maximum Entropy Regularization

What really matters is the discriminative region instead of noisy background in training set, which brings more uncertainty of depicting an image of a class. In other words, the prediction has a large entropy.

We would like to regularize the entropy of the output probability distribution $\{p_i\}_{i=1}^{C}$ to make model more general and alleviate the overfitting pain. The entropy is formulated as:

$$H\left(\mathbf{p}\right) = -\sum_{i=1}^{C} p_i \log p_i \tag{5}$$

Mathematically, the entropy $H\left(\mathbf{p}\right)$ reaches the minimum when $\mathbf{p}$ is a one-hot vector, and maximum when $\mathbf{p}$ is the uniform distribution. The former is realized automatically by the vanilla cross-entropy loss, while the latter is promising to contribute to regularizing. Hence we take the negative entropy as Maximum Entropy Regularization (MER) term which is directly imposed on the common cross-entropy loss function:

$$L_{\mathrm{MER}} = -H\left(\mathbf{p}\right) \tag{6}$$

$$L_{\mathrm{REG}} = L_{\mathrm{CE}} + \lambda L_{\mathrm{MER}} \tag{7}$$

where λ is the hyperparameter deciding the influence of MER. Intuitively, MER reduces the extreme confidence caused by cross-entropy loss. L_{CE} and L_{MER} perform like two kinds of forces that push the probability of positive class to opposite direction, as shown in Fig. 2. The subsequent part illustrates it from the perspective of gradient.

3.3   Derivative of Regularized Loss

We now consider the derivative of the regularized loss with respect to output scores $\{z_i\}_{i=1}^{C}$ which is directly related to the model, just like Eq.4.

The derivative with respect to the probability distribution is:

$$\frac{\partial L_{\mathrm{REG}}}{\partial p_i} = \begin{cases} -\frac{1}{p_i} + \lambda\left(\log p_i + 1\right), & i = y \\ \lambda\left(\log p_i + 1\right), & i \neq y \end{cases} \tag{8}$$

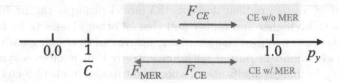

Fig. 2. Illustration of how MER term influences the convergence probability of positive class. Cross-entropy loss without MER always pushes the convergence probability to 1.0, whereas MER term pushes the probabilities to uniform distribution. They behave like two kinds of force, and the probability will finally reach to a point where the 2 forces get balanced.

According to Eq. 8, Eq. 2 and the chain rule for derivation, the derivative with respect to the score is:

$$
\frac{\partial L_{\text{REG}}}{\partial z_i} =
\begin{cases}
p_i \left(1 - \frac{1}{p_i} + \lambda \log p_i - \lambda \sum_j p_j \log p_j \right), & i = y \\
p_i \left(1 + \lambda \log p_i - \lambda \sum_j p_j \log p_j \right), & i \neq y
\end{cases}
\tag{9}
$$

By defining a cell function f:

$$
f(q) = q \left(1 + \lambda \log q - \lambda \sum_j p_j \log p_j \right)
\tag{10}
$$

we reformulate Eq. 9 as:

$$
\frac{\partial L_{\text{REG}}}{\partial z_i} =
\begin{cases}
f(p_i) - 1, & i = y \\
f(p_i), & i \neq y
\end{cases}
\tag{11}
$$

Note that Eq. 11 has the similar elegant format with Eq. 4. Differently, with $p_i \in [0, 1]$, the gradient in Eq. 11 is not always positive or negative, so the probabilities are not decreasing to 0 or increasing to 1 under more distributed scores.

3.4 Convergence Probability Distribution

Here we give the theoretical convergence probability distribution with a little strong assumption. To simplify the problem, we now only consider the probabilities instead of scores, which means the softmax operation is ignored:

$$
\min L_{\text{REG}} = -\log p_y + \lambda \sum_{i=1}^{C} p_i \log p_i
\tag{12}
$$

$$
\text{s.t.} \sum_{i=1}^{C} p_i = 1
$$

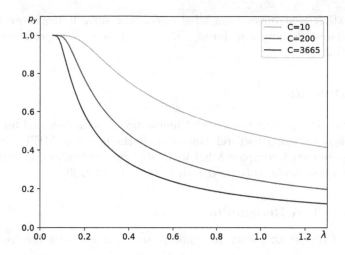

Fig. 3. The $p_y - \lambda$ curve. The convergence probability of positive class decreases as λ increases. The more classes we have, the lower probability the model converges to.

A natural idea to solve it is to convert it to an unconstrained optimization problem by lagrange multiplier:

$$\min L_{\text{lag}} = -\log p_y + \lambda \sum_{i=1}^{C} p_i \log p_i + \alpha \left(\sum_{i=1}^{C} p_i - 1 \right) \tag{13}$$

We assume that every negative class has the same position. Setting the derivatives with respect to p_i and α to 0, we get the convergence probability relation between positive class and negative class:

$$p_i = p_y e^{-\frac{1}{\lambda p_y}}, i \neq y \tag{14}$$

Furthermore, the relationship between the convergence probability and the coefficience λ can be formulated by Eq. 14 and the constrain in Eq. 12 as:

$$\lambda = \frac{m}{\log \frac{C-1}{m-1}} \tag{15}$$

where $m = \frac{1}{p_y} > 1$. Equations 14 and 15 together indicate the ideal output probability distribution when model converges for a given λ. To be more intuitive, we plot the $p_y - \lambda$ curve in Fig. 3. The convergence probability of positive class decreases monotonically with the increase of λ. When $\lambda \to +\infty$, $p_y \to \frac{1}{C}$, which is exactly a uniform distribution.

Under the assumption, MER has no difference from label smoothing [27]. They both leverage a uniform distribution to regularize, which is unrealistic for all classes. We can bridge them through the $p_y - \lambda$ curve. Also we get a deeper understanding of λ and have some guidance on how to choose a proper λ.

In fact, MER is stronger than label smoothing since it has more potential beyond the naive assumption. Former [2,19] and present experiments are all illustrating this.

4 Experiments

We make several experiments on both Chinese text recognition and fine-grained image classification using PyTorch [18] to prove the power of MER. Besides, we verify the $p_y - \lambda$ curve, compare MER with label smoothing, and investigate the effectiveness when model is trained with label corruption [3].

4.1 Chinese Text Recognition

We conduct Chinese character recognition and Chinese text line recognition respectively.

Datasets. CTW is a very large dataset of Chinese text in street view images [32]. The dataset contains 32,285 images with 1,018,402 Chinese characters from 3850 unique ones. The annotation is only in character level. The texts appear as various styles, including planar text, raised text, text under poor illumination, distant text, partially occluded text, etc.

ReCTS consists of 25,000 scene text images captured from signboards [6]. All text lines and characters are labeled. It has 440,027 Chinese characters from 4,435 unique ones and 108,924 Chinese text lines from 4,135 unique characters.

Implementation Details. For Chinese character recognition, we take it as a classification problem and use SE-ResNet50 [9] as backbone. Images are resized to the same size 32×32. The training batch size is 128. In CTW, specifically, we use both training and validation set to train with 3665 characters instead of only 1000 common characters [33]. For Chinese text line recognition, we use attention-based Encoder-Decoder as the same in ASTER [24] but without STN framework. Images are resized to 32×128. Batch size is 64.

Data augmentation is used, including changing angles in range $[-10°, 10°]$, performing perspective transformation and changing the brightness, contrast and saturation randomly. We first train a base model from scratch. Then we finetune the pretrained model with/without MER using a same training strategy to keep fair. Stochastic gradient decent (SGD) with Momentum is used for optimization and the learning rate hits a decay (from 1e−2 to 1e−5, decay rate is 0.1) if the training loss stops falling for a while. We set the weight decay as 1e−4 and momentum as 0.9. All the models are trained in one NVIDIA 1080Ti graphics card with 11 GB memory. It takes less than 12 h to reach convergence in character recognition, and about 2 days to reach convergence in text line recognition.

Table 1. Accuracy of Chinese text recognition

(a) CTW character recognition

Backbone	ResNet50 [33]	ResNet152 [33]	SE-ResNet50	
MER	✗	✗	✗	✓
Acc.	78.2	79.0	78.53	**79.24**

(b) Model accuracy with/without MER

Method	Character recognition		Text line recognition
	CTW	ReCTS	ReCTS
CE w/o MER	78.53	92.44	76.89
CE w/ MER	**79.24**	**92.74**	**77.50**

Results. Training using MER improves accuracy without any additional overhead. A model thus can be strengthened easily. As shown in Table 1, model trained with MER outperforms the one without MER on both character recognition and text line recognition. MER can even make a model outperforms another deeper one, like SE-ResNet50 and ResNet152 in Table 1(a).

To be more intuitive, we visualize the region response of character images in CTW by summing over the intermediate feature map channels and exerting minmax normalization. As shown in the top two rows of Fig. 4, the model trained without MER usually has distributed response and is prone to focus on noisy regions. By using MER, the model concentrates mainly on the text body, thus is more robust to noisy background.

4.2 Fine-Grained Image Classification

Text recognition can be regarded as a kind of fine-grained image classification since many characters have subtle inter-class but big intra-class difference. To be more general, we also verify the effectiveness of MER on the classical fine-grained image dataset: CUB-200-2011.

CUB-200-2011 contains 11,788 images of 200 types of birds, 5,994 for training and 5,794 for testing [28]. It is a typical and popular dataset for fine-grained image classification.

Implementation Details. When preprocessing the training data, we adopt random crop and random horizontal flip to augment data. Then the images are resized to 448×448. ResNet50 [7] is the backbone network whose parameters are initialized from pretrained model on ImageNet. We train the model for 80 epochs with batch size set as 8 using the Momentum SGD optimizer. Learning rate starts from 1e-3 and decays by 0.3 when the current epoch is in [40, 60, 70]. When we use MER, λ is set to [1.0, 0.5, 0.2, 0.1] empirically when the current epoch is [30, 50, 70] respectively.

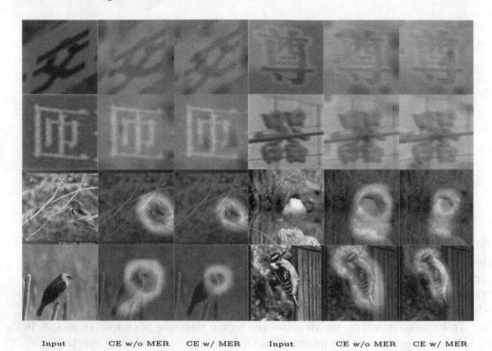

Input   CE w/o MER   CE w/ MER   Input   CE w/o MER   CE w/ MER

Fig. 4. Visualization of activation maps. Row 1–2 presents Chinese characters, and Row 3–4 presents fine-grained birds. Every triplet contains input image and attention map from model trained with/without MER. Training without MER is prone to make the model focus on broader region, including much background noise. MER regularizes the model to focus on more compact and discriminative region. Note that Row 1–2 is visualized by summing over intermediate feature map channels, whereas Row 3–4 is the class activation map.

Results. Our simple ResNet50 trained with MER also gains a lot, even outperforms the former complicated models, as shown in Table 2. Hence our model is both accurate and fast.

We visualize the activation map using the last convolutional feature map and the last linear layer weights by CAM [40]. As shown in the bottom two rows of Fig. 4, MER makes a model focus on more compact and discriminative region, and ignore the noisy background which is harmful to model generalization. In some circumstances, appearance in background or common body region (not discriminative) can make a model more confident on training set, but in this way the really discriminative region are not attended enough.

4.3   Verification of $p_y - \lambda$ Curve

We experimentally verify the theoretical convergence probability distribution described as Eq. 15 and Fig. 3.

Table 2. Accuracy on CUB-200-2011

Method	RACNN [4]	MACNN [38]	MAMC [26]	ResNet50	
MER	✗	✗	✗	✗	✓
Acc.	85.3	86.5	86.5	86.4	**87.3**

Table 3. Theoretical and experimental convergence probability of positive class (CPP) with different λ on CTW character recognition.

λ	Experimental		Theoretical
	CE loss	CPP/%	CPP/%
0.06	0.061	94.08	99.99
0.1	0.099	90.57	93.00
0.2	0.593	55.26	58.50
0.7	1.691	18.43	20.80

Training with MER on CTW character recognition, we set a fixed λ for every experiment. When the model converges, the value of cross-entropy loss in training set is used to calculate the expected experimental convergence probability of positive class (CPP):

$$p_y = e^{-L_{\mathrm{CE}}} \tag{16}$$

The theoretical convergence probability can be got directly from the curve in Fig. 3. As shown in Table 3, the experimental value is always slightly lower than the theoretical value. This is normal since there are no perfect models that can fit a complicated distribution completely. What is more, the curve is also only a proximation under an assumption that very negative class has the same position. So they are actually in accordance. As a result, the theoretical curve can be used to estimate the convergence probability distribution in training set roughly, which gives us a practical meaning of λ and may guide us to choose a proper λ.

4.4 Comparison with Label Smoothing

Since MER is very similar to label smoothing (LS), we compare their results on CUB-200-2011. LS also has a coefficience λ which means the final convergence probability of positive class (CPP) is $\left(1 - \lambda + \frac{\lambda}{C}\right)$. Hence both the two methods have the attribute that CPP decreases as λ increases.

Inspired by the theoretical relationship between CPP (p_y) and λ, we choose 4 theoretical CPPs to have 4 pairs of experiments. For each CPP, λ of LS is $1 - p_y$, and λ of MER is chosen from the curve in Fig. 3 as the previous part.

As shown in Table 4, MER always gets improvement more or less, while LS is more sensitive to λ and can even be harmful (see CPP=24). MER can achieve better accuracy than LS with their own λs. Besides, LS only works well when

Table 4. Comparison of MER and label smoothing on CUB-200-2011

Theoretical CPP/%	λ	Method	Training entropy	Acc.
24	0.76	LS	4.64	86.16
	1.0	MER	4.55	87.11
41	0.59	LS	3.90	86.56
	0.5	MER	3.71	**87.14**
77	0.23	LS	1.82	87.00
	0.2	MER	1.35	86.73
90	0.10	LS	0.92	86.61
	0.15	MER	0.36	86.50
100	0.0	w/o regularization	0.09	86.42

Table 5. Results of model trained with label corruption. Note that $\lambda = 0.0$ means the model is trained without MER.

(a) CUB-200-2011

Corruption Rate	λ		
	0.0	0.5	1.0
0.1	82.91	84.19	84.74
0.2	79.01	81.33	81.91

(b) CTW character recognition (Validation Set)

Corruption Rate	λ		
	0.0	0.3	0.6
0.1	84.12	84.61	84.58
0.2	82.63	83.20	83.46

λ is small. We argue that the negative influence of the dependence on uniform distribution can be zoomed and nonnegligible with a large λ (a small CPP), which limits the potential ability of LS. By contrast, MER is more flexible to regularize the expected probability distribution. With a same theoretical CPP, training entropy of LS is higher than MER, which also reflects the influence of uniform distribution.

4.5   Train with Label Corruption

To explore the power of MER on noisy dataset, we randomly corrupt labels of a certain proportion of training images. Label corruption is usually more harmful than feature corruption [3]. The false labels can mislead the learning process.

We find that the model trained with MER is more robust on both Chinese character recognition and fine-grained classification. In Table 5, corruption rate is the proportion of training images whose labels are randomly corrupted. The model trained without MER ($\lambda = 0.0$) suffers a lot because it is always very confident and reach to blind devotion to the given labels, even though some labels are wrong. MER regularizes a model to be less confident of labels in training set, so the learning process is less disturbed.

5   Conclusion

In this paper, we make a deep analysis on Maximum Entropy Regularization, including how MER term influence the convergence probability distribution and the learning process. MER improves generalization and robustness of a model without any additional parameters. We employ MER on both Chinese text recognition and common fine-grained classification to alleviate overfitting, and gain consistent improvement. We hope that our theoretical analysis can be useful for the further study on the regularization.

Acknowledgments. This work was supported by the National Natural Science Foundation of China (NSFC, grant No. 61733007).

References

1. Cheng, C., Huang, Q., Bai, X., Feng, B., Liu, W.: Patch aggregator for scene text script identification. In: 2019 15th International Conference on Document Analysis and Recognition (ICDAR), pp. 1077–1083. IEEE (2019)
2. Dubey, A., Gupta, O., Raskar, R., Naik, N.: Maximum-entropy fine grained classification. In: Advances in Neural Information Processing Systems, pp. 637–647 (2018)
3. Frénay, B., Verleysen, M.: Classification in the presence of label noise: a survey. IEEE Trans. Neural Networks Learn. Syst. **25**(5), 845–869 (2013)
4. Fu, J., Zheng, H., Mei, T.: Look closer to see better: recurrent attention convolutional neural network for fine-grained image recognition. In: Proceedings of the IEEE Conference on Computer Vision and Pattern Recognition, pp. 4438–4446 (2017)
5. Graves, A., Fernández, S., Gomez, F., Schmidhuber, J.: Connectionist temporal classification: labelling unsegmented sequence data with recurrent neural networks. In: Proceedings of the 23rd International Conference on Machine Learning, pp. 369–376. ACM (2006)
6. Group, M.D.: ICDAR 2019 robust reading challenge on reading Chinese text on signboard (2019). https://rrc.cvc.uab.es/?ch=12
7. He, K., Zhang, X., Ren, S., Sun, J.: Deep residual learning for image recognition. In: Proceedings of the IEEE Conference on Computer Vision and Pattern Recognition, pp. 770–778 (2016)
8. Hinton, G., Vinyals, O., Dean, J.: Distilling the knowledge in a neural network. arXiv preprint arXiv:1503.02531 (2015)
9. Hu, J., Shen, L., Sun, G.: Squeeze-and-excitation networks. In: Proceedings of the IEEE Conference on Computer Vision and Pattern Recognition, pp. 7132–7141 (2018)
10. Ioffe, S., Szegedy, C.: Batch normalization: accelerating deep network training by reducing internal covariate shift. arXiv preprint arXiv:1502.03167 (2015)
11. Jaderberg, M., Simonyan, K., Zisserman, A., et al.: Spatial transformer networks. In: Advances in Neural Information Processing Systems, pp. 2017–2025 (2015)
12. Jaynes, E.T.: Information theory and statistical mechanics. Phys. Rev. **106**(4), 620 (1957)

13. Liao, M., Zhang, J., Wan, Z., Xie, F., Liang, J., Lyu, P., Yao, C., Bai, X.: Scene text recognition from two-dimensional perspective. Proceedings of the AAAI Conference on Artificial Intelligence, vol. 33, pp. 8714–8721 (2019)
14. Liu, H., Jin, S., Zhang, C.: Connectionist temporal classification with maximum entropy regularization. In: Advances in Neural Information Processing Systems, pp. 831–841 (2018)
15. Lyu, P., Yang, Z., Leng, X., Wu, X., Li, R., Shen, X.: 2d attentional irregular scene text recognizer. arXiv preprint arXiv:1906.05708 (2019)
16. Ma, L.L., Liu, C.L.: A new radical-based approach to online handwritten Chinese character recognition. In: 2008 19th International Conference on Pattern Recognition, pp. 1–4. IEEE (2008)
17. Müller, R., Kornblith, S., Hinton, G.: When does label smoothing help? arXiv preprint arXiv:1906.02629 (2019)
18. Paszke, A., et al.: Automatic differentiation in PyTorch (2017)
19. Pereyra, G., Tucker, G., Chorowski, J., Kaiser, Ł., Hinton, G.: Regularizing neural networks by penalizing confident output distributions. arXiv preprint arXiv:1701.06548 (2017)
20. Puigcerver, J.: Are multidimensional recurrent layers really necessary for handwritten text recognition? In: 2017 14th IAPR International Conference on Document Analysis and Recognition (ICDAR), vol. 1, pp. 67–72. IEEE (2017)
21. Reed, S., Lee, H., Anguelov, D., Szegedy, C., Erhan, D., Rabinovich, A.: Training deep neural networks on noisy labels with bootstrapping. arXiv preprint arXiv:1412.6596 (2014)
22. Reeve Ingle, R., Fujii, Y., Deselaers, T., Baccash, J., Popat, A.C.: A scalable handwritten text recognition system. arXiv preprint arXiv:1904.09150 (2019)
23. Shi, B., Bai, X., Yao, C.: An end-to-end trainable neural network for image-based sequence recognition and its application to scene text recognition. IEEE Trans. Pattern Anal. Mach. Intell. **39**(11), 2298–2304 (2016)
24. Shi, B., Yang, M., Wang, X., Lyu, P., Yao, C., Bai, X.: Aster: an attentional scene text recognizer with flexible rectification. IEEE Trans. Pattern Anal. Mach. Intell. (2018)
25. Srivastava, N., Hinton, G., Krizhevsky, A., Sutskever, I., Salakhutdinov, R.: Dropout: a simple way to prevent neural networks from overfitting. J. Mach. Learn. Res. **15**(1), 1929–1958 (2014)
26. Sun, M., Yuan, Y., Zhou, F., Ding, E.: Multi-attention multi-class constraint for fine-grained image recognition. In: Ferrari, V., Hebert, M., Sminchisescu, C., Weiss, Y. (eds.) ECCV 2018. LNCS, vol. 11220, pp. 834–850. Springer, Cham (2018). https://doi.org/10.1007/978-3-030-01270-0_49
27. Szegedy, C., Vanhoucke, V., Ioffe, S., Shlens, J., Wojna, Z.: Rethinking the inception architecture for computer vision. In: Proceedings of the IEEE Conference on Computer Vision and Pattern Recognition, pp. 2818–2826 (2016)
28. Wah, C., Branson, S., Welinder, P., Perona, P., Belongie, S.: The Caltech-UCSD birds-200-2011 dataset (2011)
29. Wang, T.Q., Yin, F., Liu, C.L.: Radical-based Chinese character recognition via multi-labeled learning of deep residual networks. In: 2017 14th IAPR International Conference on Document Analysis and Recognition (ICDAR), vol. 1, pp. 579–584. IEEE (2017)
30. Wu, C., Wang, Z.R., Du, J., Zhang, J., Wang, J.: Joint spatial and radical analysis network for distorted Chinese character recognition. In: 2019 International Conference on Document Analysis and Recognition Workshops (ICDARW), vol. 5, pp. 122–127. IEEE (2019)

31. Yepes, A.J.: Confidence penalty, annealing Gaussian noise and zoneout for biLSTM-CRF networks for named entity recognition. arXiv preprint arXiv:1808.04029 (2018)
32. Yuan, T.L., Zhu, Z., Xu, K., Li, C.J., Hu, S.M.: Chinese text in the wild. arXiv preprint arXiv:1803.00085 (2018)
33. Yuan, T.L., Zhu, Z., Xu, K., Li, C.J., Mu, T.J., Hu, S.M.: A large Chinese text dataset in the wild. J. Comput. Sci. Technol. **34**(3), 509–521 (2019)
34. Zhang, H., Cisse, M., Dauphin, Y.N., Lopez-Paz, D.: mixup: beyond empirical risk minimization. arXiv preprint arXiv:1710.09412 (2017)
35. Zhang, J., Zhu, Y., Du, J., Dai, L.: Radical analysis network for zero-shot learning in printed Chinese character recognition. In: 2018 IEEE International Conference on Multimedia and Expo (ICME), pp. 1–6. IEEE (2018)
36. Zhang, X.Y., Bengio, Y., Liu, C.L.: Online and offline handwritten Chinese character recognition: a comprehensive study and new benchmark. Pattern Recogn. **61**, 348–360 (2017)
37. Zhang, X.Y., Wu, Y.C., Yin, F., Liu, C.L.: Deep learning based handwritten Chinese character and text recognition. In: Huang, K., Hussain, A., Wang, Q.F., Zhang, R. (eds.) Deep Learning: Fundamentals, Theory and Applications. Cognitive Computation Trends, vol. 2, pp. 57–88. Springer, Cham (2019). https://doi.org/10.1007/978-3-030-06073-2_3
38. Zheng, H., Fu, J., Mei, T., Luo, J.: Learning multi-attention convolutional neural network for fine-grained image recognition. In: Proceedings of the IEEE International Conference on Computer Vision, pp. 5209–5217 (2017)
39. Zhong, Z., Jin, L., Feng, Z.: Multi-font printed Chinese character recognition using multi-pooling convolutional neural network. In: 2015 13th International Conference on Document Analysis and Recognition (ICDAR), pp. 96–100. IEEE (2015)
40. Zhou, B., Khosla, A., Lapedriza, A., Oliva, A., Torralba, A.: Learning deep features for discriminative localization. In: Proceedings of the IEEE Conference on Computer Vision and Pattern Recognition, pp. 2921–2929 (2016)

An Improved Convolutional Block Attention Module for Chinese Character Recognition

Kai Zhou$^{(\boxtimes)}$, Yongsheng Zhou, Rui Zhang, and Xiaolin Wei

Meituan-Dianping Group, Beijing, China
{zhoukai03,zhouyongsheng,zhangrui36,weixiaolin02}@meituan.com

Abstract. Recognizing Chinese characters in natural images is a very challenging task, because they usually appear with artistic fonts, different styles, various lighting and occlusion conditions. This paper proposes a novel method named ICBAM (Improved Convolutional Block Attention Module) for Chinese character recognition in the wild. We present the concept of attention disturbance and combine it with CBAM (Convolutional Block Attention Module), which improve the generalization performance of the network and effectively avoid over-fitting. ICBAM is easy to train and deploy due to the ingenious design. Besides, it is worth mentioning that this module does not have any trainable parameters. Experiments conducted on the ICDAR 2019 ReCTS competition dataset demonstrate that our approach significantly outperforms the state-of-the-art techniques. In addition, we also verify the generalization performance of our method on the CTW dataset.

Keywords: Chinese recognition · Attention · Convolutional neural networks

1 Introduction

Scene text recognition is a very critical step in computer vision tasks, because text is an important information source. Despite the plenty of publicly available data, most focus on English text. As Chinese is the most widely used language in the world, recognizing Chinese has a large potential practical value. In the scene text recognition task, recognizing Chinese is a complex problem, because Chinese has a much larger character set than English and the layout of Chinese characters is usually more complicated.

Nowadays, text recognition methods have achieved distinguished success. Accordingly, methods based on deep learning have been broadly applied. Among them, there are many text line recognition schemes, CRNN [15], SCAN [22], ASTER [16], MORAN [13]. However, Chinese characters are always arranged in arbitrary forms which are difficult to handle for text line based methods. Most current recognition models remain too unstable to handle irregular layout problem (see Fig. 1).

© Springer Nature Switzerland AG 2020
X. Bai et al. (Eds.): DAS 2020, LNCS 12116, pp. 18–29, 2020.
https://doi.org/10.1007/978-3-030-57058-3_2

Fig. 1. Several complex layouts in natural images. (Green box: single Chinese character, yellow box: Chinese text line, blue box: other text line, red arrow: reading order) (Color figure online)

In order to solve the problem above, there are also character based approaches [18]. In the character based approaches, the core module is character recognition. Different from English, Chinese has a large character set and the intra-class distance is minor as some of them are extremely similar. Furthermore, Chinese characters have a wide variety of fonts, leading to a large inter-class space. These reasons above make the Chinese character recognition task very challenging. Some samples are shown in Fig. 2.

Fig. 2. Chinese character samples

It is well known that attention is very important in human visual perception system, humans use a series of partial images and selectively focus on salient parts to capture the visual structure better. Inspired by this, more and more researches incorporate attention to improve the performance of Convolutional neural networks in large-scale classification tasks, such as Residual Attention [20], SENet [6], CBAM [21], GCNet [1]. CBAM, a recently proposed method, shows the better performance compared with other attention methods. It contains spatial attention and channel attention part, as illustrated in Fig. 3.

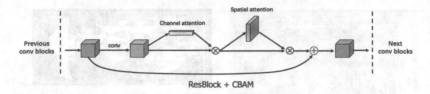

Fig. 3. CBAM integrated with a ResBlock in ResNet [4]

In the Chinese character recognition task, we incorporate CBAM to ResNet [4] and also achieve decent performance. However, we found that there are still some defects in CBAM. As we know, the human visual system's perception of topological properties takes precedence over other geometric properties. Correspondingly, the low-level features of the Chinese character images extracted by CBAM-integrated network should be clear and complete. More specifically, the edges of Chinese character extracted by our networks should be distinct. We compared the low-level features' visualization images of the Chinese characters that were identified correctly and incorrectly, and found that the low-level features of misidentified images are often unclear.

The contributions of this paper can be summarized as follows: (1) We propose the ICBAM to improve the generalization performance of the network and effectively avoid over-fitting. (2) The proposed module is very ingenious and modular design makes it easy to train and deploy in practical systems. (3) Different from dropout [5] layer, ICBAM can cooperate freely with the BN(Batch Normalization) [8] layer and improve the performance of networks, besides, it does not have any trainable parameters. (4) Experiments on ICDAR 2019 ReCTS competition dataset [11] demonstrate that our method can achieve comparable or state-of-the-art performance on the standard benchmarks, and we also verify the generalization performance of our method on the CTW dataset.

2   Related Work

Deep Convolutional Networks. After AlexNets [9] achieved record-breaking results in large scale image recognition tasks [14], convolutional networks have enjoyed a great success in recent years. With the deep convolutional networks widely used in image recognition, object detection, video analysis and image segmentation, a number of convolutional network architectures are becoming more and more powerful. Among them, VGGNets [17] are thorough evaluation of networks of increasing depth using an architecture with very small (3×3) convolution filters, which shows a larger and deeper convolutional network can significantly increase the ability of representation. Afterwards, GoogleNet [19] develops a new network architecture named "inception". This architecture improve the utilization of computing resources in the network, because "the inception" module allows for increasing the depth and width of the network while keeping the computational budget constant. As an effective network, ResNet [4] adds residual connections in the network architecture, and the residual unit makes it possible for networks to be deeper and stronger. This method has good generalization performance on other tasks such as object detection and image segmentation. Different from ResNet's identity-based skip connections, DenseNet [7] connect all layers directly with each other. This dense connectivity pattern ensures maximum information flow between layers in the network, so that, this pattern alleviate the vanishing-gradient problem and strengthen feature propagation. ResNeXt [23] proposes a simple architecture which adopts VGG and ResNets strategy of repeating layers, while exploiting the split transform-merge strategy in an easy extensible way. This strategy proposes a new concept called "cardinality" (the size of the transformations' set), which is an important factor in determining network performance. This work shows that when we design a deep convolutional network architecture, increasing cardinality is more effective.

Attention Mechanism. Recently, a broad range of works focus on attention mechanism, which uses top information to guide bottom-up feed forward process. Wang et al. propose a Residual Attention Networks [20] which generate attention-aware features by stacking attention model. This architecture makes it easier to train a residual deep convolutional network with hundreds of layers. Different from Residual Attention Networks, SENet [6] focus on finding the inter-channel relationship of convolutional networks layers. In their works, they get channel-wise attention from global average-pooled features. It's obvious that SENet do not take the spatial attention in to consideration, so SENet pays more attention to what to focus rather than where to focus. SCA-CNN [2] shows that spatial attention is also important compared with channel attention. CBAM [21] is a simple and effective attention architecture for feed-forward convolutional networks, it designs a spatial and channel-wise attention based on an efficient architecture. And it shows that using max-pooled features is better than average-pooled features in channel-wise attention. To extract model's global content features, Cao et al. exploit an novel network named GCNet [1] which is lightweight, and it could be applied to all residual blocks in the ResNet architecture. Different from the previous works, SKNet [10] is the first to explicitly focus on adaptive receptive field size of neurons by attention mechanisms. In this work, they proposed a dynamic selection mechanism in convolutional neural networks that allow each neuron to adaptively adjust its receptive field size based on multiple scales of input information. Different from data augmentation and dropout [5], the motivation for designing ICBAM is that we want to combine disturbance with attention module, helping the network extract more robust features of the Chinese characters.

3   Improved Convolutional Block Attention Module

In order to make CBAM focus on important features more accurately and suppress unnecessary features, we come up with the concept of attention disturbance and combine it with CBAM, this module slightly disturb the attention features extracted by channel and spatial attention module in training process to obtain more robust attention features. The overview of ICBAM is shown in Fig. 4.

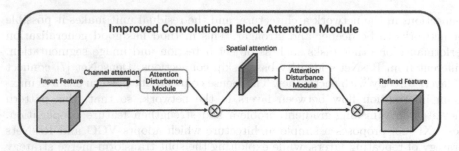

Fig. 4. The overview of ICBAM. We add attention disturbance module behind the channel and spatial attention in the training process to obtain more robust attention features at every convolutional block of deep networks. The structures of these two attention disturbance modules are identical.

The attention disturbance module is a core module in ICBAM (attention disturbance module is shown in Fig. 5). Given an intermediate attention feature map $F \in \mathbb{R}^{B \times H \times W \times C}$ as input, attention disturbance module firstly calculate a standard deviation set $S \in \mathbb{R}^{B \times 1 \times 1 \times 1}$, $s_i \in S$ and $i \in (0, 1, 2,..., B)$. Because the distributions of each image's attention features are different, to generate more suitable attention disturbance mask, we calculate the standard deviation s_i of each image's attention features. Secondly, we generate a disturbance mask $D \in \mathbb{R}^{B \times H \times W \times C}$, each image's disturbance mask $d_i \in \mathbb{R}^{1 \times H \times W \times C}$, we modify d_i as a uniform distribution and $d_i \sim \mathcal{U}(\text{-}s_i \times r, s_i \times r)$, r is attention disturbance ratio, usually $0.01 \leq r \leq 0.05$, we use 0.025 in our experiments. After getting the disturbance mask, perform the operation in Eq. 1. Where $\oplus$ denotes element-wise addition. F' is the disturbed feature, after obtaining F', the subsequent convolution process is carried out normally.

$$F' = F \oplus D \tag{1}$$

On account of the element-wise addition, the disturbance mask dimension and attention feature dimension need to be consistent. As shown in Fig. 4, the dimension of channel attention disturbance mask is $B \times 1 \times 1 \times C$, in the same vein, the dimension of spatial attention disturbance mask is $B \times H \times W \times C$.

During the training process, each step will go through the above processing, forcing the attention module to learn more robust attention features. With the attention disturbance module, an area that should have been given strong attention, if it does not get a high response, it may cause recognition errors and increase losses after disturbance.

In terms of deployment mode, channel and spatial attention module can be placed in a parallel or sequential manner, besides, these two modules can be deployed in different order in the sequential process. For the arrangement of these two attention modules in ICBAM, we learn from the best performance deployment method in CBAM, using the channel-first order in a sequential manner. We will discuss experimental results on network engineering in Sect. 4.

Because the attention disturbance module is just to make the features learned by the attention module more robust and does not contain any trainable parameters, it can be removed directly in the inference process.

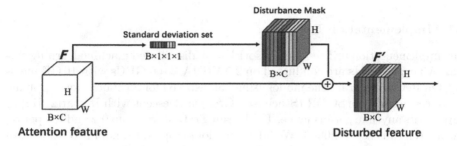

Fig. 5. The structure of Attention Disturbance Module.

4 Experiments

4.1 Dataset

ReCTS. The ICDAR 2019 Robust Reading Challenge on Reading Chinese Text on Signboard (ICDAR 2019-ReCTS) dataset is a new scene text recognition dataset which has 25,000 signboard images of Chinese and English Text. All the images are from Meituan-Dianping Group, collected by Meituan business merchains, using phone cameras under uncontrolled. ReCTS consist of Chinese characters, digits and English characters, with Chinese characters taking the largest portion. The 25,000 images are split training set contain 20,000 images and test contain 5,000 images. Moreover, 29335 character images cropped from 5000 test images, are used for ReCTS-task1 "Character Recognition in the Signboard" evaluation respectively.

CTW. CTW (Chinese Text in the Wild) is a very large dataset of Chinese text in street view images. There are 32285 images collected from more than 10 different cities in China and each image has a resolution of 2048 * 2048. The training, validation and test set contain 24290, 1597 and 3269 images. Because the origin test set does not provide character annotations, so we only evaluate our method on the validation set. In evaluation, we ignore the character size less than 20 pixel. In ReCTS dataset, they think when characters' size is less than 20 pixel especially Chinese characters, it is difficult to recognize for people. So, recognizing such a small text is meaningless.

4.2 Evaluation Metrics

We refer to the evaluation method of ReCTS-task1. The recognition accuracy is given as the metric:

$$accuracy = \frac{N_{right}}{N_{total}} \qquad (2)$$

N_{right} is the number of characters predicted correctly and N_{total} is the number of the test characters.

4.3   Implementation Details

Our implementation utilizes the TensorFlow [3] distributed machine learning system. All experiments are conducted on 2 NVIDIA V100 GPUs with 16 GB memory. On ReCTS we train the model using batchsize 64 for 25 epochs and applying warm restart scheme SGDR (Stochastic Gradient Descent with Restarts) [12] to improve its anytime performance. The learning rate starts from 0.2 and the period parameter T_0 is set to 20000. We adopt random-crop and random-rotation data augmentation technique on training images, and then resize images to 64 * 64. We verify our method on ResNet18 and ResNet50. The ICBAM is added before the last convolution layer of each residual unit. In training process, we use the SOFTMAX cross-entropy loss and set the last classify layer channels to 4487.

4.4   Experimental Results

To evaluate the effectiveness of our module and investigate the impact of different deployment forms of ICBAM on recognition accuracy, firstly, we add the attention disturbance module behind the channel attention module and the spatial attention module separately, then we add the disturbance module behind the channel attention module and the spatial attention module at the same time. All models are trained on the same training data set and use the same data augmentation scheme. We also made a horizontal comparison with other methods. Table 1 shows the performance of some CNN models trained on ReCTS.

Table 1. Base experiments on ReCTS.

Description	Accuracy
DenseNet	92.54%
ResNet18	92.75%
ResNet18 + SENet	**93.10%**

Table 2 shows the results on ReCTS test set with backbone ResNet18 and ResNet50. The results indicate that recognition accuracy can be improved by ICBAM. No matter ResNet18 or ResNet50, using ICBAM can achieve about 0.4% to 0.7% improvement.

Table 3 shows the performance of the ResNet18 and ResNet50 models with the ICBAM on CTW validation set. In order to verify the generalization performance of the ICBAM, we only use the training set of ReCTS to train the model.

2328r

Table 2. Experiments on ReCTS.

Backbone	Description	Accuracy
ResNet18	CBAM	93.56%
	ICBAM-channel attention disturbance	93.92%
	ICBAM-spatial attention disturbance	**94.22%**
ResNet50	CBAM	94.87%
	ICBAM-channel attention disturbance	**95.37%**
	ICBAM-spatial attention disturbance	95.12%

Table 3. Experiments on CTW.

Backbone	Description	Accuracy
ResNet18	CBAM	75.24%
	ICBAM-channel attention disturbance	75.66%
	ICBAM-spatial attention disturbance	**76.20%**
ResNet50	CBAM	78.58%
	ICBAM-channel attention disturbance	**80.26%**
	ICBAM-spatial attention disturbance	80.13%

Compared with baseline, ResNet18 and ResNet50 with attention disturbance mechanism have a better performance on all designed experiments. Besides, comparing with the experimental results on ReCTS, we found that the accuracy improvement on the CTW dataset is more obvious. We think, if ICBAM could improve the networks' generalization ability, the improvement of ICBAM on CTW should be greater than that on ReCTS. The improvements on ReCTS and CTW are 0.5% and 1.6% respectively. This result illustrates that attention disturbance mechanism could improve the generalization ability of networks in some ways.

Table 4. Experiments on ReCTS: add two disturbance module at the same time

Backbone	Channel disturbance ratio	Spatial disturbance ratio	Accuracy
ResNet18	0.01	0.01	92.79%
	0.02	0.02	93.02%
	0.08	0.05	93.60%
	0.01	0.05	93.76%
	0.05	0.01	**94.32%**

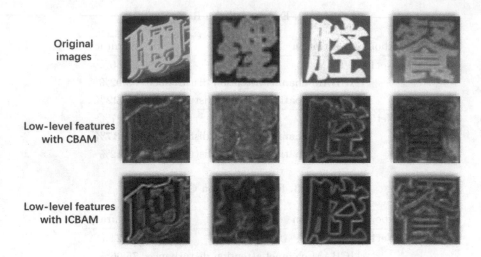

Original
images

Low-level features
with CBAM

Low-level features
with ICBAM

Fig. 6. Visualize the low-level features with CBAM and ICBAM.

Original
images

Attention
with CBAM

Attention
with ICBAM

Fig. 7. Attention visualization results.

Then we add the disturbance module behind the channel attention module and the spatial attention module at the same time. We perform 5 sets of experiments with different attention disturbance ratios, as shown in Table 4. Obviously, the results are not very stable. According to our analysis, the reason may be that if both two disturbances are superimposed, too many random factors may lead to unstable training results. This is a question worth exploring and we are still trying to explain it with quantitative experiments. Currently, some experiments on different network configurations are in progress.

On the ReCTS test set and CTW validation set, we obtain accuracy improvement from various baseline networks by plugging our tiny module, revealing the efficacy of ICBAM. We visualize some Chinese character images' low-level features extracted by CBAM and ICBAM, as shown in Fig. 6, the results show that features extracted by ICBAM is more accurate and our module suppress unnecessary features effectively.

In Fig. 7, we visualize the attention features of ResNet18 with ICBAM and ResNet18 with CBAM, it is obvious that the attention areas of ICBAM cover the character object regions better than CBAM. The experiments above show that CBAM is not robust and accurate in some cases. We conjecture that ICBAM makes network focus on important features more accurately and suppress unnecessary features, so as to help the network extract more robust features of the Chinese characters.

Through the above experiments, we verify that adding attention disturbance mechanism to CBAM can improve the performance and generalization ability of the model. On the other hand, we trained a ResNet50 model with ICBAM using ReCTS training data and 10,000 extra labeled image data, meanwhile, we extend the recognition category to 4593. Eventually, the top-3 results of ReCTS-task1 [11] are 97.37%, 97.28% and 96.12%, our method achieved state of the art result 97.10%.

5   Conclusion

We have presented the Improved Convolutional Block Attention Module (ICBAM), a novel method to improve representation and generalization ability of convolutional neural networks. From the perspective of improving attention robustness, we come up with the concept of attention disturbance and combine it with CBAM. Through the experiments, we verify the effectiveness of ICBAM, besides, we visualize the low-level features and attention features of ICBAM, the results shows that our module (ICBAM) learns what and where to emphasize or suppress and refines intermediate features effectively.

In conclusion, (1) ICBAM could improve the generalization performance of the network and effectively avoid over-fitting; (2) whether the attention disturbance module is separately added behind the channel attention module or the spatial attention module, the accuracy of recognition can be improved. However, adding attention disturbance module behind the channel attention module and the space attention module at the same time, will lead to unstable training, and it is difficult to set the hyper-parameter r; (3) this module does not have any trainable parameters, what's more, it is easy to train and deploy in practical systems.

In the future, we will analyze the influence of hyper-parameter r on network performance in more depth, and do more experiments to prove the effectiveness of ICBAM on more classification tasks. At last, we hope ICBAM become an important module of various network architectures.

References

1. Cao, Y., Xu, J., Lin, S., Wei, F., Hu, H.: GCNet: non-local networks meet squeeze-excitation networks and beyond. CoRR, abs/1904.11492 (2019)
2. Chen, L., et al.: SCA-CNN: spatial and channel-wise attention in convolutional networks for image captioning. In: The IEEE Conference on Computer Vision and Pattern Recognition (CVPR), July 2017
3. Google: Tensorflow. https://github.com/tensorflow/tensorflow
4. He, K., Zhang, X., Ren, S., Sun, J.: Deep residual learning for image recognition. In: The IEEE Conference on Computer Vision and Pattern Recognition (CVPR), June 2016
5. Hinton, G.F., Srivastava, N., Krizhevsky, A., Sutskever, I., Salakhutdinov, R.: Improving neural networks by preventing co-adaptation of feature detectors. CoRR, abs/1207.0580 (2012)
6. Hu, J., Shen, L., Sun, G.: Squeeze-and-excitation networks. In: The IEEE Conference on Computer Vision and Pattern Recognition (CVPR), June 2018
7. Huang, G., Liu, Z., van der Maaten, L., Weinberger, K.Q.: Densely connected convolutional networks. In: The IEEE Conference on Computer Vision and Pattern Recognition (CVPR), July 2017
8. Ioffe, S., Szegedy, C.: Batch normalization: accelerating deep network training by reducing internal covariate shift. CoRR, abs/1502.03167 (2015)
9. Krizhevsky, A., Sutskever, I., Hinton, G.E.: ImageNet classification with deep convolutional neural networks. In: Pereira, F., Burges, C.J.C., Bottou, L., Weinberger, K.Q. (eds.) Advances in Neural Information Processing Systems, vol. 25, pp. 1097–1105. Curran Associates Inc. (2012)
10. Li, X., Wang, W., Hu, X., Yang, J.: Selective kernel networks
11. Liu, X., et al.: ICDAR 2019 robust reading challenge on reading Chinese text on signboard (2019)
12. Loshchilov, I., Hutter, F.: SGDR: stochastic gradient descent with restarts. CoRR, abs/1608.03983 (2016)
13. Luo, C., Jin, L., Sun, Z.: A multi-object rectified attention network for scene text recognition. CoRR, abs/1901.03003 (2019)
14. Russakovsky, O., et al.. ImageNet large scale visual recognition challenge. Int. J. Comput. Vis. **115**(3), 211–252 (2015)
15. Shi, B., Bai, X., Yao, C.: An end-to-end trainable neural network for image-based sequence recognition and its application to scene text recognition. IEEE Trans. Pattern Anal. Mach. Intell. **39**(11), 2298–2304 (2017)
16. Shi, B., Yang, M., Wang, X., Lyu, P., Yao, C., Bai, X.: Aster: an attentional scene text recognizer with flexible rectification. IEEE Trans. Pattern Anal. Mach. Intell. **41**(9), 2035–2048 (2019)
17. Simonyan, K., Zisserman, A.: Very deep convolutional networks for large-scale image recognition. Computer Science (2014)
18. Song, Q., et al.: Reading Chinese scene text with arbitrary arrangement based on character spotting. In: 2019 International Conference on Document Analysis and Recognition Workshops (ICDARW), vol. 5, pp. 91–96, September 2019
19. Szegedy, C., et al.: Going deeper with convolutions. In: The IEEE Conference on Computer Vision and Pattern Recognition (CVPR), June 2015
20. Wang, F., et al.: Residual attention network for image classification. In: The IEEE Conference on Computer Vision and Pattern Recognition (CVPR), July 2017

21. Woo, S., Park, J., Lee, J.-Y., Kweon, I.S.: CBAM: convolutional block attention module. In: Ferrari, V., Hebert, M., Sminchisescu, C., Weiss, Y. (eds.) ECCV 2018. LNCS, vol. 11211, pp. 3–19. Springer, Cham (2018). https://doi.org/10.1007/978-3-030-01234-2_1
22. Wu, Y.-H., Yin, F., Zhang, X.-Y., Liu, L., Liu, C.-L.: SCAN: sliding convolutional attention network for scene text recognition. CoRR, abs/1806.00578 (2018)
23. Xie, S., Girshick, R., Dollar, P., Tu, Z., He, K.: Aggregated residual transformations for deep neural networks. In: The IEEE Conference on Computer Vision and Pattern Recognition (CVPR), July 2017

Adapting OCR with Limited Supervision

Deepayan Das$^{(\boxtimes)}$ and C. V. Jawahar

Centre for Visual Information Technology,
International Institute of Information Technology, Hyderabad, India
deepayan.das@research.iiit.ac.in, jawahar@iiit.ac.in

Abstract. Text recognition systems of today (aka OCRs) are mostly based on supervised learning of deep neural networks. Performance of these is limited by the type of data that is used for training. In the presence of diverse style in the document images (eg. fonts, print, writer, imaging process), creating a large amount of training data is impossible. In this paper, we explore the problem of adapting an existing OCR, already trained for a specific collection to a new collection, with minimal supervision or human effort. We explore three popular strategies for this: (i) Fine Tuning (ii) Self Training (ii) Fine Tuning + Self Training. We discuss details on how these popular approaches in Machine Learning can be adapted to the text recognition problem of our interest. We hope, our empirical observations on two different languages will be of relevance to wider use cases in text recognition.

Keywords: Finetuning · Semi-supervised learning · Self Training

1 Introduction and Related Works

At the beginning of the last decade, Deep Learning ushered us into a new era of artificial intelligence. Deep Neural Networks (DNNs) like CNNs [14] and RNNs [10] have shown to learn higher-order abstractions directly from raw data for various machine learning tasks. The need to hand-craft the features for tasks which earlier required domain experts has drastically declined. DNNs have established state-of-the-art results in almost all the computer vision (CV) tasks like image segmentation, object detection, pose estimation etc. Thus, with so much success, it was only natural that DNNs also forayed into one of the oldest computer vision problem of text recognition/OCR.

Motivation. One of the crucial steps towards digitizing books is the recognition of the document images using an OCR. More often than not there exists a domain gap between the data on which the OCR was trained and the books on which we want to perform recognition. In the case of digital library creation [1,2] books that come for digitization will invariably contain variations in their font-style, font-size as well as in their print quality (in case of historical books and manuscripts). Thus an OCR trained on a single source data will invariably fail across such variations. This leads to the inferior performance of the OCR module

X. Bai et al. (Eds.): DAS 2020, LNCS 12116, pp. 30–44, 2020.
https://doi.org/10.1007/978-3-030-57058-3_3

Fig. 1. (a) Word Images from the collection 1 containing clean-annotated images (b) Word Images from the collection 2 containing partially annotated noisy-degraded images.

resulting in many wrong words. Thus it becomes essential to fine-tune the OCR model on portions of books that we want to digitize. However, fine-tuning suffers from its inability to learn from unlabelled data as well as an added cost of annotation. Thus, in this paper in addition to fine-tuning, we also explore a branch of the semi-supervised approach called self-training where the OCR can learn from its own predictions, therefore bypassing the need for the creation of annotated data.

Deep Learning in Text Recognition. Traditionally, to recognize text from printed documents, sequential classifiers like Hidden Markov Models (HMMs) and graphical models like Conditional Random Fields (CRFs) have been used. These algorithms were popular since they did not require the line images to be explicitly segmented into words and characters, thus reducing the chances of error. However, due to the inability of such algorithms to retain long-term dependencies, Long Short Term Memory networks (LSTMs) emerged as the de facto choice for such sequential tasks. For recognition of handwritten text [9] was one of the earlier works to make use of LSTMs. Later it was adopted by [5,23,24] for recognition of printed text for Latin and Indic languages. More recently in [25] CNN was used as a feature extractor in the text recognition pipeline, which helped them achieve the state of the art results on various scene text datasets. Taking the same architecture forward, [6,12] showed that (CRNN) outperformed Bi-LSTM networks on printed documents. In the last few years attention-based encoder-decoder architecture [15,17] have also been proposed for character sequence prediction.

Learning with Limited Supervision. Although deep learning algorithms have become very popular, much of its success can be attributed to the availability of large quantities of clean annotated data. Collection of such datasets are both laborious and costly and it proves to be the bottleneck in applying such algorithms. A common approach to mitigate the effects of unavailability of clean and annotated data is to leverage the knowledge gained by DNNs on source tasks for whom labeled examples exist and generalize it to target task which suffers from the unavailability of labeled data. Such approaches fall under the purview of transfer learning which has been widely used in the past especially in Computer Vision (CV) tasks (object detection, classification and segmentation) under a data-constrained setting. Training models from scratch is both resource and time exhaustive. Adapting OCR models by pre-training on ImageNet data and later fine-tuning on Historical documents has been attempted in the past [27]. However, the improvement in the character recognition rate was not very significant. This may be attributed to the difference in image properties between document images and the natural images found in the ImageNet dataset.

Although, fine-tuning improves the performance of DNNs reasonably, yet it suffers from the in consequence of having to annotate a sizeable portion of the data which leads to an added cost. Additionally, fine-tuning also fails to take advantage of the vast amounts of unlabelled data that can be used to enhance the performance of machine learning models. In order to make use of the unannotated data, semi-supervised approaches like pseudo-labeling have become recently popular, where proxy labels are generated for the unannotated images. In [8, 28] the authors showed that by utilizing the unlabeled data in addition to a fraction of the labeled data, they were able to boost the performance of an existing handwriting recognition system (HWR) on a new target dataset. The authors used self-training (discussed in detail in the later sections of this paper) and were able to achieve a performance gain equivalent to a model which was finetuned on the same dataset but with full annotations.

The key contributions of this paper are as follows:

- We study the effect of fine-tuning a pre-trained OCR model on target dataset using a variety of fine-tuning approaches.
- We also, present a self-training approach for adapting OCR to the target data. We show that by combining simple regularization measures like data augmentation and dropout, we can attain improvement in accuracies close to 11% in the case of English and close to 4% in the case of Hindi dataset with no additional manual annotations.
- We also show that by combining the self-training and fine-tuning strategy we can outperform models that have been trained exclusively using the fine-tuning method.
- We empirically support our claims on the dataset for both English and Hindi and show that our proposed approach is language independent.

2   Empirical Verification Framework

Dataset. Our dataset is divided into two collections 1) Collection 1: which consists of reasonably clean-annotated images on which our OCR is trained. 2) Collection 2: which consists of partially annotated data containing noisy-degraded images. The images in the collection 1 are significantly different in terms of font style and print quality from the document images in collection 2. This can be

Table 1. Details of the data used in our work. The table describes the language of the type of collection from which the line images are used. Annotation refers to whether the data split is annotated or not. Also, the column purpose defines the role of each data split.

Collection	Language	Annotation	Purpose	#Pages	#Lines
Collection 1	English	Yes	Training OCR	1000	14K
	Hindi	Yes	Training OCR	4287	92K
Collection 2	English	Yes	Fine-tuning	50	7K
		Yes	Evaluation	200	9K
		No	Self-training	1100	18K
	Hindi	Yes	Fine-tuning	250	8K
		Yes	Evaluation	1K	20K
		No	Self-training	5K	100K

observed from Fig. 1. The various splits for each collection as well as the purpose of each split is described in Table 1. We annotate a part of collection 2 and split it into two parts (1) fine-tuning (2) evaluation. Our main objective is to transfer knowledge from collection 1 to collection 2 with limited supervision.

Evaluation. We use the character recognition rate (CRR) and word recognition rate (WRR) metrics to compare the performance of various models. CRR is the ratio of the sum of edit distance between the predicted text (pt) and ground truth (gt) to the sum of total number characters in pt and gt while WRR is defined as the number of words correctly identified, averaged over the total number of words present in the ground truth.

Implementation Details. We use a learning rate of 10^{-5} with a step scheduler. Initially we keep the batch size as 32 and the number of epochs as 100. We use early stopping criterion to avoid overfitting and the optimizer used is Adam.

3 Fine-Tuning for Text Recognition

Image representations learned with CNN on a large scale annotated data can be transferred to other visual tasks that have limited annotated training data [4]. Convolution layers pre-trained on a source task adapt better to the target task as compared to a model trained from scratch. There exist several ways to fine-tune a pre-trained model. The most common approach involves training only the last layer while keeping all the layers frozen. Alternatively, another common approach is to use pre-trained weights of a CNN as initialization, thus fine-tuning all the layers. However, such techniques have gained minimal exposure in the domain of text recognition. In one of the first attempts, [22] reported improvement in model performance when an existing model is fine-tuned on target data as opposed to training from scratch. We study the following fine-tuning approaches.

- **Full**: Using a pre-trained model as a mode of initialization and fine-tune all the layers on the target dataset. We train the model on the annotated portion of collection 2 until the loss on validation data stops decreasing.
- **Last**: Fine-tuning only the recurrent layers while keeping the convolution layers frozen.
- **Chain-thaw**: Fine-tuning one layer at a time while keeping the other layers frozen [7].
- **Unfreeze**: Another variation of chain thaw where we sequentially unfreeze the layers and fine-tune each such instance and finally fine-tune the whole network until convergence [11].

We further assess the effect of learning rate schedulers like slanted triangular learning rate ('Stlr') [11] and cosine annealing scheduler ('Cos') [18] on fine-tuning.

Table 2. Character and Word Recognition results for various finetuning methods. The first row shows the CRR and WRR for the pre-trained model. Subsequent rows contain values for models obtained after finetuning the base model.

Method	English		Hindi	
	CRR	WRR	CRR	WRR
Base model	93.65	83.40	91.63	83.95
Full	97.46	95.88	92.90	86.52
Full + Stlr	98.75	98.22	92.99	86.85
Last	96.16	90.88	92.42	85.27
Chain thaw	97.96	97.53	93.01	86.75
Unfreeze	98.02	97.75	93.03	86.96
Unfreeze + Cos	98.23	98.01	93.11	87.09
Unfreeze + Stlr	**98.79**	**98.46**	**93.20**	**87.40**

3.1   Results and Discussions

Table 2 shows the result of various fine-tuning approaches on our English and Hindi datasets. The base model refers to the pre-trained OCR models trained on the clean-annotated source data, which acts as our baseline. Due to the significant difference in the font-style and page image quality between the source and target data, the base model performs poorly as is evident from the results shown. From the experiments, we observe that (1) Fine-tuning the network consistently results in better recognition rates as opposed to training from scratch for both the datasets. (2) Fine-tuning only the recurrent layers ('Last') results in under-fitting, particularly on the English dataset. (3) We observe that fine-tuning the entire network ('Full') seems to give us more favorable results on both the datasets. We believe that this is because the target data contains minute nuances in the font style, which the pre-trained feature extractors are not able to capture well. (4) We also observe that Gradual unfreezing ('Unfreeze') approach to fine-tuning performs the best and is closely followed by 'Chain-Thaw'. 'Chain-thaw' and 'Unfreeze' methods work better than the traditional fine-tuning method since training one layer at a time helps the network to adapt better to the new domain and avoid forgetting. Also, from Fig. 3a we observe that validation loss quickly converges for 'Unfreeze' and 'Chain-thaw' techniques with 'Unfreeze' attaining the lowest validation error rate. This confirms the effectiveness of the above two approaches. (5) Additionally, learning rate schedulers consistently boost the performance for 'Full' and 'Unfreeze' methods with 'Unfreeze + Stlr' attaining the best overall accuracies. This is in support of our hypothesis that the network learns better, one layer at a time.

4   Self-training for Text Recognition

Self-training is one of the widely used approaches in semi-supervised learning. In this method, we generate the prediction for the unannotated data using a pre-trained model and then use them as pseudo-labels to train a new model [19,21]. Formally, suppose we have m labeled data (L) and n unlabelled data (U) such that $n >> m$. Additionally, L and U do not come from the same source which introduces a domain shift. M_0 is a pre-trained model trained on L. M_0 is used to generate predictions for U such that we now have images along with their predictions $[(I_0, y_0^*), (I_1, y_2^*)....(I_n, y_n^*)]$. The most confident samples are taken and added to the labelled data L. The model M_0 is then trained on the $n + p$ samples and at the end of the training, we obtain model M_1. This process is repeated for a fixed number of cycles such that at the end of each cycle, model M_{i+1} is obtained, which is then used to generate pseudo labels on the remaining $n - p$ unlabelled data samples. The process is continued until there are no more unlabelled samples or when there is no improvement in accuracy on the evaluation dataset. Figure 2 illustrates the self-training approach. We follow the same procedure for training the model using the self-training strategy on our unlabeled dataset.

Although self-training has shown success in a variety of tasks it has the downside of suffering from confirmation bias [16], where the model becomes over-confident on incorrect pseudo labels thus hindering the model's ability to learn and rectify its errors. In [28] the authors proposed to use a lexicon to verify the correctness of the pseudo-labels thereby avoiding the inclusion of any incorrect labels in the training data. However in the absence of a pre-defined vocabulary (in case of highly inflectional languages like Hindi/Tamil) it becomes essential to carefully regularize the network to alleviate the effects of confirmation bias. We use the recently proposed mixup strategy [32] to augment data which is an effective method for network regularization. Additionally, we also provide perturbations to the network using dropout regularization [26], which further mitigates the confirmation bias and enhances the model's performance. We show that by applying careful regularization measures to our self-training framework,

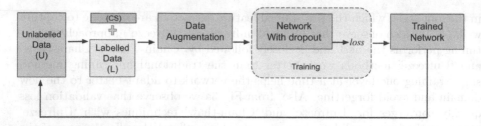

Fig. 2. A pipeline of our proposed iterative self-training approach. At each iteration model, M_i performs inference and confidence estimation on the unannotated dataset. Top k confident samples is mixed with the labelled data L and the combined samples are noised. The network (M_i) is trained on the combined samples at the end of which we obtain model M_{i+1}. The above procedure is repeated until there is no more improvement in word accuracy on test data.

we can achieve comparable accuracies to the fine-tuning approach on the Hindi dataset even though no actual labels were used. We also show that by initially training the network on pseudo-labels followed by fine-tuning on actual labels helped us achieve the best accuracies on both English and Hindi datasets.

4.1   Details

Confidence Estimation. To estimate the network confidence for each unlabelled sample in U, we take the log probability distribution given by the model M_i over the C classes where each class represents a character present in our vocabulary as shown in Eq. 1, where p_i is the probability outputted by the RNN at each time step t. We sort the predictions in decreasing order of their confidence scores and take the top 20% at the beginning of each cycle. However, due to the presence of a domain gap between the source and target data, the network tends to confuse between similar-looking characters resulting in a high probability of being assigned to the wrong character. This leads to predictions containing a considerable number of error words. When the network is trained over such confident but wrong predictions, the errors get reinforced in the subsequent trained models. This further amplifies the errors, which shows that the probability distribution can be a poor estimator for selecting the pseudo labels. To neutralize our dependence on the model's probability distribution as confidence estimator, we also take into consideration, the perplexity score of each prediction obtained by a pre-trained language model and finally take a weighted sum over both as shown by Eq. 2.

$$\text{score} = -\sum_{i=1}^{t} \log(p_i) \tag{1}$$

$$\text{score} = -\alpha \sum_{i=1}^{t} \log(p_i) + (1-\alpha)\frac{1}{m} \log P(w_1, \dots, w_m) \tag{2}$$

where, $logP(w_i, \ldots, w_t)$ is the joint probability of a sentence and α is the weight parameter that we determine empirically. The language model tends to assign a low score to predicted sentences that have error words, which lead to the overall score being low. This leads to sentences with error words to get weeded out from the confident pseudo labels, which in turn helps the model to avoid confirmation bias while training.

Prediction Ensemble. Additionally, while computing the maximum likelihood given by model M_i for each sample, we also take into consideration the probability values outputted by the earlier models i.e. $M_0, \ldots, M_i - 1$ which is given by Eq. 3 where Z_i is the model prediction at iteration i and z_{j-1} is the average of predictions for models at iteration 0 to $i - 1$. Thus, Z contains a weighted average of outputs of an ensemble of models, with recent models having greater weight than the distant models. In the first cycle of our iterative self-training method, both Z and z are zero since no previous models are available. For this reason, we specify λ to be a ramp-up function to be zero on the first iteration. The idea has been borrowed from [13] where it is used to enforce consistency regularization for semi-supervised learning.

$$Z_i = \lambda Z_i + (1 - \lambda)z_{i-1} \tag{3}$$

4.2 Regularization

Regularization plays an essential role in our design of the self-training framework. We regularize our network mainly by two ways 1) perturbing the input data and 2) perturbing network. Perturbations to the input data are done by adding Gaussian noise to the input image. In contrast, perturbations to the network are provided by adding a dropout layer, the details of which we discuss in the following sections. Earlier works [3,30] noted that providing perturbations enforced local smoothness in the decision functions of both labelled and unlabelled data. It has also the added advantage of preventing the model from getting stuck at local minima and avoid overfitting.

Gaussian Noise. We multiply the input image with a mask sampled from a binomial distribution. The mask zeros the pixel values at multiple locations, resulting in loss of information. This forces the model to become more robust while making predictions.

Mixup. In addition to Gaussian Noise, we also experiment with another type of data augmentation known as mixup [32]. Mixup creates new training samples using a weighted interpolation between two randomly sampled data points $(x_1, y_1), (x_2, y_2)$. where $\lambda \in [0,1]$ is a random number drawn from a $\beta(\alpha, \alpha)$ distribution. Mixup encourages the model towards linear behavior in between training samples. Additionally, mixup has the property of curbing confirmation bias by enforcing label smoothness by combining y_i and y_j as noted by [29]. This is especially important from the perspective of self-training strategy since we are

Table 3. Comparison of word and character recognition rates of CRNN models between baseline and our self training framework.

Model	Proposed				Baseline			
	English		Hindi		English		Hindi	
	CRR	WRR	CRR	WRR	CRR	WRR	CRR	WRR
VGG + BiLSTMs	**96.53**	**94.01**	**93.04**	**87.23**	93.65	83.40	91.63	83.95
ResNet18 + BiLSTMs	**96.76**	**94.64**	**93.22**	**87.90**	95.23	89.71	92.30	85.97

using predictions of unlabelled images as targets. In such cases, the networks, while training, tend to overfit to its predictions.

$$\hat{x} = \lambda x_i + (1 - \lambda)x_j \qquad (4)$$

$$\hat{y} = \lambda y_i + (1 - \lambda)y_j \qquad (5)$$

Weight Dropped LSTM. In addition to the above data augmentation techniques, we also apply dropout to the recurrent and fully connected network (FCN). The recurrent layers (Bi-LSTMS) is the most fundamental block in any modern OCR models. Therefore, it only makes sense to regularize them for optimal performance. Dropout is one of the most widely used approaches towards regularizing a neural network. However, naively applying dropout to the hidden state affects the RNN's ability to retain long-term dependency [31]. We use Weight Dropped LSTMs proposed in [20] which uses DropConnect on the recurrent hidden to hidden weight matrices. We also introduce dropout on the two fully connected layers after each Bi-LSTM layer with dropout probability set to 0.5. The above data augmentation and model perturbation techniques force the model to act as an ensemble which is known to yield better results than a single network in the ensemble.

We also use slanted triangular learning rates (STLR) proposed in [11] for scheduling our learning rates (LR). The authors argue that STLR enables the network to quickly converge to a suitable region in the parametric space at the start and then gradually refine its parameters.

4.3 Results and Discussions

Table 3 presents the results of our self-training strategy on two kinds of CRNN architectures where the CNN part comes from 1) VGG 2) RESNET18. We observe a significant improvement in word and character accuracies from the baselines for both the models which shows that the refinements that we suggest are model agnostic and can work under any setting. It is interesting to note that our self-training method on the Hindi base model improves the word recognition rate quite significantly and brings it at par with the 'Unfreeze + Stlr' which is the best performing finetuning approach (shown in Table 2), even though no actual labels were used to train the network. In the case of English, our proposed approach

Table 4. The breakdown effect of each regularization heuristic

Heuristics	English		Hindi	
	CRR	WRR	CRR	WRR
ST	94.22	85.12	92.13	85.41
+ STLR	94.72	86.88	92.17	85.45
+ noise	95.61	91.87	92.26	85.57
+ dropout	95.99	92.57	92.26	85.54
+ mixup	**96.48**	**93.57**	**92.57**	**86.23**

Table 5. Ablation study of various refinements

	English		Hindi	
	CRR	WRR	CRR	WRR
ST base	94.22	85.12	92.13	85.41
ST noise	95.61	91.87	92.17	85.47
ST dropout	95.22	89.28	91.91	85.27
ST mixup	96.09	91.73	92.57	86.46

also performs comparatively well with an improvement of 10% in the WRR and 3% in CRR but lags behind the 'Unfreeze' methods. This can be attributed to the difference in the number of training examples between the two datasets with Hindi being far superior in numbers. Thus, we can conclude that the self-training framework benefits from the amount of training data. Additionally, the testing accuracies for each refinement are shown in Table 4. By perturbing input images followed by data augmentation using mixup and adding weighted dropout, we systematically improve the recognition rates at both character and word level. Also, it is essential to note that the values are shown in Table 4 are for models that have been trained for only one cycle of our self-training framework and top samples were generated using sum over log probabilities.

Table 6. Character and word recognition rates at the end of each iterative cycle for both English and Hindi datasets.

Cycles	English		Hindi	
	CRR	WRR	CRR	WRR
Cycle 1	96.49	93.88	92.57	86.23
Cycle 2	96.52	93.92	92.95	87.07
Cycle 3	95.52	93.97	93.04	87.23
Cycle 4	**95.53**	**94.01**	**93.04**	**87.22**

Table 7. Character and word recognition rates for different scoring mechanisms on English dataset

Scoring	CRR	WRR
Prob-dist	96.48	93.45
Lang-model	96.01	91.56
Weighted score	**96.52**	**93.88**

4.4 Observations

To study the impact of individual regularization measures on the performance of our self-trained models we add each regularization one at a time. We then train the model with our self-training framework and check the performance for each model on the evaluation dataset. We report the accuracies in Table 5. For the above experiments we use VGG16 architecture and we run on only one cycle of our iterative self-training framework. From our experiments, we observe that (1) only by corrupting the images with Gaussian noise the self-trained attains better recognition rates compared to the base model. The improvement is rather significant especially in the case of English. (2) Similarly, in the case of dropout, WRR and CRR show an improvement of 4% and 2% respectively. However, we observe that in the case of Hindi, the performance drops. We believe that this happened due to the dropout model under-fitting on the Hindi data and believe that a lower dropout probability will easily fix the issue. We also compare the performance self-trained model with and without the *mixup* data augmentation technique. In this study, no other perturbations were provided. We observe that (4) *mixup* improves the character and word recognition rate on both the datasets, Also, from Fig. 3b we observe that the validation loss for self-training with mixup strategy converges more quickly than other regularization techniques which demonstrates the effectiveness of the mixup strategy in dealing with the confirmation bias. Additionally, we also study the effect of our iterative self-training strategy and (5) report the gradual increase in accuracy at the end of each self-training cycle in Table 6. To show the effectiveness of our proposed weighted sum approach for confidence estimation against log probability distribution and sentence probability score. From Table 7 we observe that the proposed weighted scoring achieves the best accuracies. Since no pre-trained language model was available for the Hindi language, this set of experiments were performed only on the English data. Using only the probability scores from the language model results in an inferior performance. This happens because every language models suffer from an inherent bias towards the dataset on which it was trained. Also, sentences containing proper nouns and abbreviations were given a low score in spite of being correct.

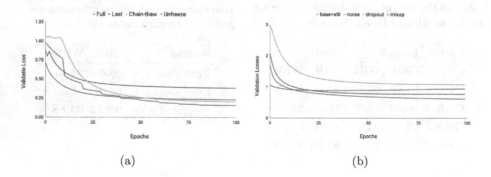

(a)   (b)

Fig. 3. (a) shows the Validation curves for different fine-tuning methods (b) Validation vs. Epochs plots for the self-training method with various regularizers on English dataset.

5   Hybrid Approach: Self-training + Fine-Tuning

Now, we combine both finetuning and the proposed self-training framework to further improve the performance of our OCR. The success of a finetuning depends on how well weights of the pre-trained models are initialized. In case the source and target tasks are very different or else if the source and target data come from very different distributions, then training the model from scratch seems to be a more viable option as compared to finetuning. In order to acclimatize the weights to the source data, we take advantage of the huge quantities of unlabelled data by employing our self-training strategy. We then follow it by finetuning the self-trained models on the limited annotated data. We show by a series of ablations that this approach outperforms the finetuning approach on both the datasets.

5.1   Results and Discussions

From Table 8 we observe that (1) in the case of Hindi, the improvement in word and character recognition rates are quite significant for our hybrid models with 2% and 1.5% improvement in word accuracy over fine-tuning and self-training methods respectively. (2) In the case of English, the improvement is significant from self-training. However, the increase in accuracies are not noticeable when compared to the fine-tuning approach. We attribute it to the fact that the character and word recognition rates had already maxed out using only fine-tuning, leaving little scope for improvement by the hybrid approach. Fine-tuning on top of self-training has the effect of canceling out any wrong prediction that the model learned during the self-training phase. It also boosts the model's confidence over uncertain predictions. Hence, we always see a spike in character and word accuracy for the hybrid approach. Figure 4a shows the validation curves for the above three approaches. We observe that the validation error rate for the hybrid approach reaches minimum within the stipulated number of epochs and it is closely followed by fine-tuning. This is in agreement with the results shown

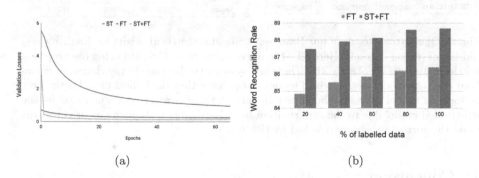

(a)   (b)

Fig. 4. (a) Validation losses vs Epochs for self-train, fine-tune and hybrid approaches on English dataset. (b) Comparison of word recognition rates between finetuning and hybrid approach when different percentages of labelled data is used

Table 8. Character and word recognition rates for fine-tuning, self-training and hybrid approach on English and Hindi datasets.

Method	English		Hindi	
	CRR	WRR	CRR	WRR
Fine-tuning	98.76	98.54	92.87	86.41
Self-train	96.53	94.01	93.04	87.23
Hybrid	**98.80**	**98.64**	**93.65**	**88.68**

in Table 8 where the hybrid method outperforms the rest of the two methods. Additionally, we also investigate the effect of the amount of data on fine-tuning. For this, we systematically increase the amount of labelled data by a factor of 20% to observe the character and word recognition rates for fine-tuning and hybrid approaches. Fig. 4b presents the comparison between the hybrid and the fine-tuning approach for different percentages of labelled data. We observe that by (3) fine-tuning the self-trained model on only a fraction of the labelled data helps us achieve better word recognition rate when compared to the models which were trained using the fine-tuning approach alone (Fig. 5).

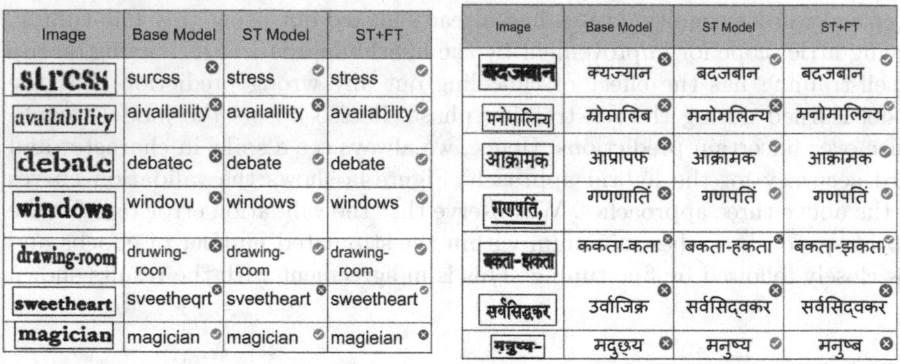

Fig. 5. Qualitative Result of our Base, self-trained and hybrid model for English (left) and Hindi (right) datasets. Here ST+FT refers to the model trained using the proposed hybrid approach. We observe that there is a systematic decrease in the character errors from the base model to the hybrid model. This shows that the hybrid model is the best performing model as it has the ability to correct the character errors incurred by the self-trained model due to the confirmation bias. Here crosses show the incorrect words while the correct words are denoted by tick marks.

6   Conclusion

In this paper, we present three different approaches for knowledge transfer between source and target data and compare the performance of each in terms of character and word recognition rates on two different datasets. We also show that simple regularization measure on self-training enables the models to learn more efficiently without getting biased towards the faulty network predictions. Additionally, we also show that combining self-training and fine-tuning can significantly boost the performance of the OCR across datasets of different languages even when the amount of labelled samples is considerably less.

References

1. Google Books. https://books.google.co.in/
2. Project Gutenberg. www.gutenberg.org
3. Arazo, E., Ortego, D., Albert, P., O'Connor, N.E., McGuinness, K.: Pseudo-labeling and confirmation bias in deep semi-supervised learning. In: ICLR (2019)
4. Bengio, Y.: Deep learning of representations for unsupervised and transfer learning. In: ICMLW, pp. 17–36 (2012)
5. Breuel, T.M., Ul-Hasan, A., Al-Azawi, M.A., Shafait, F.: High-performance OCR for printed English and Fraktur using LSTM networks (2013)
6. Dutta, K., Krishnan, P., Mathew, M., Jawahar, C.: Offline handwriting recognition on devanagari using a new benchmark dataset. In: DAS (2018)
7. Felbo, B., Mislove, A., Søgaard, A., Rahwan, I., Lehmann, S.: Using millions of emoji occurrences to learn any-domain representations for detecting sentiment, emotion and sarcasm. In: ACL (2017)
8. Frinken, V., Fischer, A., Bunke, H., Foornes, A.: Co-training for handwritten word recognition. In: ICDAR (2011)
9. Graves, A., Schmidhuber, J.: Offline handwriting recognition with multidimensional recurrent neural networks. In: NIPS (2009)
10. Hochreiter, S., Schmidhuber, J.: Long short-term memory. Neural Comput. $9(8)$, 1735–1780 (1997)
11. Howard, J., Ruder, S.: Universal language model fine-tuning for text classification (2018)
12. Jain, M., Mathew, M., Jawahar, C.: Unconstrained scene text and video text recognition for Arabic script. In: ASAR (2017)
13. Laine, S., Aila, T.: Temporal ensembling for semi-supervised learning. In: ICLR (2017)
14. LeCun, Y., Bottou, L., Bengio, Y., Haffner, P., et al.: Gradient-based learning applied to document recognition. Proc. IEEE $86(11)$, 2278–2324 (1998)
15. Lee, C.Y., Osindero, S.: Recursive recurrent nets with attention modeling for OCR in the wild. In: CVPR (2016)
16. Li, Y., Liu, L., Tan, R.T.: Certainty-driven consistency loss for semi-supervised learning. CoRR (2019)
17. Liu, W., Chen, C., Wong, K.Y.K., Su, Z., Han, J.: STAR-Net: a SpaTial attention residue network for scene text recognition. In: BMVC (2016)
18. Loshchilov, I., Hutter, F.: SGDR: stochastic gradient descent with warm restarts. In: ICLR (2017)
19. McClosky, D., Charniak, E., Johnson, M.: Effective self-training for parsing. In: ACL (2006)
20. Merity, S., Keskar, N.S., Socher, R.: Regularizing and optimizing LSTM language models. In: ICLR (2017)

21. Reichart, R., Rappoport, A.: Self-training for enhancement and domain adaptation of statistical parsers trained on small datasets. In: ACL (2007)
22. Reul, C., Wick, C., Springmann, U., Puppe, F.: Transfer learning for OCRopus model training on early printed books. CoRR (2017)
23. Sankaran, N., Jawahar, C.: Recognition of printed devanagari text using BLSTM neural network. In: ICPR (2012)
24. Sankaran, N., Neelappa, A., Jawahar, C.: Devanagari text recognition: a transcription based formulation. In: ICDAR (2013)
25. Shi, B., Bai, X., Yao, C.: An end-to-end trainable neural network for image-based sequence recognition and its application to scene text recognition. PAMI **39**(11), 2298–2304 (2016)
26. Srivastava, N., Hinton, G., Krizhevsky, A., Sutskever, I., Salakhutdinov, R.: Dropout: a simple way to prevent neural networks from overfitting. JMLR **15**(1), 1929–1958 (2014)
27. Studer, L., et al.: A comprehensive study of imagenet pre-training for historical document image analysis. In: ICDAR (2019)
28. Stuner, B., Chatelain, C., Paquet, T.: Self-training of BLSTM with lexicon verification for handwriting recognition. In: ICDAR (2017)
29. Thulasidasan, S., Chennupati, G., Bilmes, J., Bhattacharya, T., Michalak, S.: On mixup training: Improved calibration and predictive uncertainty for deep neural networks. In: NIPS (2019)
30. Xie, Q., Hovy, E., Luong, M.T., Le, Q.V.: Self-training with noisy student improves imagenet classification (2019)
31. Xu, K., et al.: Show, attend and tell: neural image caption generation with visual attention. In: ICML (2015)
32. Zhang, H., Cisse, M., Dauphin, Y.N., Lopez-Paz, D.: mixup: beyond empirical risk minimization. In: ICLR (2017)

High Performance Offline Handwritten Chinese Text Recognition with a New Data Preprocessing and Augmentation Pipeline

Canyu Xie[1], Songxuan Lai[1], Qianying Liao[1], and Lianwen Jin[1,2(✉)]

[1] College of Electronic and Information Engineering,
South China University of Technology, Guangzhou, China
`eelwjin@scut.edu.cn`
[2] SCUT-Zhuhai Institute of Modern Industrial Innovation, Zhuhai 519000, China

Abstract. Offline handwritten text recognition (HCTR) has been a long-standing research topic. To build robust and high-performance offline HCTR systems, it is natural to develop data preprocessing and augmentation techniques, which, however, have not been fully explored. In this paper, we propose a data preprocessing and augmentation pipeline and a CNN-ResLSTM model for high-performance offline HCTR. The data preprocessing and augmentation pipeline consists of three steps: training text sample generation, text sample preprocessing and text sample synthesis. The CNN-ResLSTM model is derived by introducing residual connections into the RNN part of the CRNN architecture. Experiments show that on the proposed CNN-ResLSTM, the data preprocessing and augmentation pipeline can effectively and robustly improve the system performance: On two standard benchmarks, namely the CASIA-HWDB and the ICDAR-2013 handwriting competition dataset, the proposed approach achieves state-of-the-art results with correct rates of 97.28% and 96.99%, respectively. Furthermore, to make our model more practical, we employ model acceleration and compression techniques to build a fast and compact model without sacrificing the accuracy.

Keywords: Offline Handwritten Text Recognition (HCTR) · Data preprocessing · Data augmentation · CNN-ResLSTM

1 Introduction

Handwritten Chinese text recognition (HCTR), as a challenging problem in the field of pattern recognition, has long received intensive concerns and observed steady progresses. The difficulties of HCTR are mainly from three aspects: a very large character set, diversity of writing styles and the character-touching problem.

To overcome these difficulties, researchers have proposed many solutions. Over-segmentation-based methods are very popular due to their efficiency and interpretablity. They slice the input string into sequential character segments,

© Springer Nature Switzerland AG 2020
X. Bai et al. (Eds.): DAS 2020, LNCS 12116, pp. 45–59, 2020.
https://doi.org/10.1007/978-3-030-57058-3_4

construct a segmentation-recognition lattice, and search for the optimal recognition path. For example, Zhou et al. [24] proposed a method based on semi-Markov conditional random fields, which combined candidate character recognition scores with geometric and linguistic contexts. Zhou et al. [25] described an alternative parameter learning method, which aimed at minimizing the character error rate rather than the string error rate. Wang et al. [16] proposed to train heterogeneous CNN with hierarchical supervision information obtained from the segmentation-recognition lattice. Wu et al. [18] investigated the effects of neural network language models and CNN shape models.

However, over-segmentation-based methods have their own limitations: their effectiveness highly rely on the over-segmentation step. In other words, the system would perform badly if the text is not well segmented. To solve this problem, segmentation-free methods have been proposed based on deep learning frameworks [9,11,12,19]. Messina and Louradour [9] applied Multi-Dimensional Long Short Term Memory (MDLSTM) with Connectionist Temporal Classifier (CTC) to offline HCTR. Shi et al. [11] proposed a network called Convolutional Recurrent Neural Network (CRNN) which consists of convolutional layers, recurrent layers and a transcription layer for scene text recognition. Wu et al. [19] presented a separable MDLSTM network which aimed to explore deep structure while at the same time reducing the computation efforts and resources of naive MDLSTM. As is well-known, deep learning is data driven, and data preprocessing and augmentation is very important for the system. In 2001, Vinciarelli and Luettin [14] proposed a normalization technique for cursive handwritten English words by removing the slant and slope of the words. Inspired by this work, Chen et al. [1] adopted similar method in the data preprocessing step of their system for online HCTR. However, to the best of our knowledge, no works have studied the importance of data preprocessing and augmentation for offline HCTR systematically.

In this paper, we propose a high-performance system for offline HCTR, which includes a data preprocessing, augmentation pipeline and a CNN-ResLSTM model. To prove the validity of the proposed method, we use the methods of controlling variates. Firstly, we train the baseline model with the CNN-ResLSTM (Long Short Term Memory with residual connections) model using the original training samples generated from CASIA-HWDB2.0-2.2. And then, the data augmentation and preprocessing methods including text sample generation, text sample preprocessing and text sample synthesis have been added to the training process step by step. Experiments show that the proposed pipeline can effectively and robustly improve the performance of the system. After verifying the validity of the proposed data augmentation and preprocessing pipeline, we adapt it on three models – the proposed CNN-ResLSTM model, the traditional CNN-LSTM and CNN-BLSTM models, and verify the superiority of our CNN-ResLSTM model. Finally, by combining with language model, our system outperforms the previous state-of-the-art system with a correct rate (CR) of 97.28% and an accurate rate (AR) of 96.97% on the CASIA-HWDB, and a CR of 96.99% and an AR of 96.72% on the ICDAR-2013 handwriting competition dataset. Furthermore, we accelerate and compress the CNN-ResLSTM model with Tucker decomposition [7], singular value decomposition (SVD) and adaptive drop weights (ADW) [20,22] to make it more practical for real-time applications. On the ICDAR-2013 database, the accelerated and

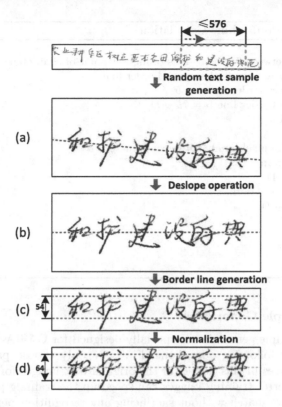

Fig. 1. Data preprocessing and augmentation pipeline.

compressed model requires only 146 ms for one text line image on average on a i7-8700K CPU with a single thread and 2.8 MB for storage, yet maintains nearly the same recognition accuracies as compared to the full model.

This paper is organized as follows: Sect. 2 introduces the data preprocessing and augmentation pipeline detailly; Sect. 3 gives the whole end-to-end recognition system; Sect. 4 presents experimental results and Sect. 5 makes a conclusion.

2   Data Preprocessing and Augmentation Pipeline

In this section, we introduce our data preprocessing and augmentation pipeline, including text sample generation, text sample preprocessing and text sample synthesis, following the process steps at training phase, as well as testing sample preprocessing.

Algorithm 1: Border Line Calculation

Input:
Difference D between $\mathcal{H}$ and Y-axis of highest point of characters
tolerable pixels K outside the upper border line
Output: Upper border line $y = b$
1   initialize upper border line b $= \mathcal{H}$ + D/2
2   **while** $D \geq 1$ **do**
3   D = D/2
4   calculate the pixel number n beyond line $y = b$
5   **if** $n < K$ **then**
6 | b = b - D
7   **else if** $n > K$ **then**
8 | b = b + D
9   **else**
10 | break
11   **end**
12 **end**

2.1   Text Sample Generation

Training text sample generation is specially designed for CASIA-HWDB2.0-2.2 dataset with the following operations. First, we only choose part (576 pixel length of the images in this paper) of the original text images for training each time, because shorter training images can speed up the training process several times for each mini-batch without sacrificing any recognition performance. As for testing, the length of the image is increased to 3,000 to contain the full text lines in the test sets. Second, we shuffle the characters within training samples to avoid the overfitting problem, because the test set of CASIA-HWDB2.0-2.2 share the same corpus with the training set (although written by different writers).

Thanks to the careful and detailed annotation of database CASIA-HWDB2.0-2.2, document images are not only segmented at text line level, but also at character level with bounding box information available. As shown by red arrow in Fig. 1, we first randomly select a character from the text samples and then extend in the right direction until length of 576 is reached. If the end of the original text line is reached while the generated text line are far smaller than length of 576, we continue to extend in the left direction. After that, we can crop the corresponding area from the original images and perform character shuffle operation with respect to their bounding box positions. Typical randomly generated training sample is shown in Fig. 1(a).

2.2   Text Sample Preprocessing

Offline handwritten text line suffers from the diversity of writing style. In this paper, we treat the offline handwritten text line images as pixels collection. Denote the pixel collection of the randomly generated training sample as follows:

$$\mathbf{I} = \{p_i | p_i = ((x_i, y_i),\ v_i),\ i = 1, \cdots, N\}, \tag{1}$$

Algorithm 2: Synthesize text samples for mixed training

Input:

isolated character set: D

total number of text samples to synthesize: N

superscript symbol set: S_p

subscript symbol set: S_b

height of target synthesized text samples: H

width of target synthesized text samples: W

Output: N synthesized text samples of shape $H \times W$

1 **for** $i = 1$ *to* N **do**
2   initialize a white background image G of shape $H \times W$
3   initialize column position $k = 5$
4   **while** 1 **do**
5   randomly fetch a character sample $A \in D$
6   get the shape of A: $h \times w$
7   **if** $h > H$ **then**
8   scale down A by scale factor $r = H/h$
9   **else if** $A \in S_p$ **then**
10   pad A to shape $85 \times w$ by padding white pixels below A
11   **else if** $A \in S_b$ **then**
12   pad A to shape $85 \times w$ by padding white pixels above A
13         $s = k + w$
14   **if** $s < W$ **then**
15   place A between columns $[k, s]$ in the center rows of G
16   generate a random integer $d \in [-2, 15]$
17             $k = s + d$
18   **else**
19   break
20   **end**
21   **end**
22   output one synthesized text sample G
23 **end**

---

where (x_i, y_i) denote the (x, y) coordinate, v_i represents the pixel value, and N is the product of image height and width. Therefore, the pixel collection of foreground characters in the text image can be represented as:

$$\mathbf{C} = \{c_j | c_j = ((x_j, y_j), \ v_j), v_j \neq 255, c_j \in \mathbf{I}\}, \tag{2}$$

where we neglect the background with pixel value of 255.

Now, given the coordinates of character pixels, we can apply de-slope operation on the generated training samples with linear function $x_j \times \hat{k} + \hat{b}$, which is derived through linear least square curve fitting approach as follows:

$$\hat{k}, \hat{b} = \arg \min_{k,b} \sum_j \|x_j \times k + b - y_j\| \tag{3}$$

Then, the de-slope operation is accomplished by transforming the Y-axis of the coordinates of character pixels as follows:

$$y'_j = (y_j - (x_j \times \hat{k} + \hat{b})) + \mathcal{H}, \quad c_j = ((x_j, y_j), \ v_j) \in \mathbf{C}, \tag{4}$$

where $\mathcal{H}$ is chosen as half the height of the normalized image, e.g. 64 in this paper. As shown in Fig. 1(b), the de-slopped text sample can now be represented as follows:

$$\mathbf{C}' = \{c'_j | c'_j = ((x_j, y'_j), \ v_j), v_j \neq 255, c_j \in \mathbf{I}\} \tag{5}$$

Next, we remove redundant blank areas in the upper and lower parts of the image, as shown in Fig. 1(c). And then, we briefly introduce *border line* of the text sample and its calculation, as shown by dash blue line in Fig. 1(c). As we all know, the height of text image are extremely unstable, depending on the highest and lowest point of the characters; thus, we need some stable measurements of the height of all text samples for normalization. In this paper, we present border line to estimate the 'height' of a text sample with the following assumption. First, a border line is a straight line parallel to the X-axis. Second, the pixels of characters between the upper and lower border line should be approximately 80%, i.e. about 20% of pixels (tolerable pixels) equally distributed outside the upper and lower border lines. In Algorithm 1, we detail the calculation of upper border line, and the lower border line is calculated in a similar way.

Finally, we rescale the text images so that distance between the upper and lower border line is exactly 64, and shift character pixels so that the border lines are mediately symmetry about picture in the vertical direction as shown in Fig. 1(d).

2.3 Text Sample Synthesis

We synthesize training samples using isolated characters of CASIA-HWDB1.0-1.2 for two reasons: (1) We want to enrich the text training set so that the distribution of the training set can cover more writing styles and habits of real writers. (2) The training set of CASIA-HWDB2.0-2.2 only has 2703 character classes that is not enough for real world applications, so we want to expand the character classes of our system to make it more practical.

At training phase, the input images to our model are gray-scale, with height and width fixed to 126 and 576, respectively. For consistency, we synthesize 750,000 text line images with the same size based on isolated characters of CASIA-HWDB1.0-1.2. The detail is described in Algorithm 2. The synthesized text samples are used together with text samples generated using aforementioned methods to create a mixed training set. It is noteworthy that, we put the isolated characters on the same horizontal line when we synthesize the sample, so we do not need to apply the preprocessing steps to the synthesized text samples.

2.4 Testing Sample Preprocessing

At testing phase, we evaluate our system on datasets described in Sect. 4. Given a testing sample image, we first convert it to gray-scale with height of 126 and

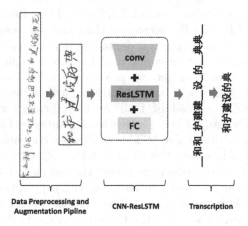

Fig. 2. Overview of our end-to-end recognition system.

Fig. 3. One ResLSTM layer in our model. It is constructed by introducing a shortcut (i.e. residual connection) into the naive LSTM layer.

width of 3,000 (3,000 is enough to contain the test samples in the test sets), then normalize it with the text sample preprocessing operations described in Sect. 2.2.

3 End-to-end Recognition System

As shown in Fig. 2, our end-to-end recognition system consists of three components. First, the raw input text image is processed by the data preprocessing and augmentation pipeline. After that, the convolutional neural network (CNN) extracts a feature sequence from the processed image and fed into the ResLSTM module to generate a probability distribution over the character dictionary for each time step. Finally, the transcription layer derives a label sequence from the distribution. The detailed setting of the proposed recognition system is described as follows.

3.1 CNN-ResLSTM Model

The offline text data has both visual (optical characters) and sequential (context) properties. Since CNN is a powerful visual feature extractor, while recurrent neural network (RNN) is good at modeling sequence, it's natural to combine CNN and RNN to explore their full potential for offline HCTR. Convolutional Recurrent Neural Network (CRNN) [11] was proposed to recognize scene text by combining CNN and Bidirectional-LSTM (BLSTM). Residual connections, the key component of ResNet [5], accelerate the convergence speed and increase network performance by reducing the degradation problem of deep networks, transferring the network to an ensemble model of many shallow models [13].

Inspired by CRNN and ResNet, we proposed a CNN-ResLSTM model by introducing residual connections into the RNN part of the CRNN structure. Besides, we use LSTM instead of BLSTM for efficiency consideration. As explained in [21], CNN-ResLSTM model has the advantages of faster convergence and better performance than naive LSTM. Our model consists of 11 convolution layers and 3 ResLSTM layers each with 512 LSTM cells. ReLU activation function is used after each convolution layer except the last one. Batch normalization is used after the last convolution layer. Dropout with drop-ratio 0.3 is introduced after each ResLSTM layer. A diagram of the ResLSTM layer is shown in Fig. 3. We illustrate the forward process of the CNN-ResLSTM in detail as follows.

Given input gray-scale image with shape of $1 \times 126 \times W$ ($channel \times height \times width$), the CNN module extracts features from the processed image and outputs feature maps with shape of $512 \times 1 \times [W/16]$. These feature maps are then converted to $[W/16]$ frames of feature vectors, with each frame containing a 512-dimensional feature vector and serving as one time step for the ResLSTM module. The ResLSTM module generates feature vector for each frame, and a fully connected (FC) layer predict probability distribution with a 7357-dimensional score vector for each frame. The first dimension of the score vector is reserved as the $blank$ symbol of Connectionist Temporal Classification (CTC) [4], and the rest 7356 dimensions correspond to our character set.

3.2   Transcription

Since our CNN-ResLSTM model does a per-frame prediction and the number of frames is usually more than the length of label sequence, we need to decode the predictions, i.e., mapping the predictions to their corresponding labels. Besides, we need to define a special loss function, so that we can train the model end-to-end. To solve this problem, we adopt CTC [4] as our transcription layer. CTC allows the CNN module and ResLSTM module to be trained jointly without any prior alignment between input image and its corresponding label sequence. As suggested in [4], since the exact positions of the labels within a certain transcription cannot be determined, CTC consider all the locations where the labels could appear, which allows a network to be trained without pre-segmented data. A detailed forward-backward algorithm to efficiently calculate the negative log-likelihood (NLL) loss between the input sequences and the target labels was described in [3].

3.3   Decoding and Language Model

Decoding the CTC trained CNN-ResLSTM model can be easily accomplished by the so-called $best\ path\ decoding$, also named $naive\ decoding$ [3]. To further increase the performance of our system, explicit language model (LM) is integrated to explore the semantic relationships between characters. By incorporating lexical constraints and prior knowledge about the language, language model can rectify some obvious semantic errors, thus improves the recognition result. In this paper, we only considered character trigram language model in experiments and a refined beam search algorithm [4] is adopted to decode our CNN-ResLSTM model.

3.4 Model Acceleration and Compression

In our previous work, [20] and [22], we have proposed a framework including ADW to compress the CNN and LSTM network, and use the SVD decomposition method to accelerate the LSTM and FC layer. For CNN acceleration, we adopt the tucker decomposition methods [7]. In this paper, we combined them together to accelerate and compress our CNN-ResLSTM model. And we also give some of our experience in accelerating and compressing the model, which will be discript in Sect. 4.3.

4   Experiments

4.1   Dataset

The training set of CASIA-HWDB [8] was used as training dataset. Specifically, the training set of CASIA-HWDB is divided into 6 subsets, in which CASIA-HWDB1.0-1.2 contain isolated characters and CASIA-HWDB2.0-2.2 contain unconstrained handwritten texts. In the training set of isolated characters, there are 3,118,477 samples of 7,356 classes. The training set of the texts contains 41,781 text lines of 2,703 classes. As mentioned before, we use the isolated characters to synthesize 750,000 text lines to enrich the training set. We evaluate the performance of our system on the test sets of CASIA-HWDB (denoted as **D-Casia**) and ICDAR-2013 Chinese handwriting recognition competition [23] Task 4 (denoted as **D-ICDAR**), which contain 10,449 text lines and 3,432 text lines respectively.

Note that, in CASIA-HWDB2.0-2.2 and the competition test set, characters outside the 7356 isolated character classes are not used in both training and testing.

4.2   Experimental Settings

We implemented our system on Caffe [6] deep learning framework. The CNN-ResLSTM model was trained with CTC criterion using mini-batch based on stochastic gradient decent (SGD) method with momentum. Momentum and weight decay were set to 0.9 and 1×10^{-4} respectively. Batch size was set to 32. The initial learning rate was set to 0.1 for the first 150,000 iterations, then reduced to 0.01 and 0.001 for two more 50,000 iterations. Inspired by [1],

Table 1. Effect of data processing pipeline (without LM)

Operations		D-Casia		D-ICDAR	
		CR (%)	AR (%)	CR (%)	AR (%)
baseline		90.28	89.75	85.07	84.50
text sample generation[a]		92.25	91.76	78.56	77.59
+shuffling characters[a,b]		90.02	89.49	81.05	80.28
+text sample preprocessing[a,b]		94.50	94.21	89.51	89.11
+text sample synthesis[a,b]	0.1	94.48	94.17	90.37	89.99
(with different ratios)	0.2	94.60	94.23	91.26	90.79
	0.3	94.88	94.44	91.67	91.08
	0.4	93.63	93.35	90.35	89.99
	0.5	**95.37**	**94.90**	**92.13**	**91.55**
	0.6	94.39	93.91	91.32	90.67
	0.7	94.07	93.57	90.95	90.31
	0.8	93.29	92.93	90.31	89.87
	0.9	92.25	91.78	89.00	88.46
	1.0	76.96	76.60	73.16	72.89

[a]In these experiments, only one CNN-ResLSTM model was used but with different data processing methods applied incrementally, which shows the effect of the proposed data preprocessing and augmentation pipeline.
[b]In the table, the '+' indicates the method and the methods above it were applied together.

we turned off the shuffle operation applied to the random generated text samples for another 50,000 iterations. Therefore, we trained the model for 300,000 iterations in total and it takes approximately 20 h to reach convergence using a single TitanX GPU.

4.3   Experimental Results

Effect of Data Processing Pipeline. We evaluated our CNN-ResLSTM model using the proposed data processing pipeline on both D-Casia and D-ICDAR, as shown in Table 1. The *baseline* refers to the CNN-ResLSTM model trained with the training set from CASIA-HWDB2.0-2.2 without using the proposed data processing pipeline.

We started our experiments using training samples generated from CASIA-HWDB2.0-2.2, but without characters shuffle operation. And then, we introduce the characters shuffle operation, text sample synthesis and synthesized text samples step by step. As we can see, when we trained the model only with the generated samples, the performance on D-CASIA is better than the baseline but much worse on D-ICDAR. The reason may be that D-ICDAR has more cursively samples, and the training set for the baseline may include more cursively

Table 2. Effects of residual connections (without LM)

Models	D-Casia		D-ICDAR	
	CR (%)	AR (%)	CR (%)	AR (%)
CNN-LSTM	93.53	93.11	89.37	88.92
CNN-BLSTM	93.87	93.60	88.73	88.27
CNN-ResLSTM	**95.37**	**94.90**	**92.13**	**91.55**

Table 3. Acceleration and compression result

Models	Storage (MB)	GFLOPs	Speed (ms)	D-CASIA		D-ICDAR	
				CR (%)	AR (%)	CR (%)	AR (%)
Baseline	61	16.57	318	**95.37**	**94.90**	**92.13**	**91.55**
Compact model	**2.8**	**4.46**	**146**	94.37	93.83	91.18	90.50

samples than the generated samples, since the length of the generated samples is shorter than the original samples. By adding character shuffle operation, we observed that the performance is improved on D-ICDAR, but surprisingly reduced on D-Casia. This is because when we only using training samples generated from CASIA-HWDB2.0-2.2, the model overfits D-Casia (D-Casia share the same corpus with the training set of CASIA-HWDB2.0-2.2) and the shuffle operation can reduce this phenomenon. By adding text samples preprocessing, we observed the performances are significantly improved, especially on D-ICDAR. This is probably because D-ICDAR has much more cursively written samples, and our text samples preprocessing operation can effectively normalize the text samples to improve recognition result. Finally, we performed experiments by randomly picking samples from the generated texts or synthesized texts with different ratios. We obtain the best model with ratio of 0.5, which we adopt in the following experiments.

Effects of Residual Connections. As shown in Table 2, to investigate the effect of the residual connections in CNN-ResLSTM model, we further constructed two models namely the CNN-LSTM and CNN-BLSTM. The CNN-LSTM was constructed by just removing the residual connections of CNN-ResLSTM, while the CNN-BLSTM was derived by replace LSTM layers of CNN-LSTM with BLSTM layers of 1024 cells (2 × 512). As shown in Table 2, the CNN-ResLSTM achieved the best results among these three models, which verifies the significance of the residual connections. It is interesting to note that the results of CNN-LSTM and CNN-BLSTM are comparable, even though BLSTM has the potential to capture contextual information from both directions.

Acceleration and Compression. To build a compact model, we firstly employed low-rank expansion method to accelerate the model. A natural idea is

Table 4. Comparison with the start-of-the-art methods

Methods	Without LM				With LM			
	D-Casia		D-ICDAR		D-Casia		D-ICDAR	
	CR (%)	AR (%)	CR (%)	AR (%)	CR (%)	AR (%)	CR (%)	AR (%)
HIT2 2013 [23]	-	-	-	-	-	-	88.76	86.73
Wang et al. [15]	-	-	-	-	91.39	90.75	-	-
MDLSTM [9]	-	-	-	83.50	-	-	-	89.40
Du et al. [2]	-	-	-	83.89	-	-	-	93.50
NA-CNN [16]	93.24	92.04	90.67	88.79	96.28	95.21	95.53	94.02
CNN-HRMELM [18]	-	-	-	-	95.88	95.95	96.20	96.32
SMDLSTM [19]	-	-	87.43	86.64	-	-	-	92.61
Wang et al. [17]	-	-	-	89.66	-	-	-	96.47
Peng et al. [10]	-	-	89.61	90.52	-	-	94.88	95.51
Ours, full model	**95.37**	**94.90**	**92.13**	**91.55**	**97.28**	**96.97**	**96.99**	**96.72**
Ours, compact model	94.37	93.83	91.18	90.50	96.85	96.22	96.75	96.38

that decomposing the CNN, LSTM and FC layers at once and then finetuning the network, but we find that the accuracy of the model would decrease a lot with such decomposing strategy. We adopt a 3-step strategy to decompose the model: (1) Decompose the convolutional layers from original CNN-ResLSTM model, keep the LSTM and FC layers intact, and only update the decomposed layers. Then we get the model named **model-1**; (2) Decompose the LSTM and FC layers from original CNN-ResLSTM model, keep the convolutional layers intact, and only update the decomposed layers. Then we get the model named **model-2**; (3) Extract the decomposed CNN layers, LSTM layers and FC layers from **model-1** and **model-2** respectively, and combine them together. Finally, we fine-tune the combined model and obtain the final decomposed model. And for the pruning part, we prune 10% connections for the layers which contain more than 100,000 parameters and clustering with 256 cluster centers.

As shown in Table 3, with about 1% accuracy loss, the storage of the model has dropped from 61 MB to 2.8 MB, which is 21.8 times smaller. And the GFLOPs is reduced from 16.57 to 4.46, which accelerate 3.7 times theoretically. In our forward implementation, the average time of a forward calculation is reduce from 318 ms to 146 ms on a i7-8700K CPU. Furthermore, the compact model could also be combined with the language model to improve the performance.

Comparison with Other Methods. As we can see in Table 4, our method achieved the best results in all measurements compared our method with other previous state-of-the-art approaches. The approaches we compared include over-segmentation-based approaches [16,18] and segmentation-free method [19] based on the CNN-RNN-CTC framework. Compared with [19], we can verify the significance of the proposed data processing pipeline and residual LSTM, which are the main differences between our method and [19]. It is worth noting that after model acceleration and compression, the result of the compact model is still outperform than other methods on dataset D-Casia and comparable on dataset D-ICDAR.

5   Conclusion

In this paper, we proposed a high-performance method for offline HCTR, which consists of a data preprocessing and augmentation pipeline and a CNN-ResLSTM model. The text images are first processed by the pipeline, and then fed into the CNN-ResLSTM for training and testing. Our data preprocessing and augmentation pipeline includes three steps: training text sample generation, text sample preprocessing, and text sample synthesis using isolated characters. The model consists of two parts: CNN part and ResLSTM part, which are jointly trained with CTC criterion. To further improve the performance of our system, we integrated a character trigram language model to rectify some obvious semantic errors. Experiments show that the performance of our CNN-ResLSTM model is improved step by step in our data preprocessing and augmentation pipeline. Compared with previous state-of-the-art approaches, our method exhibits superior performance on dataset D-Casia and D-ICDAR. Furthermore, in order to make our system more practical, we employed the model acceleration and compression method to built a compact model with small storage size and fast speed, which still outperforms than the previous state-of-the-art approaches on dataset D-Casia and have a comparable result on dataset D-ICDAR.

Acknowledgement. This research is supported in part by NSFC (Grant No. 61936003), the National Key Research and Development Program of China (No. 2016YFB1001405), GD-NSF (no. 2017A030312006), Guangdong Intellectual Property Office Project(2018-10-1).

References

1. Chen, K., et al.: A compact CNN-DBLSTM based character model for online handwritten Chinese text recognition. In: International Conference on Document Analysis and Recognition (ICDAR), pp. 1068–1073 (2017)
2. Du, J., Wang, Z.R., Zhai, J.F., Hu, J.S.: Deep neural network based hidden Markov model for offline handwritten Chinese text recognition. In: International Conference on Pattern Recognition (ICPR), pp. 3428–3433. IEEE (2016)
3. Graves, A.: Supervised Sequence Labelling. Springer, Heidelberg (2012). https://doi.org/10.1007/978-3-642-24797-2
4. Graves, A., Jaitly, N.: Towards end-to-end speech recognition with recurrent neural networks. In: Proceedings of the 31st International Conference on Machine Learning, pp. 1764–1772 (2014)
5. He, K., Zhang, X., Ren, S., Sun, J.: Deep residual learning for image recognition. In: Proceedings of the IEEE Conference on Computer Vision and Pattern Recognition, pp. 770–778 (2016)
6. Jia, Y., et al.: Caffe: convolutional architecture for fast feature embedding. In: Proceedings of the 22nd ACM International Conference on Multimedia, pp. 675–678. ACM (2014)
7. Kim, Y.D., Park, E., Yoo, S., Choi, T., Yang, L., Shin, D.: Compression of deep convolutional neural networks for fast and low power mobile applications. arXiv preprint arXiv:1511.06530 (2015)

8. Liu, C.L., Yin, F., Wang, D.H., Wang, Q.F.: Casia online and offline Chinese handwriting databases. In: International Conference on Document Analysis and Recognition (ICDAR), pp. 37–41. IEEE (2011)
9. Messina, R., Louradour, J.: Segmentation-free handwritten Chinese text recognition with LSTM-RNN. In: International Conference on Document Analysis and Recognition (ICDAR), pp. 171–175. IEEE (2015)
10. Peng, D., Jin, L., Wu, Y., Wang, Z., Cai, M.: A fast and accurate fully convolutional network for end-to-end handwritten Chinese text segmentation and recognition. In: 2019 International Conference on Document Analysis and Recognition (ICDAR), pp. 25–30. IEEE (2019)
11. Shi, B., Bai, X., Yao, C.: An end-to-end trainable neural network for image-based sequence recognition and its application to scene text recognition. IEEE Trans. Pattern Anal. Mach. Intell. **39**(11), 2298–2304 (2017)
12. Sueiras, J., Ruiz, V., Sanchez, A., Velez, J.F.: Offline continuous handwriting recognition using sequence to sequence neural networks. Neurocomputing **289**, 119–128 (2018)
13. Veit, A., Wilber, M.J., Belongie, S.: Residual networks behave like ensembles of relatively shallow networks. In: Advances in Neural Information Processing Systems, pp. 550–558 (2016)
14. Vinciarelli, A., Luettin, J.: A new normalization technique for cursive handwritten words. Pattern Recogn. Lett. **22**(9), 1043–1050 (2001)
15. Wang, Q.F., Yin, F., Liu, C.L.: Handwritten chinese text recognition by integrating multiple contexts. IEEE Trans. Pattern Anal. Mach. Intell. **34**(8), 1469–1481 (2011)
16. Wang, S., Chen, L., Xu, L., Fan, W., Sun, J., Naoi, S.: Deep knowledge training and heterogeneous CNN for handwritten Chinese text recognition. In: International Conference on Frontiers in Handwriting Recognition (ICFHR), pp. 84–89 (2017)
17. Wang, Z.R., Du, J., Wang, W.C., Zhai, J.F., Hu, J.S.: A comprehensive study of hybrid neural network hidden markov model for offline handwritten Chinese text recognition. Int. J. Doc. Anal. Recogn. (IJDAR) **21**(4), 241–251 (2018)
18. Wu, Y.C., Yin, F., Liu, C.L.: Improving handwritten Chinese text recognition using neural network language models and convolutional neural network shape models. Pattern Recogn. **65**, 251–264 (2017)
19. Wu, Y.C., Yin, F., Zhuo, C., Liu, C.L.: Handwritten Chinese text recognition using separable multi-dimensional recurrent neural network. In: International Conference on Document Analysis and Recognition (ICDAR), pp. 79–84 (2017)
20. Xiao, X., Jin, L., Yang, Y., Yang, W., Sun, J., Chang, T.: Building fast and compact convolutional neural networks for offline handwritten Chinese character recognition. Pattern Recogn. **72**, 72–81 (2017)
21. Xie, Z., Sun, Z., Jin, L., Ni, H., Lyons, T.: Learning spatial-semantic context with fully convolutional recurrent network for online handwritten Chinese text recognition. IEEE Trans. Pattern Anal. Mach. Intell. **40**(8), 1903–1917 (2017)
22. Yang, Y., et al.: Accelerating and compressing LSTM based model for online handwritten Chinese character recognition. In: International Conference on Frontiers in Handwriting Recognition (ICFHR), pp. 110–115. IEEE (2018)
23. Yin, F., Wang, Q.F., Zhang, X.Y., Liu, C.L.: ICDAR 2013 Chinese handwriting recognition competition. In: International Conference on Document Analysis and Recognition (ICDAR), pp. 1464–1470. IEEE (2013)

24. Zhou, X.D., Wang, D.H., Tian, F., Liu, C.L., Nakagawa, M.: Handwritten Chinese/Japanese text recognition using semi-Markov conditional random fields. IEEE Trans. Pattern Anal. Mach. Intell. **35**(10), 2413–2426 (2013)
25. Zhou, X.D., Zhang, Y.M., Tian, F., Wang, H.A., Liu, C.L.: Minimum-risk training for semi-Markov conditional random fields with application to handwritten Chinese/Japanese text recognition. Pattern Recogn. **47**(5), 1904–1916 (2014)

ALEC: An Accurate, Light and Efficient Network for CAPTCHA Recognition

Nan Li[✉], Qianyi Jiang, Qi Song, Rui Zhang, and Xiaolin Wei

Meituan-Dianping Group, Beijing, China
{linan21,jiangqianyi02,songqi03,zhangrui36,weixiaolin02}@meituan.com

Abstract. The CAPTCHA (Completely Automated Public Turing Test to Tell Computers and Humans Apart) is a common and effective security mechanism applied by many websites and applications. CAPTCHA recognition is an important and practical problem in text recognition research. Compared with traditional methods, DCNN (Deep Convolutional Neural Network) has achieved competitive accuracy in CAPTCHA recognition recently. However, current CAPTCHA recognition researches based on DCNN usually use conventional convolution network, which causes high computation complexity and great computing resource consumption. Aiming at the problems, we propose an **A**ccurate, **L**ight and **E**fficient network for **C**APTCHA recognition (ALEC) based on the encoder-decoder structure. The ALEC can greatly reduce the computation complexity and parameters while ensuring the recognition accuracy. In this paper, standard convolutions are replaced by depthwise separable convolutions to improve computational efficiency. The architecture utilizes group convolution and convolution channels reduction to build a deep narrow network, which reduces the model parameters and improves generalization performance. Additionally, effective and efficient attention modules are applied to suppress the background noise and extract valid foreground context. Experiments demonstrate that ALEC not only has higher speed with fewer parameters but also improves the accuracy of CAPTCHA recognition. In detail, the ALEC achieves about 4 times speed up over the standard ResNet-18 while reducing 97% parameters.

Keywords: CAPTCHA recognition · Convolutional Neural Network · Well-designed network

1 Introduction

With the development of artificial intelligence and the Internet, how to distinguish between humans and machines has become a prevalent topic. Since Luis [24] first proposed the concept of CAPTCHA (Completely Automated Public Turing Test to Tell Computers and Humans Apart), it has played an increasingly important role in the Internet security area. At present, CAPTCHA almost becomes the standard protection from attackers and is applied by many websites and applications.

© Springer Nature Switzerland AG 2020
X. Bai et al. (Eds.): DAS 2020, LNCS 12116, pp. 60–73, 2020.
https://doi.org/10.1007/978-3-030-57058-3_5

Among the many types of CAPTCHAs, such as image-based, audio-based and puzzle-based CAPTCHAs, text-based is the most convenient and commonly used type. In this paper, we focus on the text-based CAPTCHA breaking technique because it shows the importance in various ways. For example, CAPTCHA breaking can verify the security of existing CAPTCHAs then promote CAPTCHA design algorithms. Furthermore, the technology of CAPTCHA recognition can not only refresh the limit of Turing test but also inspire text recognition approaches in other fields.

The traditional CAPTCHA recognition method typically needs three steps: preprocessing, segmentation and single character recognition to deal with the low resolution, noisy, deformation and adhesive text-based CAPTCHA. But this process is rather complicated and the accuracy is relatively low. Inspired by the great success of Deep Convolutional Neural Networks (DCNN), current CAPTCHA recognition methods always prefer deep learning techniques. However, DCNN with high accuracy tends to be inefficient with respect to size and speed, which greatly restricts their applications on the computationally limited platform. Therefore, we aim to explore a highly efficient architecture specially designed for low memory resources.

In this paper, we propose an accurate yet light and efficient model named ALEC for CAPTCHA recognition. The proposed method is inspired by scene text recognition and crafted especially for CAPTCHAs. A ResNet-like backbone and Connectionist temporal classification (CTC) are implemented as the whole framework. To overcome the drawbacks of conventional DCNN, such as computation complexity and model storage, many techniques are applied for optimization. More precisely, we adopt depthwise separable convolutions, channels reduction and group convolutions to achieve the trade-off between representation capability and computational cost, the channel shuffle operation is integrated to help the information flowing across feature channels and improving accuracy. Moreover, a light-weighted attention module called Convolutional Block Attention Module (CBAM) is utilized to enhance the feature extraction ability. In summary, our main contributions are summarized as follows:

(1) We replace standard convolution with depthwise separable convolution to reduce parameters and improve recognition accuracy.
(2) We introduce Group Convolution (GConv) and channels reduction to decrease network redundancy and guarantee generalization performance.
(3) We integrate Convolutional Block Attention Module (CBAM) with our backbone that can improve accuracy with negligible overheads.
(4) Experiments conducted on two generated CAPTCHA datasets and one real-world CAPTCHA dataset verify the effectiveness and efficiency of the ALEC.

The remaining parts of the paper are organized as follows: In Sect. 2, we illustrate the related work of CAPTCHA recognition and efficient model design. In Sect. 3, we describe our ALEC method in detail. In Sect. 4, we present the details and results of experiment and in Sect. 5, we conclude the paper.

2   Related Work

CAPTCHA Recognition. Nowadays, CAPTCHA plays an important role in the multimedia security system. The idea of CAPTCHA first appeared in the paper of Moni Naor [4] who intended to design a mechanism which is easy for the user while difficult for a program or computer to solve. In 2000, as the first commercial text-based CAPTCHA was designed by Carnegie Mellon University (CMU) team, the research on CAPTCHA breaking technology and CAPTCHA security also started. At present, the frequently used CAPTCHAs [21] are text-based, image-based, audio-based, etc. Since the text-based CAPTCHA is simple to implement and easy for human users to pass, it is most wildly used in various scenarios such as hot-mail, yahoo, g-mail, QQ and so on. Some typical text-based CAPTCHA is as follows:

(a) Gimpy   (b) EZ-Gimpy   (c) PessimalPrint

Fig. 1. Samples of text-based CAPTCHA. (a) Gimpy randomly picks seven words from the dictionary and then renders a distorted image containing these words. (b) EZ-Gimpy contains a single word generated by various fonts and deformations. (c) The main idea of PessimalPrint is that low-quality text images are legible to human readers while still challenging optical character recognition (OCR).

In recent years, as the CAPTCHA design techniques become more robust by introducing large character set, distortion, adhesion, overlap or broken contours, various CAPTCHA recognition frameworks also come into being. The traditional methods usually locate a single number or character area in an image by segmentation and identify it. For example, Rabih et al. [15] dealt with the segmentation by fuzzy logic algorithm based on edge corners. Yan et al. [28] proposed an efficient segmentation method for Microsoft CAPTCHA and recognized it by multiple classifiers. In [9], two new segmentation techniques called projection and middle-axis point separation were proposed with line cluttering and character warping. All in all, these systems mentioned above are with multiple stages, they are originally designed for a specific type of CAPTCHA and each module is optimized independently. As a result, errors compound between modules can be significant and the systems tend to behave poorly in terms of generalization performance.

In the last several years, most of the text-based CAPTCHA uses Crowded Characters Together (CCT) which remarkably reduces the success rate of the segmentation process. Therefore, some alternative promising approaches resort to the high level features [27], such as deep learning [2,3,16]. The commonly used deep learning models in the CAPTCHA recognition are CNN, RNN, and so forth. [4] utilized a model consisting of two convolutional layers to learn image features. A fixed number of Softmax layers were introduced to predict each character of the fixed-length CAPTCHA. For the variable-length of CAPTCHA, they adopt a series of recurrent layers where the number of layers equals the maximum number of possible characters in CAPTCHA. Jing et al. [25] designed a modified DenseNet and achieved excellent performance on the CAPTCHA datasets of 9th China University Student Service Outsourcing Innovation and Entrepreneurship Competition. [12] chosen a combination of CNNs followed by the RNN trained on synthetic data and successfully broke the real-world CAPTCHA currently used by Facebook and Wikipedia. [29] applied the RNN in CAPTCHA recognition to avoid gradient vanishing problems and it can keep the long context in the network. [17] proposed a novel decoding approach based on the multi-population genetic algorithm and used the two-dimensional RNN to obtain relative information of both the horizontal and vertical context.

Efficient Models. As neural networks achieve remarkable success in many visual recognition tasks, the expensive computation and intensive memory become an obstacle to the deployment in low memory devices and applications with strict latency requirements. The idea of reducing their storage and computational cost has been a hot issue and tremendous progress has been made in this area. In the aspect of model simplifying, the common methods include model compression and efficient model design. Model compression [7,10,23,31] is to compress original model so that the network has fewer parameters yet little accuracy reduction. But model compression usually causes accuracy reduction and the reduction of model size is limited. The efficient model design represents a new neural network architecture that is specifically tailored for some desired computing ranges or resource-constrained environments. In recent years, the increasing needs for running deep neural networks on multiple devices encourage the study of various efficient model designs. Many states of the art models involved such an idea. For example, GoogLeNet considered improving utilization of the computing resources inside the network which ensures less computational growth while increasing the depth and width of the network [22]. MobileNet introduced a streamline based architecture with depthwise separable convolutions and global hyper-parameters to build light weight deep neural networks [8]. Afterward, an improvement of MobileNet called MobileNetv2 came into being [18], it is based on an inverted residual structure where the non-linearities are removed in the narrow layers and the shortcut connections are between the thin bottleneck layers. Another efficient model design family called ShuffleNet [30] and ShuffleNetv2 [14] utilize new operations such as group convolution and channel shuffle to improve efficiency and derive four guidelines for efficient network design.

3   Method

As shown in Fig. 2, the proposed model named ALEC is an encoder-decoder structure in which the encoder is a residue-like network and the decoder is implemented by the CTC [19]. In Sect. 3.1, an introduction about the overall

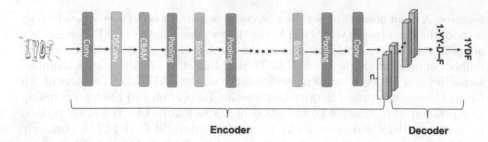

Fig. 2. Overview of the network architecture. The architecture consists of two parts: 1) encoder, which extracts feature sequence from the input image; 2) decoder, which generates final predicted sequence. Specifically, 'n' denotes the number of classes.

architecture of the ALEC is given. In Sect. 3.2, we describe the details of repeated building blocks in the ALEC.

3.1 Model Architecture

A residue-like network is used to extract feature representation from input CAPTCHA image, which compresses 2D image into 1D feature map and maintains the width. And then the CTC utilizes extracted features to produce the predicted sequence.

Residual Structure Model. This research utilizes residual structure as the backbone of the CAPTCHA recognition network to extract features. With the plain network depth increasing, accuracy gets saturated and then degrades rapidly which is called degenerative problem. The residual structure model [6] involves 'shortcut connections' to address the degradation problem. Shortcut connections are those skipping one or more layers. Skipping effectively simplifies the network by using fewer layers in the initial training stages. Since there are fewer layers to propagate through, residual learning speeds the convergence by reducing the impact of vanishing gradients. In comparison with the plain network, the residual network reformulates the layers as learning residual functions instead of learning unreferenced functions. The ResNets structure has several compelling advantages: these residual networks are easier to optimize and can gain accuracy from considerably increased depth.

Connectionist Temporal Classification. The CAPTCHA recognition requires to translate the extracted feature maps to the prediction of sequences of labels. The crucial step is to transform the network outputs into a conditional probability distribution over label sequences. The network can then be used as a classifier by selecting the most probable label sequence. We adopt the conditional probability defined in the Connectionist Temporal Classification (CTC) layer [5] which is proposed by Graves et al. The advantage of this structure is that it does not require explicit segmentation.

The CTC output layer includes the number of labels plus one unit where the extra unit represents the probability of observing a 'blank' or no label. Given an input sequence, the CTC layer outputs the probabilities of all possible ways of aligning all possible label sequences. Knowing that one label sequence can be represented by different alignments, the conditional probability distribution over a label sequence can be found by summing the total probabilities of all possible alignments. Given an input sequence x of length T, the probability of outputting π is represented by the product of the probability of each element of π, $y_{\pi_t}^t$ is interpreted as the probability of observing label k at time t and L' is the dataset of possible characters plus blank.

$$p(\pi|x) = \prod_{t=1}^{T} y_{\pi_t}^t, \forall \pi \in L'^T \tag{1}$$

The conditional probability of a given labeling l is the sum of the probabilities of all the paths π corresponding to it and the output of the classifier should be the most probable labeling for the input sequence:

$$p(l|x) = \sum_{\pi \in B^{-1}(l)} p(\pi|x) \tag{2}$$

$$h(x) = \arg \max_{l \in L \leq T} p(l|x). \tag{3}$$

One of the distinctive properties of CTC is that there are no parameters to be learned for decoding. Therefore, this addresses our target is to ensure the efficiency of encoder feature extraction network. In the next subsection, we will describe the details of block structure in ALEC.

3.2   Blocks in the ALEC

We improve the blocks in the ALEC to accelerate prediction and decrease model size while maintaining the recognition accuracy. As shown in Fig. 3, standard convolution is replaced by depthwise separable convolution and group convolution followed by channel shuffle operation. The CBAM introduced in the ALEC boosts performance with slight computation cost.

Depthwise Separable Convolution (DSConv). To decrease network redundancy and improve effectiveness, the depthwise separable convolutions are applied in our method to replace the standard convolutions. Depthwise separable convolution [20] divides the standard convolution into two parts: depthwise convolution and pointwise convolution. The depthwise convolutions and pointwise convolutions are illustrated in Fig. 7 where M is the number of input channels, N is the number of output channels and Dk * Dk is the kernel size. Each depthwise convolution filter only convolutes for per specific input channel. The pointwise convolutions combine multi-channel output of depthwise convolutions to create new features with the kernel size of 1 * 1. Taking the 3 * 3 convolution kernel used in the residual network as an example, in theory, depthwise separable convolutions can improve the efficiency by 9 times.

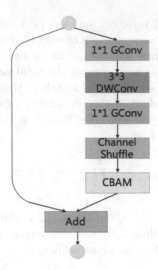

Fig. 3. Block structure in the ALEC. The block combines depthwise convolution, group convolution, channel shuffle and CBAM.

Group Convolution (GConv). Moreover, we combine group convolution with pointwise convolution as pointwise group convolution [30] in order to decrease network redundancy and guarantee generalization performance. GConv [11] is used to group the input feature map and convolute each group of features. We combine GConv with pointwise convolution from unit 2 to unit 4. Because the channel number of pointwise convolutions in unit 1 is less, the GConv operation is not performed. In the first pointwise convolution output of each block, the number of channels is extended to the same as the block output channel. In our experiment, we divided pointwise convolution into two groups.

There is no interaction between GConv. The output feature of GConv only comes from half of the input features. The input and output channels of different groups are not related, which will harm the feature expression and weight learning of convolution network. Therefore, we use channel shuffle in each block before adding the attention module which conduces feature interleaving and fusion.

Convolutional Block Attention Module (CBAM). Most of the well-designed CAPTCHAs are interfered by various kinds of noise including spots, curves or grids, therefore, we integrate the attention mechanism which plays an important role in feature extraction. Attention not only tells where to focus, but it also improves the representation of interests. [26] proposed a plug-and-play module for pre-existing base CNN architectures called Convolutional Block Attention Module.

This architecture consists of two attention modules: channel attention and spatial attention. Given an intermediate feature map, it first produces two spatial context features by computing an average pooling operation F_{avg}^c and a max-pooling operation F_{max}^c simultaneously. F_{avg}^c describes the global features and

F_{max}^c gathers distinctive object features. The channel attention map is forwarded by a shared network of multi-layer perceptron (MLP) with both descriptors.

$$M_c(F) = \sigma(MLP(F_{avg}^c) + MLP(F_{max}^c)) \qquad (4)$$

Similarly to the channel attention map, the spatial one is obtained by first generating a 2D average-pooled feature F_{avg}^s and a 2D max-pooled feature F_{max}^s across the channel axis. Afterward, those are concatenated and convolved by a standard convolution layer to produce the spatial feature map. As shown in Fig. 4, these two complementary attention modules are placed in the sequential arrangement.

$$M_a(F) = \sigma(f^{3\times3}([F_{avg}^s; F_{max}^s])) \qquad (5)$$

CBAM learns channel attention and spatial attention separately. By separating the attention generation process for a 3D feature map, it has much less computational and parameter overhead. Moreover, it can be plug at any convolutional block at many bottlenecks of the network.

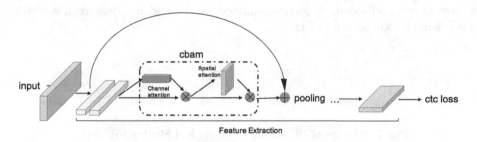

Fig. 4. An overview of the CBAM in framework.

4 Experiments

In this section, we conduct extensive experiments on 3 benchmarks to verify the efficiency and effectiveness of the proposed method. In Sect. 4.1, an introduction about training and testing datasets are given. In Sect. 4.2, we describe the implementation details of the experiments. Finally, Detailed experimental results and comparison of different configurations are presented in Sect. 4.3.

4.1 Datasets

We utilize two different public CAPTCHA generators and a collection of real-world CAPTCHA to build our datasets.

For the generated CAPTCHA dataset, we generate 20000 images as the trainset and 2000 images for testing and validation respectively. The generated CAPTCHA images consist of 4 randomly selected alphanumeric characters including upper and lower cases. The generated annotations are case insensitive.

As for the real-world CAPTCHA, we collect various types of CAPTCHA via websites and randomly shuffle these images to construct a mixed CAPTCHA trainset of over 200000 samples then 20000 images for test and validation respectively. The distribution of the number of each CAPTCHA is approximately equal. The annotations are case insensitive.

Hsiaoming[1] is a captcha library that generates image-based, text-based and audio-based CAPTCHAs. We only use text-based CAPTCHA with random color, curves and dots as noises. The text font is set to be DroidSansMono.

Skyduy[2] is another python-based public CAPTCHA generator with dense points background noise and a random color curve. The text font is randomly chosen from "FONT_HERSHEY_COMPLEX", "FONT_HERSHEY_SIMPLEX" and "FONT_ITALIC". Some samples are shown in Fig. 5.

Real-World CAPTCHA. As real-world CAPTCHA exhibits a lot of variation in their design, in order to verify the robustness and effectiveness of our proposed framework, we collect 24 different types of real-world CAPTCHA via web access. These samples cover various kinds of CAPTCHA features, including hollow shapes, adhesion, distortion, unfixed length and interference lines, etc. Some samples are shown in Fig. 6.

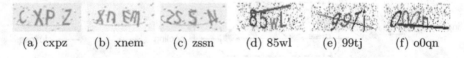

(a) cxpz   (b) xnem   (c) zssn   (d) 85wl   (e) 99tj   (f) o0qn

Fig. 5. Samples of Hsiaoming (a)(b)(c) and Skyduy (d)(e)(f)

4.2 Implementation Details

The network configurations in the experiments are summarized in Fig. 8. The ALEC modifies the standard convolutions of the first unit to a combination of pointwise convolutions and depthwise convolutions. For the rest of the units, group convolutions along with depthwise separable convolutions are adopted. Moreover, the CBAM module is embedded in every ResNet block to introduce spatial and channel attention. The input image size is fixed as 75×32. Some image process such as random rotation of -3 to $3°$, elastic deformation, random contrast, brightness, hue and saturation are applied for data augmentation. CTC output layer is used to produce the prediction sequence.

The network is trained with stochastic gradient descent with warm restarts learning rate strategy (SGDR) [13], setting the minimum learning rate η^i_{min} to 0, the maximum learning rate η^i_{max} to 0.05. T_{cur} accounts for the number of epochs performed since the last start, T_i is a prefixed constant and is set to be 15000.

---

[1] https://github.com/lepture/captcha.
[2] https://github.com/skyduy/CAPTCHAgenerator.

Fig. 6. Samples of real-world CAPTCHA from websites

$$\eta_t = \eta^i_{min} + \frac{1}{2}(\eta^i_{max} - \eta^i_{min})(1 + \cos(\frac{T_{cur}}{T_i}\pi)) \tag{6}$$

We implement the network with the Tensorflow framework [1] and experiments are carried out on a workstation with a 2.20 GHz Intel (R) Xeon (R) CPU, 256 GB RAM and a NVIDIA Tesla V100 GPU. The batchsize is set to 64 and it takes about three hours to train the CAPTCHA recognition model on each dataset.

4.3   Experimental Results

To demonstrate the effectiveness of ALEC and every utilized module, we conducted several experiments on generated CAPTCHA and real-world CAPTCHA. The model accuracy, parameter numbers and inference time are given in Table 1 and in Table 2.

Comparison Between Standard Convolution and DWConv. In Table 1, we see that depthwise separable convolutions improve the accuracy by 1.14% compared to the standard convolutions on Hsiaoming and 0.94% on Skyduy with about 68% reduction for parameters and about 2.7 times actual speedup.

The Effect of GConv. Group convolution can reduce the computation complexity by transforming full-channel convolutions to group-channel convolutions. From the results, we see that models with group convolution can reduce parameter scale by about 48% while the accuracy only reduces 0.32% on Hsiaoming and 0.55% on Skyduy with the speed almost unchanged.

The Effect of Channel Number. We compare ResNet-18 models with two different channel settings, namely the light and standard. The channel number of the light version is only a quarter of the standard version. Experiment shows that the former can acquire 0.87% accuracy improvement with about 80% parameter reduction on Hsiaoming. This modification of channel number can significantly reduce the number of parameters and improve the computational efficiency. Moreover, CAPTCHA recognition tends to get over-fitting easily, this structure allows to lower the capacity of the model and achieve better generalization performance.

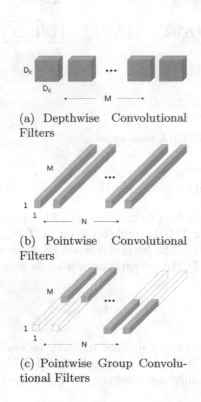

(a) Depthwise Convolutional Filters

(b) Pointwise Convolutional Filters

(c) Pointwise Group Convolutional Filters

	ALEC
Input	$75 \times 32 \times 3$
Conv	$[3, 1, 32]$
Unit1	Conv:$[1, 1 \times 1, 64]$ DWConv:$[3, 1 \times 1, 64]$ Conv:$[1, 1 \times 1, 64]$ $\times 2$ CBAM MaxPooling:$[2, 2 \times 2, 64]$
Unit2	GConv:group:2,f:96 DWConv:$[3, 1 \times 1, 96]$ GConv:group:2,f:96 Channel Shuffle $\times 2$ CBAM MaxPooling:$[2, 2 \times 1, 96]$
Unit3	GConv:group:2,f:128 DWConv:$[3, 1 \times 1, 128]$ GConv:group:2,f:128 Channel Shuffle $\times 2$ CBAM MaxPooling:$[2, 2 \times 1, 128]$
Unit4	GConv:group:2,f:160 DWConv:$[3, 1 \times 1, 160]$ GConv:group:2,f:160 Channel Shuffle $\times 2$ CBAM MaxPooling:$[2, 2 \times 1, 160]$
Conv	$[2, 1 \times 1, 192]$

Fig. 7. The standard convolution is replaced by two layers: depthwise convolution in (a) and pointwise convolution in (b). Pointwise group convolution in (c) combines pointwise convolution with group convolution.

Fig. 8. Network configurations of ALEC. The kernel size, pooling size, stride and channels are shown in brackets with number of layers.

Combination with Attention. CAPTCHA usually uses disturbance of background and other noises to increase the difficulty of recognition. To solve this problem, we utilize an attention mechanism called CBAM which can be integrated into convolution neural networks with negligible overheads. As shown in Table 1, models with CBAM only increases about 14 K parameters but improves the accuracy by 0.70% on Hsiaoming and 0.22% on Skyduy.

We also conduct experiments on the real-world CAPTCHA dataset. As shown in Table 2, the experimental results are consistent with generated CAPTCHA, the ALEC acquires 0.25% accuracy improvement which also verifies the validation of our method. With these modules mentioned above, the ALEC can achieve better efficiency and significant performance on both generated and real

Table 1. Comparison for different network structure. In the last column we report running time in milliseconds (ms) for a single core of the Intel (R) Xeon (R) CPU E5-2650 v4 @ 2.20GHz. "Standard" indicates standard ResNet-18.

Backbone	CBAM	DWConv	GConv	Hsiaoming accuracy	Skyduy accuracy	Params	CPU time
Standard				92.35%	90.45%	12.22M	228 ms
Standard		√		93.40%	91.30%	3.90M	85 ms
Light				93.15%	90.70%	2.04M	114 ms
Light	√			93.80%	90.90%	2.05M	121 ms
Light		√		94.00%	91.10%	0.62M	**54 ms**
Light	√	√		**94.10%**	**91.55%**	0.64M	60 ms
Light	√	√	√	93.80%	91.05%	**0.33M**	59 ms

Table 2. On the real-world CAPTCHA dataset, we compare the ALEC with standard ResNet-18.

Backbone	CBAM	DWConv	GConv	Accuracy
Standard				96.14%
Light				96.32%
Light	√			**96.55%**
Light	√	√		**96.55%**
Light	√	√	√	96.38%

datasets. Moreover, with fewer parameters, the ALEC can be deployed on more kinds of platforms, such as mobile phones and embedded devices.

5 Conclusion

In this paper, we propose an accurate, light and efficient network for CAPTCHA recognition called ALEC which integrates depthwise separable convolution and group convolution. Moreover, effective and efficient attention modules are applied to suppress the background noise and extract valid foreground context. All these properties make ALEC an excellent approach for CAPTCHA recognition. Comprehensive experiments demonstrate that the ALEC achieves superior accuracy with higher speed and fewer parameters, compared with standard convolution networks. Actually, the significance of this research is not only for CAPTCHA recognition but also can be applied to the related researches of text recognition. In the future, we will study CAPTCHA recognition on improving accuracy and speeding up.

References

1. Abadi, M., et al.: TensorFlow: a system for large-scale machine learning. In: 12th USENIX Symposium on Operating Systems Design and Implementation (OSDI 16), pp. 265–283 (2016)
2. El Ahmad, A.S., Yan, J., Tayara, M.: The robustness of Google CAPTCHA's. Computing Science, Newcastle University (2011)
3. El Ahmad, A.S., Yan, J., Marshall, L.: The robustness of a new captcha. In: Proceedings of the Third European Workshop on System Security, pp. 36–41. ACM (2010)
4. Garg, G., Pollett, C.: Neural network captcha crackers. In: 2016 Future Technologies Conference (FTC), pp. 853–861. IEEE (2016)
5. Graves, A., Fernández, S., Gomez, F., Schmidhuber, J.: Connectionist temporal classification: labelling unsegmented sequence data with recurrent neural networks. In: Proceedings of the 23rd International Conference on Machine Learning, pp. 369–376. ACM (2006)
6. He, K., Zhang, X., Ren, S., Sun, J.: Deep residual learning for image recognition. In: Proceedings of the IEEE Conference on Computer Vision and Pattern Recognition, pp. 770–778 (2016)
7. Hou, L., Yao, Q., Kwok, J.T.: Loss-aware binarization of deep networks. arXiv preprint arXiv:1611.01600 (2016)
8. Howard, A.G., et al.: MobileNets: efficient convolutional neural networks for mobile vision applications. arXiv preprint arXiv:1704.04861 (2017)
9. Huang, S.-Y., Lee, Y.-K., Bell, G., Zhan-he, O.: An efficient segmentation algorithm for captchas with line cluttering and character warping. Multimedia Tools Appl. 48(2), 267–289 (2010)
10. Kim, Y.-D., Park, E., Yoo, S., Choi, T., Yang, L., Shin, D.: Compression of deep convolutional neural networks for fast and low power mobile applications. arXiv preprint arXiv:1511.06530 (2015)
11. Krizhevsky, A., Sutskever, I., Hinton, G.E.: ImageNet classification with deep convolutional neural networks. In: Advances in Neural Information Processing Systems, pp. 1097–1105 (2012)
12. Le, T.A., Baydin, A.G., Zinkov, R., Wood, F.: Using synthetic data to train neural networks is model-based reasoning. In: 2017 International Joint Conference on Neural Networks (IJCNN), pp. 3514–3521. IEEE (2017)
13. Loshchilov, I., Hutter, F.: SGDR: stochastic gradient descent with warm restarts. arXiv preprint arXiv:1608.03983 (2016)
14. Ma, N., Zhang, X., Zheng, H.-T., Sun, J.: ShuffleNet V2: practical guidelines for efficient CNN architecture design. In: Ferrari, V., Hebert, M., Sminchisescu, C., Weiss, Y. (eds.) Computer Vision – ECCV 2018. LNCS, vol. 11218, pp. 122–138. Springer, Cham (2018). https://doi.org/10.1007/978-3-030-01264-9_8
15. Al Nachar, R., Inaty, E., Bonnin, P.J., Alayli, Y.: Breaking down captcha using edge corners and fuzzy logic segmentation/recognition technique. Secur. Commun. Netw. 8(18), 3995–4012 (2015)
16. Qing, K., Zhang, R.: A multi-label neural network approach to solving connected captchas. In: 2017 14th IAPR International Conference on Document Analysis and Recognition (ICDAR), vol. 1, pp. 1313–1317. IEEE (2017)
17. Rui, C., Jing, Y., Hu, R., Huang, S.: A novel LSTM-RNN decoding algorithm in captcha recognition. In: 2013 Third International Conference on Instrumentation, Measurement, Computer, Communication and Control, pp. 766–771. IEEE (2013)

18. Sandler, M., Howard, A., Zhu, M., Zhmoginov, A., Chen, L.-C.: MobileNetV2: inverted residuals and linear bottlenecks. In: Proceedings of the IEEE Conference on Computer Vision and Pattern Recognition, pp. 4510–4520 (2018)

19. Shi, B., Bai, X., Yao, C.: An end-to-end trainable neural network for image-based sequence recognition and its application to scene text recognition. IEEE Trans. Pattern Anal. Mach. Intell. **39**(11), 2298–2304 (2015)

20. Sifre, L., Mallat, S.: Rigid-motion scattering for image classification. Ph. D. dissertation (2014)

21. Singh, V.P., Pal, P.: Survey of different types of captcha. Int. J. Comput. Sci. Inf. Technol. **5**(2), 2242–2245 (2014)

22. Szegedy, C., et al.: Going deeper with convolutions. In: Proceedings of the IEEE Conference on Computer Vision and Pattern Recognition, pp. 1–9 (2015)

23. Tulloch, A., Jia, Y.: High performance ultra-low-precision convolutions on mobile devices. arXiv preprint arXiv:1712.02427 (2017)

24. von Ahn, L., Blum, M., Hopper, N.J., Langford, J.: CAPTCHA: using hard AI problems for security. In: Biham, E. (ed.) EUROCRYPT 2003. LNCS, vol. 2656, pp. 294–311. Springer, Heidelberg (2003). https://doi.org/10.1007/3-540-39200-9_18

25. Wang, J., Qin, J.H., Xiang, X.Y., Tan, Y., Pan, N.: Captcha recognition based on deep convolutional neural network. Math. Biosci. Eng **16**(5), 5851–5861 (2019)

26. Woo, S., Park, J., Lee, J.-Y., Kweon, I.S.: CBAM: convolutional block attention module. In: Ferrari, V., Hebert, M., Sminchisescu, C., Weiss, Y. (eds.) ECCV 2018. LNCS, vol. 11211, pp. 3–19. Springer, Cham (2018). https://doi.org/10.1007/978-3-030-01234-2_1

27. Yan, J., El Ahmad, A.S.: Breaking visual captchas with Naive pattern recognition algorithms. In: Twenty-Third Annual Computer Security Applications Conference (ACSAC 2007), pp. 279–291. IEEE (2007)

28. Yan, J., El Ahmad, A.S.: A low-cost attack on a Microsoft captcha. In: Proceedings of the 15th ACM Conference on Computer and Communications Security, pp. 543–554. ACM (2008)

29. Zhang, L., Zhang, L., Huang, S.G., Shi, Z.X.: A highly reliable captcha recognition algorithm based on rejection. Acta Automatica Sinica **37**(7), 891–900 (2011)

30. Zhang, X., Zhou, X., Lin, M., Sun, J.: ShuffleNet: an extremely efficient convolutional neural network for mobile devices. In: Proceedings of the IEEE Conference on Computer Vision and Pattern Recognition, pp. 6848–6856 (2018)

31. Zhu, M., Gupta, S.: To prune, or not to prune: exploring the efficacy of pruning for model compression. arXiv preprint arXiv:1710.01878 (2017)

A Benchmark System for Indian Language Text Recognition

Krishna Tulsyan, Nimisha Srivastava, Ajoy Mondal[✉], and C. V. Jawahar

Centre for Visual Information Technology,
International Institute of Information Technology, Hyderabad, India
krishna.tulsyan@research.iiit.ac.in,
{nimisha.srivastava,ajoy.mondal,jawahar}@iiit.ac.in

Abstract. The performance various academic and commercial text recognition solutions for many languages world-wide has been satisfactory. Many projects now use the OCR as a reliable module. As of now, Indian languages are far away from this state, which is unfortunate. Beyond many challenges due to script and language, this space is adversely affected by the scattered nature of research, lack of systematic evaluation, and poor resource dissemination. In this work, we aim to design and implement a web-based system that could indirectly address some of these aspects that hinder the development of OCR for Indian languages. We hope that such an attempt will help in (i) providing and establishing a consolidated view of state-of-the-art performances for character and word recognition at one place (ii) sharing resources and practices (iii) establishing standard benchmarks that clearly explain the capabilities and limitations of the recognition methods (iv) bringing research attempts from a wide variety of languages, scripts, and modalities into a common forum. We believe the proposed system will play a critical role in further promoting the research in the Indian language text recognition domain.

Keywords: Indian language · Text detection and recognition ·
Ground truth · Evaluation platform · Online benchmark system

1 Introduction and Related Work

Text recognition solutions are becoming more and more data driven in recent years [18]. Machine learning algorithms have emerged to be the central component of the OCR systems [41,42]. This is true in most areas of perception and language processing. The quality of solutions is often measured based on the empirical performance on popular benchmarks. It is observed in the history that: (i) establishing proper benchmarks has brought a community together to solve a specific problem with the objective performance systematically improving with time and with growth in the community; (ii) with the performance of the solution becoming *"satisfactory"*, newer challenges are thrown to the community. This trend has been true in some of the OCR problems, such as scene text recognition. However, Indian language OCR research has not yet adapted to

© Springer Nature Switzerland AG 2020
X. Bai et al. (Eds.): DAS 2020, LNCS 12116, pp. 74–88, 2020.
https://doi.org/10.1007/978-3-030-57058-3_6

this well-known model of research and development. Beyond technical, there are many social challenges still left out in this space.

There has been a convergence of methods for recognizing text in printed, handwritten, and natural scenes. Due to the success of deep learning-based formulations [1,16,29], advances in one modality of input (e.g., printed) influence the formulations in other modalities. We believe that this has possibly been the most impactful technical trend that can unite and advance the research and developments in Indian languages. Research groups that worked on OCR alone had a specific focus on a language or script. We observe it the world over, that research groups often work only on one of the modalities, i.e., only on one of the scene text, handwritten or printed documents. Given that there are more than 20 official languages and hundreds of unofficial languages, the number of research groups that work in this area is clearly deficient.

There has been significant research in developing highly accurate OCR solutions [28,37]. Most of these techniques are driven by the availability of a large amount of data. Unfortunately, creating standard datasets and sharing them across this community has yet not penetrated. This work is also an attempt to bring data and representations into a common format for future use.

Having dynamic leader boards or performance stats has been a way to keep up-to-date on the status of research, and know the harder challenges to be focused. Many open platforms have emerged in document image analysis and also in general machine learning. Some of the available open platforms related to this domain are EU's catch-all repository[1], Github[2] for code sharing and Kaggle[3] for hosting research related contests. The robust reading competition (RRC) platform [19–21,26,36] has been a driving force behind many of the big challenges in ICDARs. However, RRC platform is too broad for our purpose.

This paper proposes a novel benchmark system for Indian language text recognition. Using this system, we hope to tackle some of the inherent challenges evident in the domain of Indian language text recognition. Though this paper does not propose any algorithm, the proposed system could be very important for solving many of the open problems and furthering the research and development of this domain.

2 Indian Language Text Recognition: Practical Challenges

In recent years, there has been significant research in the domain of text recognition in Indian languages [17]. There have been many attempts to create OCRs for recognition in Indian languages like Bangla [5,31,34], Hindi [3,4,14], Tamil [1,30], and Kannada [2,37]. As there are many languages and numerous scripts in India, we have several challenges in developing state-of-the-art text recognition platform across all these use cases.

---

[1] https://www.openaire.eu/faqs.

[2] https://github.com/.

[3] https://www.kaggle.com/.

2.1 Lack of People

India has as many as 23 official languages [7]. Though many of these languages share common linguistic and grammatical structures, their underlying scripts remain very different. Furthermore, the non-standardization of Indian language fonts and their rendering scheme has made the development of a multilingual OCR very challenging.

Moreover, only a limited number of researchers are working on text segmentation and recognition tasks in Indian languages. This small group is not sufficient for exploring text recognition across Indian languages due to the diversity in languages (e.g., Hindi, Bangla, Tamil, Urdu, etc.), modalities (e.g., printed, historical, scene text, etc.), and tasks (e.g., text localization, word recognition, etc.).

2.2 Lack of Data

Researchers report results on their own datasets, which in most cases, are not available publicly. Text recognition of Indian languages is an emerging domain that it only recently gained the much-needed traction. Hence, there is a distinct shortage of standard datasets. Indian language consists of many scripts, among of them, only two scripts of Devanagari [24,35], and Bangla [8,24] have any substantial dataset associated with them. This lack of dataset is a serious concern as it results in subpar performance in most of the modern machine learning techniques like Neural Networks (RNN, CNN) [16], Long Short-Term Memory (LSTM) [6] and Support Vector Machines (SVM) [10] because most of these modern techniques are heavily data-driven.

The vast scope of this domain further compounds the issue. There exist multiple modalities for each of the languages like scanned documents, born-digital images, natural scene images, and text in videos. Also, most of the datasets in this domain are not available publicly, and those that do, are scattered and are individual attempts. Another significant issue is that there is no central community-driven attempt to track and benchmark the different datasets in this domain.

2.3 Challenges in Evaluation

Most of the modern OCR techniques use two primary evaluation criteria to evaluate text recognition tasks. Character Error Rate (CER) is a character metric that is based upon the Levenshtein distance [25], which is the minimum number of single-character edit operations (insertions, deletions, and substitutions) required to change the given the word to another. Word Error Rate (WER) [22] is a word metric that is also based on Levenshtein distance the same as the character metric but at the word level, i.e., a minimum number of single-word operations required to change one text to another.

In the case of Indian language, CER and WER fail to accurately represent and evaluate all aspects of the text recognition method in Indic script.

Fig. 1. A visual illustration some Indian scripts.

2.4 Challenges from Language and Scripts

Large variations are observed in the Indic scripts when compared to the Latin scripts, that resulted in many challenges in the development of general text recognition for Indian languages. There are as many as 23 official languages [7] spoken and written in India and 12 different underlying scripts [11] which results in a significant variation, further increasing the complexity of the task. Furthermore, the lack of any standard Indian language fonts and differences in their rendering schemes has made the development of multilingual OCR very challenging.

Let us consider Bangla script [32] as an example. It is primarily used for Bengali, Assamese, and Manipuri languages. Bangla script contains 11 vowels, whereas the number of consonants is 39. As depicted in Fig. 1, Bengali language shows a horizontal line running across the characters and words, which is commonly referred to as *Shirorekha*. In some cases, a set of consonants is followed by another consonant, which results in the formation of a new character that has an orthographic shape and is called a *compound* character.

Due to the presence of compound and *modified* characters (the shape of a consonant is changed when followed by a vowel, hence, a modified character), the number of distinct characters possible in Bangla [40] (roughly 400) is far higher than Latin scripts (62 different characters in English) hence, making text recognition for Bangla script is challenging when compared to English.

3 Contributions

The key factors behind building this benchmarking system are:

(i) Introduce some amount of standardization in the research in Indian language text recognition and bring the research community together on a single platform.

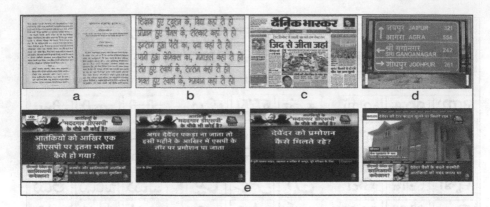

Fig. 2. A visual illustration of different modalities of text in Hindi language—(a) printed text, (b) handwritten text, (c) newspaper, (d) scene text and (e) text in few frames of a video.

(ii) Given the lack of structured resources, the portal aims to give a platform to the research community to collaborate, share, and standardize datasets to different benchmark tasks, thereby hosting challenges, workshops, and community events.

We hope that this system brings a unification in the research done in silos till now, and the community collaboratively grows. With this background, we have developed a system that is:

(a) Scalability with respect to the task, language, modality, and dataset by offering flexibility to the community to propose new tasks or new evaluation methods for existing tasks.
(b) Verification of over-fitting of the methods to the specific datasets through online testing.
(c) Common platform for logging research outcomes against the datasets/tasks and comparison of results to the state-of-the-art.

Integration of Large Community: Due to large variations in scripts in Indian languages [38] (twenty-three major scripts; Fig. 1 illustrates few), the algorithm designed for one does not work for the other. The communities of researchers in a particular language produce results on their datasets, which in most cases, are not available publicly. At the same time only a limited number of datasets [8, 24, 35] are available and various other research communities which are working on the detection, segmentation and recognition tasks in different modalities of documents (e.g., scanned [23, 33], scene text [27, 29], video text [12], camera captured [15], etc.) produce independent results that cannot be benchmarked and compared. Figure 2 visually illustrates the various modalities of Hindi text documents. Our system addresses these challenges by providing a common systematic evaluation and a benchmarking platform to quantify and track the progress in the domain of text detection, segmentation, and recognition from a variety of sources. The sources are crafted for the different tasks and approved by our panel of eminent researchers in Indian languages, thus providing the desired standardization.

The variations in task, image modality and language specific to the needs of Indian Languages are limited in the existing systems such as Kaggle, Robust Reading Competition [19–21, 26, 36], Pascal VOC[4] [13], Cityscapes[5] [9] when compared to our system.

Flexibility in Design: Due to the scale and the diversity of the research, it is becoming increasingly challenging for one owner to develop and maintain the system, introduce new datasets and challenges, organize workshops for all the different languages and scripts. We have designed our system in a way such that the fraternity gets the flexibility to contribute to the evolution of the system. The existing systems for other languages are more rigid in their design. We provide access to the full system or a small designated part of the system to any research community on this eld to update/improve the system. The research community can propose to (i) modify existing tasks, (ii) integrate new tasks, evaluation metrics and datasets, (iii) host a new challenge, and (iv) organize a workshop. All the proposals will be reviewed by a panel and included in the system upon approval. This exibility makes our system scalable as the research community grows and fosters constructive discussions and collaborations.

Online Evaluation: The existing popular web-based evaluation tools such as Kaggle, Robust Reading Competition, Pascal VOC, Cityscapes, and much more deal with only offline evaluation on the submitted results. They have no control over the overfitting[6] of submitted methods on the test datasets. Our system is different from the existing systems since we added the capability to evaluate the results online in real-time.

Since our system provides the test datasets to the users, there is a chance to overfit the submitted methods to the specified dataset. The online evaluation feature is incorporated into our portal to minimize the possibility of overfitting any submitted method on the test dataset. With this feature, we can detect and alert the users, if the results for a particular submitted method overfits the test dataset by maintaining and checking the presented technique against other random sample images that are not part of the offline test dataset. With this facility, any registered user can establish a connection to our server for a specified duration by sending a request to the server. The server sends test images to the user, one at a time, and receives the result for the individual image after running the method on that image based on the user's request. Figure 3 is a brief illustration of the working of the online testing facility.

---

[4] http://host.robots.ox.ac.uk:8080/.
[5] https://www.cityscapes-dataset.com/.
[6] Overfitting is a modeling error that occurs when a function is too closely fit a limited set of data points.

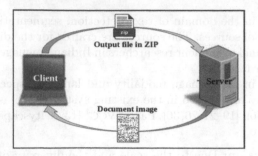

Fig. 3. A visual illustration of online evaluation. The server sets a connection to the client for a duration based on user requests. The server sends test images one at a time to the user and receives the corresponding result.

After the process is done, i.e., all the images from online test dataset are processed and evaluated using corresponding evaluation measure for the task, the system attempts to detect the overfitting of the submitted method by comparing the evaluation results for online and offline datasets. As specified, the online dataset is a super-set of an offline dataset with some random variations of the same type added.

The system calculates the probability of overfitting by measuring the response of the submitted method on the random images of the online dataset compared to the result of the specified technique on the offline dataset. Ideally, if the method is viable (not overfitting), then the difference between the result accuracy of the offline and online datasets should be minimal. On the other hand, this difference is more significant than a specified threshold. For example, in the offline dataset, the accuracy (mAP) of the method is reported to be 94%. In the case of an online dataset, the same approach has the accuracy (mAP) dropped to 62%. Then it is clear that this method is overfitting on the offline dataset.

Other than overfitting, this system can also be used as an intermediary, that gathers and compiles various stats related to processing, submission, and evaluation of the methods on a particular dataset. For example, with this online testing system, we can gather the submission method's time to process each test image. This data can be interpreted by the user to do some analysis, which offers further insights into the workings and efficiency of the submitted method.

4   Design, Functionality and Implementation

4.1   Design

The benchmark system portal has the following pages to navigate to—**Home**, **Login**, **Resource**, **Workshop/Challenge** and **Task**. Figure 4 highlights the functionalities of these heads. We subsequently explain the details available, capabilities and workflow of the various components in the system through these heads.

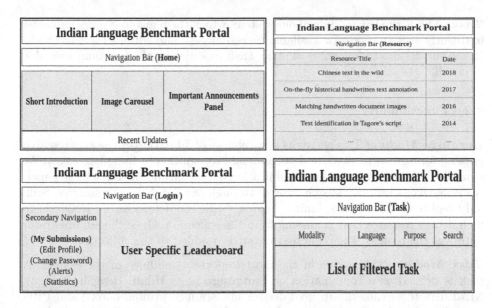

Fig. 4. Visual illustration of the functionalities of our benchmark system.

Home: This is the launch page for the benchmark system that provides consolidated information and statistics around the key indicators and trends (most used datasets, the highest number of submissions, etc.). It also highlights news, ashes, information on upcoming events, challenges, and any other essential information that the research community may need to know. The underlying idea is to eventually make the landing page as the go-to page for the community to get information on all the research related trends, any recent breakthroughs, and key information in the space.

Login: Users need to register themselves on the benchmark system. After logging into the system, the users can download datasets, test their algorithms, and participate in the workshops/challenges. The user prole page gives all the information on the activities done by the user on the system like datasets downloads, submissions made, the relative position of the users submission on the leaderboard, etc.

Resource: Here, we refer to most of the available literature on segmentation, detection, and recognition tasks in Indian languages. The users can search the extensive collection of resources based on keywords and lters. Going forward, we intend to crowd-source all the available literature or links to the online support for the same in the Indian language research space.

Workshop/Challenge: This page hosts all the information on workshops and challenges that the community plans to host. Registered users can participate in these workshops and challenges and present their work.

Task: This is the key module of our system, which the registered users will most frequently visit to check for available tasks/datasets and submit their methods. In the subsections to follow, we explain **Task** module and the functions available in detail.

4.2 Functionality

Resource Module: This serves as a warehouse of all resources (mostly research papers) for text segmentation, detection, and recognition in Indic languages. We store the title, abstract, authors, publication date, URL and BibTex for the paper. An anonymous user can search/browse through our database of resources. Still, only registered users can submit a request (with the details of the resource) to include any resource that is missing from the collection. Once the administrator approves the request, the requested resource is added to the collection.

Task Module: This is one of the most important modules of the system. A task is defined as a combination of **Language** (e.g., Hindi, Bengali, Telugu), **Modality** (e.g., camera captured documents, scanned printed books) and **Purpose** (text localization, text recognition) and represents a problem for which the user designs an appropriate method to achieve the goal. Each task in the system represents a unique research problem (e.g., word detection in Hindi scanned book, line recognition in Tamil scene images). Every task may have one or more associated test datasets available.

The benchmark system supports either multiple versions of the same dataset or an entirely new dataset for each task. Each submission in the task is evaluated against performance measures indicated for the particular task, and the result obtained is stored in the database. The users have the option to request new evaluation metrics or to make changes to the existing ones.

Submission Module: The user can use the associated dataset as a test/benchmarking set and submit the obtained results against any task. A user can have multiple submissions against a single task. Each submission is identified by a unique identifier (ID). Every time a user uploads his/her results for a *submission*, the system evaluates the results against the ground truth for the dataset as per the evaluation criterion defined for the task. The evaluation results are then displayed at an appropriate position in the leaderboard. The user can choose to keep his submission either **public** or **private**. In case, submission is marked as **public**. The submission results are available for visitors to see on the leaderboard. In case of submission being marked as **private**, the results are displayed only for this particular user.

To check for overfitting of an algorithm developed by the user, **online** evaluation has been proposed. The online verification system allows the user to verify his/her algorithm dynamically by calling the interfacing function (API) to get one test image at a time and submit the output generated immediately back into the system. Once the entire dataset is iterated upon, and the system receives the output. The results are evaluated as per the metrics defined for the task and displayed at an appropriate position in the leaderboard. It is noted that additionally, the test dataset in the **online** mode is a super-set of the dataset in the **offline** mode for the same task. Figure 3 displays the **online** submission process.

The user gets to see the relative position of his result in the public leaderboard on his/her profile page under the **My Submission** tab. The user can also see the statistics of his/her submissions under the **Stats** tab on his profile page.

Evaluation of Results: As the number and variations of the tasks hosted on the system continue to grow, we faced a different challenge in managing and organizing the evaluation metrics for each task. We aim to design the system so that each task can have multiple evaluation measures and are easily configured/ modified by the administrator. The algorithm developed for a particular task is evaluated by comparing the results submitted with the corresponding ground truths stored in the system. For each task, multiple standard measures are considered and may differ from one task to the other. The user can propose other evaluation measures that may be relevant to the task by writing to the system administrator. The proposal goes to a review board before getting added to the task.

Leaderboard: It shows the relative performance of the method for a particular task and evaluation criteria against a dataset. Every task has its separate leaderboard, where ordered results are shown for every evaluation criteria, thereby ranking the methods based on their performance. Users can reorder the displayed ordered list by selecting from the evaluation measures. The user can choose to see the leaderboard corresponding to the **online** or **offline** way of evaluation (as described in the submission module) or for the different dataset versions available for a task.

4.3   System Implementation

The benchmark system uses Django[7]/Python for back-end functionality and HTML/CSS for front-end. The system's dynamic components include resource search, Task display page, Leaderboard, Online/Offline Submissions, and User Authentication. The system is planned in such a way that every individual module of the system can act as an optional app/plugin separately. The system is designed with a standard model, view, and controller (MVC) architecture in mind. There are five major models for the project as follows:

- **User Model:** All the corresponding details of the registered users, are stored in this model (e.g., being full name, username, email, date of birth, affiliation, etc.).
- **Task Model:** This is a central model that stores the details of each defined task. Some of the tasks' attributes include the name of the task, description, modality, language, the purpose of the task, etc. The task model is connected to the dataset model by a one-to-many relationship.

---

[7] https://www.djangoproject.com/.

- **Dataset Model:** The following model is used to store the details of the available test datasets. Each dataset is inexplicably linked to a single corresponding task model. The model also saves the ground truth zip/file for the dataset.
- **Submission Model:** This model is used to store the details of the user's submissions.
- **Resource Model:** This model is used to store the metadata for the essential resources (like articles published in various journals and conferences).

Our designed system is highly modular and consists of a collection of various tasks. Each of these tasks is independent of others and can be added or removed by the supervisor of the system independently or based on the user or an administration panel's suggestions without affecting the rest of the system. The supervisor or admin has the freedom to add a new task/dataset or modify/delete the existing tasks/datasets using the admin portal with the approval from an advisory committee. It will comprise of eminent researchers from the fraternity. The administration panel is only accessible to the supervisor and a selected number of staff accounts that correspond to the administration panel.

Workflow: Any anonymous user (without registration) can only view the latest news, upcoming events, overall statistics flushed on the Homepage, resource search page, and public leaderboard corresponding to a Task. They are not allowed to download the test datasets for any task. On the other hand, registered users have access to a lot more features. Currently, the system supports both username and email-based registration, and the user alone can login into the website using username.

- Request addition of new resources by contacting the supervisor or a staff member.
- Download the dataset of any available task.
- Upload result corresponds to the dataset of a particular task, in case of offline submission.
- Upload result of a particular task through API, in case of online submission.
- View the submitted results on either public or private leaderboard.
- View the submitted results on his/her profile page.
- Edit and view the profile associated with the user.
- Modify/delete the submitted results.

5   Current Status

Task/Dataset Results: Currently, the system has the following 5 different tasks/datasets against which any user can upload the submissions.

(i) **Line detection for Bangla printed books**: The objective of this task is to localize individual text lines presents on a page. We use the Tesseract [39] line detector as a baseline. Quantitative scores of the results obtained using this method are: (Hmean = 98.9%, Recall = 97.9%, Precision = 100%, AP = 97.9%). Figure 5 (First Row) displays the evaluation result for this task on our leaderboard.

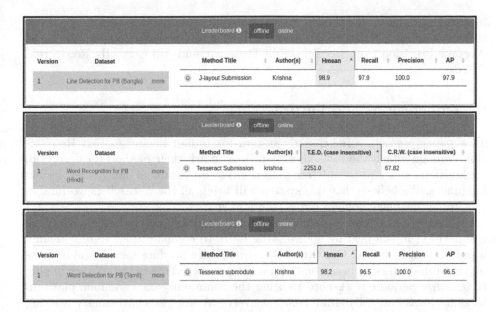

Fig. 5. A visual illustration of Leaderboard containing results of various tasks in Indian languages. **First Row:** result of text line detection for Bangla printed books. **Second Row:** result of word recognition for Hindi printed books. **Third Row:** result of word detection for Tamil printed books.

(ii) **Word recognition for Hindi printed books:** Given a set of cropped images of Hindi words, the objective of this task is to recognize those words correctly. We use the Tesseract OCR [39] as a baseline for this task. The evaluation results for this method are: (Correctly Recognized Words = 67.82%, Total Edit distance = 2251). Figure 5 (Second Row) displays the evaluation result for this task on our leaderboard.

(iii) **Word detection for Tamil printed books:** The objective of this task is to localize each word on a page. Again we use a sub-module of Tesseract OCR [39] as a baseline to evaluate this task. The quantitative scores of results obtained using this method are: (Hmean = 98.2%, Recall = 96.5%, Precision = 100%, AP = 96.5%). Figure 5 (Third Row) displays the evaluation result for this task on our leaderboard.

(iv) **Block detection for Telugu printed books:** The objective of this task is to predict the bounding box of each block present on a page.

(v) **End-to-End recognition for Hindi printed books:** The objective of the task is to localize and recognize the words present in a page of Hindi printed books.

The evaluation measures available for the detection tasks are Recall, Precision, Hmean (F-score), and Average Precision (AP). In the case of recognition task, two evaluation measures are used—(i) Total Edit Distance and (ii) Correctly Recognized Words. Both of these measures are evaluated in a case insensitive manner. For the end-to-end recognition task, three evaluation measures (as a combination of detection and recognition measures) are used - (i) Average Precision (AP), (ii) Total Edit Distance, and (iii) Correctly Recognized Words.

6 Conclusions and Future Work

This paper presents a Benchmark System for Indian Language Text Recognition through which the researchers can benchmark their method for detection, recognition, and end-to-end recognition tasks in document images for Indian languages. We believe that this system will bring all the researchers working in the Indian language domain into a common space.

We expect that this system will foster a sense of cooperation and healthy competition in the field by allowing the users to compare their results against a well-known standard. Our target is to eventually produce meaningful insights on the state-of-the-art based on the statistics collected over time. We are hoping to organize periodic workshops to bring the community on a common platform to share ideas and collaborate constructively. We are also continuously working towards improving and adding exciting new functionalities in the system. We target to introduce challenges to the system.

References

1. Achanta, R., Hastie, T.J.: Telugu OCR framework using deep learning. ArXiv (2015)
2. Ashwin, T.V., Sastry, P.S.: A font and size-independent OCR system for printed Kannada documents using support vector machines. Sadhana **27**, 35–38 (2002)
3. Bansal, V., Sinha, R.: A complete OCR for printed Hindi text in Devanagari script. In: ICDAR (2001)
4. Bansal, V., Sinha, R.M.K.: A complete OCR for printed Hindi text in Devanagari script. In: ICDAR (2001)
5. Basu, S., Das, N., Sarkar, R., Kundu, M., Nasipuri, M., Basu, D.K.: Handwritten Bangla alphabet recognition using an MLP based classifier. CoRR (2012)
6. Breuel, T.M.: High performance text recognition using a hybrid convolutional-LSTM implementation. In: ICDAR (2017)
7. Chandramouli, C., General, R.: Census of India 2011. Government of India, Provisional Population Totals, New Delhi (2011)
8. Chaudhuri, B.B.: A complete handwritten numeral database of Bangla – a major Indic script. In: IWFHR (2006)
9. Cordts, M., et al.: The Cityscapes dataset for semantic urban scene understanding. In: CVPR (2016)
10. Das, N., Das, B., Sarkar, R., Basu, S., Kundu, M., Nasipuri, M.: Handwritten Bangla basic and compound character recognition using MLP and SVM classifier. ArXiv (2010)
11. Datta, A.K.: A generalized formal approach for description and analysis of major Indian scripts. IETE J. Res. (1984)

12. Dutta, K., Mathew, M., Krishnan, P., Jawahar, C.V.: Localizing and recognizing text in lecture videos. In: ICFHR (2018)
13. Everingham, M., Van Gool, L., Williams, C.K., Winn, J., Zisserman, A.: The Pascal visual object classes (VOC) challenge. IJCV **88**, 303–338 (2010)
14. Gaur, A., Yadav, S.: Handwritten Hindi character recognition using k-means clustering and SVM. ISETTLIS (2015)
15. Gupta, V., Rathna, G.N., Ramakrishnan, K.: Automatic Kannada text extraction from camera captured images. In: MCDES, IISc Centenary Conference (2008)
16. Jain, M., Mathew, M., Jawahar, C.V.: Unconstrained OCR for Urdu using deep CNN-RNN hybrid networks. In: ACPR (2017)
17. Jomy, J., Pramod, K.V., Kannan, B.: Handwritten character recognition of south Indian scripts: a review. CoRR (2011)
18. Jordan, M.I., Mitchell, T.M.: Machine learning: trends, perspectives, and prospects. Science **349**, 255–260 (2015)
19. Karatzas, D., Gómez, L., Nicolaou, A., Rusiñol, M.: The robust reading competition annotation and evaluation platform. In: DAS (2018)
20. Karatzas, D., et al.: ICDAR 2015 competition on robust reading. In: ICDAR (2015)
21. Karatzas, D., et al.: ICDAR 2013 robust reading competition. In: ICDAR (2013)
22. Klakow, D., Peters, J.: Testing the correlation of word error rate and perplexity. Speech Commun. **38**, 19–28 (2002)
23. Krishnan, P., Sankaran, N., Singh, A.K., Jawahar, C.: Towards a robust OCR system for Indic scripts. In: DAS (2014)
24. Kumar, A., Jawahar, C.V.: Content-level annotation of large collection of printed document images. In: ICDAR (2007)
25. Levenshtein, V.: Binary codes capable of correcting deletions, insertions and reversals. Soviet Physics Doklady **10**, 707–710 (1966)
26. Lucas, S.M., Panaretos, A., Sosa, L., Tang, A., Wong, S., Young, R.: ICDAR 2003 robust reading competitions. In: ICDAR (2003)
27. Mathew, M., Jain, M., Jawahar, C.V.: Benchmarking scene text recognition in Devanagari, Telugu and Malayalam (2017)
28. Mathew, M., Singh, A.K., Jawahar, C.V.: Multilingual OCR for Indic scripts. In: DAS (2016)
29. Nag, S., et al.: Offline extraction of Indic regional language from natural scene image using text segmentation and deep convolutional sequence. ArXiv (2018)
30. Negi, A., Bhagvati, C., Krishna, B.: An OCR system for Telugu. In: ICDAR (2001)
31. Omee, F.Y., Himel, S.S., Bikas, M.A.N.: A complete workflow for development of Bangla OCR. CoRR (2012)
32. Pal, U., Chaudhuri, B.: Indian script character recognition: a survey. Pattern Recogn. **37**, 1887–1899 (2004)
33. Sankaran, N., Jawahar, C.V.: Recognition of printed Devanagari text using BLSTM neural network (2012)
34. Sarkar, R., Das, N., Basu, S., Kundu, M., Nasipuri, M., Basu, D.K.: Word level script identification from Bangla and Devanagri handwritten texts mixed with Roman script. CoRR (2010)
35. Setlur, S., Kompalli, S., Ramanaprasad, V., Govindaraju, V.: Creation of data resources and design of an evaluation test bed for Devanagari script recognition. In: WPDS (2003)
36. Shahab, A., Shafait, F., Dengel, A.: ICDAR 2011 robust reading competition challenge 2: reading text in scene images. In: ICDAR (2011)
37. Sheshadri, K., Ambekar, P.K.T., Prasad, D.P., Kumar, R.P.: An OCR system for printed Kannada using k-means clustering. In: ICIT (2010)

38. Sinha, R.M.K.: A journey from Indian scripts processing to Indian language processing. IEEE Ann. Hist. Comput. **31**, 8–31 (2009)
39. Smith, R.: An overview of the Tesseract OCR engine. In: ICDAR (2007)
40. Stiehl, U.: Sanskrit-kompendium. Economica Verlag (2002)
41. Ye, Q., Doermann, D.S.: Text detection and recognition in imagery: a survey. IEEE Trans. Pattern Anal. Mach. Intell. **37**, 1480–1500 (2015)
42. Zhu, Y., Yao, C., Bai, X.: Scene text detection and recognition: recent advances and future trends. Front. Comput. Sci. (2015)

Classification of Phonetic Characters by Space-Filling Curves

Valentin Owczarek[1(✉)], Jordan Drapeau[1], Jean-Christophe Burie[1],
Patrick Franco[1], Mickaël Coustaty[1], Rémy Mullot[1], and Véronique Eglin[2]

[1] Laboratoire L3i, University of La Rochelle,
17042 La Rochelle Cedex 1, France
{valentin.owczarek1,jordan.drapeau,jean-christophe.burie,patrick.franco,
mickael.coustaty,remy.mullot}@univ-lr.fr
[2] Université de Lyon, CNRS, INSA-Lyon, LIRIS, UMR5205,
69621 Villeurbanne, France
veronique.eglin@insa-lyon.fr

Abstract. Ancient printed documents are an infinite source of knowledge, but digital uses are usually complicated due to the age and the quality of the print. The Linguistic Atlas of France (ALF) maps are composed of printed phonetic words used to locate how words were pronounced over the country. Those words were printed using the Rousselot-Gillieron alphabet (extension of Latin alphabet) which bring character recognition problems due to the large number of diacritics. In this paper, we propose a phonetic character recognition process based on a space-filling curves approach. We proposed an original method adapted to this particular data set, able to finely classify, with more than 70% of accuracy, noisy and specific characters.

Keywords: Space-filling curves · Image classification · Phonetic alphabet · Linguistic maps

1 Introduction

The historical heritage largely contributes to the culture of each country around the world. This legacy generally appears as historical documents or ancient maps where graphical elements are often present. In this paper, we more specifically take interest in specific graphical documents named linguistic maps. Those maps transcribe the way a language is spoken in each area and help to understand the evolution of the language over time. In our research, we consider the Linguistic Atlas of France (ALF)[1] which is an atlas created between 1896 and 1900, then printed and published between 1902 and 1910. The ALF is an influential dialect atlas which presents an instantaneous picture of the dialect situation of France at the end of the 19th century, published in 35 booklets, bringing together in 13 volumes, which represents 1920 geolinguistic maps. The Swiss linguist Jules Gilliéron and the French businessman Edmond Edmont carry out the surveys for the ALF by travelling by rail, car and on foot through the 639 survey points

---

[1] Maps dataset available at http://lig-tdcge.imag.fr/cartodialect5.

© Springer Nature Switzerland AG 2020
X. Bai et al. (Eds.): DAS 2020, LNCS 12116, pp. 89–100, 2020.
https://doi.org/10.1007/978-3-030-57058-3_7

of the Gallo-Romanic territory to spread the investigations as widely as possible. Thanks to its data homogeneously transcribed using the Rousselot-Gilliéron alphabet and published in a raw form on its maps, the ALF can be assimilated to a first-generation atlas which gathers more than one million reliable lexical data, which inspired many other linguistic atlas in Europe.

The ALF maps are mainly composed of four kinds of information: names of French departments (always surrounded by a rectangle), survey point numbers (identification of a city where a survey has been done), words in phonetics (pronunciation of the word written in Rousselot-Gilliéron phonetic alphabet), and borders. An illustration of these components is given in Fig. 1. Note that each map gather the different pronunciations of a given word in a single map. For example, the Fig. 1 represents a sample of the map made for the word "balance".

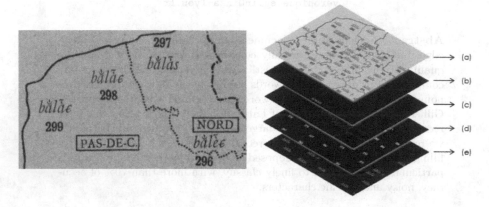

Fig. 1. Left: Two French departments names and some survey point numbers with their phonetic words, surrounded by a few borders. Right: layer separation (a) Map; (b) French departments names; (c) Borders; (d) Survey point numbers; (e) Words in phonetics.

Our research aims at automatically extracting the ALF information and generating maps with selected elements (currently, this process is done manually and it takes weeks to build a single map). In a previous work [5], we proposed to separate each type of information into layers (see Fig. 1) in order to prepare data for a subsequent analysis. Based on these results, this paper focus on the classification of characters in the phonetic layer.

2   Dataset Specifications

Edmond Edmont and Jules Gilliéron use in the ALF the phonetic notation developed and broadcast by Abbé Rousselot and Jules Gilliéron himself. The conventions that define the Rousselot-Gilliéron alphabet are recorded in the "Revue des patois gallo-romans (no. 4 1891, p. 5–6)" and repeated in the maps intelligence notice that accompanies the ALF. This alphabet is mainly made up of the letters of the Gallo-Roman languages (like French), on which diacritics (accents and notations) may be placed to symbolize more faithfully the way of pronouncing this letter or a part of the word (lemma). There are 1920 maps in the ALF which have all been written uniformly with the Rousselot-Gilliéron alphabet for the transcription of phonetic words. The Fig. 2 shows an example of a word transcribed into phonetic.

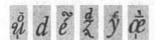

Fig. 2. An example of a phonetic word written in Rousselot-Gilliéron alphabet that we can find on the maps. Here is the transcription of the town name "L'Hermitage".

An inventory of the different characters used on all the maps has been made. The protocol used was to insert a character as new in the inventory if the diacritic was different. The superposition of diacritics being possible, for a given basic character (a, e, ...), the number of variations of some characters can be important. For example, the basic character "e" offers a range of 60 variations. Each variation is considered as a character of the inventory. From this inventory, a dataset has been created. It consists of an image of each character of the inventory found in the ALF maps. We chose to extract only one image per character because finding one representation of some specific characters, among the 1920 maps, is quite complex. Finding a second representation would have required a lot of effort. To date, there is no search tool within the maps, so the work has been done manually. This work brought together a collection of 251 different characters images (181 vowels, 61 consonants, 9 legend symbols) (Fig. 3).

Fig. 3. Some samples of character images that can be found in the dataset.

Note that in this alphabet, 389 characters have been listed, but only 251 of them were found printed on the various maps. The images (from this dataset) have been extracted directly from the maps, which also bring a lot of noise to them. Indeed, noises are either related to the maps (textures degradations like holes, ink smudges, partially erased or slightly rotated characters) or created when scanning them (artifacts, low resolution, blur). This is why our dataset is a reduced dataset that shows wide disparities of low image quality (Fig. 4).

(a)   (b)   (c)   (d)

Fig. 4. Some samples of character images that can be found in the dataset, showing some noises: (a) ink smudge, (b) hole, (c) blur, (d) low resolution.

The image annotation consists in associating a class to each thumbnail (a class is an index in the latin alphabet) and its transcription in Rousselot-Gilliéron alphabet. The Table 1 shows a sample of the correspondence file of the dataset for the character "a".

Table 1. A sample of the correspondence file of the dataset.

Filename	Class	Transcription	Image
image1.tif	a	à̧	 ǎ
image10.png	a	ā	 ā
image100.tif	a	a̠	 ɑ
image110.tif	e	è̄	 ē

The transcription of accented character (phonetic) in the correspondence file was made possible thanks to the *Symbola* font, which includes almost the entire Rousselot-Gilliéron alphabet. The font is available online and also provided with the dataset. Sometimes a problem of diacritics overlay can be observed depending on the editor which the font is used, but this phenomenon doesn't have impact on the transcription result.

The dataset is available online and can be accessed for free[2]. It includes images of the characters before and after pre-processing, its correspondence file (filename, class, transcription), the previously presented *Symbola* font, and some learning logs.

All experiments that are proposed on this dataset are presented in the next sections. We present here our first attempt to cluster those letters and to map them to their transcription.

---

[2] The phonetics characters images dataset is available here: http://l3i-share.univ-lr. fr/datasets/Dataset_CharRousselotGillerion.zip.

3   Related Work

The principle of image classification is based on a fast search for object similarity among a large collection of already identified objects. In the literature, there is several works to specify this similarity.

A few of them [4, 7] use the matching of some basic properties of the image like colors, textures, shapes, or edges. However, the maps from our project are ancient printed documents. This means that the time-degradation (variable texture, colors) and the printing made with the tools of that time (variable shapes, edges) make it difficult to exploit this kind of maps with these approaches.

Another solution is to use local invariant features which is becoming more and more popular these days, like SIFT [10] or SURF [2]. In our case, Rousselot-Gilliéron alphabet brings also classification problems, because of these diacritics, compared to Latin languages. Here again, our maps are printed with the tools of that time with ink. This means that we never have two purely identical letters (and mostly diacritics) on all the data that we have for comparison purposes which is not suitable.

Recently, in image classification Artificial Neural Network (ANN) are widely used, more precisely Convolutional neural network (CNN) [3] which is inspired by human vision. The ANN is composed of specific layer computing results from small part of the image. The main drawback of this kind of technique is the complexity of training. In our case, the network will not converge efficiently, because of the lack of training images, and high variety in the characters (root and diacritics). However, there are methodologies based on a few numbers of samples or inclusive zero shot called one-shot learning. These methods are based on transfer learning or on a mapping between several representation spaces, and in our case, there is no consistent base offers similar characteristics to the studied characters (root and diacritics).

In opposite, we proposed a new technique based on space-filling curve. This technique takes advantage of the automatic discovery of key points and compact representation of class model. These features match our requirements and we will then explore them in the following of this paper.

4   Space-Filling Curve for Characters Classification

4.1   An Overview of Space-Filling Curves

Space-Filling curves (SFCs) are historically a mathematical curiosity, they are continuous non-differentiable functions. The first one was discovered by G. Peano in 1890 [13]. One year later D. Hilbert have proposed a different curve [8], commonly called Hilbert curve, this curve is originally defined for dimension $D = 2$, but multidimensional are defined using the Reflected Binary Gray code (RBG). The Fig. 5, shown the Hilbert space-filling curve for dimension $D = 2$ and order $n = 1, 2, 3$.

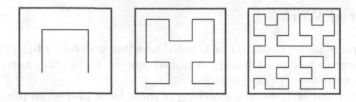

Fig. 5. Hilbert space-filling curve, in dimension $D = 2$ for $n = 1$, 2, 3. Left: the Hilbert curve at order $n = 1$. Center: the Hilbert curve at order $n = 2$, construct by subdividing the curve at order $n = 1$. Right: the Hilbert curve at order $n = 3$.

In this paper, S_n^D is the space-filling curve function transforming a D dimensional point into an integer called index with the order n by the multidimensional Hilbert curve and the inverse function by $\bar{S}_n^D$.

The main property of the Hilbert curve is the neighborhood preserving, the fact that two close D dimensional points separated by a short distance have high probability to be associate to indices by S_n^D separated by a short distance i.e:

$$S^D(\mathbf{X}_i) = I_i, \quad \bar{S}^D(I_i) = \mathbf{X}_i$$
$$S^D(\mathbf{X}_j) = I_j, \quad \bar{S}^D(I_j) = \mathbf{X}_j \qquad (1)$$
$$m(\mathbf{X}_i, \mathbf{X}_j) = \varepsilon_D, \quad m(I_i, I_j) = \varepsilon_1$$

where m is a distance function, with ε_D and ε_1 small. According to [6,11], the curve conserving the best the locality is the multidimensional extension of the Hilbert curve and for that reason was used in this paper. Complementary information on Space-filling curves can be find in [14] with more applications in [1].

Along years, many applications are drawn on SFCs, for example image storing and retrieving [15], derivative-free global optimization [9] and image encryption [12]. Here SFC are used in a new framework to recognize the characters of the Rousselot-Gilliéron phonetic alphabet.

4.2   Characters Image Classification

When dealing with characters image classification with the proposed approach, several points have to be met. Firstly, images have to be pre-processed in order to standardize them, for example changing the color scheme to black and white (binarization), deleting small noisy components, resizing the images. Secondly, the classification technique has to be translation and rotation invariant or not very sensitive.

In this section, the SFC character images classification is explained with a focus on the pre-processing, and the translation and rotation sensitivity.

Data Pre-processing. As mentioned in Sect. 2, images are degraded, so we propose to standardize the images with five steps presented in Fig. 6:

1. Transform the color images to black and white:
   - Divided each pixel of the image by the standard deviation,
   - Then set to white or black each pixel depending if the value is lower or bigger than the mean pixel value.
2. Delete small black components to reduce noise and black and white transformation artefact.
3. Find the Region of Interest (ROI) as the smallest box embracing the black pixels.
4. Center the main component by adding white padding.
5. Resize the image.

As result of this pre-processing, 64×64 pixels images are obtained with gray scale color scheme and main component centered.

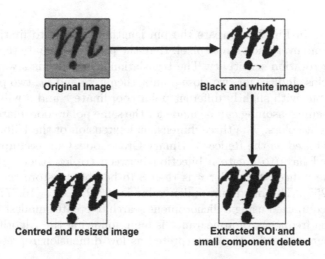

Fig. 6. Data pre-processing process from the original to the resized and centered black and white image.

Training the SFC Classifier. As shown in Eq. 1, it is possible to transform a data point into an index using the SFC function. But to represent an image, we need as much indices as pixels. In this work, an image is represented by a distribution histogram where the bins length is not regular. Those are calculated using the distribution of all indices produced by all images of a training data set. The irregularity in the bins length permit to adapt the classifier to the particularities of the dataset. Once the bins are known, we can create for each image an histogram and then merge all histograms for a given character into one unique using the statistic mean. To summarize, the training part create two linked objects: The irregular bins length and one histogram by classes of character.

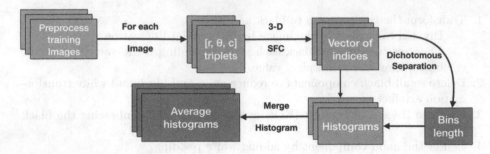

Fig. 7. Training of a classifier based on space-filling curve. Two object is learn: the average histograms and the bin length. For each image, a vector of indices is created, then transformed into an histogram using the irregular bins. The training ends with the creation of an average histogram for each classes.

As mention in Fig. 7, to create the bin length you need to firstly compute every index from every non-white pixel. Here the polar coordinate (r, θ) are used to ensure low rotation sensitivity. The translation sensitivity is always assured: SFC assigns close indices for two close points, then if x_a and x_b, two points, with same color c but with slightly different polar coordinate r and θ will have close indices. The same reasoning can be made for the same polar coordinates but with slightly different colors. The three-dimensional extension of the Hilbert-curve is used which is based on the Reflected-Binary-Gray-Code. Our assumption is that the curve used have to create a bijection between triplet (theta, rho, c), the pixels and the indices: the order n is then 8 to be able to store the maximum value of c(=255). The maximum index value is $(2^n D) - 1 (= 16.777.215)$. The bin length is computed using a dichotomous search to create smallest histograms bins where the frequency of appearance is high, following the Algorithm 1.

The histogram can be then interpreted as low dimension representation of each image.

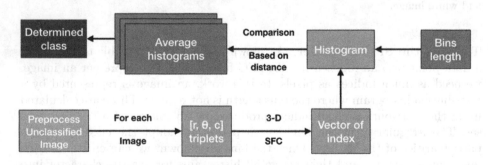

Fig. 8. Classification of phonetic character based on Space-filling curve. The average histograms and the bin length must be determined by training (cf. Fig. 7). The comparison between the histogram and the average histograms is compute with cosine distance.

Algorithm 1. Algorithm use to create the bins of the histograms based on a dichotomous separation.

1: **DichotomousBinsLength**   *in:* list of indices: tabI, the order: n, the dimension: D, threshold: k   *out:* list of bins: res

2: res : [], initBin = $[0, 2^{nD} - 1]$

3: res.append(initBin)

4: **repeat**

5:     lenRes = len(res)

6:     newRes = []

7:     **for all** bin in res **do**

8:         **if** countIndices(tabI, bin) > k **then**

9:             a = bin[0], b = bin[1]

10:             newRes.append($[a, a + \frac{b-a}{2}]$)

11:             newRes.append($[a + \frac{b-a}{2} + 1, b]$)

12:         **else**

13:             newRes.append(bin)

14:         **end if**

15:     **end for**

16:     res = newRes

17: **until** len(res) == lenRes

18: **return** res

Classification of Character. The classification process is similar to k-means where the distribution histograms can be seen as centroids. Then, the class of a new image is the class of the nearest centroids. Given the sparsity of dimensional space ($D \approx 300$), the cosine distance is used to compare two histograms. The Fig. 8 shows the process of classification:

1. A new image is pre-process (cf. Fig. 6)
2. The image is transformed to a histogram with irregular bins, previously compute in the training step (cf. Fig. 7)
3. The determined class is the class corresponding to the nearest mean distribution histogram.

In this section, a framework able to classify noisy and low quality images is presented. The input data are standardized to a square binarized images where the root character is centered. The SFC-classification is based on an automatic selection of zone-of-interest using indices and reduce to variable length bins histograms. The method is specialized by learning the bins length of these histograms. This task is done on the training dataset using dichotomous separation. The method is slightly robust to translation and rotation through the use of polar coordinates and SFC, making the classification process well adapted to the dataset presented in Sect. 2.

4.3 Results

As described earlier, the SFC based classifier use a training and test principle. This results in having to set aside all the characters that are present only once in the dataset, because with only one image we won't be able to train and test the model. As a security measure, any image class that does not have at least four images will be set aside for these tests. The Table 2 shows the repartition of used character within the dataset.

Table 2. Number of images by class of character

Class of characters	a	c	d	e	g	i	n	o	oe	ou	r	u	z	Total
Number of images	42	4	4	35	4	18	5	25	17	25	7	19	4	209

The tests were therefore carried out on a total of two hundred and nine character images after subtracting the small classes. To evaluate the robustness of the model, three different sets of tests were carried out with for each class of character a random selection, where 75% are assigned for training and 25% for test. The results are presented in Table 3.

Table 3. Portion of images for the training and the tests, and the result accuracy of classification for the three experiments.

Exp	NbTrainImg	NbTestImg	%Train	%Test	Accuracy
1	152	57	73%	27%	71.9%
2	156	53	74%	26%	77.3%
3	158	51	75%	25%	72.5%

Concerning the results, the proposed method is able to detect slightly more than 70% on each experiment. These results are understandable because the standardization of the images reduce the variance of the numerous noises present on the images of the basic dataset, and thus it offers a certain stability, but on the other hand it is not perfect either. Indeed, the standardization sometimes goes:

- cut letters during binarization,
- cause a zoom effect of the character, to which the model is sensitive,
- add or not add a padding if a diacritic is present.

We also notice that classes with too few images give us variable recognition rates as can be expected with learning models. The Table 4 perfectly illustrates this last point by showing the accuracy of classification for each character inside each experiment.

Results can still be improved. On the one hand, a next step could be to increase the number of images per class by capturing more occurrences of a character, or by generating slightly different images by adding few noises on the one already extracted. On the other hand, the pre-processing that seeks to

Table 4. For each test set, this table shows the accuracy of classification for each classes of characters.

	Classes of characters												
	a	c	d	e	g	i	n	o	oe	ou	r	u	z
Exp1	64%	0%	0%	78%	0%	80%	0%	57%	100%	100%	100%	60%	100%
Exp2	91%	0%	0%	67%	0%	100%	100%	50%	75%	83%	100%	100%	100%
Exp3	90%	100%	100%	56%	0%	100%	100%	83%	25%	83%	100%	40%	100%

standardize images by binarizing, removing noises and putting them at the same resolution, generates defects that should be fixed, because some generate missing diacritics or sections of characters.

5 Conclusion

This paper presents an original approach for classifying phonetics characters images. The proposed method consists in the representation of characters by irregular histograms, created with space-filling curve, able to differentiate characters between them. Note that our method is able to correctly classify 70% of our phonetic characters images dataset subset.

The obtained results are quite encouraging because they prove that from few data, which are not perfect, we are able to obtain reliable performance. An interesting strategy could be, for example, to inject its result in another algorithm which will focus to identify the diacritics in an optimized way. This may be a final alternative solution to build an OCR for this kind of phonetic alphabet.

However, these results will have to be compared with related works presented in Sect. 3, and then can still be improved as described in the Sect. 4.3. The SFC technique here consider only the color of the pixels and the position (polar coordinate) to produce an index, it will be thoughtful to use nearest pixels to increase the quality of the characterization.

To go further, it might also be interesting to use the SFC with phonetic words images. This will result in the formation of groups of similar words, and by definition, isoglosses (regions where people say exactly the same thing) that are of real value to dialectologists.

Acknowledgment. This work is carried out in the framework of the ECLATS project and supported by the French National Research Agency (ANR) under the grant number ANR-15-CE38-0002.

References

1. Bader, M.: Space-Filling Curves. Springer, Heidelberg (2013). https://doi.org/10.1007/978-3-642-31046-1

2. Bay, H., Ess, A., Tuytelaars, T., Gool, L.V.: Speeded-up robust features (surf). Comput. Vis. Image Underst. **110**(3), 346–359 (2008). Similarity Matching in Computer Vision and Multimedia
3. Chang, O., Constante, P., Gordon, A., Singaña, M.: A novel deep neural network that uses space-time features for tracking and recognizing a moving object. J. Artif. Intell. Soft Comput. Res. **7**(2), 125–136 (2016)
4. Cheng, Y.C., Chen, S.Y.: Image classification using color, texture and regions. Image Vis. Comput. **21**(9), 759–776 (2003)
5. Drapeau, J., et al.: Extraction of ancient map contents using trees of connected components. In: Fornés, A., Lamiroy, B. (eds.) GREC 2017. LNCS, vol. 11009, pp. 115–130. Springer, Cham (2018). https://doi.org/10.1007/978-3-030-02284-6_9
6. Faloutsos, C., Roseman, S.: Fractals for secondary key retrieval. In: Proceedings of the Eighth ACM SIGACT-SIGMOD-SIGART Symposium on Principles of Database Systems, PODS 1989, pp. 247–252. ACM (1989)
7. Fredembach, C., Schröder, M., Süsstrunk, S.: Region-Based Image Classification for Automatic Color Correction, January 2003
8. Hilbert, D.: Ueber die stetige Abbildung einer Line auf ein Flächenstück. Math. Ann. **38**(3), 459–460 (1891)
9. Lera, D., Sergeyev, Y.D.: GOSH: derivative-free global optimization using multi-dimensional space-filling curves. J. Glob. Optim. **71**(1), 193–211 (2018)
10. Lowe, D.G.: Distinctive image features from scale-invariant keypoints. Int. J. Comput. Vision **60**(2), 91–110 (2004)
11. Moon, B., Jagadish, H.V., Faloutsos, C., Saltz, J.H.: Analysis of the clustering properties of the Hilbert space-filling curve. IEEE Trans. Knowl. Data Eng. **13**(1), 124–141 (2001)
12. Murali, P., Sankaradass, V.: An efficient space filling curve based image encryption. Multimedia Tools Appl. **78**(2), 2135–2156 (2018). https://doi.org/10.1007/s11042-018-6234-8
13. Peano, G.: Sur une courbe, qui remplit toute une aire plane. Math. Ann. **36**, 157–160 (1890). https://doi.org/10.1007/BF01199438
14. Sagan, H.: Space-Filling Curves. Springer, New York (1994). https://doi.org/10.1007/978-1-4612-0871-6
15. Song, Z., Roussopoulos, N.: Using Hilbert curve in image storing and retrieving. Inf. Syst. **27**(8), 523–536 (2002)

Document Image Processing

Self-supervised Representation Learning on Document Images

Adrian Cosma[1,2(✉)], Mihai Ghidoveanu[1,3], Michael Panaitescu-Liess[1,3], and Marius Popescu[1,3]

[1] Sparktech Software, Bucharest, Romania
{adrian.cosma,mihai.ghidoveanu,michael.panaitescu}@sparktech.ro,
popescunmarius@gmail.com
[2] University Politehnica of Bucharest, Bucharest, Romania
[3] Faculty of Mathematics and Computer Science, University of Bucharest, Bucharest, Romania

Abstract. This work analyses the impact of self-supervised pre-training on document images in the context of document image classification. While previous approaches explore the effect of self-supervision on natural images, we show that patch-based pre-training performs poorly on document images because of their different structural properties and poor intra-sample semantic information. We propose two context-aware alternatives to improve performance on the Tobacco-3482 image classification task. We also propose a novel method for self-supervision, which makes use of the inherent multi-modality of documents (image and text), which performs better than other popular self-supervised methods, including supervised ImageNet pre-training, on document image classification scenarios with a limited amount of data.

Keywords: Self-supervision · Pre-training · Transfer learning · Document images · Convolutional neural networks

1 Introduction

A document analysis system is an important component in many business applications because it reduces human effort in the extraction and classification of information present in documents.

While many applications use Optical Character Recognition systems (OCR) to extract text from document images and directly operate on it, documents often have an implicit visual structure. Helpful contextual information is given by the position of text in a page and, generally, the page layout. Reports containing tables and figures, invoices, resumes, and forms are difficult to process without considering the relationship between layout and textual content.

As such, while there are efforts in dealing with the visual structure in documents leveraging text [31,47], relevant-sized datasets are mostly internal, and

---

M. Ghidoveanu and M. Panaitescu-Liess—Equal contribution.

© Springer Nature Switzerland AG 2020
X. Bai et al. (Eds.): DAS 2020, LNCS 12116, pp. 103–117, 2020.
https://doi.org/10.1007/978-3-030-57058-3_8

privacy concerns inhibit public release. Moreover, labeling of such datasets is an expensive and time-consuming process.

This is not the case for natural images. Natural images are prevalent on the internet, and large-scale annotated datasets are publicly available. The ImageNet database [9] contains 14M annotated natural images with 1000 classes and has powered many advances in computer vision and image understanding through training of high capacity convolutional neural networks (CNNs). ImageNet also provides neural networks with the ability to transfer the information to other unrelated tasks like object detection and semantic segmentation [21]. A neural network pre-trained on ImageNet has substantial performance gains compared to a network trained from scratch [49].

However, it was shown that pre-training neural networks with large amounts of noisily labeled images [46] substantially improves the performance after fine-tuning on the main classification task. This is indicative of a need to make use of a large corpus of partially labeled or unlabeled data. Moreover, modern methods of leveraging unlabeled data have been developed [11,33], by creating a pretext task, in which the network is under self-supervision, and afterwords fine-tuning on the main task.

Unlike natural image datasets, document datasets are hard to come by, especially fully annotated ones, and have only a fraction of the scale of ImageNet [15,29]. However, unlabeled documents are easily found online in the form of e-books and scientific papers.

Qualitatively, document images are very different from natural images, and therefore using a pre-trained CNN on ImageNet for fine-tuning on documents is questionable. Document images are also structurally different from natural images, as they are not invariant to scaling and flips. It has been shown that models trained on ImageNet often generalize poorly to fine-grained classification tasks on classes that are poorly represented in ImageNet [27]. While there are classes that are marginally similar to document images (i.e. menus, websites, envelopes), they are vastly outnumbered by other natural images. Moreover, models that are pre-trained on RVL-CDIP [15] dataset have a much better performance on document classification tasks with a limited amount of data [26].

Self-supervision methods designed for document images have received little attention. As such, there is a clear need for learning more robust representations of documents, which make use of large, unlabeled document image datasets.

This paper makes the following contributions to the field of document understanding:

1. We make a quantitative analysis of self-supervised methods for pre-training convolutional neural networks on document images.
2. We show that patch-based pre-training is sub-optimal for document images. To that end, we propose improved versions of some of the most popular methods for self-supervision that are better suited for learning structure from documents.
3. We propose an additional self-supervision method which exploits the inherent multi-modality (text and visual layout) of documents and show that the

representations they provide are superior to pre-training on ImageNet, and subsequently better than all other self-training methods we have tested in the context of document image classification on Tobacco-3482 [28].

4. We make a qualitative analysis of the filters learned through our multi-modal pre-training method and show that they are similar to those learned through direct supervision, which makes our method a viable option for pre-training neural networks on document images.

2 Related Work

2.1 Transfer Learning

One of the requirements of practicing statistical modeling is that the training and test data examples must be independent and identically distributed (i.i.d.). Transfer learning relaxes this hypothesis [43]. In computer vision, most applications employ transfer learning through fine-tuning a model trained on the ImageNet Dataset [9]. Empirically, ImageNet models do transfer well on other subsequent tasks [21,27], even with little data for fine-tuning. However, it only has marginal performance gains for tasks in which labels are not well-represented in the ImageNet dataset.

State-of-the-art results on related tasks such as object-detection [39] and instance segmentation [16] are improved with the full ImageNet dataset used as pre-training, but data-efficient learning still remains a challenge.

2.2 Self-supervision

Unsupervised learning methods for pre-training neural networks has sparked great interest in recent years. Given the large quantity of available data on the internet and the cost to rigorously annotate it, several methods have been proposed to learn general features. Most modern methods pre-train models to predict pseudo-labels on pretext tasks, to be fine-tuned on a supervised downstream task - usually with smaller amounts of data [22].

With its roots in natural language processing, one of the most successful approaches is the skip-gram method [32], which provides general semantic textual representations by predicting the preceding and succeeding tokens from a single input token. More recent developments in natural language processing show promising results with models such as BERT [38] and GPT-2 [10], which are pre-trained on a very large corpus of text to predict the next token.

Similarly, this approach has been explored for images, with works trying to generate representations by "context prediction" [11]. Authors use a pretext task to classify the relative position of patches in an image. The same principle is used in works which explore solving jigsaw puzzles as a pretext task [7,33]. In both cases, the intuition is that a good performance on the patch classification task is directly correlated with a good performance on the downstream task, and with the network learning semantic information from the image.

Other self-supervision methods include predicting image rotations [14], image colorization [48] and even a multi-task model with several tasks at once [12]. Furthermore, exemplar networks [13] are trained to discriminate between a set of surrogate classes, to learn transformation invariant features. A more recent advancement in this area is Contrastive Predictive Coding [17], which is one of the most performing methods, for self-supervised pre-training.

An interesting multi-modal technique for self-supervision leverages a corpus of images from Wikipedia and their description [3]. The authors pre-train a network to predict the topic probabilities of the text description of an image, thereby leveraging the language context in which images appear.

Clustering techniques have also been explored [6,20,34] - by generating a classification pretext task with pseudo-labels based on cluster assignments. One method for unsupervised pre-training that makes very few assumptions about the input space is proposed by Bojanowski et al. [5]. This approach trains a network to align a fixed set of target representations randomly sampled from the unit sphere.

Interestingly, a study by Kolesnikov et al. [25] demonstrated that there is an inconsistency between self-supervision methods and network architectures. Some network architectures are better suited to encode image rotation, while others are better suited to handle patch information. We argue that this inconsistency also holds for datasets. These techniques show promising results on natural images, but very little research is devoted to learning good representations for document images, which have entirely different structural and semantic properties.

2.3 Document Analysis

The representation of document images has a practical interest in commercial applications for tasks such as classification, retrieval, clustering, attribute extraction and historical document processing [36]. Shallow features for representing documents [8] have proven to be less effective compared to deep features learned by a convolutional neural network. Several medium-scale datasets containing labeled document images are available, the ones used in this work being RVL-CDIP [15] and Tobacco-3482 [28]. For classification problems on document images, state-of-the-art approaches leverage domain knowledge of documents [1], combining features from the header, the footer and the contents of an image. Layout-methods are used in other works [2,24,47] to make use of both textual information and their visual position in the image for use in extracting semantic structure.

One study by Kölsch et al. [26] showed that pre-training networks using the RVL-CDIP dataset is better than pre-training with ImageNet in a supervised classification problem on the Tobacco-3482 dataset. Still, training from scratch is far worse than with ImageNet pre-training [41].

3   Methods

For our experiments, we implemented several methods for self-supervision and evaluated their performance on Tobacco-3482 document image classification task, where there is a limited amount of data. We implemented two Context-Free Networks (CFN), relative patch classification [11], and solving jigsaw puzzles [33], which are patch-based, and, by design, are not using the broader context of the document. We also trained a model to predict image rotations [14], as a method that could intuitively make use of the layout and an input-agnostic method developed by Bojanowski et al. [5], which forces the model to learn mappings to the input image to noise vectors that are progressively aligned with deep features. We propose variations to context-free solving of jigsaw puzzles and to rotation prediction, which improves performance. We propose Jigsaw Whole, which is a pretext task to solve jigsaw puzzles, but with the whole image given as input, and predicting flips, which is in the same spirit of predicting rotations, but better suited for document images.

We also developed a method that makes use of the information-rich textual modality: the model is tasked to predict the topic probabilities of the text present in the document using only the image as an input. This method is superior to ImageNet pre-training.

3.1   Implementation Details

Given the extensive survey of [44], we used the document images in grayscale format, resized to a fixed size of 384×384. Images are scaled so that the pixels fall in the interval $(-0.5, 0.5)$, by dividing by 255 and subtracting 0.5 [23].

Shear transformations or crops are usually used to improve the performance and robustness of CNNs on document images [44]. We intentionally don't use augmentations during training or evaluation to speed up the process and lower the experiment's complexity.

InceptionV3 [42] architecture was used in all our experiments because of its popularity, performance and availability in common deep learning frameworks.

3.2   Jigsaw Puzzles

In the original paper for pre-training with solving Jigsaw puzzles [33], the authors propose a Context-Free Network architecture, with nine inputs, each being a crop from the original image. There, the pretext task is to reassemble the crops into the original image by predicting the permutation.

This is sensible for natural images, which are invariant to zooms: objects appear at different scales in images, and random crops could contain information that is useful when fine-tuning on the main task.

On the other hand, document images are more rigid in their structure. A random crop could end up falling in an area with blank space, or in the middle of a paragraph. Such crops contain no information for the layout of the document, and their relationship is not clear when processed independently. Moreover, text size changes relative to the crop size. As such, when fine-tuning, the text size is significantly smaller relative to the input size, which is inconsistent with the pretext task.

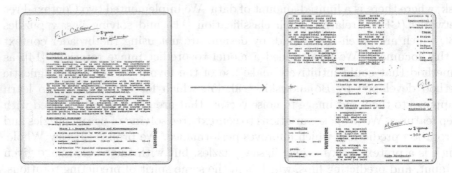

Fig. 1. Input image for our Jigsaw Whole method. The scrambled image is given as is to a single network, without using siamese branches, such that context and layout information is preserved.

We propose a new way of pre-training by solving jigsaw puzzles with convolutional networks, by keeping the layout of the document visible to the model. After splitting the image into nine crops and shuffling them, we reassemble them into a single puzzle image. An example of the model input is exemplified in Fig. 1. We name this variation **Jigsaw Whole**. The intuition is that the convolutional network will better learn semantic features by leveraging the context of the document as a whole.

In order to obtain the final resolution, we resized the initial image at 384×384 pixels and then split it into nine crops of 128×128 pixels each. Using jitter (as recommended by Noroozi et al. [33]) of 10 pixels, results in a resolution of 118×118 pixels for each of the nine patches.

As described by Noroozi et al. [33], we chose only 100 out of $9! = 362880$ possible permutations of the nine crops. Those were selected using a greedy algorithm to approximate the maximal average hamming distance between pairs of permutations from a set.

3.3   Relative Position of Patches

Using a similar 3×3 grid as in the previous method, and based on Doersch et al. [11], we implemented a siamese network to predict which is the position of a patch relative to the square in the center. The model has two inputs (the crop in the middle and one of the crops around it), and after the Global Average Pooling layer from InceptionV3, we added a fully-connected layer with 512 neurons, activated with a rectified linear unit and then a final fully-connected layer with eight neurons and softmax activation. For fine-tuning, we kept the representations created after the Global Average Pooling layer, ignoring the added fully-connected layer. To train the siamese network, we resized all the images to 384 × 384 pixels, then we created a grid with nine squares of 128 × 128 pixels each, and using jitter of 10 pixels, we obtained the input crops of 118 × 118.

Note that jitter is used both in solving jigsaw puzzles and predicting the relative position of patches, in order to prevent the network from solving the problem immediately by analysing the margins of the crops only (in which case it does not need to learn any other structural or semantic features).

Similar to solving jigsaw puzzles, predicting the relative position of patches suffer from the same problems of having too little context in a patch. We show that these methods perform poorly on document image classification.

3.4   Rotations and Flips

A recent method for self-supervision proposed by Gidaris et al. [14] is the simple task of predicting image rotations. This task works for natural images quite well, since objects have an implicit orientation, and determining the rotation requires semantic information about that object. Documents, on the other hand, have only one orientation - upright. We pre-train our network to discriminate between 4 different orientations (0°, 90°, 180° and 270°). It is evident that discriminating between (0°, 180°) pair and (90°, 270°) pair is trivial as the text lines are positioned differently. We argue that this is a shortcut for the model, and in this case, the task is not useful for learning semantic or layout information.

Instead, we propose a new method, in the same spirit, by creating a pre-text task that requires the model to discriminate between different flips of the document image. This way, the more challenging scenarios from the rotations methods are kept (in which text lines are always horizontal), and we argue that this forces the model to learn layout information or more fine-grained text features in order to discriminate between flips. It is worth noting that this method does not work in the case of natural images, as they are invariant to flips, at least across the vertical axis. In our experiments, we named this variation **Flips**.

3.5   Multi-modal Self-supervised Pre-training

While plain computer vision methods are used with some degree of success, many applications do require textual information to be extracted from the documents. Be it the semantic structure of documents [47], or extracting attributes from financial documents [24] or table understanding [19,31,37], the text modality present in documents is a rich source of information that can be leveraged to obtain better document representations. We assume that the visual document structure is correlated with the textual information present in the document. Audebert et al. [2] use textual information to jointly classify documents from the RVL-CDIP dataset with significant results. Instead of jointly classifying, we explore self-supervised representation learning using text modality.

The text is extracted by an OCR engine [40] making resulting text very noisy - many words have low document frequency due to OCR mistakes. While this should not be a problem given the large amount of data in the RVL-CDIP dataset, we do clean the text by lower-casing it, replacing all numbers with a single token and discarding any non-alpha-numeric characters.

Text Topic Spaces. Using textual modality to self-train a neural network was used by Gomez et al. [3] by exploiting the semantic context present in illustrated Wikipedia articles. The authors use the topic probabilities in the text as soft labels for the images in the article. Our approach is similar - we extract text from the RVL-CDIP dataset and analyse it using Latent Dirichlet Allocation [4] to extract topics. The CNN is then trained to predict the topic distribution, given only the image of the document. Different from the approach proposed by Gomez et al. [3], there is a more intimate and direct correspondence between the document layout and its text content.

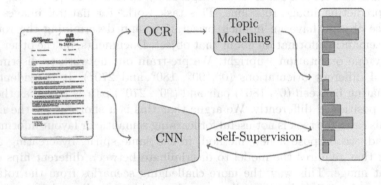

Fig. 2. General methodology for multi-modal self-supervision. The text from documents is extracted using an OCR engine, and then each of our method for topic modelling are used to generate topic probabilities. The neural network is then tasked to predict the topic probabilities using only the document image.

In this topic modeling method, we used soft-labels, as it was shown to improve performance in the context of knowledge distillation [18]. Soft-labels are also robust against noise [45], and have shown to increase performance for large-scale semi-supervised learning [46]. Figure 2 depicts the general overview this method. Our intuition is that documents that are similar in topic spaces should also be similar in appearance. In our experiments, we named this self-supervision method **LDA Topic Spaces**.

4   Experiments

For the pre-training phase of the self-supervised methods, we used the training set from RVL-CDIP [15]. RVL-CDIP is a dataset consisting of 400.000 grayscale document images, of which 320.000 are provided for training, 40.000 for validation, and the remaining 40.000 for testing. The images are labeled into 16 classes, some of which are also present in Tobacco-3482. Naturally, during our self-supervised pre-training experiments, we discard the labels. During the evaluation, we used the pre-trained models as feature extractors, and compute feature vectors for each image. We then trained a logistic classifier using L-BFGS [30]. As shown by Kolesnikov et al. [25], a linear model is sufficient for evaluating the quality of features. Images used in the extraction phase come from the Tobacco-3482 dataset and are pre-processed exactly as during training. For partitioning the dataset, we used the same method as in [1,15,23,26] for consistency and for a fair comparison with other works. We used Top-1 Accuracy as a metric, and we trained on a total of 10 to 100 images per class (with ten images increment), randomly sampled from Tobacco-3482. Testing was done on the rest of the images. We ran each experiment 10 times to reduce the odds of having a favourable configuration of training samples. Our evaluation scheme is designed for testing the performance in a document image classification setting with a limited amount of data.

In the particular case of LDA, we varied the number of topics and trained three models tasked to predict the topic probabilities of 16, 32, and 64 topics. By using this form of soft-clustering in the topic space, the model benefited from having a finer-grained topic distribution.

In our experiments, we used two supervised benchmarks: a model pre-trained on ImageNet and a model pre-trained on RVL-CDIP. The supervised pre-training methods have an obvious advantage, due to the high amount of consistent and correct information present in annotations. Consistent with other works [26], supervised RVL-CDIP pre-training is far superior.

5   Results

In Fig. 3, we show some of the more relevant methods in the evaluation scheme. Supervised RVL-CDIP is, unsurprisingly, the most performing method, and our self-supervised multi-modal approach has a significantly higher accuracy overall when compared to supervised ImageNet pre-training. Features extracted from patch-based methods and methods which rely only on layout information are not discriminative enough to have higher accuracy. This is also consistent with the original works [5,11,33] in which self-supervised pre-training did not provide a boost in performance compared to the supervised baseline.

Relative Patches and Jigsaw Puzzles have only modest performance. Both these methods initially used a "context-free" approach. Our variation of Jigsaw Puzzles - Jigsaw Whole - works around this by actually including more context. Receiving the entire document helps the network to learn features that are relevant for the layout. Features learned this way are more discriminative for the classes in Tobacco-3842. In the case of Relative Patches, there is no sensible way to include more context in the input, as stitching together two patches changes the aspect ratio of the input.

In Fig. 3, we present the mean accuracy for 100 samples per class on all methods. We also implemented the work of Bojanowski et al. [5], to pre-train

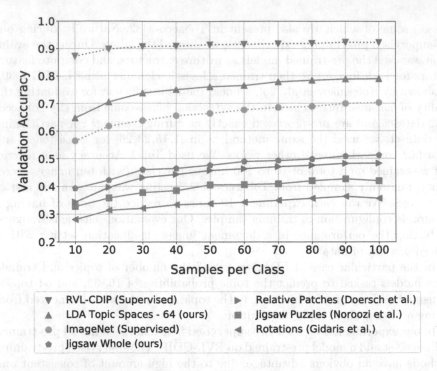

Fig. 3. Performance on fine-tuning for some of the more relevant methods on different sample sizes from Tobacco-3482. Our proposed methods have higher accuracy than previous attempts on natural images. Multi-modal self-supervision using LDA (LDA Topic Spaces) with 64 topics is significantly higher than supervised ImageNet pretraining and much higher than the other self-supervised methods we tested.

a model by predicting noise. This method is very general and assumes very little of the inputs. We discovered that it was better than predicting rotations. The features extracted by predicting rotations were, in fact, the weakest, as this task is far too easy for a model in the case of document images. Our variation, predicting flips, provides a much harder task, which translates into a better performance on the classification task (Table 1).

In the case of topic modeling, we argue that the boost in performance is due to the high correlation between the similarity in the topic space and the similarity in the "image" space. The 16 classes in RVL-CDIP are sufficiently dissimilar between them in terms of topics (i.e., news article, resume, advertisement), and each class of documents has a specific layout. Surprisingly, LDA with 16 topics was the weakest. A finer-grained topic distribution helped the model in learning more discriminative features.

Table 1. Results on our pre-training experiments. We also implemented Noise as Targets (NAT) [5] as an input-agnostic method for self-supervision. For LDA, we tried multiple number of topics and have decided upon 64 before experiencing diminishing returns. All presented methods employ self-supervised pre-training, unless otherwise specified.

Method	Accuracy %
Rotations (Gidaris et al.)	35.63 ± 2.74
Noise as Targets (Bojanowski et al.)	40.85 ± 1.58
Jigsaw Puzzles (Noroozi et al.)	41.95 ± 1.71
Relative Patches (Doersch et al.)	48.28 ± 1.69
Flips (ours)	$\mathbf{50.52 \pm 1.93}$
Jigsaw Whole (ours)	$\mathbf{51.19 \pm 2.06}$
ImageNet pre-training (Supervised)	70.3 ± 1.5
LDA Topic Spaces - 16 (ours)	75.75 ± 1.47
LDA Topic Spaces - 32 (ours)	76.22 ± 1.72
LDA Topic Spaces - 64 (ours)	$\mathbf{78.96 \pm 1.42}$
RVL-CDIP pre-training (Supervised)	92.28 ± 0.8

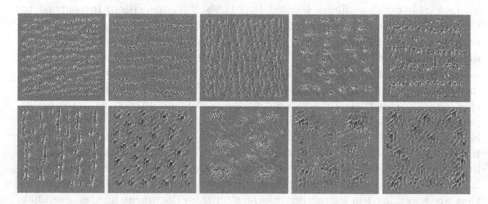

Fig. 4. Gradient ascent visualization of filters learned through supervised RVL pre-training. Patterns for text lines, columns and paragraphs appear.

5.1 Qualitative Analysis

For the qualitative analysis, we compare filters learned through LDA self - supervision with those learned through RVL-CDIP supervised pre-training. In Figs. 4 and 5, we show gradient ascent visualizations learned by both methods - a randomly initialized input image is progressively modified such that the mean activation of a given filter is maximized. The filters shown are from increasing depths in the InceptionV3 network, from conv2d_10 to conv2d_60.

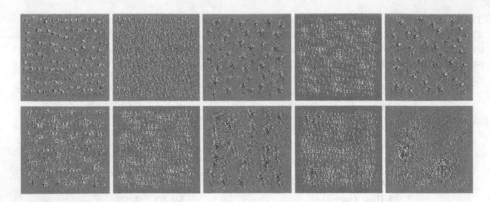

Fig. 5. Gradient ascent visualization of filters learned by LDA pre-training. Activation patterns that correspond to words and paragraphs emerge. More distinctively, patterns that appear to resemble words are more frequent than in the supervised setting.

In Fig. 4 there are clear emerging patters that correspond to text lines, columns and paragraphs - general features that apply to a large subset of documents.

In Fig. 5, we show filters learned by LDA self - supervision. In contrast to the "gold-standard" filter learned by RVL-CDIP supervision, the patterns that emerge here are frequently more akin to what appears to be words. This is a direct consequence of the way LDA constructs topics, in which a topic is based on a bag-of-words model. Our neural network, therefore, has a high response in regions corresponding to particular words in the image. The features learned this way are nonetheless discriminative for document images, as patterns for paragraphs and text columns still emerge. Another particularity of these filters is that they are noisier than those learned by direct supervision. This likely results from the soft and noisy labels generated by the topic model, and due to imperfect text extracted by the OCR engine. Naturally, features learned from ImageNet pre-training are qualitatively different and more general from filters that are specialized in extracting information from document images. See Olah et al. [35] for a comprehensive visualization of InceptionV3 trained on ImageNet.

6   Conclusions

We have explored self-supervision methods that were previously introduced in the realm of natural images and showed that document images have a more rigid visual structure, which makes patch-based methods not work as well on them. To that end, we proposed slight alterations that exploit documents' visual properties: self-supervised pre-training by predicting flips and by solving jigsaw puzzles with the whole layout present in the input.

Documents are inherently multi-modal. As such, by extracting text from the document images, we developed a method to pre-train a network, in a self-supervised manner, to predict topics generated by Latent Dirichlet Allocation.

This method outperforms the strong baseline of supervised pre-training from ImageNet. We also show that the features learned this way are closely related to those learned through direct supervision on RVL-CDIP, making this method a viable method for pre-training neural networks on document images.

Acknowledgements. We want to express our appreciation for everyone involved at Sparktech Software, for fruitful discussions, and much-needed suggestions. In particular, we would like to thank Andrei Manea, Andrei Sbârcea, Antonio Bărbălău, and Andrei Iușan.

References

1. Afzal, M.Z., Kölsch, A., Ahmed, S., Liwicki, M.: Cutting the error by half: investigation of very deep CNN and advanced training strategies for document image classification. CoRR abs/1704.03557 (2017). http://arxiv.org/abs/1704.03557
2. Audebert, N., Herold, C., Slimani, K., Vidal, C.: Multimodal deep networks for text and image-based document classification. CoRR abs/1907.06370 (2019). http://arxiv.org/abs/1907.06370
3. i Bigorda, L.G., Patel, Y., Rusiñol, M., Karatzas, D., Jawahar, C.V.: Self-supervised learning of visual features through embedding images into text topic spaces. CoRR abs/1705.08631 (2017). http://arxiv.org/abs/1705.08631
4. Blei, D.M., Ng, A.Y., Jordan, M.I.: Latent dirichlet allocation. J. Mach. Learn. Res. **3**, 993–1022 (2003). http://dl.acm.org/citation.cfm?id=944919.944937
5. Bojanowski, P., Joulin, A.: Unsupervised learning by predicting noise. In: Proceedings of the 34th International Conference on Machine Learning, ICML 2017, vol. 70, pp. 517–526. JMLR.org (2017). http://dl.acm.org/citation.cfm?id=3305381.3305435
6. Caron, M., Bojanowski, P., Mairal, J., Joulin, A.: Leveraging large-scale uncurated data for unsupervised pre-training of visual features. CoRR abs/1905.01278 (2019). http://arxiv.org/abs/1905.01278
7. Cruz, R.S., Fernando, B., Cherian, A., Gould, S.: Deeppermnet: visual permutation learning. CoRR abs/1704.02729 (2017). http://arxiv.org/abs/1704.02729
8. Csurka, G., Larlus, D., Gordo, A., Almazán, J.: What is the right way to represent document images? CoRR abs/1603.01076 (2016). http://arxiv.org/abs/1603.01076
9. Deng, J., Dong, W., Socher, R., Li, L.J., Li, K., Fei-Fei, L.: ImageNet: a large-scale hierarchical image database. In: CVPR 2009 (2009)
10. Devlin, J., Chang, M., Lee, K., Toutanova, K.: BERT: pre-training of deep bidirectional transformers for language understanding. CoRR abs/1810.04805 (2018). http://arxiv.org/abs/1810.04805
11. Doersch, C., Gupta, A., Efros, A.A.: Unsupervised visual representation learning by context prediction. CoRR abs/1505.05192 (2015). http://arxiv.org/abs/1505.05192
12. Doersch, C., Zisserman, A.: Multi-task self-supervised visual learning. CoRR abs/1708.07860 (2017). http://arxiv.org/abs/1708.07860
13. Dosovitskiy, A., Springenberg, J.T., Riedmiller, M.A., Brox, T.: Discriminative unsupervised feature learning with convolutional neural networks. CoRR abs/1406.6909 (2014). http://arxiv.org/abs/1406.6909
14. Gidaris, S., Singh, P., Komodakis, N.: Unsupervised representation learning by predicting image rotations. CoRR abs/1803.07728 (2018). http://arxiv.org/abs/1803.07728

15. Harley, A.W., Ufkes, A., Derpanis, K.G.: Evaluation of deep convolutional nets for document image classification and retrieval. In: International Conference on Document Analysis and Recognition (ICDAR) (2015)
16. He, K., Gkioxari, G., Dollár, P., Girshick, R.B.: Mask R-CNN. CoRR abs/1703.06870 (2017). http://arxiv.org/abs/1703.06870
17. Hénaff, O.J., Razavi, A., Doersch, C., Eslami, S.M.A., van den Oord, A.: Data-efficient image recognition with contrastive predictive coding. CoRR abs/1905.09272 (2019). http://arxiv.org/abs/1905.09272
18. Hinton, G., Vinyals, O., Dean, J.: Distilling the knowledge in a neural network. In: NIPS Deep Learning and Representation Learning Workshop (2015). http://arxiv.org/abs/1503.02531
19. Holecek, M., Hoskovec, A., Baudis, P., Klinger, P.: Line-items and table understanding in structured documents. CoRR abs/1904.12577 (2019). http://arxiv.org/abs/1904.12577
20. Hsu, K., Levine, S., Finn, C.: Unsupervised learning via meta-learning. CoRR abs/1810.02334 (2018). http://arxiv.org/abs/1810.02334
21. Huh, M., Agrawal, P., Efros, A.A.: What makes imagenet good for transfer learning? CoRR abs/1608.08614 (2016). http://arxiv.org/abs/1608.08614
22. Jing, L., Tian, Y.: Self-supervised visual feature learning with deep neural networks: a survey. CoRR abs/1902.06162 (2019). http://arxiv.org/abs/1902.06162
23. Kang, L., Kumar, J., Ye, P., Li, Y., Doermann, D.: Convolutional neural networks for document image classification. In: 2014 22nd International Conference on Pattern Recognition, pp. 3168–3172, August 2014. https://doi.org/10.1109/ICPR.2014.546
24. Katti, A.R., et al.: Chargrid: towards understanding 2D documents. CoRR abs/1809.08799 (2018). http://arxiv.org/abs/1809.08799
25. Kolesnikov, A., Zhai, X., Beyer, L.: Revisiting self-supervised visual representation learning. CoRR abs/1901.09005 (2019). http://arxiv.org/abs/1901.09005
26. Kölsch, A., Afzal, M.Z., Ebbecke, M., Liwicki, M.: Real-time document image classification using deep CNN and extreme learning machines. CoRR abs/1711.05862 (2017). http://arxiv.org/abs/1711.05862
27. Kornblith, S., Shlens, J., Le, Q.V.: Do better imagenet models transfer better? CoRR abs/1805.08974 (2018). http://arxiv.org/abs/1805.08974
28. Kumar, J., Ye, P., Doermann, D.: Structural similarity for document image classification and retrieval. Pattern Recognit. Lett. **43**, 119–126 (2014). https://doi.org/10.1016/j.patrec.2013.10.030
29. Li, M., Cui, L., Huang, S., Wei, F., Zhou, M., Li, Z.: Tablebank: table benchmark for image-based table detection and recognition. arXiv preprint arXiv:1903.01949 (2019)
30. Liu, D.C., Nocedal, J.: On the limited memory BFGS method for large scale optimization. Math. Program. **45**(1), 503–528 (1989). https://doi.org/10.1007/BF01589116
31. Liu, X., Gao, F., Zhang, Q., Zhao, H.: Graph convolution for multimodal information extraction from visually rich documents. In: Proceedings of the 2019 Conference of the North American Chapter of the Association for Computational Linguistics: Human Language Technologies, vol. 2 (Industry Papers), pp. 32–39. Association for Computational Linguistics, Minneapolis (2019). https://doi.org/10.18653/v1/N19-2005. https://www.aclweb.org/anthology/N19-2005
32. Mikolov, T., Sutskever, I., Chen, K., Corrado, G., Dean, J.: Distributed representations of words and phrases and their compositionality, vol. abs/1310.4546 (2013). http://arxiv.org/abs/1310.4546

33. Noroozi, M., Favaro, P.: Unsupervised learning of visual representations by solving jigsaw puzzles. CoRR abs/1603.09246 (2016). http://arxiv.org/abs/1603.09246
34. Noroozi, M., Vinjimoor, A., Favaro, P., Pirsiavash, H.: Boosting self-supervised learning via knowledge transfer. CoRR abs/1805.00385 (2018). http://arxiv.org/abs/1805.00385
35. Olah, C., Mordvintsev, A., Schubert, L.: Feature visualization. Distill (2017). https://doi.org/10.23915/distill.00007. https://distill.pub/2017/feature-visualization
36. Oliveira, S.A., Seguin, B., Kaplan, F.: dhsegment: a generic deep-learning approach for document segmentation. CoRR abs/1804.10371 (2018). http://arxiv.org/abs/1804.10371
37. Qasim, S.R., Mahmood, H., Shafait, F.: Rethinking table parsing using graph neural networks. CoRR abs/1905.13391 (2019). http://arxiv.org/abs/1905.13391
38. Radford, A., Wu, J., Child, R., Luan, D., Amodei, D., Sutskever, I.: Language models are unsupervised multitask learners (2019)
39. Ren, S., He, K., Girshick, R.B., Sun, J.: Faster R-CNN: towards real-time object detection with region proposal networks. CoRR abs/1506.01497 (2015). http://arxiv.org/abs/1506.01497
40. Smith, R.: An overview of the tesseract OCR engine. In: Proceedings of Ninth International Conference on Document Analysis and Recognition (ICDAR), pp. 629–633 (2007)
41. Studer, L., et al.: A comprehensive study of imagenet pre-training for historical document image analysis. CoRR abs/1905.09113 (2019). http://arxiv.org/abs/1905.09113
42. Szegedy, C., Vanhoucke, V., Ioffe, S., Shlens, J., Wojna, Z.: Rethinking the inception architecture for computer vision. CoRR abs/1512.00567 (2015). http://arxiv.org/abs/1512.00567
43. Tan, C., Sun, F., Kong, T., Zhang, W., Yang, C., Liu, C.: A survey on deep transfer learning. CoRR abs/1808.01974 (2018). http://arxiv.org/abs/1808.01974
44. Tensmeyer, C., Martinez, T.: Analysis of convolutional neural networks for document image classification (2017)
45. Thiel, C.: Classification on soft labels is robust against label noise. In: Lovrek, I., Howlett, R.J., Jain, L.C. (eds.) KES 2008. LNCS (LNAI), vol. 5177, pp. 65–73. Springer, Heidelberg (2008). https://doi.org/10.1007/978-3-540-85563-7_14
46. Yalniz, I.Z., Jégou, H., Chen, K., Paluri, M., Mahajan, D.: Billion-scale semi-supervised learning for image classification. CoRR abs/1905.00546 (2019). http://arxiv.org/abs/1905.00546
47. Yang, X., Yümer, M.E., Asente, P., Kraley, M., Kifer, D., Giles, C.L.: Learning to extract semantic structure from documents using multimodal fully convolutional neural network. CoRR abs/1706.02337 (2017). http://arxiv.org/abs/1706.02337
48. Zhang, R., Isola, P., Efros, A.A.: Colorful image colorization. CoRR abs/1603.08511 (2016). http://arxiv.org/abs/1603.08511
49. Zhuang, F., et al.: A comprehensive survey on transfer learning (2019)

ACMU-Nets: Attention Cascading Modular U-Nets Incorporating Squeeze and Excitation Blocks

Seokjun Kang$^{(\boxtimes)}$ ⓘ, Brian Kenji Iwana ⓘ, and Seiichi Uchida ⓘ

Department of Advanced Information Technology, Kyushu University,
Fukuoka, Japan
{seokjun.kang,brian,uchida}@human.ait.kyushu-u.ac.jp

Abstract. In document analysis research, image-to-image conversion models such as a U-Net have been shown significant performance. Recently, cascaded U-Nets research is suggested for solving complex document analysis studies. However, improving performance by adding U-Net modules requires using too many parameters in cascaded U-Nets. Therefore, in this paper, we propose a method for enhancing the performance of cascaded U-Nets. We suggest a novel document image binarization method by utilizing Cascading Modular U-Nets (CMU-Nets) and Squeeze and Excitation blocks (SE-blocks). Through verification experiments, we point out the problems caused by the use of SE-blocks in existing CMU-Nets and suggest how to use SE-blocks in CMU-Nets. We use the Document Image Binarization (DIBCO) 2017 dataset to evaluate the proposed model.

Keywords: U-Net · DIBCO · Image binarization

1 Introduction

Recently, semantic segmentation models such as U-Net [19] and Fully Convolutional Network (FCN) [12] have been widely used in document analysis research. Due to the excellent image-to-image conversion performance of these models, many variants are proposed in the International Conference of Document Analysis and Recognition (ICDAR), International Conference on Frontiers of Handwriting Recognition (ICFHR), and International Workshop on Document Analysis Systems (DAS) [2,22]. Among these models, U-Net has shown strong pixel localization performance and are used in various areas as well as document analysis.

However, U-Nets are usually trained to achieve one specific purpose and additional processes are sometimes required to solve complex tasks. For example, to binarize historical document images, pre-processing such as background noise removal, non-document area removal, and image quality improvement are typical as well as a binarization process. Therefore, Cascading Modular U-Nets (CMU-Nets) were proposed in ICDAR 2019 [10], which utilize modularized U-Nets that perform different tasks.

© Springer Nature Switzerland AG 2020
X. Bai et al. (Eds.): DAS 2020, LNCS 12116, pp. 118–130, 2020.
https://doi.org/10.1007/978-3-030-57058-3_9

CMU-Nets are consist of pre-trained U-Net modules to conduct the specific tasks required for document image binarization. Next, modularized U-Nets are cascaded for conducting a document image binarization task. To enhance the performance of CMU-Nets, the modular U-Nets have inter-module skip-connections so that features can be transferred between modules. Since pre-trained U-Net modules are trained by using sufficient external datasets in CMU-Nets, it has the advantage of being able to train with only limited historical document images. However, since CMU-Nets use an original U-Net, there is a memory limitation of available U-Nets to design CMU-Nets. Therefore, only four U-Nets were used for document image binarization in CMU-Nets. In this paper, we improve the inter-module connections by applying the Squeeze-and-Excitation block (SE-block) [6].

Many studies increase the performance of neural networks using the attention mechanism [9,23]. The attention mechanism aims to enhance NN's performance with a relatively tiny architectural change of NN. In particular, unlike the previous attention mechanism, the SE-block focuses on improving the representation performance of the network by modeling channel-wise relationships. The SE-block can be attached directly to anywhere in various types of networks. Therefore, many researchers have tried to utilize SE-blocks with their model [20].

This paper tries to find out how we use SE-blocks in CMU-Nets through three steps. First, we apply the SE-block to simply stacked U-Nets and CMU-Nets, respectively. Next, we propose optimal architecture by analyzing the problems of CMU-Nets with SE-blocks. Finally, we utilize the DIBCO 2017 dataset to verify the performance improvement of document image binarization of CMU-Nets with the proposed method. Through these evaluation steps, we can define our contribution as follow.

- We experimentally demonstrate the value of SE-blocks in CMU-Nets and conventional stacked U-Nets to verify how SE-blocks can contribute to model performance.
- We propose the optimal architecture by using SE-blocks in CMU-Nets.
- We evaluate the proposed method by using the DIBCO 2017 dataset.

The remaining of this paper is organized as follows. Section 2 reviews related work for the document image binarization model by using FCN and U-Nets and the variants by using attention mechanism and the SE-block. Section 3 provides detailed information on the proposed model. Section 4 shows experiment results analysis by comparing the conventional CMU-Nets and stacked U-Nets. Finally, Sect. 5 is the conclusion section.

2   Related Work

In this section, we briefly mention related work on document image binarization systems by adopting FCN and U-Net and Artificial Neural Networks (ANN) with SE-blocks.

2.1 Document Image Binarization with FCN and U-Net

In recent times, document analysis systems by using FCN [14,25] and U-Net [3,11] are suggested. Since FCN and U-Net have robust image-to-image conversion ability, many researchers adapt these models to their document image binarization systems. In DIBCO 2017, six of the 18 participating teams proposed ANNs-based document binarization systems, of which three teams adopted FCN and U-Net [17]. The top four teams' systems of DIBCO 2017 were suggested by using ANNs. The winner's system of DIBCO 2017 utilized a single U-Net. The winner used effective data augmentation techniques and optimization tasks for network training.

Aside from these competitions, U-Net-based document binarization systems have been introduced. Huang et al. improved the document binarization performance by combining local and global predict results using three U-Nets [7]. Their model adopts a partially stacked U-Nets that extract global and local-level features by applying different patch sizes. He and Schomaker used single U-Net to learn the degradation of document images rather than traditional pixel-wise training and prediction methods [5]. They created degraded uniform images and trained the model through iterative work. Accurate binarization image was output by obtaining a global or local threshold value by utilizing single U-Net.

2.2 ANNs with the Attention Networks

The concept of the attention mechanism was introduced by Graves [4] and Bahdanau et al. [1]. Since then, attention mechanisms have been applied to various tasks such as image classification [24,26] and semantic segmentation [8,18].

Wang et al. demonstrated robust image classification performance by applying attention mechanisms to a deep convolutional neural network (CNN) [23]. Wang et al. trained each module to detect different types of attention by stacking multiple attention modules, and used residual attention structure to prevent performance degradation of stacked attention modules. Oktay et al. proposed Attention U-Net by applying the stacked residual attention mechanism proposed by Wang et al. to U-Net. Oktay et al. demonstrated robust semantic segmentation performance by attaching attention gates to all expanding paths of U-Net.

SE-block, is suggested by Hu et al., complemented the existing attention mechanism [6]. SE-block contributed to network performance improvement by extracting important information of network through squeeze operation and performing feature re-calibration through excitation operation as shown in Fig. 1. Rundo et al. demonstrated robust semantic segmentation performance by applying SE-blocks inside U-Net [20]. Rundo et al. added SE-blocks to the existing U-Net contracting path and used SE-blocks as a way to concatenating by adding them in the upsampling process of expanding path. Zhu et al. proposed 3D SE U-Net for fast and fully automated anatomical segmentation. [28] Zhu et al. used SE-blocks to replace some of the max-pooling and convolution filters in U-Net to transfer features between layers.

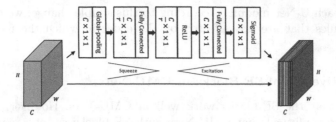

Fig. 1. Brief description of SE-block. SE-block is consist of squeeze and excitation operation. SE-block can extract features by using global pooling and calibrate these features by using fully connected layers.

3   Proposed Method

In this paper, we use two strategies to design the proposed model. The first strategy is to modify the conventional training method when U-Net modules in CMU-Nets are trained. We verify how much the revised method affects document image binarization performance. The second strategy is to analyze the effect of SE-blocks on conventional inter-module skip-connections in CMU-Nets. According to this analysis steps, we design the proper architecture of individual U-Net modules and overall CMU-Nets.

3.1   Preparation Process of the Proposed Method

CMU-Nets has two major differences from stacked U-Nets. The first is that CMU-Nets adopt skip-connections between connected modules for robust feature transfer. In CMU-Nets, the proposed connection method between modules is called by inter-module skip-connections. Inter-module skip-connections can transfer pixels' localization information to the next modules and prevent excessive interference from the previously used U-Net modules. Second, to solve the problem of insufficient document image quantity, CMU-Nets adopted pre-trained U-Net modules by using external datasets for training. In general, when training NN models by using external datasets, final models' performance sometimes can show worse results. Therefore, many researchers avoid this situation, they usually utilize data augmentation methods for solving the lack of training data. However, in specific cases such as document analysis, there are too insufficient training data for optimizing NN models, hence CMU-Nets use pre-trained U-Net modules. CMU-Nets are combined with these modules to create cascading U-Nets for a specific purpose such as an image binarization task. Therefore, in CMU-Nets, the process of making and combining pre-trained U-Net modules is very important for the whole model training process.

In conventional CMU-Nets, only gray-scaled images were used for training U-Net modules. However, the actual input images used in each module are the output images of the other modules as well as the gray-scaled images. Therefore, we set gray-scaled, dilated, eroded, and histogram-equalized images as training images for each U-Net module. According to this simple change, we can create U-Net modules that are less affected by the order in which the modules are arranged.

3.2   Effectiveness of SE-Blocks in CMU-Nets

We expect that the SE-blocks work well in CMU-Nets. However, according to Fig. 2, we confirmed that CMU-Nets with SE-blocks as inter-module skip-connections cannot give a big performance boost. To find out the reason for the degradation, we use randomly initialized cascaded U-Nets to see if the degradation is occurring in typical stacked U-Nets structures. And then, we inserted SE-blocks to replace the existing inter-module skip-connections. According to Fig. 3, We found better performance improvement results than randomly initialized cascaded U-Nets compared to CMU-Nets. Since CMU-Nets use pre-trained U-Net modules, it is judged that the effect of the SE-block is much smaller, thus the performance change is less. In particular, cascaded U-Nets, rather than single U-Net, require much more parameters, thus the stability of the model training due to the addition of SE-blocks is insufficient compared to randomly initialized cascaded U-Nets. Therefore, we confirmed that the existing method of adding SE-blocks is not suitable for CMU-Nets using pre-trained U-Net modules.

3.3   Attention CMU-Nets with SE-Blocks

Before utilizing SE-blocks in CMU-Nets, we confirmed two major considerations to design attention CMU-Nets. The first is that CMU-Nets is a model created by combining pre-trained U-Net modules. SE-blocks must be trained from the initial model training process to learn which features of the input image are required for accurate image-to-image conversion tasks. However, since CMU-Nets use pre-trained U-Net modules, thus the training process of the SE-blocks must be completed during the fine-tuning process. If we overload the training process of the SE-blocks intentionally enough, we can face a risk of over-fitting. The second is that each U-Net module used in CMU-Nets has its role (specific task). Among the used U-Net modules in CMU-Nets, there is just one module that performs image binarization. The others are designed to remove noise and improve image quality. Therefore, the features of the previous module extracted by the SE-block are not unconditionally required by the next module or image binarization module.

The proposed attention CMU-Nets is applied seven SE-blocks to the conventional CMU-Nets for reflecting these two considerations. To enhance the performance between connected U-Net modules, we inject three SE-blocks in the inter-module skip-connection. When we inject SE-blocks in all inter-module skip-connections, we can confirm the performance degradation. Therefore, we build three SE-blocks in the first connections between U-Net modules. We focused on Global Average Pooling (GAP) within the SE-block. SE-block uses GAP to compress global information into channel descriptors. Our pre-trained U-Net also conducts a global process on the image, thus the proposed attention CMU-Nets

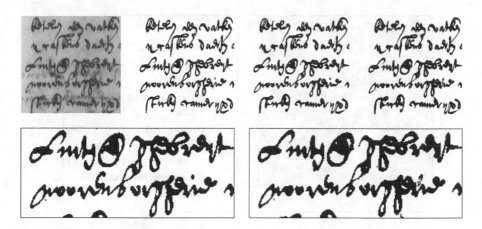

Fig. 2. The output images of CMU-Nets with SE-blocks instead of the inter-module skip-connections. Images from upper left to lower right correspond to the original image, ground truth image, the output image of CMU-Nets, the output image of CMU-Nets with SE-blocks, the enlarged result image of CMU-Nets, and the enlarged result image of CMU-Nets with SE-blocks. The output image of CMU-Nets with SE-blocks is thicker than the output image of CMU-Nets.

can secure the improved performance by adding SE-blocks in the first inter-module connections.

Also, we inject single SE-blocks in the first convolutional steps of the pre-trained U-Net for improving the performance of single U-Net. Since our pre-trained U-Nets perform global processing such as histogram-equalization and image binarization, we utilized SE-blocks' global feature extraction capabilities for improving the performance of pre-trained U-Nets. Therefore, instead of the first max-pooling of the U-Net module, we used a SE-block to extract the global information of the image than at the first stage of the module. We tried to inject SE-blocks after each max-pooling layer. However, there is no significant improvement by adding additional SE-blocks. We designed the usage method of SE-blocks as shown in Fig. 4 to ensure that global information is effectively delivered.

We considered adding SE-blocks to the expanding path as the model proposed by Rundo et al. [20]. However, each U-Net module except for the binarization U-Net module in CMU-Nets performs the pre-processing task in the document image binarization task. In CMU-Nets, the task of each pre-trained U-Net module is important, but consistent one pre-processing task should be accomplished for transferring appropriate information between modules. In particular, this pre-processing should be used to avoid excessive interference with the binarization module.

124   S. Kang et al.

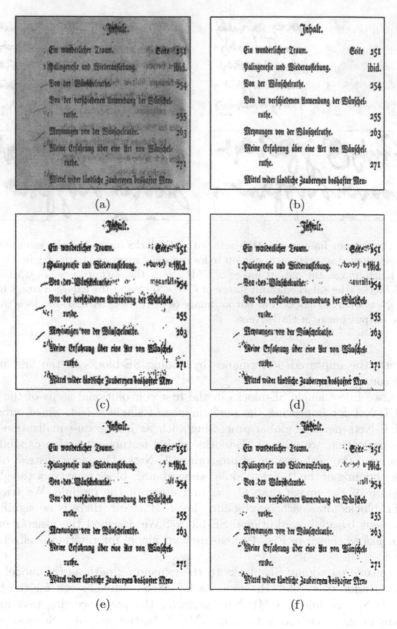

(a)   (b)   (c)   (d)   (e)   (f)

Fig. 3. The comparative output images between the empty cascading U-Nets and CMU-Nets. Images from upper left to lower right correspond to the original image, the ground truth image, the output image of empty cascading U-Nets without SE-blocks, the output image of empty cascading U-Nets with SE-blocks, the output image of CMU-Nets without SE-blocks, and the output image of CMU-Nets with SE-blocks.

4   Experimental Results

4.1   Dataset

In this experiment, two datasets are utilized for model training and fine-tuning. To compare CMU-Nets and the proposed method properly, we train the proposed method on the same conditions. First, we use The COCO-Text dataset for pre-training the U-Net modules. However, according to Sect. 3.1, we mention that dilated, eroded, and histogram-equalized images are used with gray-scaled imaged for training each U-Net module. Therefore, a total of 10,000 images were acquired, with 2,500 images in each class. If input and output images are the same results (i.e. input image: dilated, output image: dilated), relevant images substitute gray-scaled images.

Also, for fine-tuning, we use the previous DIBCO and H-DIBCO datasets. The total number of all DIBCO and H-DIBCO images are 116 images. Since these images are insufficient for fine-tuning, 100 images from each DIBCO and H-DIBCO are made by cropping the original images. Therefore, we can procure a total of 800 DIBCO and H-DIBCO images for fine-tuning. The dataset of DIBCO 2017 is used for the test. The size of our all images is resized to 256×256.

4.2   Experimental Environment

In this experiment, each U-Net module is trained by 40,000 iterations using batches of 4 for pre-trained U-Net modules. The fine-tuning is conducted 10,000 iterations. Since all conditions should be equaled with conventional CMU-Nets, the used U-Net module is based on the original U-Net. We applied 3×3 convolution filters and 2×2 max-pooling filters. We fixed the value of reduction ratio (r) in SE-block by 16. The number of used convolutional layers in each U-Net module is 13. We used Adam optimizer with an initial learning rate of 0.0001 and a sigmoid activation function in the output layer.

4.3   Evaluation Criteria

To evaluate the proposed method, we use the DIBCO criteria. There are four criteria, F-measure (FM), pseudo-F-measure (F_{ps}) [15], Peak Signal-to-Noise Ratio (PSNR), and Distance Reciprocal Distortion measure (DRD) [13]. These criteria were designed for performance evaluation of image binarization. Since our target is document image binarization, we utilized these criteria for model evaluation.

4.4   Result Analysis

We demonstrated the comparative results among the DIBCO 2017 winner's, CMU-Nets, and the proposed method as shown in Fig. 5. Also, we showed the quantitative results with Table 1. In Fig. 5, our proposed method shows robust document image binarization results compared with the others. We can find out

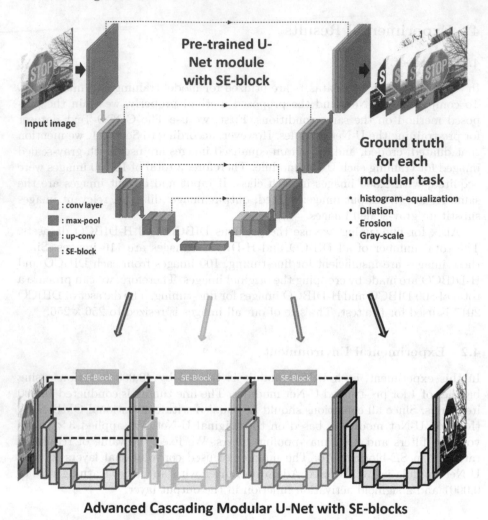

Advanced Cascading Modular U-Net with SE-blocks

Fig. 4. Concept of the proposed method. First, we procure four kinds of training samples. Second, we inject SE-block in the first pooling layer of U-Net. Third, we make pre-trained U-Net modules. Finally, we cascade the pre-trained U-Net modules with the proposed SE-blocks scheme.

that the proposed method controls global and local features well according to erased noisy parts and fore-ground pixels' thickness. Indeed, since conventional CMU-Nets use dilation and erosion U-Net modules, there is a tiny difference of fore-ground pixels' thickness compared with the ground-truth image. However, because of the SE-blocks that can calibrate the necessary features, we can erase more noisy parts.

In Table 1, we obtained the three highest values compared with the others. We obtained the value of FM of 92.19 and F_{ps} of 94.07. In particular, the value

Fig. 5. Comparison results with the DIBCO 2017 winner's, CMU-Nets, and the proposed method. Images in the columns from left to right correspond to the original images, ground truth images, the result images of the DIBCO 2017 winner, the result images of CMU-Nets, and the result images of the proposed method.

Table 1. Comparative results on DIBCO 2017

Method	FM	F_{ps}	PSNR	DRD
Otsu [16]	77.73	77.89	13.85	15.54
Sauvola and Pietikäinen [21]	77.11	84.1	14.25	8.85
Zhao et al. [27]	90.73	92.58	17.83	3.58
Competition winner [17]	91.04	92.86	**18.28**	3.40
CMU-Nets [10]	91.57	93.55	15.85	2.92
The proposed method	**92.19**	**94.07**	17.13	**2.69**

of PSNR is up by 8.08%. According to these results, the proposed scheme is proper for CMU-Nets.

Also, we conducted a comparative evaluation related to the usage of changed training samples. The revised training method is obtained higher value in all criteria in Table 2. However, the difference of the values between the proposed method without the revised training method and the proposed method is not large value. According to this result, if CMU-Nets are designed for more complex tasks, our proposed method can be considered to avoid a problem with the order of arranged U-Net modules.

Table 2. Results of the proposed method with the revised training method on DIBCO 2017

Method	FM	F_{ps}	PSNR	DRD
Conventional training method	92.07	93.86	16.89	2.71
Revised training method	**92.19**	**94.07**	**17.13**	**2.69**

5   Conclusion

In this paper, we suggest two things in CMU-Nets, the revised training method for making pre-trained U-Net modules, and the proper usage of SE-blocks in CMU-Nets. According to the quantitative and qualitative results, our proposed method can be considered if someone wants to make the model by using CMU-Nets. We also find out that simple usage of SE-blocks in cascading U-Nets can not make dramatic performance improvement. We hope that our proposed method will be applied not only to document binarization task but also to the other document analysis tasks.

Acknowledgement. This work was supported by JSPS KAKENHI Grant Number JP17K19402 and JP17H06100.

References

1. Bahdanau, D., Cho, K., Bengio, Y.: Neural machine translation by jointly learning to align and translate. arXiv preprint arXiv:1409.0473 (2014)
2. Fink, M., Layer, T., Mackenbrock, G., Sprinzl, M.: Baseline detection in historical documents using convolutional u-nets. In: 2018 13th IAPR International Workshop on Document Analysis Systems (DAS), pp. 37–42, April 2018. https://doi.org/10.1109/DAS.2018.34
3. Fink, M., Layer, T., Mackenbrock, G., Sprinzl, M.: Baseline detection in historical documents using convolutional u-nets. In: 2018 13th IAPR International Workshop on Document Analysis Systems (DAS), pp. 37–42, April 2018
4. Graves, A.: Generating sequences with recurrent neural networks. arXiv preprint arXiv:1308.0850 (2013)
5. He, S., Schomaker, L.: DeepOtsu: document enhancement and binarization using iterative deep learning. Pattern Recognit. **91**, 379–390 (2019). https://doi.org/10.1016/j.patcog.2019.01.025
6. Hu, J., Shen, L., Sun, G.: Squeeze-and-excitation networks. In: Proceedings of the IEEE Conference on Computer Vision and Pattern Recognition, pp. 7132–7141 (2018)
7. Huang, X., Li, L., Liu, R., Xu, C., Ye, M.: Binarization of degraded document images with global-local u-nets. Optik **203**, 164025 (2020). https://doi.org/10.1016/j.ijleo.2019.164025
8. Huang, Z., Wang, X., Huang, L., Huang, C., Wei, Y., Liu, W.: CCNet: criss-cross attention for semantic segmentation. In: Proceedings of the IEEE International Conference on Computer Vision, pp. 603–612 (2019)

9. Jaderberg, M., Simonyan, K., Zisserman, A., et al.: Spatial transformer networks. In: Proceedings of the Conference on Neural Information Processing Systems, pp. 2017–2025 (2015)
10. Kang, S., Uchida, S., Iwana, B.K.: Cascading modular u-nets for document image binarization. In: International Conference on Document Analysis and Recognition (ICDAR), pp. 675–680 (2019)
11. Lee, J., Hayashi, H., Ohyama, W., Uchida, S.: Page segmentation using a convolutional neural network with trainable co-occurrence features. In: International Conference on Document Analysis and Recognition (ICDAR) (2019)
12. Long, J., Shelhamer, E., Darrell, T.: Fully convolutional networks for semantic segmentation. In: Proceedings of the 2015 IEEE Conference on Computer Vision and Pattern Recognition, pp. 3431–3440. IEEE, June 2015. https://doi.org/10.1109/cvpr.2015.7298965
13. Lu, H., Kot, A., Shi, Y.: Distance-reciprocal distortion measure for binary document images. IEEE Signal Process. Lett. **11**(2), 228–231 (2004). https://doi.org/10.1109/lsp.2003.821748
14. Meier, B., Stadelmann, T., Stampfli, J., Arnold, M., Cieliebak, M.: Fully convolutional neural networks for newspaper article segmentation. In: 2017 14th IAPR International Conference on Document Analysis and Recognition (ICDAR), vol. 1, pp. 414–419, November 2017. https://doi.org/10.1109/ICDAR.2017.75
15. Ntirogiannis, K., Gatos, B., Pratikakis, I.: Performance evaluation methodology for historical document image binarization. IEEE Trans. Image Process. **22**(2), 595–609 (2013). https://doi.org/10.1109/tip.2012.2219550
16. Otsu, N.: A threshold selection method from gray-level histograms. IEEE Trans. Syst. Man Cybern. **9**(1), 62–66 (1979). https://doi.org/10.1109/tsmc.1979.4310076
17. Pratikakis, I., Zagoris, K., Barlas, G., Gatos, B.: ICDAR2017 competition on document image binarization (DIBCO 2017). In: Proceedings of the 2017 International Conference on Document Analysis and Recognition, pp. 1395–1403. IEEE, November 2017. https://doi.org/10.1109/icdar.2017.228
18. Ren, M., Zemel, R.S.: End-to-end instance segmentation with recurrent attention. In: Proceedings of the IEEE Conference on Computer Vision and Pattern Recognition, pp. 6656–6664 (2017)
19. Ronneberger, O., Fischer, P., Brox, T.: U-Net: convolutional networks for biomedical image segmentation. In: Navab, N., Hornegger, J., Wells, W.M., Frangi, A.F. (eds.) MICCAI 2015. LNCS, vol. 9351, pp. 234–241. Springer, Cham (2015). https://doi.org/10.1007/978-3-319-24574-4_28
20. Rundo, L., et al.: USE-Net: incorporating squeeze-and-excitation blocks into U-Net for prostate zonal segmentation of multi-institutional MRI datasets. arXiv preprint arXiv:1904.08254 (2019)
21. Sauvola, J., Pietikäinen, M.: Adaptive document image binarization. Pattern Recognit. **33**(2), 225–236 (2000). https://doi.org/10.1016/s0031-3203(99)00055-2
22. Tensmeyer, C., Martinez, T.: Document image binarization with fully convolutional neural networks. In: Proceedings of the 2017 International Conference on Document Analysis and Recognition, pp. 99–104. IEEE, November 2017. https://doi.org/10.1109/icdar.2017.25
23. Wang, F., et al.: Residual attention network for image classification. In: Proceedings of the IEEE Conference on Computer Vision and Pattern Recognition, pp. 3156–3164 (2017)
24. Woo, S., Park, J., Lee, J.Y., So Kweon, I.: CBAM: convolutional block attention module. In: Proceedings of the European Conference on Computer Vision (ECCV), pp. 3–19 (2018)

25. Xu, Y., He, W., Yin, F., Liu, C.: Page segmentation for historical handwritten documents using fully convolutional networks. In: 2017 14th IAPR International Conference on Document Analysis and Recognition (ICDAR), vol. 1, pp. 541–546, November 2017. https://doi.org/10.1109/ICDAR.2017.94
26. Yang, Z., Nevatia, R.: A multi-scale cascade fully convolutional network face detector. In: Proceedings of the 2016 International Conference on Pattern Recognition, pp. 633–638. IEEE, December 2016. https://doi.org/10.1109/icpr.2016.7899705
27. Zhao, J., Shi, C., Jia, F., Wang, Y., Xiao, B.: Document image binarization with cascaded generators of conditional generative adversarial networks. Pattern Recognit. **96**, 106968 (2019). https://doi.org/10.1016/j.patcog.2019.106968
28. Zhu, W., et al.: AnatomyNet: deep 3D squeeze-and-excitation U-nets for fast and fully automated whole-volume anatomical segmentation. bioRxiv, p. 392969 (2018)

Dewarping Document Image by Displacement Flow Estimation with Fully Convolutional Network

Guo-Wang Xie[1,2], Fei Yin[2], Xu-Yao Zhang[1,2], and Cheng-Lin Liu[1,2,3(✉)]

[1] School of Artificial Intelligence, University of Chinese Academy of Sciences,
Beijing 100049, People's Republic of China
[2] National Laboratory of Pattern Recognition,
Institute of Automation of Chinese Academy of Sciences,
95 Zhongguancun East Road, Beijing 100190, People's Republic of China
xieguowang2018@ia.ac.cn, fyin@nlpr.ia.ac.cn, {xyz,liucl}@nlpr.ia.ac.cn
[3] CAS Center for Excellence of Brain Science and Intelligence Technology,
Beijing, People's Republic of China

Abstract. As camera-based documents are increasingly used, the rectification of distorted document images becomes a need to improve the recognition performance. In this paper, we propose a novel framework for both rectifying distorted document image and removing background finely, by estimating pixel-wise displacements using a fully convolutional network (FCN). The document image is rectified by transformation according to the displacements of pixels. The FCN is trained by regressing displacements of synthesized distorted documents, and to control the smoothness of displacements, we propose a Local Smooth Constraint (LSC) in regularization. Our approach is easy to implement and consumes moderate computing resource. Experiments proved that our approach can dewarp document images effectively under various geometric distortions, and has achieved the state-of-the-art performance in terms of local details and overall effect.

Keywords: Dewarping document image · Pixel-wise displacement · Fully convolutional network · Local Smooth Constraint

1 Introduction

With the popularity of mobile devices in recent years, camera-captured document images are becoming more and more common. Unlike document images captured by flat scanners, camera-based document images are more likely to deform due to multiple factors such as uneven illumination, perspective change, paper distortion, folding and wrinkling. This makes the processing and recognition of document image more difficult. To reduce the effect of distortion in processing of document images, dewarping approaches have been proposed to estimate the distortion and rectify the document images.

© Springer Nature Switzerland AG 2020
X. Bai et al. (Eds.): DAS 2020, LNCS 12116, pp. 131–144, 2020.
https://doi.org/10.1007/978-3-030-57058-3_10

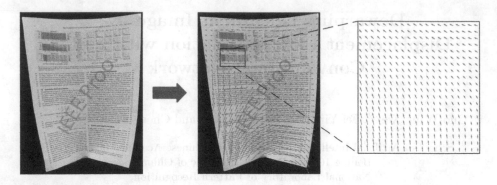

Fig. 1. Our approach regards the document image as a field of displacement flow, which represents the displacements of pixels for transforming one image into another for rectification.

Traditional approaches estimate the 3D shape of document images using auxiliary hardware [16,25] or the geometric properties and visual cues of the document images [5,18,20]. Some approaches [8,12] restrict the page surface to be warped as a cylinder for simplifying the difficulty of 3D reconstruction. Then using raw images and the document shape computer flattened image to correct the distortions. These methods require specific hardware, external conditions or strong assumptions which restrict their generality. For improving the generality of dewarping model, the methods in [15] and [6] use deep neural networks to regress the dewarping function from deformed document image by using 2D and 3D supervised information of the warping respectively. Li et al. [11] considered that it was not possible to accurately and efficiently process the entire image, and proposed patch-based learning approach and stitch the patch results into the rectified document by processing in the gradient domain. Although these methods have obtained promising performance in rectification, further research is needed to deal with situations of more difficult distortions and background.

In this paper, we propose a novel framework to address the difficulties in both rectifying distorted document image and removing background finely. We view the document image is a field of displacement flow, such that by estimating pixel-wise displacements, the image can be transformed to another image accordingly. For rectifying distorted documents, the displacement flow is estimated using a fully convolutional network (FCN), which is trained by regressing the ground-truth displacements of synthesized document images. The FCN has two output branches, for regressing pixel displacements and classifying foreground/background. We design appropriate loss functions for training the network, and to control the smoothness of displacements, we propose Local Smooth Constraint (LSC) for regularization in training. Figure 1 shows the effect of displacement flow and image transformation.

Compared with previous methods based on DNNs, our approach is easy to implement. It can process a whole document image efficiently in moderate computation complexity. The design of network output layers renders good effect in both rectifying distortion and removing background. The LSC in regularization makes the rectified image has smooth shape and preserves local details well. Experiments show that our approach can rectify document images and various contents and distortions, and yields state of the art performance on real-world dataset.

2   Related Works

A lot of techniques for rectifying distorted document have been proposed in the literature. We partitioned them into two groups according to whether deep learning is adopted or not.

Non-deep-Learning-Based Rectification. Prior to the prevalence of deep learning, most approaches rectified document image by estimating the 3D shape of the document images. For reconstructing the 3D shape of document image, many approaches used auxiliary hardware or the geometric properties of the document images to compute an approximate 3D structure. Zhang et al. [25] utilized a more advanced laser range scanner to reconstruct the 3D shape of the warped document. Meng et al. [16] recovered the document curl by using two structured laser beams. Tsoi et al. [19] used multi-view document images and composed together to rectify document image. Liang et al. [12] and Fu et al. [8] restricted the page surface to be warped as a cylinder to simplify the difficulty of 3D reconstruction. Moreover, some techniques utilized geometric properties and visual cues of the document images to reconstruct the document surface, such as illumination/shading [5,20], text lines [14,18], document boundaries [2,3] etc. Although these method can handle simple skew, binder curl, and fold distortion, it is difficult for complicated geometric distortion(i.e., document suffer from fold, curve, crumple and combinations of these etc.) and changeable external conditions (i.e., camera positions, illumination and laying on a complex background etc.)

Deep-Learning-Based Rectification. The emergence of deep learning inspires people to investigate the deep architectures for document image rectification. Das et al. [7] used a CNN to detect creases of document and segmented document into multiple blocks for rectification. Xing et al. [23] applied CNN to estimate document deformation and camera attitude for rectification. Ramanna et al. [17] removed curl and geometric distortion of document by utilizing a pix2pixhd network (Conditional Generative Adversarial Networks). However, these method were only useful for simple deformation and monotone background. Recently, Ma et al. [15] proposed a stacked U-Net which was trained end-to-end to predict the forward mapping for the warping. Because of the generated dataset is quite different from the real-world image, [15] trained on its dataset has worse generalization when tested on real-world images. Das and Ma et al. [6] think dewarping model was not always perform well when trained by the synthetic training dataset only used 2D deformation, so they created a Doc3D dataset which has multiple types of pixel-wise document image ground truth by using both real-world document and rendering software. Meanwhile, [6] proposed a dewarping network and refinement network to correct geometric and shading of document images. Li et al. [11] generated training dataset in the 3D space and use rendering engine to get the finer, realistic details of distorted document image. They proposed patch-based learning approach and stitch the patch results into the rectified document by processing in the gradient domain, and a illumination correction network used to remove the shading. Compared to prior approaches, [6,11] cared more about the difference between the generated training dataset and the real-world testing dataset, and focused on generating more realistic training dataset to improve generalization in real-world images. Although these results are amazing, the learning and expression capability of deep neural network was not fully explored.

3   Proposed Approach

Our approach uses a FCN with two output branches for predicting pixel displacements and foreground/background classification. In dewarping, the foreground pixels are mapped to the rectified image by interpolation according to the predicted displacements.

3.1   Dewarping Process

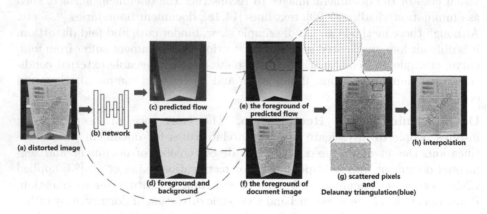

Fig. 2. Illustration of the process of dewarping document image. An input distorted document is first fed into network to predict the pixel-wise displacements and foreground/background classification. When performing rectification, Delaunay triangulation is applied for interpolation in all scattered pixels.

Figure 2 illustrates the process of dewarping document image in our work. We predict the displacement and the categories (foreground or background) at pixel-level by applying two tasks in FCN, and then remove the background of the input image, and mapped the foreground pixels to rectified image by interpolation according to the predicted displacements. The cracks maybe emerge in rectified image when using a forward mapping interpolation. Therefore, we construct Delaunay triangulations in all scattered pixels and then using interpolation [1].

For facilitating implementation, we resize the input image into 1024×960 (zooming in or out along the longest side and keeping the aspect ratio, then filling zero for padding.) in our work. Although smaller input image requires less computing, some information may be lost or unreadable when the distorted document image has small text, picture etc. To trade-off between computational complexity and rectification effect, the document pixels are mapped to rectified image of the same size as the original image, and all the pixels in rectified image are filled by interpolation. We adjust the mapping size as follows:

$$I = F(\lambda \cdot \Re; I_{HD}), \tag{1}$$

where λ is the scaling factor of zooming in or out, $\Re$ is the map of displacement prediction, I_{HD} is the high-resolution distorted image which has same size as $\lambda \cdot \Re$, F is the linear interpolation and I is the rectified image with higher resolution. As shown in Fig. 3, we can implement this method when computing and storage resources are limited.

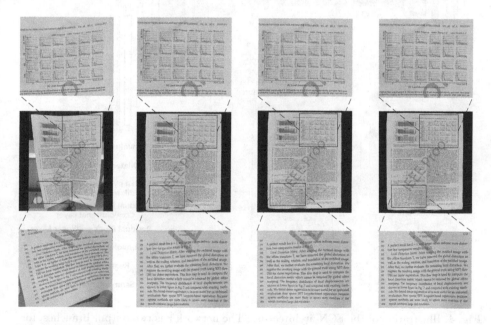

Fig. 3. Results of the different resolution by zooming the flow of displacement. Column 1: Original distorted image, Column 2: Initial ($\times 1$) rectified image, Column 3: $\times 1.5$ rectified image, Column 4: $\times 2$ rectified image

3.2 Network Architecture

In this section, we introduce the architecture of neural network as shown in Fig. 4. For improving the generalization in real-world images, instead of focusing on the vulnerable visual cues, such as illumination/shading, clear text etc., our network architecture infer the displacement of entire document from the image texture layout. Compared with [6,15], our method can simplifies the difficulty of rectification because our model need not to predict the global position of each pixel in flatten image. Different from [11], our approach can take into account both local and global distortion.

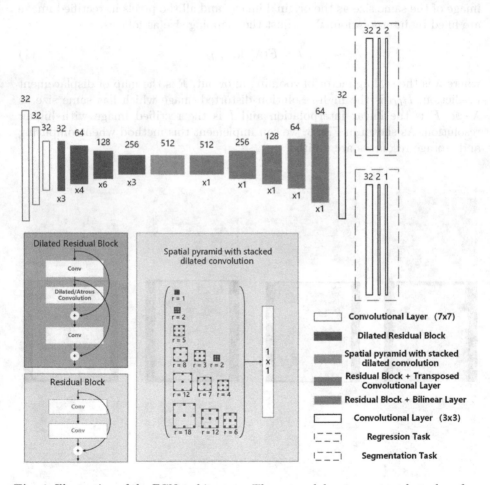

Fig. 4. Illustration of the FCN architecture. The network has two output branches, for pixel displacements prediction and foreground/background classification, respectively.

We adopt an auto-encoder structure and add group normalization (separating the channels into 32 groups) and ReLU after each convolution. To trade-off between computational complexity and rectification effect, the encoder extract local feature by using three convolutional layers with three strides of 1, 2, 2 and 7×7 kernels. Inspired by the architecture from [4,21], We design a dilated residual block which fuse local and dilated semantic by utilizing general convolution, dilated convolution with rate $= 3$ and residual connection. In this way, we can extract denser and larger receptive field distortion feature. After that, we use one spatial pyramid with stacked dilated convolution to encode global high-level semantic information by parallel and cascaded manners. Distortion feature extractor reduces the spatial resolution and obtain the global feature maps, then we gradually recover the displacement of the entire image (the raw resolution) from the spatial feature by using residual block [9] with transposed convolutional layer or bilinear layer.

We use multi-task manner to rectify the document image and separate the foreground and background. The regression task applies group normalization and PReLU after each convolution except for the last layer, and the segmentation task applies group normalization and ReLU after each convolution except for the last layer which adds a sigmoid layer.

3.3 Loss Functions

We train the deep neural network by defining four loss function as a guide to regress the compact and smooth displacement and separate the foreground and background.

The segmentation loss we use in this work is the standard cross entropy loss, which is defined as:

$$L_B = -\frac{1}{N} \sum_{i}^{N} \left[y_i \cdot log\left(\hat{p}_i\right) + \left(1 - y_i\right) \cdot log\left(1 - \hat{p}_i\right) \right], \tag{2}$$

where N is the number of elements in flow, y_i and $\hat{p}_i$ respectively denote the ground-truth and predicted classification.

We optimize the network by minimizing the L1 element-wise loss which measures the distance of pixel-displacement of the foreground between the predicted flow and the ground-truth flow. We formulate L_D function as follows:

$$L_D = \frac{1}{N_f} \sum_{i}^{N_f} \|\Delta D_i - \Delta \hat{D}_i\|_1, \tag{3}$$

where N_f is the elements of foreground which is specified by ground-truth. D_i and $\hat{D}_i$ denote the pixel-displacement in ground-truth and output value of regression network, respectively.

Although the network can be trained by measuring the pixel-wise error between the generated flow and the ground-truth, it's difficult to make model obey the continuum assumption between pixels as shown in Fig. 5. To keep the vary continuously from one point to another in a local, we propose a Local

Smooth Constraint (LSC). In a local region, the LSC expects the predicted displacement trend to be as close to the ground-truth flow as possible. The displacement trend represents the relative relationship between a local region and its central point, which can be defined as:

$$\delta = \sum_{j=1}^{k}(\Delta D_j - \Delta D_{center}), \tag{4}$$

where k is the number of elements in a local region. In our work, we define the local region as a 3×3 rectangle, and ΔD_{center} represents the center of rectangle. To speed up the calculation, we apply a 2D convolution with a strides of 1 and 3×3 kernel. We formulate LSC as follows:

$$\begin{aligned} L_{LSC} &= \frac{1}{N_f} \sum_i^{N_f} \|\delta_i - \hat{\delta}_i\|_1 \\ &= \frac{1}{N_f} \sum_i^{N_f} \| \sum_{j=1}^{k}(\Delta D_j - \Delta D_i) - \sum_{j=1}^{k}(\Delta \hat{D}_j - \Delta \hat{D}_i)\|_1 \\ &= \frac{1}{N_f} \sum_i^{N_f} \| \sum_{j=1}^{k}(\Delta D_j - \Delta \hat{D}_j) - k \times (\Delta D_i - \Delta \hat{D}_i)\|_1, \end{aligned} \tag{5}$$

where $(\Delta D_i - \Delta \hat{D}_i)$ is the distance of the pixel-displacement between the predicted flow and the ground-truth flow, and $\sum_{j=1}^{k}$ can be calculated by using convolution which has square kernels with weight of 1.

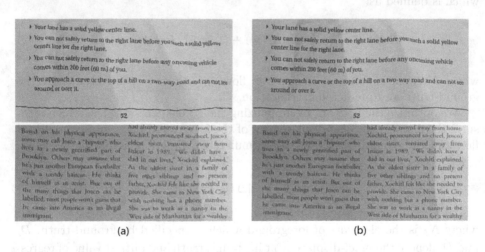

Fig. 5. Results of using Local Smooth Constraint (LSC) after 8 epoch of training. (a) Without using LSC. (b) Using LSC.

Different form general multi-task learning which be applied for assisting with one of the tasks by learning task relationships, our two tasks will be merged for dewarping. We utilize the cosine distance measures the loss in orientation between ground-truth and combined output value of two branch network.

$$L_{cos} = 1 - \cos\theta = 1 - \frac{1}{N} \sum_{i}^{N} \frac{\Delta D_i \cdot \Delta \hat{D}_i}{\|\Delta D_i\| \|\Delta \hat{D}_i\|} \quad (6)$$

These losses are defined as a linear combination:

$$L = L_B + \alpha L_D + \beta L_{LSC} + \gamma L_{cos}, \quad (7)$$

where α, β and γ are weights associated to L_D, L_{LSC} and L_{cos}, respectively.

3.4   Training Details

In our work, the resolution of input data is 1024×960. We train our model on the synthetic dataset of 80,000 images, and none of the documents used in the challenging benchmark dataset proposed by Ma et al. [6] are used to create the synthetic data for training. The network is trained with Adam optimizer [10]. We set the batch size of 6 and learning rate of 2×10^{-4} which reduced by a factor of 0.5 after each 10 epochs. Our method can produce satisfactory rectified document image in about 30 epochs. We set the hyperparameters as $\alpha = 0.1$, $\beta = 0.01$ and $\gamma = 0.05$.

4   Experiments

4.1   Datasets

Our networks are trained in a supervised manner by synthesizing the distorted document image and the rectified ground-truth. Recently, [6,11] generate training dataset in the 3D space to obtain more natural distortion or rich annotations, and use rendering engine to get the finer, realistic details of distorted document image. Although these methods are beneficial for generating more realistic training dataset to improve generalization in real-world images, none of synthetic algorithm could simulate the changeable real-world scenarios. On the other hand, our approach is content-independent which simplifies the difficulty of the dewarping problem and has better performance on rough or unreadable training dataset (as shown in Fig. 6). Experiments proved that our method still maintains the ability of learning and generalization on real-world images, although the generated training dataset is quite different from the real-world dataset.

   For faster and easier synthesizing training dataset, we directly generate distorted document image in 2D mesh. We warp the scanned document such as receipts, papers and books etc., and then using two functions proposed by [15] to change the distortion type, such as folds and curves. Meanwhile, we augment the synthetic images by adding various background textures and jitter in the HSV color space. We synthesized 80K images which have the same height and

(a) 512x480   (b) 1024x960   (c) 2048x1920

Fig. 6. Resolution of the synthetic images. The higher the resolution, the more informa-tion is retained. With minor modifications to the model, any resolution can be applied to the training.

width (i.e., 1024×960). Moreover, our ground-truth flow has three channels. For the first two channels, we define the displacement (Δx, Δy) at pixel-level which indicate how far each pixel have to move to reach its position in the undis-torted image as the rectified Ground-truth. For the last channel, we represent the foreground or background by using the categories (1 or 0) at pixel-level.

4.2 Experimental Setup and Results

We train our network on a synthetic dataset and test on the Ma et al. [15] bench-mark dataset which has various real-world distorted document images. We run our network and post-processing on a NVIDIA TITAN X GPU which processes 10 input images per batch and Intel(R) Xeon(R) CPU E5-2650 v4 which rectifies distorted image by using forward mapping in multiprocessing, respectively. Our implementation takes around 0.67 to 0.72 s to process a 1024×960 image.

Table 1. Comparison of different methods on the Ma et al. [15] benchmark dataset which has various real-world distorted document images. DocUnet was proposed by Ma et al. [15]. DewarpNet was proposed by Das and Ma et al. [6] recently, and Dewarp-Net(ref) is DewarpNet combined with the refinement network to adjust for illumination effects.

Method	MS-SSIM	LD
DocUnet [15]	0.41	14.08
DewarpNet [6]	0.4692	8.98
DewarpNet (ref) [6]	**0.4735**	8.95
Our	0.4361	**8.50**

We compare our results with Ma et al. [15] and Das and Ma et al. [6] on the real-world document images. Compared with previous method, our proposal can rectify various distortions while removing background and replace it to transpar-ent (the visual comparison is shown in Fig. 7). As shown in Fig. 8, our method addresses the difficulties in both rectifying distorted document image and remov-ing background finely. For visually view the process, we don't crop the redundant boundary and retain the original corrected state (No post-cropping, the black edge is the background).

We use two quantitative evaluation criteria as [15] which provided the code with default parameters. One of them is Multi-Scale Structural Similarity (MS-SSIM) [22] which evaluate the global similarity between the rectified document images and scanned images in multi-scale. The other one is Local Distortion (LD) [24] which evaluate the local details by computing a dense SIFT flow [13]. The quantitative comparisons between MS-SSIM and LD are shown in Table 1. Because our approach is more concerned with how to expand the distorted image than whether the structure is similar, we demonstrate state-of-the-art performance in the quantitative metric of local details.

Fig. 7. Results on the Ma et al. [15] benchmark dataset. Row 1: Original distorted images, Row 2: Results of Ma et al. [15], Row 3: Results of Das and Ma et al. [6], Row 4: Results of our method, Row 5: Scanned images.

As shown in Fig. 5, Local Smooth Constraint (LSC) can keep the vary continuously from one point to another in a local. With modifying the hyperparameter of L_D and L_{LSC}, the effect is similar between the loss functions that measure the element-wise mean squared error and the mean element-wise absolute value difference. In our implementation, the loss function of cosine distance is functionally similar to the pixel-displacement distance, however it can significantly improve the convergence speed. Results in Table 2 show ablation experiments

142   G.-W. Xie et al.

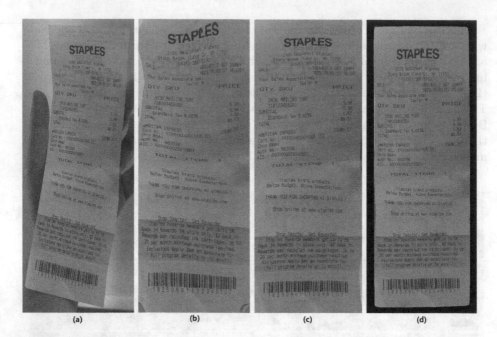

Fig. 8. Results in details. (a) Original distorted images. (b) Ma et al. [15]. (c) Das and Ma et al. [6]. (d) Our.

Table 2. Effect of hyperparameters about the pixel-displacement and cosine distance. α and γ represent the hyperparameter of L_D and L_{cos}, respectively.

Loss function	MS-SSIM	LD
$\alpha = 0.1, \gamma = 0.01$	**0.4434**	8.72
$\alpha = 0.1, \gamma = 0.05$	0.4361	**8.50**
$\alpha = 0.1, \gamma = 0.1$	0.4422	8.76
$\alpha = 0.01, \gamma = 0.05$	0.4389	8.70
$\alpha = 1, \gamma = 0.05$	0.4319	9.14

when we change the hyperparameters of the pixel-displacement and cosine distance. We find that our approach achieves the state-of-the-art performance in terms of local details and overall effect when we set $\alpha = 0.1, \gamma = 0.05$, although it's not best in the global similarity (MS-SSIM). Our network is more concerned with expanding the distortions in local and global components, which is no excessive pursuit of the global similarity between the rectified document images and scanned images. Therefore, we can yield the smoother rectified document image and achieves the better performance in overall effect on the various real-world distorted document images.

5   Conclusion

In this paper, we presented a novel framework for rectifying distorted document images using a fully convolutional network for pixel-wise displacement flow estimation and foreground/background classification, so as to address the difficulties in both rectifying distortions and removing background finely. We define a Local Smooth Constraint (LSC) based on the continuum assumption to make the local pixels and global structure of the rectified image to be compact and smooth. Our approach shows the ability of learning and generalization from imperfect virtual training dataset, even if the synthesized dataset is quite different from the real-world dataset. Although our approach has better tradeoff between computational complexity and rectification effect, the edge of partially rectified image was still not neat enough and the speed of calculation still need to be further improved. In the future, we plan to enhance the performance and get rid of the post-processing steps.

Acknowledgements. This work has been supported by National Natural Science Foundation of China (NSFC) Grants 61733007, 61573355 and 61721004.

References

1. Amidror, I.: Scattered data interpolation methods for electronic imaging systems: a survey. J. Electron. Imaging **11**, 157–76 (2002)
2. Brown, M.S., Tsoi, Y.C.: Geometric and shading correction for images of printed materials using boundary. IEEE Trans. Image Process. **15**(6), 1544–1554 (2006)
3. Cao, H., Ding, X., Liu, C.: A cylindrical surface model to rectify the bound document image. In: Proceedings Ninth IEEE International Conference on Computer Vision, pp. 228–233. IEEE (2003)
4. Chen, L.C., Papandreou, G., Schroff, F., Adam, H.: Rethinking atrous convolution for semantic image segmentation. arXiv preprint arXiv:1706.05587 (2017)
5. Courteille, F., Crouzil, A., Durou, J.D., Gurdjos, P.: Shape from shading for the digitization of curved documents. Mach. Vis. Appl. **18**(5), 301–316 (2007)
6. Das, S., Ma, K., Shu, Z., Samaras, D., Shilkrot, R.: Dewarpnet: single-image document unwarping with stacked 3D and 2D regression networks. In: Proceedings of the IEEE International Conference on Computer Vision, pp. 131–140 (2019)
7. Das, S., Mishra, G., Sudharshana, A., Shilkrot, R.: The common fold: utilizing the four-fold to dewarp printed documents from a single image. In: Proceedings of the 2017 ACM Symposium on Document Engineering, pp. 125–128. ACM (2017)
8. Fu, B., Wu, M., Li, R., Li, W., Xu, Z., Yang, C.: A model-based book dewarping method using text line detection. In: Proceedings of 2nd International Workshop on Camera Based Document Analysis and Recognition, Curitiba, Barazil, pp. 63–70 (2007)
9. He, K., Zhang, X., Ren, S., Sun, J.: Deep residual learning for image recognition. In: Proceedings of the IEEE Conference on Computer Vision and Pattern Recognition, pp. 770–778 (2016)
10. Kingma, D.P., Ba, J.: Adam: a method for stochastic optimization. arXiv preprint arXiv:1412.6980 (2014)

11. Li, X., Zhang, B., Liao, J., Sander, P.V.: Document rectification and illumination correction using a patch-based cnn. ACM Trans. Graph. **38**(6), 1–11 (2019)
12. Liang, J., DeMenthon, D., Doermann, D.: Geometric rectification of camera-captured document images. IEEE Trans. Pattern Anal. Mach. Intell. **30**(4), 591–605 (2008)
13. Liu, C., Yuen, J., Torralba, A.: Sift flow: dense correspondence across scenes and its applications. IEEE Trans. Pattern Anal. Mach. Intell. **33**(5), 978–994 (2010)
14. Liu, C., Zhang, Y., Wang, B., Ding, X.: Restoring camera-captured distorted document images. Int. J. Doc. Anal. Recogn. **18**(2), 111–124 (2015)
15. Ma, K., Shu, Z., Bai, X., Wang, J., Samaras, D.: Docunet: document image unwarping via a stacked u-net. In: Proceedings of the IEEE Conference on Computer Vision and Pattern Recognition, pp. 4700–4709 (2018)
16. Meng, G., Wang, Y., Qu, S., Xiang, S., Pan, C.: Active flattening of curved document images via two structured beams. In: Proceedings of the IEEE Conference on Computer Vision and Pattern Recognition, pp. 3890–3897 (2014)
17. Ramanna, V., Bukhari, S.S., Dengel, A.: Document image dewarping using deep learning. In: International Conference on Pattern Recognition Applications and Methods (2019)
18. Tian, Y., Narasimhan, S.G.: Rectification and 3D reconstruction of curved document images. In: Proceedings of the IEEE Conference on Computer Vision and Pattern Recognition, pp. 377–384. IEEE (2011)
19. Tsoi, Y.C., Brown, M.S.: Multi-view document rectification using boundary. In: Proceedings of the IEEE Conference on Computer Vision and Pattern Recognition, pp. 1–8. IEEE (2007)
20. Wada, T., Ukida, H., Matsuyama, T.: Shape from shading with interreflections under a proximal light source: distortion-free copying of an unfolded book. Int. J. Comput. Vision **24**(2), 125–135 (1997)
21. Wang, P., et al.: Understanding convolution for semantic segmentation. In: IEEE Winter Conference on Applications of Computer Vision, pp. 1451–1460. IEEE (2018)
22. Wang, Z., Simoncelli, E.P., Bovik, A.C.: Multiscale structural similarity for image quality assessment. In: The Thirty-Seventh Asilomar Conference on Signals, Systems and Computers, vol. 2, pp. 1398–1402. IEEE (2003)
23. Xing, Y., Li, R., Cheng, L., Wu, Z.: Research on curved Chinese document correction based on deep neural network. In: International Symposium on Computational Intelligence and Design, vol. 2, pp. 342–345. IEEE (2018)
24. You, S., Matsushita, Y., Sinha, S., Bou, Y., Ikeuchi, K.: Multiview rectification of folded documents. IEEE Trans. Pattern Anal. Mach. Intell. **40**(2), 505–511 (2017)
25. Zhang, L., Zhang, Y., Tan, C.: An improved physically-based method for geometric restoration of distorted document images. IEEE Trans. Pattern Anal. Mach. Intell. **30**(4), 728–734 (2008)

Building Super-Resolution Image Generator for OCR Accuracy Improvement

Xujun Peng[1(✉)] and Chao Wang[2]

[1] ISI, University of Southern California, Marina del Rey, CA, USA
xpeng@isi.edu
[2] LinkedMed Co., Ltd., Spreadtrum Center, Zuchongzhi Road, Shanghai, China
chao.wang@linkedmed.com.cn

Abstract. Super-resolving a low resolution (LR) document image can not only enhance the visual quality and readability of the text, but improve the optical character recognition (OCR) accuracy. However, even despite the ill-posed nature of image super-resolution (SR) problem, how do we treat the finer details of text with large upscale factors and suppress noises and artifacts at the same time, especially for low quality document images is still a challenging task. Thus, in order to boost the OCR accuracy, we propose a generative adversarial network (GAN) based framework in this paper, where a SR image generator and a document image quality discriminator are constructed. To obtain high quality SR document image, multiple losses are designed to encourage the generator to learn the structural properties of texts. Meanwhile, the quality discriminator is trained based on a relativistic loss function. Based on the proposed framework, the obtained SR document images not only maintain the details of textures but remove the background noises, which achieve better OCR performance on the public databases. The source codes and pre-trained models are available at https://gitlab.com/xujun. peng/doc-super-resolution.

Keywords: Structural similarity · GAN · Super-resolution · OCR

1 Introduction

Nowadays, hand-held image capturing devices are prevalent. As a result, massive amount of document images are obtained in our day life, which urges the OCR techniques to facilitate the information retrieval from these rich resources. However, most OCR systems are built upon the document images with high qualities and resolutions, which cannot be guaranteed for hand-held captured images under the uncontrolled environments. Generally, document images acquired by hand-held devices are easily degraded by different types of blurs and noises. This degradation is even worse if it takes into account the fact that the text resolution is low. This phenomenon can be observed in Fig. 1(a), where the OCR performance is damaged by a noisy/blurry low resolution document image.

X. Bai et al. (Eds.): DAS 2020, LNCS 12116, pp. 145–160, 2020.
https://doi.org/10.1007/978-3-030-57058-3_11

Document image super resolution, which aims to restore the high resolution (HR) document image from one or more of its LR counterparts by reducing the noises and maintaining the sharpness of strokes, can not only enhances the document image's perceptual quality and readability, but can be used as an effective tool to improve the OCR accuracy [26]. However, as can be seen from the image capturing process that is illustrated in Eq. 1:

$$\mathbf{y} = (\mathbf{x} \otimes \mathbf{k}) \downarrow_s + \mathbf{n},\tag{1}$$

where $\mathbf{y}$ is the observed LR image, $\mathbf{x} \otimes \mathbf{k}$ is the convolutional operation between unknown HR image (normally the noise free document image) and degradation kernel, $\downarrow_s$ is the downsampling operation with scale of s and $\mathbf{n}$ is the additive white Gaussian noises introduced by capturing devices, there are infinite HR images $\mathbf{x}$ satisfying Eq. 1 given LR image $\mathbf{y}$, such that it is always infeasible to find the proper and efficient mapping from LR image to HR image [2,40].

If no *a priori* knowledge is available, the vision system cannot reliably distinguish between overlaps caused by two different objects in the scene and overlaps caused by a single self-occluding object. A flat rigid object supported by and totally occluding another smaller object may be recognized as a large box-shaped object. Similarly, a flat nonrigid object supported in the middle by a smaller object may be recognized as convex, while if it is supported at the edges by more than one object, it may be recognized as concave.	If no *a priori* knowledge is available, the vision system cannot reliably distinguish between overlaps caused by two different objects in the scene and overlaps caused by a single self-occluding object. A flat rigid object supported by and totally occluding another smaller object may be recognized as a large box-shaped object. Similarly, a flat nonrigid object supported in the middle by a smaller object may be recognized as convex, while if it is supported at the edges by more than one object, it may be recognized as concave.
(a) Original LR image	(b) Restored SR image

Fig. 1. The portion of degraded LR document image and the corresponding SR image produced by the proposed system. The character error rate (CER) obtained by tesseract OCR engine for the original LR image is 27.04%, whilst the CER for the generated SR image is 5.59%.

In order to tackle this challenging problem, many researches have been conducted which can be roughly categorized as two main groups: interpolation based and learning based methods. Interpolation based SR methods, such as bicubic interpolation or Markov random field (MRF) based smoothing [18], are simple and efficient but cannot provide plausible results, especially with large upscale factors. The learning based SR systems learn the mapping between LR images/patches and the corresponding HR images/patches and apply the learnt correlations to infer the HR image. Recently, deep learning approaches have produced more favorable results for computer vision tasks, which also dominated the research direction of SR.

In this paper, we focus our research on SR problem for document images to improve the OCR accuracy of them. To achieve this goal, we propose a generative adversarial network (GAN) based framework for SR where the generator is applied to produce the HR document images based on its corresponding LR images, and the discriminator which is trained by a relativistic loss function is employed to distinguish LR and HR images. In Fig. 1(b), a SR document image generated by the proposed system is demonstrated from which a higher OCR accuracy is obtained comparing to its LR counterpart. In summary, our main contributions are three folds:

- We propose to use multiscale structural similarity (M-SSIM) loss instead of mean squared error (MSE) loss to train the super-resolution document image generator, which can effectively capture the structural properties for text in the document images;
- Inspired by the transformer employed in the machine translation (MT) system [36], the spatial attention layer is applied in the proposed generator to boost super-resolution performance;
- Based on the GAN, an end-to-end document image super-resolution system is designed, which is not depended on the annotated or aligned low-resolution/high-resolution image pairs for training but improves the OCR accuracy on public datasets.

We organize the rest of this article as follows. In Sect. 2, we introduce the related work for image SR and document image OCR accuracy improvements, which is followed by the proposed document image SR approach in Sect. 3. The experiments setup and analysis of this research is covered in Sect. 4. And we conclude our work in Sect. 5.

2   Related Work

2.1   OCR Improvements

Generally, in order to improve the performance of OCR, three types of approaches are carried out. One trend is to use different kinds of preprocessing methods to improve the quality of document images which include many simple manipulations, such as noise removal, image enhancement, deskew, dewarping, etc. [1,3,8,34].

Another types of approaches aim to improve the recognition capability of OCR. The early OCR engines are mostly segmentation-based, which require a sophisticated segmentation algorithm to guarantee the performance of OCR [33]. However, in most applications, it is hard or impossible to segment text line into single character, especially for images with low quality or handwritten text. Thus, instead of relying on models for individual characters, segmentation-free OCR engines consider entire text lines as sequential signals and encode them into a single model. For example, Decerbo et al. used hidden markov models (HMM) to model the handwritten text using a slide window strategy to convert 2D text line image into 1D signal [4,5]. One potential problem of HMM based OCR is that it uses hand-crafted features, which requires domain knowledge to design the features and degrades the performance. So the modern OCR engines tend to utilize recurrent neural network (RNN) combined with CNN to automatically extract features for text line image [29].

The other researches attempt to apply post-processing techniques to correct OCR's outputs. In [11], Jean-Caurant *et al.* proposed to use lexicographical-based similarity to re-order the OCR outputs and build a graph connecting similar named entities to improve the OCR performance for names. By employing ivector features to estimate the OCR accuracy for each text line, Peng *et al.* rescored the OCR outputs and used a lattice to correct the outputs [27]. In [39], Xu and Smith designed an approach that detected duplicated text in OCR's outputs and performed a consensus decoding combined with a language model to improve the OCR's accuracy. Inspired by the idea from machine translation, Mokhtar *et al.* applied a neural machine translation approach to convert OCR's initial outputs to final corrected outputs [20].

2.2 Document Image Super-Resolution

As one of the powerful image processing methods, which can effectively enhance the image's quality, image super-resolution (sometimes it is also called image restoration) has been a mainstream research direction for a long time. Recently, document image super-resolution also gains increasing attentions from research community. In [22], a selective patch processing scheme was proposed by Nayef *et al.*, where the patches with high variance were reconstructed by learned model but other patches were interpolated by bicubic approach to ensure the efficiency and accuracy. To learn the mapping relation between noisy LR document patches and HR patches, Walha *et al.* designed an textual image resolution enhancement framework using both online and offline dictionaries [37].

More the state-of-the-art document super-resolution approaches prefer deep learning based frameworks. In ICDAR2015 Competition on Text Image Super-Resolution [28], Dong *et al.* modified a SR convolutional neural network (CNN), which was originally applied for natural image SR, for the document image SR and won the top one in this competition [7]. Based on the similar idea, Su *et al.* employed SRGAN [17] on the document images to improve the accuracy of OCR [35]. SRGAN was also used by Lat and Jawahar in their work, where the SR document image generated by the GAN was combined with bilinear interpolated image to obtain the final SR document image [16]. To avoid the loss of textual details from denoising steps prior to the restoration, Sharma *et al.* suggested a noise-resilient SR framework to boost OCR performance, where super-resolution and denoising were performed simultaneously based on stacked sparse de-noising auto-encoder and coupled deep convolutional auto-encoder [31]. This idea was also applied in Fu *et al.*'s approach, where multiple detail-preserving networks were stacked to extract detailed features from a set of LR document images during the up-scaling process [9]. In order to overcome the severe degradation problem for historical documents, Nguyen *et al.* proposed a character atten-tion GAN to restore degraded characters in old documents so OCR engine can improve its accuracy [24]. In [21], Nakao *et al.* trained two SR-CNNs on char-acter dataset and ImageNet dataset separately and combined the outputs from these two networks to obtain the SR document images. Unlike other approaches that only image based criteria were used to guide the training of neural network,

Sharma *et al.* introduced an end-to-end DNN framework which integrated the document image preprocessing and OCR into a single pipeline to boost OCR performance [32]. In this framework, they trained a GAN for de-noising which was followed by a deep back projection network (DBPN) for SR task and a bidirectional long short term memory (BLSTM) for OCR.

3 Proposed Approach

3.1 Overall System Architecture

In this paper, our aim is to improve the OCR accuracy of the document image by solving the SR problem. To this end, a generative adversarial network based super-resolution system is proposed, where the input and output images have the same size but with different resolution. Although in real applications, lots of degraded LR document images are available, it is hard to find the corresponding HR images for the supervised training. Thus, in this work, a two-stage training scheme is carried out, where in the first stage, a small amount of aligned training samples were used to pre-train the CNN based generator, and large amount of training samples of LR document images were used to train the generator with the help of the discriminator in the second stage.

In Fig. 2, the overall architecture of the proposed SR system is illustrated, where the proposed super-resolution document image generator $\mathcal{G}_\theta$ is composed by a feature extractor and a image generator, which is connected by the visual attention layers. In the framework of GAN, a discriminator $\mathcal{D}_\xi$ is also designed in our work which distinguishes the generated super-resolution document images from high resolution images. To train the proposed super-resolution image generator, the structural similarity loss is applied to guide the training for both generator and discriminator, which is described in Sect. 3.4 in details.

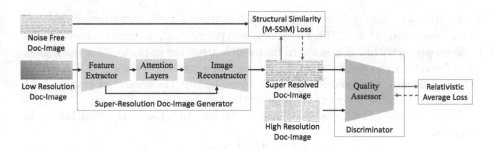

Fig. 2. An illustration of the proposed super-resolution document image generation system based on GAN.

3.2 Image Generator $\mathcal{G}_\theta$

To super resolve the LR images, many deep learning approaches use a stack of fully connected convolutional layers (such as ResNet block [10]) to reconstruct the images, where the networks' inputs are normally enlarged LR images with $4\times$ upscaling factor by bicubic interpolation and each convolutional layer has the same input size [13,17].

Unlike these methods which do not employ internal up-sampling and down-sampling for CNN layers, in our work, we apply the idea of encoder-decoder, which is widely used for semantic labeling [25], to accomplish the super-resolution task. In this architecture, the encoder is used to extract features of image in different scales and decoder is utilized to reconstruct the super resolved images. Particularly, we employ a set of CNN layers along with the batch normalization and LeakyReLU layers to construct the feature extractor. To avoid using pooling layer for down-sampling operation, we apply the stride of 2 for each CNN layer in feature extractor. To build the image reconstructor, the same amount of transposed convolutional layers are used, which forms a mirrored version of feature extractor. To overcome the vanishing gradient problem that is easily happened for the very deep neural networks, the shortcuts of corresponding layers between feature extractor and image reconstructor are applied, which effectively utilizes the extracted features in the image reconstruction process [30].

Additionally, to encourage the proposed SR document image generator to focus on those features contributing more for the details of SR image, the visual self-attention layers are introduced and applied to connect the feature extractor and image reconstructor. Originated from machine translation (MT) [36], self-attention technique has been successfully applied for many image/vision tasks [41]. Formally, given an image or feature map $\mathbf{x}$ whose size is $n \times n$, the self-attention model can be expressed as:

$$A(\mathbf{q}, \mathbf{k}, \mathbf{v}) = \mathrm{softmax}\left(\mathbf{q}(\mathbf{x})\mathbf{k}(\mathbf{x})^{\mathbf{T}}\right)\mathbf{v}(\mathbf{x}), \tag{2}$$

where $\mathbf{q}(\cdot), \mathbf{k}(\cdot)$ and $\mathbf{v}(\cdot)$ are vectorized features of $\mathbf{x}$ in three feature spaces, whose sizes are $n^2 \times 1$. Theoretically, $\mathbf{q}(\mathbf{x})\mathbf{k}(\mathbf{x})^T$ calculates the correlations between pixels within the image to find which areas are important for the image, which is used to re-weight $\mathbf{v}(\mathbf{x})$. In our implementation, we use 3×3 convolutions to calculate $\mathbf{q}(\mathbf{x})$ and $\mathbf{k}(\mathbf{x})$ with reduced feature channels, but employ 1×1 convolution to compute $\mathbf{v}(\mathbf{x})$ to balance the accuracy and efficiency.

Based on the obtained self-attention, the output of self-attention layer can be calculated by:

$$\mathbf{o} = \mathbf{h}\left(M\big(A(\mathbf{q}, \mathbf{k}, \mathbf{v})\big)\right) + \mathbf{x}, \tag{3}$$

where $M(\cdot)$ is the operation to convert $A(\mathbf{q}, \mathbf{k}, \mathbf{v})$ to a $n \times n$ matrix and $\mathbf{h}(\cdot)$ is the 1×1 convolution operation.

The structure of the proposed SR document image generator is demonstrated in Fig. 3.

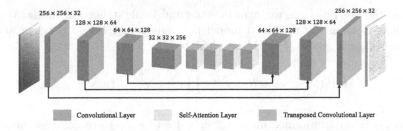

256 × 256 × 32
128 × 128 × 64
64 × 64 × 128
32 × 32 × 256
256 × 256 × 32
128 × 128 × 64
64 × 64 × 128

Convolutional Layer   Self-Attention Layer   Transposed Convolutional Layer

Fig. 3. The structure of the proposed SR document image generator $\mathcal{G}_\theta$.

3.3   Quality Assessor $\mathcal{D}_\xi$

Under the framework of GAN, the discriminator $\mathcal{D}_\xi$ is designed to differentiate LR document images from HR document images, which guide the training of image generator $\mathcal{G}_\theta$ based on their visual qualities. In this work, the quality assessor $\mathcal{D}_\xi$ is built by 4 ResNet blocks, where each ResNet block is followed by a LeakyReLU layer and a max-pooling layer for down-sampling. The last three layers for the quality assessor $\mathcal{D}_\xi$ are three fully connected layers whose size are 1024, 512 and 1, respectively.

To train the $\mathcal{D}_\xi$, the document images produced by the proposed image generator $\mathcal{G}_\theta$ and the ideally HR images are fed into the discriminator for training, where the dissimilarity or loss between these two type of images are calculated and used to train image generator $\mathcal{G}_\theta$ consequently.

3.4   Loss Functions for Training

Given a LR document image $\mathbf{y}$ and its HR counterpart $\mathbf{x}$, the goal of training for image generator $\mathcal{G}_\theta$ is to search the optimized parameter θ such that:

$$\bar{\theta} = \arg\min_\theta KL\left(\mathcal{G}_\theta(\mathbf{y})\|\mathbf{x}\right), \tag{4}$$

where $KL(\cdot)$ is the Kullback-Leibler divergence which measures the dissimilarity between super resolved image $\mathcal{G}_\theta(\mathbf{y})$ and HR image $\mathbf{x}$, and $\uparrow_s$ means the upscaling operation with factor s. Ideally, Eq. 4 is minimized when SR document image $\mathbf{y} \uparrow_s$ has the same distribution as HR image $\mathbf{x}$ w.r.t. resolution/quality. In the scenario of GAN, the Kullback-Leibler divergence between these two types of images are calculated by maximizing the object function of quality assessor $\mathcal{D}_\xi$.

Although GAN can learn the SR image generator $\mathcal{G}_\theta$ by using large amount of unlabeled data, it can be saturated easily because of GAN's large capacity and huge search space. Thus, in this work, we carry out a two-phases training scheme to train the $\mathcal{G}_\theta$. In the first training phase, we apply the supervised training to obtain the initial parameter θ for $\mathcal{G}_\theta$ such that it is close to the optimized $\mathcal{G}_\theta$, where we collect a small amount of HR images along with their LR counterparts for training. Unlike the conventional approaches that mean squared error (MSE) is used as the loss function, we introduce the multiscale structural similarity (M-SSIM) loss to train the proposed model [38]. Given two image patch $\mathbf{x}$ and $\mathbf{y}$ with the same size, their M-SSIM is calculated by:

$$\text{M-SSIM}(\mathbf{x}, \mathbf{y}) = l(\mathbf{x}, \mathbf{y})^{\alpha} \cdot \prod_{j=1}^{M} c_j(\mathbf{x}, \mathbf{y})^{\beta_j} s_j(\mathbf{x}, \mathbf{y})^{\gamma_j}, \tag{5}$$

where $l(\cdot, \cdot)$ is the luminance measure, $c_j(\cdot, \cdot)$ and $s_j(\cdot, \cdot)$ are contrast measure and structure measure with scale j, the hyper-parameters α, β and γ control the influence of these three measures. The detailed definition of these measure can be found in [38].

Based on M-SSIM, we define the M-SSIM loss between two image patches $\mathbf{x}$ and $\mathbf{y}$ as:

$$\mathcal{L}_{M-SSIM} = \frac{1 - \text{M-SSIM}(\mathbf{x}, \mathbf{y})}{2} \tag{6}$$

As can be seen from Eq. 5 that the advantage of M-SSIM loss over MSE loss is that MSE only calculates the absolute errors by comparing the differences between pixels, but M-SSIM loss concerns more on the perceived quality by using the statistics between two images.

In the second training phase, large amount of unlabeled training samples are used with the combination of LS-GAN [19] and Ra-GAN [12] to train the generator $\mathcal{G}_\theta$, where the losses for generator $\mathcal{G}_\theta$ and discriminator $\mathcal{D}_\xi$ are defined by:

$$\min \mathcal{L}_{\mathcal{G}_\theta} = \frac{1}{2}\mathbb{E}_{\mathbf{y} \sim \mathbb{P}(\mathbf{y})} \left[\left(\delta\big(\mathcal{D}_\xi(\mathcal{G}_\theta(\mathbf{y})) - \mathbb{E}_{\mathbf{x} \sim \mathbb{P}(\mathbf{x})}[\mathcal{D}_\xi(\mathbf{x})]\big) - a \right)^2 \right] \tag{7}$$

$$\min \mathcal{L}_{\mathcal{D}_\xi} = \frac{1}{2}\mathbb{E}_{\mathbf{x} \sim \mathbb{P}(\mathbf{x})} \left[\left(\delta\big(\mathcal{D}_\xi(\mathbf{x}) - \mathbb{E}_{\mathbf{y} \sim \mathbb{P}(\mathbf{y})}[\mathcal{D}_\xi(\mathcal{G}_\theta(\mathbf{y}))]\big) - b \right)^2 \right]$$
$$+ \frac{1}{2}\mathbb{E}_{\mathbf{y} \sim \mathbb{P}(\mathbf{y})} \left[\left(\delta\big(\mathcal{D}_\xi(\mathcal{G}_\theta(\mathbf{y})) - \mathbb{E}_{\mathbf{x} \sim \mathbb{P}(\mathbf{x})}[\mathcal{D}_\xi(\mathbf{x})]\big) - c \right)^2 \right] \tag{8}$$

where $\mathbb{E}[\cdot]$ means the expectation, $\delta(\cdot)$ is the Sigmoid function, $\mathbf{x}$ is the HR training image and $\mathbf{y}$ is the LR training sample, and we take $a = 1, b = 1$ and $c = 0$ in our work.

As we can see from Eq. 7 and Eq. 8, unlike the conventional losses designed for GAN, Ra-GAN uses relative average difference from the discriminator between HR and LR training samples within the training batch, which provides more stable training process. And the use of LS-GAN helps the proposed system avoid saturation during the GAN's training phase.

4   Experiments and Results

4.1   Datasets and Evaluation Metrics

In order to train the proposed super-resolution document generator, SmartDoc-QA dataset [23] and B-MOD dataset [14] were employed.

SmartDoc-QA dataset contains a total of 4260 camera captured document images whose size are 3096 × 4128. All these images were obtained based on 30 noise-free documents through two types of mobile phone cameras with different types of distortions and degradations, such as uneven lighting, out-of-focus blur, etc. In our experiment, we selected 90% images from this set for the supervised training in the first training phase, and the remaining 10% images were used for tuning purpose.

Prior to the supervised training, the captured images for SmartDoc-QA dataset were rectified by using RANSAC based perspective transformation, where the matching points between each LR image and the corresponding HR image were extracted by SURF features.

In B-MOD dataset, the document images were captured using the same manner as the SmartDoc-QA dataset but with larger number of samples, where 2,113 unique pages from random scientific papers were collected and photographed, which caused a total of 19725 images obtained by using 23 different mobile devices. In this dataset, although the rectified document images were available but the correspondences between these rectified images and their HR counterparts were not provided. Thus, in our experiments, we used the HR and LR images in this dataset to accomplish the unsupervised training for the GAN.

Because there is no public benchmark dataset available for the document image super-resolution task, we used SoC dataset [15] to evaluate the performance of the proposed document image super-resolution system. In SoC dataset, a total of 175 document images were captured from 25 "ideally clean" documents with different focus lengths. For each image, the ground truth text was also provided.

In Fig. 4, sample images from these three datasets are demonstrated.

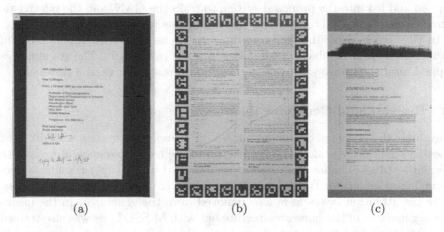

(a)   (b)   (c)

Fig. 4. Sample images from the training and evaluation datasets. (a) Rectified sample image from SmartDoc-QA dataset. (b) Rectified training image from B-MoD dataset, where the markers around the image are used for rectification. (c) Raw image from the SoC dataset for evaluation.

To measure the performance of image super-resolution, peak signal-to-noise ratio (PSNR) is a widely used metric where higher value means better performance. However, PSNR only considers the difference between the pixel values at the same positions but ignores human visual perception, which causes this approach to provide poor performance in representing the quality of SR images. Furthermore, as we are more interested in improving the OCR performance through the SR document image and expect the resolved images not only have higher resolution but contain less noises than the original LR images, thus we use character error rate (CER) of OCR outputs to measure the proposed document image super-resolution system in this work.

4.2   System Training

As described in Sect. 3.2 and Sect. 3.3, we train the proposed super-resolution generator in two training phases.

In the first training phase, the supervised training was carried out, where the training data from SmartDoc-QA dataset was employed. To train the generator $\mathcal{G}_\theta$, we randomly cropped small patches from the rectified camera captured document images to feed into the generator. The patches of the same location from the corresponding noise-free images were also cropped and used as the ground truths to guide the training of generator $\mathcal{G}_\theta$. In our experiment, the patches' size were 256×256.

In the second training phase, the rectified camera captured images from B-MOD dataset were used as the LR training samples, and 2,113 clean images from this dataset, along with other 3,554 in-house noise free document images were utilized as the HR training samples. These unlabeled training images were randomly cropped and fed into the proposed system to train the GAN, and the relativistic average losses were used to guide the training of generator $\mathcal{G}_\theta$ and $\mathcal{D}_\xi$.

In this experiment, the Adam optimizer was used in both supervised and unsupervised training phases with initial learning rate of 10^3. The training was stopped until the improvement of training loss was smaller than a threshold.

4.3   Evaluation of Generator $\mathcal{G}_\theta$

To evaluate the proposed document image super-resolution system with different losses and neural network architectures, we trained multiple generator $\mathcal{G}_\theta$ with different settings. The baseline system we obtained was a super-resolution document image generator trained using supervised training only with MSE loss, where the attention layers were also removed from the generator. In the meantime, a generator of the same architecture but with M-SSIM loss was also trained based on supervised training strategy. In the third SR document image generator, we added the attention layers into the network as shown in Fig. 3. Based on the third generator, we retrained it in the framework of GAN and obtained the final SR document image generator.

To assess the performance of these different systems, the 175 images from SoC dataset were applied. By considering these degraded images as enlarged images from its LR counterparts, we used them as the inputs of SR image generator $\mathcal{G}_\theta$ directly to produce the SR document images. With the obtained SR images, Tesseract OCR engine was used to retrieve their transcripts and the CERs were also calculated. In Table 1, the CERs of the original image and the images generated from the SR system we implemented are listed. It can be observed that the CER of the original degraded document images are as high as 22.03%. By using CNN based SR image generator, the CERs of the produced document images are decreased to 17.00%. With the introduced M-SSIM loss and attention layers, we can see the CERs of those restored images are decreased further to 14.95%. From the last row of this table, it can be seen that the images generated by the SR system that was trained based on GAN achieve the lowest CER 14.40%, which shows the effectiveness of the proposed M-SSIM loss, attention layers and the GAN training scheme for the SR system.

Table 1. OCR performances for different SR document image generators trained in our work, where the original image is taken as the system input.

Approach	CER (%)
Original image	22.03
Base Generator w/o attention layers + MSE Loss	17.00
Base Generator w/o attention layers + M-SSIM Loss	15.93
Generator w/attention layers + M-SSIM Loss	14.95
Generator w/attention layers + Ra-GAN	14.40

4.4   Comparison with the State-of-the Arts

To measure the OCR performance of the proposed SR document image generating system and other state-of-the-art SR approaches, a comparison experiment was implemented where the images in the SoC dataset were downsampled by the factor of 4 initially. Then, those down-sized images were super resolved by using BiCubic interpolation, SRCNN super-resolution [6], SRGAN super-resolution [17], and the proposed super-resolution approach, respectively. In this experiment, we retrained SRGAN using the same training data from SmartDoc-QA and B-MOD datasets. Need to note that because the proposed SR document image generator $\mathcal{G}_\theta$ takes the input whose size is the same as the output, we upsampled the down-sized images using the nearest neighbor interpolation prior to send them to the proposed generator. The obtained SR document images were transcribed by using Tesseract OCR engine, whose CERs are reported in Table 2.

Table 2. Comparison between the proposed SR approach and the state-of-the-art methods w.r.t. OCR performances, where the original image is downsampled 4 times before sending as the system input.

Approach	CER (%)
Original image	22.0
BiCubic	31.3
SRCNN [6]	30.3
SRGAN [17]	27.7
Proposed	22.3

As can be seen from this table, for document images which were downsampled by factor of 4, the resolved images obtained by the proposed methods achieved almost the same OCR accuracy as the original images. But the images produced by other methods, even they were visually close to the original image still had worse OCR performance.

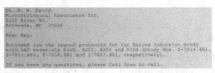

(a) Original document image

(b) Upsampled image by BiCubic interpolation   (c) SR image by SRCNN [6]

(d) SR image by SRGAN [17]   (e) SR image by the proposed system

Fig. 5. Output SR images by different approaches.

In Fig. 5, the original document image and the restored SR images by different approaches are also illustrated. We can observe from this figure that unlike the SR images generated by other methods, which still have lots of background noises and uneven/bad lighting effects, the image produced by the proposed SR method removes all the lighting effects and provides clean backgrounds.

The main reason for this advantage is that the proposed SR system uses GAN to train the generator $\mathcal{G}_\theta$ which does not rely on HR/LR image training pairs. So we can collect large amount of ideally noise free document as the targets to guide the training for discriminator and generator, which makes the obtained generator can produce high quality images with less noises.

5   Conclusion

In this work, we propose an end-to-end neural network for SR document image generation, where the image generator is composed by a encoder and a decoder, which are connected by the proposed attention layers. To obtain the SR document generator, a two training phase strategy is implemented, where in the first stage, the supervised training is carried out. And in this stage, a M-SSIM loss is designed which effectively improves the performance of SR w.r.t. OCR accuracy. In the second training phase, a combination of LS-GAN and Ra-GAN is applied, which largely relaxes the restriction for the training samples. So we can collect a large number of unlabeled training samples, including ideally noise free images, to train the proposed SR system. The experimental results show the effectiveness of the proposed SR methods based on the OCR performances. In the future, we will explore other loss functions, including OCR based loss functions to guide the training of SR image generator.

References

1. Agrawal, M., Doermann, D.: Stroke-like pattern noise removal in binary document images. In: 2011 International Conference on Document Analysis and Recognition, pp. 17–21 (2011)
2. Anwar, S., Khan, S., Barnes, N.: A deep journey into super-resolution: a survey. arXiv preprint arXiv:1904.07523 (2019)
3. Caner, G., Haritaoglu, I.: Shape-DNA: effective character restoration and enhancement for Arabic text documents. In: 2010 20th International Conference on Pattern Recognition, pp. 2053–2056 (2010)
4. Cao, H., Natarajan, P., Peng, X., Subramanian, K., Belanger, D., Li, N.: Progress in the Raytheon BBN Arabic offline handwriting recognition system. In: 2014 14th International Conference on Frontiers in Handwriting Recognition, pp. 555–560 (2014)
5. Decerbo, M., Natarajan, P., Prasad, R., MacRostie, E., Ravindran, A.: Performance improvements to the BBN Byblos OCR system. In: Eighth International Conference on Document Analysis and Recognition (ICDAR 2005), vol. 1, pp. 411–415 (2005)
6. Dong, C., Loy, C.C., He, K., Tang, X.: Image super-resolution using deep convolutional networks. IEEE Trans. Pattern Anal. Mach. Intell. **38**(2), 295–307 (2016)
7. Dong, C., Zhu, X., Deng, Y., Loy, C.C., Qiao, Y.: Boosting optical character recognition: a super-resolution approach. CoRR abs/1506.02211 (2015). http://arxiv.org/abs/1506.02211

8. Fawzi, M., et al.: Rectification of camera captured document images for camera-based OCR technology. In: 2015 13th International Conference on Document Analysis and Recognition (ICDAR), pp. 1226–1230 (2015)
9. Fu, Z., et al.: Cascaded detail-preserving networks for super-resolution of document images. In: 15th IAPR International Conference on Document Analysis and Recognition (ICDAR), pp. 240–245 (2019)
10. He, K., Zhang, X., Ren, S., Sun, J.: Deep residual learning for image recognition. In: 2016 IEEE Conference on Computer Vision and Pattern Recognition (CVPR), pp. 770–778 (2016)
11. Jean-Caurant, A., Tamani, N., Courboulay, V., Burie, J.: Lexicographical-based order for post-OCR correction of named entities. In: 2017 14th IAPR International Conference on Document Analysis and Recognition (ICDAR), vol. 1, pp. 1192–1197 (2017)
12. Jolicoeur-Martineau, A.: The relativistic discriminator: a key element missing from standard GAN. In: International Conference on Learning Representations (2019). https://openreview.net/forum?id=S1erHoR5t7
13. Kim, J., Lee, J.K., Lee, K.M.: Deeply-recursive convolutional network for image super-resolution. In: 2016 IEEE Conference on Computer Vision and Pattern Recognition (CVPR), pp. 1637–1645 (2016)
14. Kiss, M., Hradis, M., Kodym, O.: Brno mobile OCR dataset. In: 15th IAPR International Conference on Document Analysis and Recognition (ICDAR), pp. 1352–1357 (2019)
15. Kumar, J., Ye, P., Doermann, D.: A dataset for quality assessment of camera captured document images. In: Camera-Based Document Analysis and Recognition, pp. 113–125 (2014)
16. Lat, A., Jawahar, C.V.: Enhancing OCR accuracy with super resolution. In: 2018 24th International Conference on Pattern Recognition (ICPR), pp. 3162–3167 (2018)
17. Ledig, C., et al.: Photo-realistic single image super-resolution using a generative adversarial network. In: 2017 IEEE Conference on Computer Vision and Pattern Recognition (CVPR), pp. 105–114 (2017)
18. Lu, J., Min, D., Pahwa, R.S., Do, M.N.: A revisit to MRF-based depth map super-resolution and enhancement. In: 2011 IEEE International Conference on Acoustics, Speech and Signal Processing (ICASSP), pp. 985–988 (2011)
19. Mao, X., Li, Q., Xie, H., Lau, R.Y.K., Wang, Z., Smolley, S.P.: Least squares generative adversarial networks. In: 2017 IEEE International Conference on Computer Vision (ICCV), pp. 2813–2821 (2017)
20. Mokhtar, K., Bukhari, S.S., Dengel, A.: OCR error correction: state-of-the-art vs an NMT-based approach. In: 2018 13th IAPR International Workshop on Document Analysis Systems (DAS), pp. 429–434 (2018)
21. Nakao, R., Iwana, B.K., Uchida, S.: Selective super-resolution for scene text images. In: 15th IAPR International Conference on Document Analysis and Recognition (ICDAR), pp. 401–406 (2019)
22. Nayef, N., Chazalon, J., Gomez-Krämer, P., Ogier, J.: Efficient example-based super-resolution of single text images based on selective patch processing. In: 2014 11th IAPR International Workshop on Document Analysis Systems, pp. 227–231 (2014)
23. Nayef, N., Luqman, M.M., Prum, S., Eskenazi, S., Chazalon, J., Ogier, J.: SmartDoc-QA: a dataset for quality assessment of smartphone captured document images - single and multiple distortions. In: 2015 13th International Conference on Document Analysis and Recognition (ICDAR), pp. 1231–1235 (2015)

24. Nguyen, K.C., Nguyen, C.T., Hotta, S., Nakagawa, M.: A character attention generative adversarial network for degraded historical document restoration. In: 15th IAPR International Conference on Document Analysis and Recognition (ICDAR), pp. 420–425 (2019)
25. Noh, H., Hong, S., Han, B.: Learning deconvolution network for semantic segmentation. In: 2015 IEEE International Conference on Computer Vision (ICCV), pp. 1520–1528 (2015)
26. Ohkura, A., Deguchi, D., Takahashi, T., Ide, I., Murase, H.: Low-resolution character recognition by video-based super-resolution. In: 10th International Conference on Document Analysis and Recognition, pp. 191–195 (2009)
27. Peng, X., Cao, H., Natarajan, P.: Boost OCR accuracy using iVector based system combination approach. In: Document Recognition and Retrieval XXII, vol. 9402, pp. 116–123 (2015)
28. Peyrard, C., Baccouche, M., Mamalet, F., Garcia, C.: ICDAR2015 competition on text image super-resolution. In: 2015 13th International Conference on Document Analysis and Recognition (ICDAR), pp. 1201–1205 (2015)
29. Rawls, S., Cao, H., Kumar, S., Natarajan, P.: Combining convolutional neural networks and LSTMS for segmentation-free OCR. In: 2017 14th IAPR International Conference on Document Analysis and Recognition (ICDAR), vol. 1, pp. 155–160 (2017)
30. Ronneberger, O., Fischer, P., Brox, T.: U-Net: convolutional networks for biomedical image segmentation. In: Navab, N., Hornegger, J., Wells, W.M., Frangi, A.F. (eds.) MICCAI 2015. LNCS, vol. 9351, pp. 234–241. Springer, Cham (2015). https://doi.org/10.1007/978-3-319-24574-4_28
31. Sharma, M., Ray, A., Chaudhury, S., Lall, B.: A noise-resilient super-resolution framework to boost OCR performance. In: 14th IAPR International Conference on Document Analysis and Recognition (ICDAR), pp. 466–471 (2017)
32. Sharma, M., et al.: An end-to-end trainable framework for joint optimization of document enhancement and recognition. In: 15th IAPR International Conference on Document Analysis and Recognition (ICDAR), pp. 59–64 (2019)
33. Smith, R., Antonova, D., Lee, D.S.: Adapting the tesseract open source OCR engine for multilingual OCR. In: Proceedings of the International Workshop on Multilingual OCR, pp. 1:1–1:8 (2009)
34. Stamatopoulos, N., Gatos, B., Pratikakis, I., Perantonis, S.J.: A two-step dewarping of camera document images. In: 2008 The Eighth IAPR International Workshop on Document Analysis Systems, pp. 209–216 (2008)
35. Su, X., Xu, H., Kang, Y., Hao, X., Gao, G., Zhang, Y.: Improving text image resolution using a deep generative adversarial network for optical character recognition. In: 15th IAPR International Conference on Document Analysis and Recognition (ICDAR), pp. 1193–1199 (2019)
36. Vaswani, A., et al.: Attention is all you need. In: Advances in Neural Information Processing Systems 30, pp. 5998–6008 (2017)
37. Walha, R., Drira, F., Lebourgeois, F., Garcia, C., Alimi, A.M.: Handling noise in textual image resolution enhancement using online and offline learned dictionaries. Int. J. Doc. Anal. Recognit. (IJDAR) 21(1), 137–157 (2018)
38. Wang, Z., Simoncelli, E.P., Bovik, A.C.: Multiscale structural similarity for image quality assessment. In: The Thirty-Seventh Asilomar Conference on Signals, Systems Computers, 2003, vol. 2, pp. 1398–1402 (2003)
39. Xu, S., Smith, D.: Retrieving and combining repeated passages to improve OCR. In: 2017 ACM/IEEE Joint Conference on Digital Libraries (JCDL), pp. 1–4 (2017)

40. Yang, W., Zhang, X., Tian, Y., Wang, W., Xue, J.H.: Deep learning for single image super-resolution: a brief review. arxiv abs/1808.03344 (2018)
41. Zhang, H., Goodfellow, I., Metaxas, D., Odena, A.: Self-attention generative adversarial networks. In: Proceedings of the 36th International Conference on Machine Learning. Proceedings of Machine Learning Research, vol. 97, pp. 7354–7363. PMLR (2019)

Faster Glare Detection on Document Images

Dmitry Rodin[1,2]([✉]) [iD], Andrey Zharkov[1,2][iD], and Ivan Zagaynov[1,2][iD]

[1] R&D Department, ABBYY Production LLC, Moscow, Russia
{d.rodin,andrew.zharkov,ivan.zagaynov}@abbyy.com
[2] Moscow Institute of Physics and Technology (National Research University),
Moscow, Russia

Abstract. Glare on images is caused by a bright light source or its reflection. This effect can hide or corrupt important information on a document image, negatively affecting the overall quality of text recognition and data extraction. In this paper, we present a new method for glare detection which is based on our prior work on barcode detection, where we used a convolutional neural network (CNN) with consecutive downsampling and context aggregation modules to extract relevant features. In this paper, we propose a similar CNN to create a segmentation map for image glare. The proposed model is fast, lightweight, and outperforms previous approaches in both inference time and quality. We also introduce a new dataset of 687 document images with glare. The glare regions on the document images were marked and the dataset itself has been made publicly available. All of the measurements and comparisons discussed in this paper were performed using this dataset. We managed to obtain an F_1-measure of 0.812 with an average run time of 24.3 ms on an iPhone XS. The experimental data is available at https://github.com/RodinDmitry/glare.

Keywords: Machine learning · Semantic segmentation · Glare detection · Image analysis

1 Introduction

Glare occurs if a scene contains a bright light source or a reflection of a bright light source. In the case of document images, glare is typically caused by a reflection of a bright light source, which means that glare is determined by the properties of the light source and the reflective material, resulting in very high in-class variance in shape, size, and the dynamic range of the glare region.

Detecting glare is an important problem in optical character recognition and document quality assessment, because glare may hide important information needed for text detection and recognition. By locating the glare region, we can assess the expected quality of recognition results and predict the loss of information. This may be useful in systems that extract information from document

© Springer Nature Switzerland AG 2020
X. Bai et al. (Eds.): DAS 2020, LNCS 12116, pp. 161–167, 2020.
https://doi.org/10.1007/978-3-030-57058-3_12

fields based on a template. In the case of mobile devices, real-time glare detection can inform the user about the expected quality of text recognition and help him/her to take a photo that is better suitable for text recognition purposes.

Currently, neural networks significantly outperform traditional algorithms in many different tasks. However, most of the state-of-the-art glare detection methods, are based on classic approaches [2,5]. In our opinion, this situation is due to a lack of publicly available datasets that can be used in glare detection research. As part of our research, we created a new dataset containing images of documents of multiple types affected by glare. The documents were printed on different kinds of materials and present glare regions varying in shape and size. To detect glare on document images, we propose a novel solution based on [7].

2   Related Work

One approach was proposed by Andalibi et al. [2]. It is primarily intended to detect glare on footage from a camera placed on a moving car and consists in building saturation, intensity, and contrast maps and calculating features based on those maps. As the approach proposed in [2] is mainly concerned with glare caused by direct light sources, it is not well suited for document images, where glare is typically caused by reflections of light sources, and where the shape of glare regions is affected by document material, document shape, and reflection angles.

Another approach was proposed by Singh et al. [5] for detecting glare in natural night scenes. As [2] above, this approach addresses glare caused by direct light sources.

3   Network Structure

The main focus of our research was to detect the approximate position of the glare region with low latency and to present the result to the end-user of the mobile device, preventing him/her from creating a document photo unsuitable for text recognition purposes.

Delineating the boundaries of a glare region is a complicated task due to a large variety of possible shapes. As the approach proposed in [7] proved capable of detecting barcodes of varying shapes, we adopted it for glare detection with several modifications.

Our convolutional neural network consists of two main parts: a downsampling module and the context aggregation network module (CAN-module). The downsampling module consists of four blocks of convolution layers with stride 1 or 2 and kernel size 3, separated by padding layers. All the downsampling blocks except the first have 16 filters. In the first block, the first convolution uses 8 filters to produce a more lightweight model. The downsampling module reduces the image size by a factor of 16 on each side, so the model makes predictions based on super-pixels.

The CAN-module consists of 5 dilated convolutions with 24 filters and an output layer. In all the layers except the last, ReLU is used as the activation function. The glare region is often the local maximum in the adjacent part of the image. To account for this and to be able to ignore the luminance of the image, we use an instance normalization layer [6]. The resulting network structure is presented in Table 1.

Table 1. Network Description

Layer	Parameters	Filters
Instance normalization	–	–
Zero padding	Size: 1	–
Convolution	Kernel: 3×3; stride: 2	8
Convolution	Kernel: 3×3	16
Zero padding	Size: 1	–
Convolution	Kernel: 3×3; stride: 2	16
Convolution	Kernel: 3×3	16
Zero padding	Size: 1	–
Convolution	Kernel: 3×3; stride: 2	16
Convolution	Kernel: 3×3	16
Zero padding	Size: 1	–
Convolution	Kernel: 3×3; stride: 2	16
Convolution	Kernel: 3×3	16
Convolution	Kernel: 3×3; dilation: 1	24
Convolution	Kernel: 3×3; dilation: 2	24
Convolution	Kernel: 3×3; dilation: 4	24
Convolution	Kernel: 3×3; dilation: 8	24
Convolution	Kernel: 3×3; dilation: 1	24
Convolution	Kernel: 3×3	1

4   Description of the Dataset

Our dataset contains 687 images split into two subsets: a training subset of 550 images and a test subset of 137 images. The resolution of the images varies from 1080p to 4K. The data set contains the following types of documents: 380 magazine pages, 198 document pages, and 109 business cards and IDs. All of the images have glare. All of the images have normal luminance. Document content occupies at least 60% of each image. The glare region occupies 5.8% of the image on average, but it can take up to 40% of on some images. All of the business cards and IDs in the dataset are fictional and do not contain any personal data of real people.

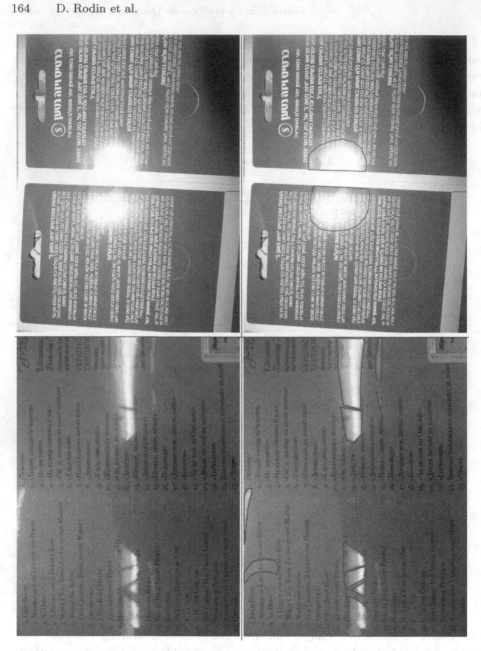

Fig. 1. Left: example of glare from the dataset; Right: corresponding markup.

Figure 1 shows some images from the dataset with particularly complex glare shapes. Figure 2 provides information about the relative glare area distribution in the dataset.

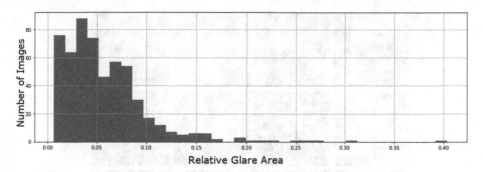

Fig. 2. Relative glare area distribution in the dataset

5 Experiments

This section provides the details of the experiments conducted on the dataset. Subsection 5.1 describes the results achieved and compares them with those obtained using other approaches.

Our objective will be achieved if we prevent the user from taking a photo with glare. For this reason, we use recall and F-measure as our main metrics. To calculate precision and recall, we divide the image into blocks of 16×16 pixels. A block is marked as containing glare if at least 25% of its area is glare. The output of the network has the same size as the obtained glare map, so we calculate precision, recall, and F_1-measure block by block.

In the experiments, we aimed to maximize the recall and F-measure, whilst keeping the number of parameters to a minimum and keeping inference time below 40 ms. We used the Tensorflow [1] framework and an Adam [3] optimizer with a learning rate of 0.001, $\beta_1 = 0.99$, $\beta_2 = 0.999$. For the loss function, we used weighted cross-entropy.

5.1 Results

In our experiments, we managed to achieve an F_1-measure of 0.812, which is higher than in previous attempts. Different approaches are compared in Table 3. As can be seen from the table, our network is much lighter than other networks. We also ported our solution to mobile platforms and measured the run time. The results are listed in Table 2. We also collected result heatmaps from a set of images and combined them in Fig. 3. While the Rodin and Orlov [4] approach often produces noisy outputs on document edges, our solution often captures regions larger than the actual glare.

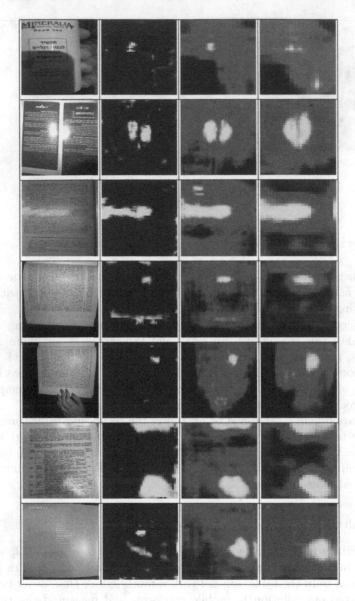

Fig. 3. Network outputs compared (columns left to right: original image, Rodin and Orlov, U-net-like architecture, CAN-module network)

Table 2. Run Times Compared

Method	Device	Total time, ms	Runs	Run time
Rodin and Orlov [4]	iPhone 8	5,300	100	53.0
Rodin and Orlov [4]	iPhone XS	3,729	100	37.3
CAN-module	iPhone 8	3,738	100	37.4
CAN-module	iPhone XS	2,434	100	24.3

Table 3. Metrics compared

Method	F_1	Precision	Recall	Total parameters
Rodin and Orlov [4]	0.740	0.744	0.736	405,473
U-Net [4]	0.798	0.767	0.841	323,489
CAN-module	0.812	0.795	0.836	34,865

6 Conclusion

In this paper, we proposed a new method for glare detection and introduced a dataset of document images with glare that can be successfully used in glare detection research. The proposed method proved to be fast (24 ms on an iPhone XS), lightweight (34,865 parameters), and capable of achieving an F_1-measure of 0.812. The proposed model outperforms its predecessors and is capable of running on mobile devices in real time. In future research, other neural architectures may be explored to achieve better results.

References

1. Abadi, M., et al.: Tensorflow: large-scale machine learning on heterogeneous distributed systems. CoRR abs/1603.04467 (2015)
2. Andalibi, M., Chandler, D.M.: Automatic glare detection via photometric, geometric, and global positioning information (2017)
3. Kingma, D.P., Ba, J.: Adam: a method for stochastic optimization. CoRR abs/1412.6980 (2015)
4. Rodin, D., Orlov, N.: Fast glare detection in document images. In: International Conference on Document Analysis and Recognition Workshops (ICDARW), September 2019. https://doi.org/10.1109/icdarw.2019.60123. http://dx.doi.org/10.1109/ICDARW.2019.60123
5. Singh, M., Tiwari, R.K., Swami, K., Vijayvargiya, A.: Detection of glare in night photography. In: 2016 23rd International Conference on Pattern Recognition (ICPR), pp. 865–870, Decembre 2016. https://doi.org/10.1109/ICPR.2016.7899744
6. Ulyanov, D., Vedaldi, A., Lempitsky, V.: Instance normalization: the missing ingredient for fast stylization (2016)
7. Zharkov, A., Zagaynov, I.: Universal barcode detector via semantic segmentation. In: 15th International Conference on Document Analysis and Recognition (ICDAR), Sydney, Australia, pp. 837–843, September 2019. https://doi.org/10.1109/ICDAR.2019.00139

Camera Captured DIQA with Linearity and Monotonicity Constraints

Xujun Peng[1](✉) and Chao Wang[2]

[1] ISI, University of Southern California, Marina del Rey, CA, USA
xpeng@isi.edu
[2] LinkedMed Co., Ltd., Spreadtrum Center, Zuchongzhi Road, Shanghai, China
chao.wang@linkedmed.com.cn

Abstract. Document image quality assessment (DIQA), which predicts the visual quality of the document images, can not only be applied to estimate document's optical character recognition (OCR) performance prior to any actual recognition, but also provides immediate feedback on whether the documents meet the quality requirements for other high level document processing and analysis tasks. In this work, we present a deep neural network (DNN) to accomplish the DIQA task, where a Saimese based deep convolutional neural network (DCNN) is employed with customized losses to improve system's capability of linearity and monotonicity to predict the quality of document images. Based on the proposed network along with the new losses, the obtained DCNN achieves the state-of-the-art quality assessment performance on the public datasets. The source codes and pre-trained models are available at https://gitlab.com/xujun.peng/DIQA-linearity-monotonicity.

Keywords: Document image · Siamese network · Quality assessment

1 Introduction

Mobile phones are ubiquitous nowadays in our life, which provide a convenient method to capture and digitalize document images to facilitate higher level document processing, recognition and analysis tasks, such as document classification, OCR, information retrieval, etc. However, most document recognition and retrieval systems expect document images with consistent and high qualities, which are not naturally to be satisfied under the uncontrolled circumstances. Normally, the camera-captured document images are easily suffering from motion blur and out-of-focus blur during acquisition process. Physical noises caused by capture devices and environments, such as salt-and-pepper noise, bad/uneven illuminations, are also common factors to degrade the qualities of obtained images [21]. Thus, it is essential to assess the image's perceived quality prior to passing them into the pipeline of processing, which can avoid unnecessary computation and ensure the systems are user friendly.

Typically, based on the availability of original reference image, DIQA can be categorized into two main groups: full-reference (FR) assessment and no-reference (NR) assessment [10,18,30]. By comparing document images with the

© Springer Nature Switzerland AG 2020
X. Bai et al. (Eds.): DAS 2020, LNCS 12116, pp. 168–181, 2020.
https://doi.org/10.1007/978-3-030-57058-3_13

corresponding references, FR assessment approaches achieve higher accuracy for predicting the qualities of document images. For example, Alaei used second order of Hast deviations to calculate the similarity map between reference document images and distorted images to assess the quality of them [1]. However, the lack of references makes it not applicable for most cases. So, NR based assessment attracts more interests from both research community and industry, and has wide range of applications. As no information of the reference is obtainable and only images' own properties can be used for the assessment, NR assessment is considered as the most challenging problem for DIQA.

To accomplish the NR assessment for document images, hand-crafted features were investigated by many pioneer researches. In [4], Blando *et al.* used the amount of white speckle, character fragments and the number of white connected components along with their sizes to predict the OCR accuracy of document image, which is considered as an important indicator of document image's quality. Similarly, Souza *et al.* proposed to employ the statistics of connected components and the font size to measure the document quality, which were utilized to guide the filter selection to improve the document image's visual quality [26]. In order to assess the quality of typewritten document images and restore them, Cannon *et al.* used similar features, such as background speckle, touching characters, and broken characters to measure the quality of the documents [6]. By using stroke's gradient and average height-width ratio, Peng *et al.* estimated the document image's quality based on the Support Vector Regression (SVR), which was trained depended on normalized OCR scores [22]. In [12], Kumar *et al.* developed a DIQA approach, which applied difference of differences in grayscale values of a median-filtered image (ΔDoM) as the indicator of document image's edge sharpness. In the same manner, Bui *et al.* combined numerous features to predict the sharpness of images, which was applied to the camera-captured document images [5].

As can be seen from these literature that most conventional hand-crafted features based DIQA approaches are heavily relied on the deep understanding and domain knowledge of different types of document image's degradation. Hence, to alleviate the burden of finding appropriate feature representations of document images heuristically, learning based approaches are proposed for DIQA. So, Ye and Doermann applied a localized soft-assignment and max-pooling technique to encode document representations, which were utilized to estimate document's OCR accuracy [29]. In [20], Peng *et al.* suggested a latent Dirichlet allocation (LDA) based algorithm to learn the "quality" topics from each document and map these topics to the corresponding OCR scores for quality assessment. To effectively learn the codebook that can represent the document image's quality, Peng *et al.* also proposed a discriminative sparse representation approach for DIQA [21]. To the same end, Xu *et al.* also employed a codebook to distinguish high quality and low quality document images [27]. But unlike the existing approaches that only first-order statistics of clusters for codebook were calculated, their system learned high order properties of the codebook for quality assessment. In [2], Alaei *et al.* built a DIQA system which learned a more informative visual codebook where representative bag-of-visual (BoW) words were extracted from structure features. By combining document image restoration and quality assessment tasks together, Garg and Chaudhury designed an EM based framework to iteratively learn the parameters that enhance the image's quality and estimate the quality of the image [7]. In [28], the pairwise rank learning approach is proposed to incorporated into the optimization framework for image quality assessment.

Instead of using OCR scores to guide the training of DIQA systems, many researchers attempted to apply other metrics as the indicator for document image quality. Emphasizing on the out-of-focus blur, Rai *et al.* constructed a dataset where each document image was captured under a controlled environment and the blur degree of each image was quantitatively computed based on the radius of circle of confusion from the camera [24]. Then, they used blur degree as the groundtruth to train the DIQA systems. In their other work, Rai *et al.* proposed the idea of spatial frequency response (SFR) to quantify the document image's quality, from which four slanted edges were placed on all sides of the reference document image and the blurs were calculated based on the degradation of slanted edges [23]. This SFR based dataset along with images' blur scores can be used to train DIQA systems, consequently. By using surrogate modeling [8], Singh *et al.* trained a model that can predict a vector value for different types of quality metrics given small amount of training samples, such that the instant on-the-fly performance feedback for a particular quality assessment algorithm on the image can be evaluated [25].

Nowadays, thanks to the great success of DNN and its generalization abilities, researchers have pushed the performance of many computer vision tasks to new heights. Consequently, recent researches attempted to use representations produced by DNNs to address the problem of DIQA. Convolutional neural networks (CNNs), which are widely used for different types of recognition problems nowadays, were also applied in [11] to assess the document image's quality, where small patches from the document images were extracted and sent to a CNN to predict the OCR score of the document. Not simply relied on the CNN for feature extraction, Li *et al.* developed a reinforcement learning approach to integrate CNN into the recurrent neural network (RNN) for camera based DIQA [16]. In this method, the attention model was applied to select patches which can boost system's accuracy. In [17], Lu and Dooms fine-tuned the Alex-Net and transformed the knowledge from natural scene images to the document images for DIQA.

In this paper, we proposed a ResNet [9] based Siamese network [31] with customized losses, which aimed to extract more valuable representations from document image and build a stronger classifier for DIQA. We organize the rest of this article as follows. We introduce the proposed DIQA approach in Sect. 2, including the DNN based quality recognizer and the proposed losses. Our experimental setup, results and analysis are covered in Sect. 3. And we conclude our work in Sect. 4.

2   Proposed Methodology

2.1   Motivation and Network Architecture

Unlike the conventional DIQA methods where hand-crafted features and heuristic rules are designed by developers or learnt from bag-of-visual (BoV) words, CNN based learning approaches provide an unified framework to extract deep representation of images and accomplish the classification tasks. Normally, the DNN based DIQA approaches assume that the degradation within a single document image is evenly distributed, such that the overall quality score of the image can be calculated by averaging the qualities from a bunch of random sampled patches in the image [11,14–16]. Based on the same assumption, in the training phase, the random selected training patches from the same document image are considered as independent and identically distributed, which are consequently assigned to a single quality score, ignoring the different quality characteristics between them. However, in real applications, the degradation and blurs are not consistent across the areas in the document image, as shown in the Fig. 1 and Fig. 2. In these two sample document images, we can see that texts in the same image might suffer different level of out-of-focus blurs and uneven illuminations. Therefore, the discriminant capability of the DIQA models is damaged if we treat all training patches within the same document image indiscriminately.

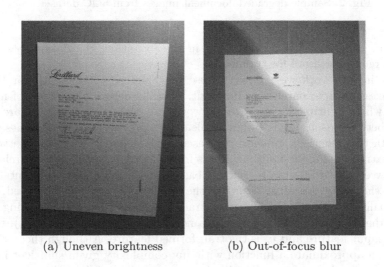

(a) Uneven brightness   (b) Out-of-focus blur

Fig. 1. Sample degraded document images from SmartDoc-QA dataset.

One intuitive but infeasible approach to overcome this problem is to feed the entire document image or patches that cover most part of the image into the DIQA model simultaneously to obtain the quality score, which is inapplicable because normally the camera captured document images' size is too large to

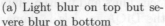

(a) Light blur on top but se-   (b) Bottom area has better
vere blur on bottom   quality than top part

Fig. 2. Sample degraded document images from SOC dataset.

exceed the capacity of the model, and in the meantime the database size for DIQA is relatively small for training.

Thus, in this work we propose to use the Siamese network [31] for DIQA which takes multiple patches as its input. Siamese network is a kind of neural network where multiple identical subnetworks are constructed which share their weights between each other. In our implementation, we build the Siamese network with two identical subnetworks based on ResNet [9] which are served as feature extractors for the multiple inputs from the same image and the relationship between these inputs are learned based on the obtained representations. Figure 3 shows the overall structure of the Siamese network we employed.

Normally, a CNN based feature extractor is constructed by stacking each layer to the next layer which serves as a non-linear function: $y_o = f(y_i)$, where y_i is the input feature and y_o is the transformed feature. Although theoretically a CNN can approximate a function with any complexity given sufficient layers, it turns out that its performance will start to saturate and degrade at particular point because of the gradient vanishing problem [3,9]. To overcome this problem, ResNet introduces an identity matrix to transmit the input feature to next layers directly and avoid data vanishing problem. Formally, the structure of ResNet can be expressed as: $y_o = f(y_i) + id(y_i)$, where $id(y_i)$ is the identity function in the form of $id(y) = y$. This identity connection enables the ResNet to learn residual representation effectively.

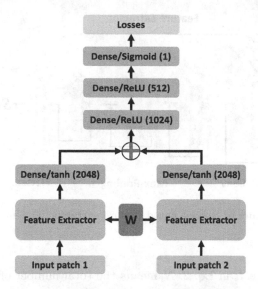

Fig. 3. The overall architecture of Siamese network used in our work.

In our implementation, we apply a multi-layers ResNet whose input shape is 128 × 128 to act as feature extractor in Siamese network. The network is constructed by 4 residual network blocks and pooling layers, where each residual block has three stacked CNN layers and one short-path layer. The output filters of each residual block are 32, 64, 128 and 256. After the last residual block, max pooling layer is applied and the features are flattened. The architecture of the network for feature extractor we designed is briefly shown in Fig. 4.

The extracted 1D features from two inputs are passed following dense layers separately, and concatenated and fed into three dense layers consequently, where the last layer uses the sigmoid activation, as shown in Fig. 3.

2.2   Losses

Generally, for the linear prediction tasks or regression tasks, l_1 or l_2 norm is used as the evaluation metric, which is also applied to guide the training procedures. But for the DIQA, the two widely used metrics are Pearson Correlation (PLCC) and Spearman's Rank-Order Correlation (SROCC), which are defined by:

$$\rho_p = \frac{\sum_{i=1}^{N}(x_i - \bar{x})(y_i - \bar{y})}{\sqrt{\sum_{i=1}^{N}(x_i - \bar{x})^2}\sqrt{\sum_{i=1}^{N}(y_i - \bar{y})^2}} \tag{1}$$

and

$$\rho_s = 1 - \frac{6\sum_{i=1}^{N}(u_i - v_i)^2}{N(N^2 - 1)}, \tag{2}$$

where x_i and y_i are the predicted quality score and ground truth of test image I_i in Eq. 1, and u_i and v_i denote the predicted and true rank order of quality

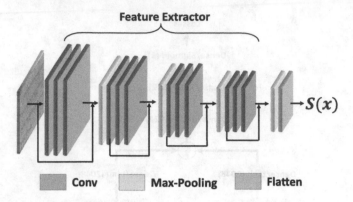

Fig. 4. The architecture of ResNet based feature extractor implemented in our work for DIQA.

scores for test image I_i in Eq. 2. N means the total number of test images for both equations.

From these two equations, we can see that PLCC measures the linearity between the predicted quality scores and ground truths, but SROCC quantitatively measure the monotonicity relationships between them. In order to encourage the proposed DIQA system to learn the document image qualities with respect to these two metrics, we define the losses in this work to train the Siamese network based DIQA model, which include two components: linearity loss and monotonicity loss.

Linearity Loss. The linearity loss which is designed to force the proposed system to learn the linear relationship between predict quality and ground truth, is simply defined by inversing power of PLCC (ρ_p): $\mathcal{L}_l = \frac{1}{2}\rho_p^{-2}$.

By letting $a_i = x_i - \bar{x}$, $b_i = y_i - \bar{y}$, $K = \sum_i b_i^2$ and $c_i = a_i b_i$, $\mathcal{L}_l$ can be expressed as:

$$\mathcal{L}_l = \frac{1}{2} \frac{\sum_i a_i^2 \sum_i b_i^2}{(\sum_i c_i)^2} = \frac{K}{2} \frac{\sum_i a_i^2}{(\sum_i c_i)^2} \tag{3}$$

To allow the backpropagation of the Pearson loss through the network, the gradient of the $\mathcal{L}_l$ with respect to a weight parameter w_k in the model can be calculated by:

$$\frac{\partial \mathcal{L}_l}{\partial w_k} = \frac{K}{2} \left(\frac{1}{(\sum_i c_i)^2} \frac{\partial \sum_i a_i^2}{\partial w_k} - \frac{\sum_i a_i^2}{(\sum_i c_i)^4} \frac{\partial (\sum_i c_i)^2}{\partial w_k} \right)$$

$$= K \left(\frac{1}{(\sum_i c_i)^2} \sum_i a_i \frac{\partial a_i}{\partial w_k} - \frac{\sum_i a_i^2}{(\sum_i c_i)^3} \sum_i \frac{\partial c_i}{\partial w_k} \right), \tag{4}$$

where $\frac{\partial a_i}{\partial w_k}$ and $\frac{\partial c_i}{\partial w_k}$ can be further computed by:

$$\frac{\partial a_i}{\partial w_k} = \frac{\partial (x_i - \bar{x})}{\partial w_k} = \frac{\partial (x_i - \sum_j x_j / N)}{\partial w_k}$$

$$= \frac{\partial x_i}{\partial w_k} - \frac{1}{N} \sum_j \frac{\partial x_j}{\partial w_k} \tag{5}$$

and

$$\frac{\partial c_i}{\partial w_k} = \frac{\partial a_i b_i}{\partial w_k} = b_i \frac{\partial a_i}{\partial w_k}$$

$$= b_i \left(\frac{\partial x_i}{\partial w_k} - \frac{1}{N} \sum_j \frac{\partial x_j}{\partial w_k} \right). \tag{6}$$

By taking Eq. 5 and Eq. 6 into Eq. 4, it can be seen that the gradient of $\mathcal{L}_l$ can pass through the network to update the parameters without blocking.

Monotonicity Loss. Because the Spearman's rank order correlation is not differentiable, we propose a pairwise loss to enhance the capability of monotonicity for the proposed DIQA model, which is defined by:

$$\mathcal{L}_m = \frac{1}{2N^2} \sum_{i=1}^{N} \sum_{j=1}^{N} ((x_i - x_j) - (y_i - y_j))^2, \tag{7}$$

where we expect the distance between the pair of predicted document images' quality scores is close to their ground truths' difference.

Similar to the linearity loss, we can calculate the gradient of $\mathcal{L}_m$ with respect to the parameter w_k of the neural network according to:

$$\frac{\partial \mathcal{L}_m}{\partial w_k} = \frac{1}{2N^2} \sum_i \sum_j \frac{\partial ((x_i - x_j) - (y_i - y_j))^2}{\partial w_k}$$

$$= \frac{1}{N^2} \sum_i \sum_j (x_i - x_j - y_i + y_j) \frac{\partial (x_i - x_j - y_i + y_j)}{\partial w_k}$$

$$= \frac{1}{N^2} \left(\sum_i (x_i - y_i) \frac{\partial x_i}{\partial w_k} + \sum_j (x_j - y_j) \frac{\partial x_j}{\partial w_k} \right.$$

$$\left. - \sum_i \sum_j (x_i - y_i) \frac{\partial x_j}{\partial w_k} - \sum_i \sum_j (x_j - y_j) \frac{\partial x_i}{\partial w_k} \right) \tag{8}$$

This provides the proposed monotonicity loss a mechanism that allows the gradient to flow back to the input of the network to update the model's parameters.

Based on the proposed linearity loss and monotonicity loss, we can have the total loss for the proposed DIQA system:

$$\mathcal{L} = (1 - \alpha)\mathcal{L}_l + \alpha\mathcal{L}_m, \tag{9}$$

where α is a hyper-parameter controlling the influence from monotonicity loss.

3 Experiments and Results

3.1 Datasets

In our experiments, we conducted the proposed Siamese network based DIQA paradigm and evaluate it on two public datasets: SmartDoc-QA [19] and SOC [13].

In SmartDoc-QA set, a total of 4260 images were captured based on 30 clean documents through two types of mobile phone cameras with different types of distortion, where the size of each image was 3096×4128. For each image, two OCR scores were obtained by two different OCR engines and we calculated the average character error rate (CER) as the quality score for each of them. In this experiment, we randomly selected 10% images from SmartDoc-QA set as a development set, 10% images as the test set and the remaining 80% images from this dataset as the training set. The development set was used to find the best value for hyper-parameter α for the monotonicity loss and the optimal iteration for training.

SOC dataset contained a total of 175 high resolution images whose size are 1840×3264. All these images were captured from 25 "ideally clean" documents with cameras. For each degraded image, the quality score was also computed based on the average CER from three provided OCR scores. For this dataset, we randomly selected 140 images for the training and the remaining 35 images for evaluation.

The sample images from these two datasets can be seen in Fig. 1 and Fig. 2.

3.2 Network's Training and Evaluation

In this work, we trained different DIQA models for two datasets separately, but tuned the hyper-parameters based on the selected development set from SmartDoc-QA set as described before.

Prior to the network's training, we randomly cropped 32 training samples from the training images to construct the training batches, where each sample's size is 128×128. To avoid including pure black or white background patches into the training batch which cannot provide any discriminant information for the DIQA, the pre-processing approach introduced in [11] was employed, where the patches were only sampled from areas with high variances.

In order to compare the performance of the proposed losses and conventional losses for DIQA, we first trained two benchmark systems for both SmartDoc-QA and SOC datasets by using the proposed ResNet based Siamese network with mean squared error (MSE) loss. For each epoch in the training phase, 60 batches were used and we trained the neural network using 800 epochs. The proposed network was trained based on SGD optimizer with Nesterov momentum $\rho = 0.9$, decay $\epsilon = 1e - 06$ and learning rate 0.001.

As we discussed in Sect. 2.2, PLCC (ρ_p) and SROCC (ρ_s) were used to measure the DIQA performance for these two benchmark systems, where we received $\rho_p = 0.914$, $\rho_s = 0.752$ on SmartDoc-QA's test set and $\rho_p = 0.951$, $\rho_s = 0.835$ on SOC's test set. From these numbers, we can find that with the MSE loss, which encourages the network to predict quality scores as close as the ground truths but ignores their orders, the proposed Siamese network obtains the state-of-the-art PLCC performances. But the SROCC which measures the rank capability of the DIQA system is relatively low for MSE loss based Siamese network.

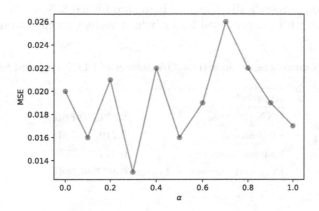

Fig. 5. Mean Squared Error (MSE) with different α for monotonicity loss on development set for SmartDoc-QA corpus.

To evaluate the effectiveness of the proposed losses, we re-trained two networks with the proposed losses. But as we can see from Eq. 9 that there is a hyper-parameter α which controls the weight of monotonicity loss, thus a tuning experiment was carried out on the SmartDoc-QA's development set to find the optimal α for the system. In this tuning process, we iterated α from 0 to 1.0 to train 11 different models and assessed the MSE values for each model on the development set. In Fig. 5, these MSE values are illustrated with respect to the corresponding α. It can be observed that the best DIQA performance was obtained with $\alpha = 0.3$. Therefore, we re-trained the Siamese networks with $\alpha = 0.3$ for proposed losses on both SmartDoc-QA and SOC datasets, where the same training scheme as the benchmark systems were taken.

In Table 1, we compared the DIQA performances of the designed Siamese network based DIQA system with the proposed losses, along with our benchmark systems and other two state-of-the-art approaches on SmartDoc-QA dataset. From this table, we can see that the proposed DIQA model outperformed other systems with large margin for PLCC measure. For the SROCC measure, the relatively low value for ρ_s is mainly because of the unbalanced OCR scores for the dataset [16]. By plotting the predicted OCR scores and the ground truths for the test set in Fig. 6, we can find that more than half of document images' CERs are 0, which causes majority of mis-ranking for these images and hence degrades the SROCC performances. Although ρ_s of the proposed system is surpassed by the method described in [16] on SmartDoc-QA set, by comparing the benchmark system, we can still see the big improvement for ρ_s which shows the proposed losses can provide more monotonicity capability than other conventional losses for the DIQA system.

In Table 2, the PLCC and SROCC of the proposed system and other existing methods were also compared on SOC dataset. From this table, it can be easily seen that the proposed network with losses has superior DIQA performance with respect to ρ_s. Specifically, the ρ_s is increased from 0.835 for the benchmark system to 0.936 with the proposed losses, which verifies its effectiveness.

Table 1. The comparisons on SmartDoc-QA dataset of PLCC (ρ_p) and SROCC (ρ_s).

Approach	ρ_p	ρ_s
RNN [16]	0.814	**0.865**
Li et al. [15]	0.719	0.746
Siamese baseline	0.914	0.752
Proposed Siamese + Losses	**0.927**	0.788

Table 2. The comparisons on SOC dataset of PLCC (ρ_p) and SROCC (ρ_s).

Approach	ρ_p	ρ_s
CNN [11]	0.950	0.898
LDA [20]	N/A	0.913
Sparse representation [21]	0.935	0.928
RNN [16]	0.956	0.916
CG-DIQA [14]	0.906	0.857
Li et al. [15]	0.914	0.857
Lu et al. [17]	**0.965**	0.931
Siamese baseline	0.951	0.835
Proposed Siamese + Losses	0.957	**0.936**

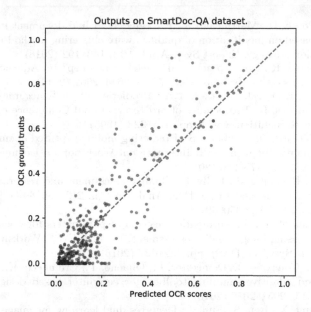

Fig. 6. Predicted OCR scores and ground truths on SmartDoc-QA dataset.

4   Conclusion

In this paper, we present a Siamese network based DIQA system, where ResNet is applied as a feature extractor. Unlike the conventional approaches that only take a single sampled patch from the image to predict and average the quality scores, based on this architecture, multiple patches from the same degraded document image can be fed into the neural network simultaneously, which not only learns the quality property of each single patch but the relations between them. The baseline DIQA system based on this network achieves the state-of-the-art performances. Furthermore, in order to increase the linearity and monotonicity capability of the proposed system, we propose the novel losses with respect to the PLCC and SROCC measures. Relied on the proposed losses, the Siamese network based DIQA system outperforms other existing approaches for both PLCC and SROCC measures on the public datasets, which shows the effectiveness of the proposed method.

In our future work, the new architecture of neural network which can learn more comprehensive correlations between patches within the image will be explored.

References

1. Alaei, A.: A new document image quality assessment method based on hast derivations. In: 15th International Conference on Document Analysis and Recognition (ICDAR), pp. 1244–1249 (2019)

2. Alaei, A., Conte, D., Martineau, M., Raveaux, R.: Blind document image quality prediction based on modification of quality aware clustering method integrating a patch selection strategy. Expert Syst. Appl. **108**, 183–192 (2018)
3. Ba, J., Caruana, R.: Do deep nets really need to be deep? In: Advances in Neural Information Processing Systems, vol. 27, pp. 2654–2662 (2014)
4. Blando, L.R., Kanai, J., Nartker, T.A.: Prediction of OCR accuracy using simple image features. In: Proceedings of 3rd International Conference on Document Analysis and Recognition, vol. 1, pp. 319–322 (1995)
5. Bui, Q.A., Molard, D., Tabbone, S.: Predicting mobile-captured document images sharpness quality. In: 13th IAPR International Workshop on Document Analysis Systems (DAS), pp. 275–280 (2018)
6. Cannon, M., Hochberg, J., Kelly, P.: Quality assessment and restoration of typewritten document images. Int. J. Doc. Anal. Recogn. **2**(2), 80–89 (1999). https://doi.org/10.1007/s100320050039
7. Garg, R., Chaudhury, S.: Automatic selection of parameters for document image enhancement using image quality assessment. In: 12th IAPR Workshop on Document Analysis Systems (DAS), pp. 422–427 (2016)
8. Gorissen, D., Couckuyt, I., Demeester, P., Dhaene, T., Crombecq, K.: A surrogate modeling and adaptive sampling toolbox for computer based design. J. Mach. Learn. Res. **11**, 2051–2055 (2010)
9. He, K., Zhang, X., Ren, S., Sun, J.: Deep residual learning for image recognition. In: IEEE Conference on Computer Vision and Pattern Recognition (CVPR), pp. 770–778 (2016)
10. Jiang, X., Shen, L., Yu, L., Jiang, M., Feng, G.: No-reference screen content image quality assessment based on multi-region features. Neurocomputing **386**, 30–41 (2020)
11. Kang, L., Ye, P., Li, Y., Doermann, D.: A deep learning approach to document image quality assessment. In: 2014 IEEE International Conference on Image Processing (ICIP), pp. 2570–2574 (2014)
12. Kumar, J., Chen, F., Doermann, D.: Sharpness estimation for document and scene images. In: Proceedings of the 21st International Conference on Pattern Recognition (ICPR2012), pp. 3292–3295 (2012)
13. Kumar, J., Ye, P., Doermann, D.: A dataset for quality assessment of camera captured document images. In: Iwamura, M., Shafait, F. (eds.) CBDAR 2013. LNCS, vol. 8357, pp. 113–125. Springer, Cham (2014). https://doi.org/10.1007/978-3-319-05167-3_9
14. Li, H., Zhu, F., Qiu, J.: CG-DIQA: no-reference document image quality assessment based on character gradient. In: 24th International Conference on Pattern Recognition (ICPR), pp. 3622–3626 (2018)
15. Li, H., Zhu, F., Qiu, J.: Towards document image quality assessment: a text line based framework and a synthetic text line image dataset. In: 15th International Conference on Document Analysis and Recognition (ICDAR), pp. 551–558 (2019)
16. Li, P., Peng, L., Cai, J., Ding, X., Ge, S.: Attention based RNN model for document image quality assessment. In: 14th IAPR International Conference on Document Analysis and Recognition (ICDAR), vol. 01, pp. 819–825 (2017)
17. Lu, T., Dooms, A.: A deep transfer learning approach to document image quality assessment. In: 15th International Conference on Document Analysis and Recognition (ICDAR), pp. 1372–1377 (2019)
18. Ma, K., Liu, W., Liu, T., Wang, Z., Tao, D.: dipIQ: blind image quality assessment by learning-to-rank discriminable image pairs. IEEE Trans. Image Process. **26**(8), 3951–3964 (2017)

19. Nayef, N., Luqman, M.M., Prum, S., Eskenazi, S., Chazalon, J., Ogier, J.: SmartDoc-QA: a dataset for quality assessment of smartphone captured document images - single and multiple distortions. In: 13th International Conference on Document Analysis and Recognition (ICDAR), pp. 1231–1235 (2015)
20. Peng, X., Cao, H., Natarajan, P.: Document image OCR accuracy prediction via latent Dirichlet allocation. In: 13th International Conference on Document Analysis and Recognition (ICDAR), pp. 771–775 (2015)
21. Peng, X., Cao, H., Natarajan, P.: Document image quality assessment using discriminative sparse representation. In: 12th IAPR Workshop on Document Analysis Systems (DAS), pp. 227–232 (2016)
22. Peng, X., Cao, H., Subramanian, K., Prasad, R., Natarajan, P.: Automated image quality assessment for camera-captured OCR. In: 18th IEEE International Conference on Image Processing, pp. 2621–2624 (2011)
23. Rai, P.K., Maheshwari, S., Gandhi, V.: Document quality estimation using spatial frequency response. In: IEEE International Conference on Acoustics, Speech and Signal Processing (ICASSP), pp. 1233–1237 (2018)
24. Rai, P.K., Maheshwari, S., Mehta, I., Sakurikar, P., Gandhi, V.: Beyond OCRs for document blur estimation. In: 14th IAPR International Conference on Document Analysis and Recognition (ICDAR), vol. 01, pp. 1101–1107 (2017)
25. Singh, P., Vats, E., Hast, A.: Learning surrogate models of document image quality metrics for automated document image processing. In: 13th IAPR International Workshop on Document Analysis Systems (DAS), pp. 67–72 (2018)
26. Souza, A., Cheriet, M., Naoi, S., Suen, C.Y.: Automatic filter selection using image quality assessment. In: Proceedings of Seventh International Conference on Document Analysis and Recognition, vol. 1, pp. 508–512 (2003)
27. Xu, J., Ye, P., Li, Q., Liu, Y., Doermann, D.: No-reference document image quality assessment based on high order image statistics. In: IEEE International Conference on Image Processing (ICIP), pp. 3289–3293 (2016)
28. Xu, L., Li, J., Lin, W., Zhang, Y., Zhang, Y., Yan, Y.: Pairwise comparison and rank learning for image quality assessment. Displays 44, 21–26 (2016)
29. Ye, P., Doermann, D.: Learning features for predicting OCR accuracy. In: Proceedings of the 21st International Conference on Pattern Recognition (ICPR2012), pp. 3204–3207 (2012)
30. Ye, P., Doermann, D.: Document image quality assessment: a brief survey. In: 12th International Conference on Document Analysis and Recognition, pp. 723–727 (2013)
31. Zagoruyko, S., Komodakis, N.: Learning to compare image patches via convolutional neural networks. In: IEEE Conference on Computer Vision and Pattern Recognition (CVPR), pp. 4353–4361 (2015)

Background Removal of French University Diplomas

Tanmoy Mondal[1]([✉]), Mickaël Coustaty[2], Petra Gomez-Krämer[2], and Jean-Marc Ogier[2]

[1] Zenith, INRIA, Montpellier, France
tanmoy.mondal@inria.fr
[2] L3i, University of La-Rochelle, La Rochelle, France
{mickael.coustaty,petra.gomez,jean-marc.ogier}@univ-lr.fr

Abstract. Separation of foreground text from noisy or textured background is an important preprocessing step for many document image processing problems. In this work we focus on decorated background removal and the extraction of textual components from French university diploma. As far as we know, this is the very first attempt to resolve this kind of problem on French university diploma images. Hence, we make our dataset public for further research, related to French university diplomas. Although, this problem is similar to a classical document binarization problem, but we have experimentally observed that classical and recent state of the art binarization techniques fail due to the different complexity of our dataset. In French diplomas, the text is superimposed on decorated background and there is only a small difference of intensities between the character borders and the overlapped background. So, we propose an approach for the separation of textual and non-textual components, based on Fuzzy C-Means clustering. After obtaining clustered pixels, a local window based thresholding approach and the Savoula binarization technique is used to correctly classify pixels, into the category of text pixels. Experimental results show convincing accuracy and robustness of our method.

Keywords: Text/graphics separation · French university diplomas · Document binarization · Fuzzy C-Means clustering

1 Introduction

Separation of text foreground from page background is an important processing step for the analysis of various kind of documents e.g. historical, administrative, scanned and camera captured documents. This separation enables various document analysis algorithms (such as character recognition, localization of stamps, logos, signatures etc.) to focus only on the written text without the disturbing background noise. If texts and graphics are located in separate regions in the page, localizing the region of texts and graphics is easier in comparison with texts superimposed on background graphics [1].

© Springer Nature Switzerland AG 2020
X. Bai et al. (Eds.): DAS 2020, LNCS 12116, pp. 182–196, 2020.
https://doi.org/10.1007/978-3-030-57058-3_14

In this paper, we are focusing on the proper extraction of all the text components from French university diplomas, which contain a complex background. The objective is to secure the university diplomas and to reduce the possibilities of fraud. The core idea is to encode a unique and secret number in the diploma while printing it. This encoding is done by changing the shape of certain characters and each modified character shape is associated with a specific digit (this association is only known to some specific people of the university). Hence, later to check the authenticity of any scanned student's diploma, we a need proper and undeformed extraction of text characters which is required for decoding the secret and unique number, associated with each diploma. All French university diplomas (around 300 thousand per year) are printed on the same decorated and authenticate thick paper/parchment (in French called *parchemin*), only and strictly fabricated by the National Printing House (in French called *Imprimerie Nationale*). So, any French university diploma has the same background and only the foreground text changes for different universities/institutes (moreover scanned diplomas will diversify due to scanning effects). Moreover, the text fonts, size and text styles also vary for different universities/institutes. In every diploma, each individual student's information e.g. name, date-of-birth, place-of-birth etc. changes also. All these variations together, makes the binarization process harder. This complex background is highly superimposed with the foreground textual components which makes it complex to separate the textual components from the background (see Fig. 2a for an example of diploma). This problem can also be seen as separation of graphical background from the textual foreground. On the other hand, this problem has a high resemblance with historical degraded document binarization. Historical degradation includes nonuniform background, stains, faded ink, ink bleeding through the page and uneven illumination [2]. The state of the art text/graphics separation approaches do not perform well in our case because unlike general text/graphics separation data e.g. maps, floor plans etc. [1], the texts in diploma images are fully superimposed upon the decorated colored background.

There exist several categories of binarization techniques, which are often evaluated on the dataset of popular document image binarization contest (DIBCO) [11]. But these categories of binarization approaches have several constraints. Many common binarization approaches are based on the calculation of local or global thresholds and image statistics, which doesn't work for our case (see experimental Section in 4). So as a contribution of this paper, we propose a new algorithm for the separation of text components from graphical background. The algorithm starts by structured forest based fast gradient image formation followed by textual region selection by using image masking for the initial filtering of textual components. Then Fuzzy C-Means clustering is applied to separate textual and non-textual components. As the clustering may result in some deformation of the textual components (characters), hence we also a propose character reconstruction technique by using a local window based thresholding approach and the Savoula binarization algorithm [12]. This technique helps to correctly classify some extra pixels into the category of text pixels, which helps to reconstruct/recover the missing character pixels. As far as we know, this is the very initial proposed attempt of removing background to obtain texts from French university diplomas.

The remainder of this paper is organized as follows. First, Sect. 2 summarizes the work related to text/graphics segmentation in general. Then, Sect. 3 provides an overview of the method proposed. Subsequently, experimental results are described and analyzed in Sect. 4. Finally, Sect. 5 concludes the paper and gives an overview of future work.

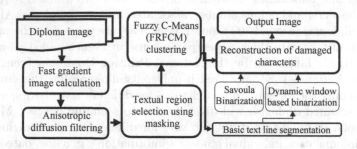

Fig. 1. The proposed algorithm architecture. The portions in bold are the main steps of the algorithm while the two non-bold blocks are the auxiliary steps needed for the "reconstruction of damaged characters" block.

2   Related Work

By definition, text binarization means the labeling of each pixel in the image as text or background which has a good resemblance with our problem. Due to the unavailability of any previous work, we have outlined here some of the related work from the domain of document image (mainly historical) binarization. The existing binarization techniques in the literature can be categorized into two principal categories: learning free and learning based approaches. Several binarization techniques have been proposed in the past decade, but very recently the trend in document binarization tends to machine learning based methods, mostly relying on deep learning based techniques.

Most deep learning based image binarization algorithms use convolutional neural networks (CNN), e.g. by Tensmeyer and Martinez [13], or variations thereof, such as the deep supervised network (DSN) approach proposed by Vo et al. [14]. Due to the requirement of sufficient amount of training data and specially in our case, where no ground truth exists, the training based approaches are not useful. Moreover, recently the learning free approaches [8] have shown high potential and comparable accuracy with respect to learning based approaches. In [2], a local binarization method based on a thresholding with dynamic and flexible windows is proposed. Jia et al. [8] proposed an approach based on structural symmetry of pixels (SSP) from text strokes for binarization. Using the combination of SSP and Fuzzy C-Means (FCM) clustering (the FRFCM technique [7]), Mondal et al. [10] also proposed an improved and fast binarization technique for historical document images.

While deep learning methods are getting more and more attention in the community, FCM [7] based models are among the popularly used methods for image segmentation. We avoided to use any learning based method (e.g. deep neural networks) for background removal due to the non availability of a pixel level ground truth (GT). It is also highly cumbersome and expensive to generate such GT for our high resolution experimental data set. These methods rely on the fuzzy set theory which introduce fuzziness for the belongingness of each image pixel to a certain class. Such clustering technique, named FRFCM [7], is used as one of the main processing step for our proposed algorithm. So, here we also provide a brief background and state of the art methods on FCM techniques to justify our choice. It is superior to hard clustering as it has more tolerance to ambiguity and retains better original image information. As FCM only considers gray-level information without considering the spatial information, it fails to segment images with complex texture and background or images corrupted by noise. So, the FCM algorithm with spatial constraint (FCM_S) [7] was proposed, which incorporates spatial information in the objective function. However, FCM_S is time consuming because the spatial neighbor term is computed in each iteration. To reduce the execution time, two modified versions, named as FCM_S1 and FCM_S2 [7] were proposed. These algorithms employ average filtering and median filtering to obtain the spatial neighborhood information in advance. However, both FCM_S1 and FCM_S2 are not robust to Gaussian noise, as well as to a known noise intensity.

The Enhanced FCM (EnFCM) [7] algorithm is an excellent technique from the viewpoint of low computational time as it performs clustering based on gray-level histograms instead of pixels of a summed image. However, the segmentation result by EnFCM is only comparable to that produced by FCM_S. To improve the segmentation results, the Fast Generalized FCM (FGFCM) [7] was proposed. It introduces a new factor as a local similarity measure, which guarantees both noise immunity and a detailed preservation of image segmentation. Along with that it also removes the requirement of empirical parameter α in EnFCM and performs clustering on gray-level histograms. But FGFCM needs more parameter settings than EnFCM. The robust Fuzzy Local Information C-Means clustering algorithm (FLICM) is introduced in [7], which is free from parameter selection. This algorithm replaces the parameter α of EnFCM by incorporating a novel fuzzy factor into the objective function to guarantee noise immunity and image detail preservation. Although FLICM overcomes the image segmentation performance, but the fixed spatial distance is not robust to different local information of images. Hence, we use a significantly fast and robust algorithm based on morphological reconstruction and membership filtering (FRFCM) [7] for image segmentation.

The aforementioned state-of-the-art binarization methods are developed for the binarization of historical document images, by testing some of them, we observed unsatisfying results. In the following section, we propose a new method for separating the textual foreground from textured background.

3   Proposed Method

In this section, the proposed algorithm is explained and its overall architecture is shown in Fig. 1. At first, the structural random forest based gradient image formation technique is applied on the image $\mathcal{I}(x, y)$ (see Fig. 2a).

3.1   Structural Forest Based Fast Gradient Image Formation

A real time gradient computation for the objective of edge detection is proposed by Dollár et al. [4], which is faster and more robust to texture presence than current state-of-the-art approaches. The gradient computation (see Fig. 2b) is performed based on the present structure in local image patches and by learning both an accurate and computationally fast edge detector.

3.2   Anisotropic Diffusion Filtering

An anisotropic diffusion filter [6] is applied on the gradient image (see Sect. 3.1) to reduce image noises (due to scanning) without removing significant contents of the image e.g. edges, lines and other details (see Fig. 2c).

3.3   Selection of Textual Regions from the Original Image

After filtering the image, edges are detected by the Canny edge detector (see Fig. 2d). Only better gradient images are obtained by Dollár et al. [4] (refer to Sect. 3.1), but to generate the binary edges we have applied the Canny edge detection algorithm. The high and low thresholds of Canny are computed automatically as ϕ (equal to Otsu's threshold value) and $0.5 \times \phi$. The idea was not to use hand-crafted thresholds.

The edge image is then dilated by using a 7×7 rectangular kernel[1] to connect the broken edges, caused by the Canny algorithm (see Fig. 2e). It can be observed that it contains holes and gaps inside the character shapes (see Fig. 2e). These holes/gaps are filled by applying the well known flood fill[2] based hole filling approach (see Fig. 2f). Let us denote this hole filled image as $\mathcal{H}(x, y)$. Now the original gray scale image pixels from the gray scale image corresponding to this hole filled image are simply obtained by creating a blank image of the same size as the original image and initialized by 255 (let's say $\mathcal{T}(x, y)$; see Fig. 2g)). This process corresponds to applying a mask[3] using image $\mathcal{H}(x, y)$. Now, a clustering

---

[1] Considering the high image resolution, we chose a 7×7 kernel size. However, this parameter should be adapted to fit other dataset requirements.

[2] https://docs.opencv.org/2.4/modules/imgproc/doc/miscellaneous_transformations.html?highlight=floodfill#floodfill.

[3] The background is white (255) and the foreground is black (0).

technique is applied to the masked image $\mathcal{T}(x,y)$ for classifying into text pixels and background pixels.

$$\mathcal{T}(x,y) = \begin{cases} \mathcal{I}(x,y); & \text{if } \mathcal{H}(x,y) = 0. \\ 255; & \text{otherwise.} \end{cases} \tag{1}$$

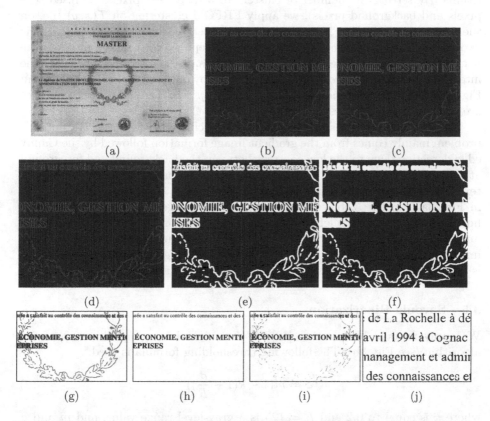

Fig. 2. Various preprocessing steps: (a) Original gray image (b) Gradient image obtained by [4] (c) Anisotropic filtered image (d) Detected edges by Canny (e) Dilated images after Canny edge detection (f) Hole filled image after dilation (g) Selected textual regions $\mathcal{T}(x,y)$ (h) Only sure text pixels after clustering (i) Sure and confused pixels after clustering (j) Existing issue of character deformation after clustering (sure text pixels only).

3.4 Fuzzy C-Means (FRFCM) Clustering Algorithm

Following to the aforementioned state of the art on FCM clustering mentioned in Sect. 2, Lei and Jia et al. [7] proposed the FRFCM algorithm with a low computational cost which can achieve good segmentation results for various types

of images and can also achieve a high segmentation precision. It employs morphological reconstruction (MR) to smooth the images in order to improve noise
immunity and image detail preservation. Results obtained by various aforementioned clustering techniques on a small cropped image of 822×674 are shown
in Table 1. It can be seen that FRFCM has an impressive computational time.

Therefore, FRFCM is faster than other aforementioned improved FCM algorithms. By setting the number of clusters to 3 (sure text pixels, confused text
pixels and background pixels), we apply FRFCM clustering on $T(x, y)$ (i.e. on
the image shown in Fig. 2g), which gives a clustered image (see Fig. 2i). The
two deeper intensities are visible in Fig. 2i except white, which represents the
background. Among these two deeper intensities, the darker one represents the
intensities of *sure text pixels* (the image formed by only these pixels are shown in
Fig. 2h) and the other ones represent *confused pixels*. The following techniques
are applied to qualify these pixels as texts or background.

Some character shapes are deformed due to clustering (see Fig. 2j), but this
problem mainly comes from the gradient image formation followed by the Canny
edge detection. Due to the low contrast between the background and foreground
at these specific regions of the image, the gradient image fails to clearly signify
text regions followed by the failure of Canny's edge detection. This is shown
in Fig. 3a (top left: zoomed portion of dilated image after Canny, top right:
hole filled image, bottom left: textual region separation, bottom right: clustered
image). Note that even before performing the dilation operation on the Canny
image, some portions of the image were already missing. The following technique
is applied to recover the missing portions and to reconstruct the image.

3.5 Savoula Binarization

We apply Savoula binarization [12] on the original image at this step of the
algorithm (see Fig. 3b). The following thresholding formula is used:

$$S = m(1 - k(1 - \frac{\sigma}{R}))$$ (2)

where k is equal to 0.2 and $R = 125$ is a gray-level range value, and m and σ
represent the mean and standard deviation of image window respectively.

3.6 Basic and Fast Text Line Segmentation

Text lines are roughly segmented based on a horizontal projection of the binarized image, obtained by the FRFCM clustering (the sure text pixels). After
obtaining the heights of each text line, a simple but efficient approach to calculate the average of these values are described in Algorithm 1. This technique
is helpful for removing outliers from the set of values which helps to calculate

Table 1. Time required (in seconds) by the different clustering algorithms on a cropped diploma image.

Method name	Time required	Method name	Time required
FCM_S1	232.15	FGFCM_S2	2.23
FCM_S2	201.14	ENFSCM	0.74
FCM_M1	3.07	FLICM	662.25
FGFCM	2.15	FRFCM	0.27
FGFCM_S1	1.79		

a better average value.[4] The values are stored in $Arr[items]$ and the intelligent average is obtained in the variable $AvgVal$.[5]

Algorithm 1: INTELLIGENT AVERAGING

Input: Arr[items]
Output: AvgVal
1 Divide the elements in 10 equal spaced bins (we use the *hist* function of Matlab)
2 From the bin centers of each bin, get the lower and upper limits of each bin
3 Sort the bins based on the number of elements existing in each bin
4 Only consider the two top bins
5 Choose those values which belong to these two top bins (using the upper and lower limits of each bin). Let's call them refined values (denoted as $Refined[items]$)
6 Calculate the mean value (**M**) and the standard deviation (**S**) from these refined values
7 Now again obtain the re-refined values by verifying if it belongs to:
   $(M - S) \leq Refined[items] \leq (M - S)$
8 Finally calculate the mean value from these re-refined values

---

3.7 Window Based Image Binarization

We apply a fast window based binarization technique on the original gray image. This technique is inspired by the method in [2]. From the previous step of text line segmentation, we get the very first (*textStart*) and the last image row (*textEnd*), which contains text. As we already have the average height of text lines ($AvgVal$) so by using this information, we divide the considered region

---

[4] Please note that, maybe for our case of diploma images, as there are not many text lines, this approach may not show a high difference in results, compared to classical averaging, but when there are more data, this technique shows better performance.
[5] We use the *hist* function of the Matlab toolbox to calculate a histogram of values, divided into 10 bins.

(region between *textStart* and *textEnd*) into equal sized horizontal stripes. Now, each horizontal stripe is divided into 20 equal parts[6] and each of these parts called (*window*) on which the binarization is performed by using the following equations:

$$\sigma_{adaptive} = \left[\frac{\sigma_W - \sigma_{min}}{\sigma_{max} - \sigma_{min}} \right] \times max_{Intensity} \tag{3}$$

$$\mathcal{T}_W = \mathcal{M}_W - \frac{\mathcal{M}_W \times \sigma_W}{(\mathcal{M}_W + \sigma_W)(\sigma_{adaptive} + \sigma_W)} \tag{4}$$

where $\sigma_{adaptive}$ is the gray-level value of the adaptive standard deviation of the window. σ_W is the standard deviation of the window, σ_{min} and σ_{max} are the minimum and maximum standard deviation of all the windows (20 here), $max_{Intensity}$ is the maximum intensity value of the horizontal stripe, $\mathcal{M}_W$ is the mean value of the pixel intensities of the window and $\mathcal{T}_W$ is the threshold of the window. The obtained binarized image is denoted by $WinImg^{binary}(x, y)$ (see Fig. 3c). It can be seen that although the simple Savoula binarization has comparatively performed well, but it fails to remove the centered decorated background (called *couronne* in French), whereas the binarized image from the window based image binarization is prominently keeping most of the background decorated portions. So, none of these can be directly used for our case.

3.8   Reconstruction of Deformed Characters

We propose a technique for the reconstruction of deformed characters with the help of the Savoula and the window based binary image. This reconstruction is required as the shape of the characters should remain intact for proper decoding of the secret number, as mentioned in the introduction (see Sect. 1). The obvious text pixels (sure text pixels), obtained from the fuzzy clustering are copied into a the new image (called $\mathcal{K}(x, y)$). Now, for each confused pixel (line 7 in Algorithm 2) in the fuzzy clustered image, we check first if the same pixel is a foreground pixel in the Savoula image (Υ) and also in the window based binary image (χ). If they are foreground pixels (line 8 in Algorithm 2), then we traverse at the neighborhood of this pixel with the window size of $winSz \times winSz$ and count the number of such pixels which are foreground both in $\Upsilon(x, y)$ and $\chi(x, y)$. If this number exceeds 20% of the window size ($winSz \times winSz$) (line 17 in Algorithm 2), then we mark this pixel as a text pixel, otherwise it is marked as a background pixel. The value of $winSz$ is taken as (*stroke width/2*). The *stroke width* is calculated by using the approach presented in [3]. The recovered and background separated image is shown in Fig. 3d and a zoomed portion is shown in Fig. 3e.

---

[6] Considering the image resolution, we choose this value as 20. The objective is to get small textual regions for local binarization. We have not properly tested this chosen value so the performance can be slightly improved by the proper choice of this value by evaluating on some test dataset.

Algorithm 2: CHARACTER RECONSTRUCTION

Input: γ, β, $MRows$, $NCols$
Output: $\mathcal{H}_{smooth}$

1   $maskSz \leftarrow 5$ ▷ $mask\ size\ for\ reconstruction$
2   $winSz = maskSz/2$
3   $combi\_Imag \leftarrow \gamma$
4   $sum \leftarrow 0$
5   **for** $iR \leftarrow 0$ **to** $(M-1)$ **do**
6   **for** $jC \leftarrow 0$ **to** $(N-1)$ **do**
      /* Verifying the confused pixels in clustered image   */
7   **if** $(\beta(iR,jC) \neq 0)$ & $(\beta(iR,jC) \neq 255)$ **then**
       /* check if this pixel is text in both in savoula and
       windowed binary image   */
8   **if** $(\Upsilon(iR,jC) = 0)$ & $(\chi(iR,jC) = 0)$ **then**
9        $sCnt \leftarrow 0;\ x1 \leftarrow 0;\ y1 \leftarrow 0$
10   **for** $k \leftarrow -winSz$ **to** $winSz$ **do**
11   **for** $c \leftarrow -winSz$ **to** $winSz$ **do**
        /* counting the number of strong pixels at the
        surrounding   */
12   **if** $(\Upsilon(y1,x1) == 0)$ & $(\chi(y1,x1) == 0)$ **then**
13         $sCnt++$
14
15   **if** $(sCnt \geq ((0.2 \times winSz^2) - 1))$ **then**
16        $combi\_Imag(iR,jC) = 0$
17        $\beta(iR,jC) = 0$
18
19
20

21   **γ (sure text pixels), β (sure and confused text pixels), $MRows$ (Total number of image rows), $NCols$ (Total number of image columns).

---

4   Results and Discussion

In this section, we will present and discuss the obtained results on a dataset of university diplomas.

4.1   Dataset

To evaluate the proposed algorithm, we have created the French university diploma dataset described in the following.

4.1.1   Diploma Data Set

A total of 40 images were scanned in color and gray level (denoted as Dip_Col and Dip_Gray in Table 2) by a Fujitsu scanner in 300 (20 images) and 600 (20 images) DPI (of size is 7000 × 5000 pixels approx.).

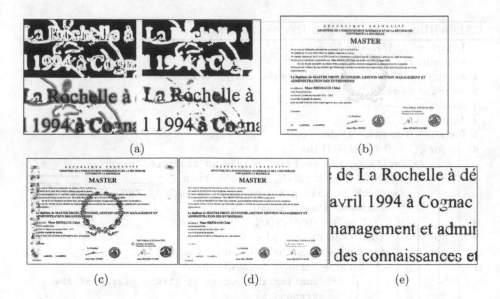

(a)   (b)

(c)   (d)   (e)

Fig. 3. (a) Reasons of character deformation (b) The binary image by the Savoula technique (c) The binary image by the dynamic window based technique (d) The text separated image after reconstruction (e) Zoomed portion after image reconstruction.

4.1.2   Ground Truth Preparation

Due to no prior availability of a dataset, we have created a dataset and a corresponding GT. Due to the high image size, it is cumbersome to create the GT manually. So, we adapted a semi-manual approach by manually selecting the range of the gray-level threshold for foreground/text pixels.[7] Then the obtained images are manually corrected by the GIMP software.[8]

4.2   Evaluation Protocol

We have used the *F-Measure (FM)*, the *Peak Signal to Noise Ratio (PSNR)* and the Distance Reciprocal Distortion Metric (DRD) as evaluation metrics (the same as in the DIBCO competition [11]). It can be seen from Table 2, that the performance of our technique is quite promising and achieves an average accuracy (F-Measure) of 93% on four parts of the complete data set. The accuracies could be further improved by a better preparation of the GT (mainly by a proper removal of the background pixels because our algorithm classifies correctly the

---

[7] Data and available at: https://github.com/tanmayGIT/University_Diploma_BackGd_Removal.git.

[8] These documents are real *University of La Rochelle, France* diplomas of students. We avoided to generate fake diplomas as it is forbidden by the law. For the privacy purpose, the confidential information e.g. name, date of birth, unique number of diploma etc. were manually removed from the gray image (for color images, it was first converted into gray).

Table 2. Results on the university diploma dataset. The best results of each metric are highlighted in bold.

Dataset	No. of images	Method name	F-Measure (%)(↑)	PSNR (↑)	DRD(↓)
300_Col	10	Proposed	**89.89**	**20.90**	**5.56**
		Niblack	10.26	2.03	501.80
		Savoula	87.70	19.98	7.90
		Wolf-Jolin	83.90	18.60	11.22
		Bataineh [2]	68.88	14.80	30.24
		Gatos [5]	84.74	18.80	10.75
		Mondal [10]	84.40	19.05	9.39
		Jia [9] (for 10 images)	71.88	15.48	25.84
300_Gray	10	Proposed	**93.62**	**22.90**	**2.99**
		Niblack	10.61	1.85	486.97
		Savoula	91.98	21.91	4.70
		Wolf-Jolin	89.40	20.50	6.56
		Gatos [5]	91.08	21.23	5.55
		Bataineh [2]	74.59	15.78	23.54
		Mondal [10]	86.88	19.89	7.19
		Jia [9] (for 8 images)	87.91	19.93	8.07
600_Col	10	Proposed	**92.05**	**21.91**	**3.59**
		Niblack	10.83	2.09	446.60
		Savoula	89.36	20.58	5.77
		Wolf-Jolin	85.88	19.16	8.40
		Bataineh [2]	68.96	14.71	27.59
		Gatos [5]	85.36	18.91	9.07
		Mondal [10]	88.48	20.48	5.86
		Jia [9] (for 8 images)	85.12	19.26	8.32
600_Gray	10	Proposed	**95.14**	**24.04**	**1.76**
		Niblack	10.96	1.90	420.37
		Savoula	92.12	21.89	4.03
		Wolf-Jolin	89.63	20.50	5.59
		Bataineh [2]	74.92	15.75	20.44
		Gatos [5]	91.30	21.25	4.57
		Mondal [10]	89.33	20.76	4.94
		Jia [9] (for 6 images)	83.60	18.70	11.05

* For the comparison purpose, only training free best performing methods are chosen from literature.

background pixels). It can be seen from Fig. 4a that the GTs are not properly cleaned/corrected, which affects the statistical result. The proposed technique performed better than the classical (e.g. Niblack, Savoula and Wolf-Jolin's [11]) binarization techniques. However, among these three classical binarization techniques, the Savoula binarization performed comparatively better than the others. But as mentioned before, these classical binarization techniques are mainly

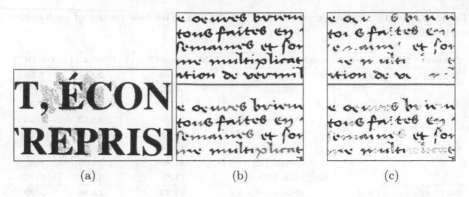

(a)   (b)   (c)

Fig. 4. (a) The issues with the prepared GT (b) *Top:* One example of GT image from the DIBCO dataset, *Bottom:* A clustered image (sure and confused text pixels) (c) *Top:* Only sure text pixels from the clustered image, *Bottom:* After the image reconstruction.

unable to remove the centered decorated background (i.e. the *couronne*) (see the Savoula binarization results in Fig. 3b and 5a).

We also compared the proposed approach with some more recent state of the art binarization techniques. A dynamic window based local thresholding approach is proposed by Bataineh et al. [2]. The comparative results shown in Table 2 of this algorithm are not very promising. It can be seen in Fig. 3c that this technique is unable to properly remove the decorated background graphics. Our proposed technique has also outperformed the popularly known binarization technique by Gatos et al. [5], whose statistical accuracy is close to the one of Savoula et al. [11] (see Table 2).

Using structural symmetry of pixels (SSP), we recently proposed a Fuzzy C-Means clustering based binarization technique [10]. Although the accuracy shown in Table 2 is good, but our new method outperforms it by reasonable margin. As shown in Fig. 5b, the previous approach misses character pixels which deforms the shape of characters and it is also unable to properly remove the background *couronne*. Another SSP based technique is proposed by Jia et al. [9] which has listed some of the best accuracies on the DIBCO data sets [11]. This technique also failed to properly remove the decorated background from the diplomas[9] (see Fig. 5c and Table 2). Our algorithm also has outperformed this technique with a suitable margin.

The computational time of the different algorithms is presented in Table 3. Due to the implementation facility, we have implemented the prototype of this algorithm using Matlab and C++.[10] In comparison with other state-of-the-art binarization algorithms e.g. Niblack [11], Gatos [5] etc., our technique takes more

---

[9] We have used the author's provided executable, but due to some unresolved memory related errors of the OpenCV library, we were unable to run it on some images.

[10] In future, we plan to make an optimal implementation of complete algorithm in C++ using the OpenCV library.

Table 3. Time required (in seconds) by the different algorithms on a diploma image of 7016 × 4960 pixels.

Method name	Time required	Method name	Time required
Niblack [11]	3.03 (C++)	Savoula [11]	3.06 (C++)
WolfJolin [11]	3.06 (C++)	Bataineh [2]	1.03 (C++)
Gatos [5]	3.21 (C++)	Mondal* [10]	421 + 33.43 + 1031 = 1485.43 (Matlab and C++)
Jia [9]	128 (.exe provided by authors)	**Proposed****	35.8 + 14 + 5.8 = 55.6 (Matlab and C++)

*Mondal et al.: Background estimation (421 s in Matlab) + gradient computation (33.43 s in C++) + remaining portion of the algorithm (1031 s in Matlab).
**Proposed: Sect. 3.1, 3.2, 3.3 (35.8 s in C++) + Sect. 3.4, 3.6 (14 s in Matlab) + Sect. 3.5, 3.7, 3.8 (5.8 s in C++).

computational time, but it is able to outperform these state-of-the-art algorithms in terms of accuracy. While even using Matlab (slower than C++ or the Python language) for implementing one portion of our algorithm, it is a lot more faster than the algorithm of Jia et al. [9] [11] and our previous work [10].

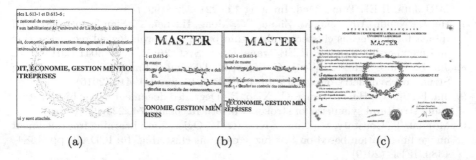

(a)   (b)   (c)

Fig. 5. (a) The result of Savoula binarization (b) Some results of our previous binarization technique [10] (c) The binarization result by Jia et al. [9]

5   Conclusion

In this paper, we have proposed a technique to remove the background and to obtain texts from French university diplomas. Our proposed method performs satisfactory well on diploma images (see Table 3). All experiments are done by using an Intel i7-8850H CPU, 32 GB RAM (in C++ and Matlab R2018a). It can be concluded that the proposed approach performs satisfactorily well on diploma images compared to many other state of the art techniques.

---

[11] We have used the optimized executable, provided by the authors.

Acknowledgment. This work is supported by the Nouvelle-Aquitaine Region and the European Union under the project "Sécurisation et Authentification des Diplômes" in the "programme opérationnel FEDER/FSE 2014–2020" (grant number P2017-BAFE-46), and by the ANR LabCom IDEAS (ANR-18-LCV3-0008). We highly thank the anonymous reviewers for their comprehensive reviews.

References

1. Ahmed, S., Liwicki, M., Dengel, A.: Extraction of text touching graphics using SURF. In: 10th IAPR International Workshop on Document Analysis Systems, pp. 349–353. IEEE, Mar 2012
2. Bataineh, B., Abdullah, S.N.H.S., Omar, K.: An adaptive local binarization method for document images based on a novel thresholding method and dynamic windows. Pattern Recogn. Lett. **32**(14), 1805–1813 (2011)
3. Bolan, S., Shijian, L., Tan, C.L.: A robust document image binarization technique for degraded document images. IEEE Trans. Image Process. **22**(4), 1408–1417 (2013)
4. Dollar, P., Zitnick, C.L.: Structured forests for fast edge detection. In: ICCV (2013)
5. Gatos, B., Pratikakis, I., Perantonis, S.: Adaptive degraded document image binarization. Pattern Recognit. **39**(3), 317–327 (2006)
6. Gerig, G., Kbler, O., Kikinis, R., Jolesz, F.A.: Nonlinear anisotropic filtering of MRI data. IEEE Trans. Med. Imaging **11**, 221–232 (1992)
7. He, L., Nandi, A.K., Jia, X., Zhang, Y., Meng, H., Lei, T.: Significantly fast and robust fuzzy c-means clustering algorithm based on morphological reconstruction and membership filtering. IEEE Trans. Fuzzy Syst. **26**(5), 3027–3041 (2018)
8. Jia, F., Shi, C., He, K., Wang, C., Xiao, B.: Document Image Binarization Using Structural Symmetry of Strokes. In: ICFHR-2016, pp. 411–416 (2016)
9. Jia, F., Shi, C., He, K., Wang, C., Xiao, B.: Degraded document image binarization using structural symmetry of strokes. Pattern Recognit. **74**, 225–240 (2018)
10. Mondal, T., Coustaty, M., Gomez-Krämer, P., Ogier, J.m.: Learning free document image binarization based on fast fuzzy c-means clustering. In: ICDAR, pp. 1384–1389. IEEE (2019)
11. Pratikakis, I., Zagoris, K., Barlas, G., Gatos, B.: ICDAR2017 Competition on Document Image Binarization (DIBCO 2017). ICDAR 1(Dibco), pp. 1395–1403 (2018)
12. Sauvola, J., Pietikäinen, M.: Adaptive document image binarization. Pattern Recognit. **33**, 225–236 (2000)
13. Tensmeyer, C., Martinez, T.: Document Image Binarization with Fully Convolutional Neural Networks. In: ICDAR, vol. 1, pp. 99–104 (2018)
14. Vo, Q.N., Kim, S.H., Yang, H.J., Lee, G.: Binarization of degraded document images based on hierarchical deep supervised network. Pattern Recognit. **74**, 568–586 (2017)

Segmentation and Layout Analysis

The Benefits of Close-Domain Fine-Tuning for Table Detection in Document Images

Ángela Casado-García$^{(\boxtimes)}$, César Domínguez(ID), Jónathan Heras(ID), Eloy Mata(ID), and Vico Pascual(ID)

Department of Mathematics and Computer Science,
University of La Rioja, Logroño, Spain
{angela.casado,cesar.dominguez,jonathan.heras,
eloy.mata,vico.pascual}@unirioja.es

Abstract. A correct localisation of tables in a document is instrumental for determining their structure and extracting their contents; therefore, table detection is a key step in table understanding. Nowadays, the most successful methods for table detection in document images employ deep learning algorithms; and, particularly, a technique known as *fine-tuning*. In this context, such a technique exports the knowledge acquired to detect objects in natural images to detect tables in document images. However, there is only a vague relation between natural and document images, and fine-tuning works better when there is a close relation between the source and target task. In this paper, we show that it is more beneficial to employ fine-tuning from a closer domain. To this aim, we train different object detection algorithms (namely, Mask R-CNN, RetinaNet, SSD and YOLO) using the TableBank dataset (a dataset of images of academic documents designed for table detection and recognition), and fine-tune them for several heterogeneous table detection datasets. Using this approach, we considerably improve the accuracy of the detection models fine-tuned from natural images (in mean a 17%, and, in the best case, up to a 60%).

Keywords: Table detection · Deep learning · Transfer learning · Fine-tuning

1 Introduction

Tables are widely present in a great variety of documents such as administrative documents, invoices, scientific papers, reports, or archival documents among

This work was partially supported by Ministerio de Economía y Competitividad [MTM2017-88804-P], Ministerio de Ciencia, Innovación y Universidades [RTC-2017-6640-7], Agencia de Desarrollo Económico de La Rioja [2017-I-IDD-00018], and the computing facilities of Extremadura Research Centre for Advanced Technologies (CETA-CIEMAT), funded by the European Regional Development Fund (ERDF). CETA-CIEMAT belongs to CIEMAT and the Government of Spain.

© Springer Nature Switzerland AG 2020
X. Bai et al. (Eds.): DAS 2020, LNCS 12116, pp. 199–215, 2020.
https://doi.org/10.1007/978-3-030-57058-3_15

others; and, therefore, techniques for table analysis are instrumental to automatically extract relevant information stored in a tabular form from several sources [7]. The first step in table analysis is *table detection*—that is, determining the position of the tables in a document—and such a step is the basis to later determine the internal table structure and, eventually, extract semantics from the table contents [7].

Table detection methods in digital born documents, such as readable PDFs or HTML documents, employ the available meta-data included in those documents to guide the analysis by means of heuristics [28]. However, table detection in image-based documents, like scanned PDFs or document images, is a more challenging task due to high intra-class variability—that is, there are several table layouts that, in addition, are highly dependent on the context of the documents—low inter-class variability—that is, other objects that commonly appear in documents (such as figures, graphics or code listing among others) are similar to tables—and the heterogeneity of document images [8]. These three issues make difficult to design rules that are generalisable to a variety of documents; and, this has led to the adoption of machine learning techniques, and, more recently, deep learning methods.

Currently, deep learning techniques are the state of the art approach to deal with computer vision tasks [32]; and, this is also the case for table detection in document images [21,23,34]. The most accurate models for table detection have been constructed using *fine-tuning* [29], a *transfer learning* technique that consists in re-using a model trained in a source task, where a lot of data is available, in a new target task, with usually scarce data. In the context of table detection, the fine-tuning approach has been applied due to the small size of table detection datasets that do not contain the necessary amount of images required to train deep learning models from scratch. In spite of its success, this approach has the limitation of applying transfer learning from natural images, a distant domain from document images. This makes necessary the application of techniques, like image transformations [11], to make document images look like natural images.

In this work, we present the benefits of applying transfer learning for table detection from a close domain thanks to the LaTeX part of the TableBank dataset [23], a dataset that consists of approximately 200 K labelled images of academic documents containing tables—a number big enough to train deep learning models from scratch. Namely, the contributions of this work are the following:

- We analyse the accuracy of four of the most successful deep learning algorithms for object detection (namely, Mask-RCNN, RetinaNet, SSD and YOLO) in the context of table detection in academic documents using the TableBank dataset.
- Moreover, we present a comprehensive study where we compare the effects of fine-tuning table detection models from a distant domain (natural images) and a closer domain (images of academic documents from the TableBank dataset), and demonstrate the advantages of the latter approach. To this aim, we employ the 4 aforementioned object detection architectures and 7

heterogeneous table detection datasets containing a wide variety of document images.

– Finally, we show the benefits of using models trained for table detection on document images to detect other objects that commonly appear in document images such as figures and formulas.

As a by-product of this work, we have produced a suite of models that can be employed via a set of Jupyter notebooks (documents for publishing code, results and explanations in a form that is both readable and executable) [22] that can be run online using Google Colaboratory [5]—a free Jupyter notebook environment that requires no setup and runs entirely in the cloud avoiding the installation of libraries in the local computer. In addition, the code for fine-tuning the models is also freely available. This allows the interested readers to adapt the models generated in this work to detect tables in their own datasets. All the code and models are available at the project webpage https://github.com/holms-ur/fine-tuning.

The rest of this paper is organised as follows. In the next section, we provide a brief overview of the methods employed in the literature to tackle the table detection task. Subsequently, in Sect. 3, we introduce our approach to train models for table detection using fine-tuning, as well as the setting that we employ to evaluate such an approach. Afterwards, we present the obtained results along with a thorough analysis in Sect. 4, and the tools that we have developed in Sect. 5. Finally, the paper ends with some conclusions and further work.

2   Related Work

Since the early 1990s, several researchers have tackled the task of table detection in document images using mainly two approaches: rule-based techniques and data-driven methods. The former are focused on defining rules to determine the position of lines and text blocks to later detect tabular structures [16,19,38]; whereas, the latter employ statistical machine learning techniques, like Hidden Markov models [6], a hierarchical representation based on the MXY tree [3] or feature engineering together with SVMs [20]. However, both approaches have drawbacks: rule-based methods require the design of handcrafted rules, that do not usually generalise to several kinds of documents; and, machine learning methods require manual feature engineering to decide the features of the documents that are feed to machine learning algorithms. These problems have been recently alleviated by using deep learning methods.

Nowadays, deep learning techniques are the state of the art approach to deal with table detection. The reason is twofold: deep learning techniques are robust for different document types; and, they do not need handcrafted features since they automatically learn a hierarchy of relevant features using convolutional neural networks (CNNs) [14]. Initially, hybrid methods combining rules and deep-learning models were suggested; for instance, in [15] and [27], CNNs were employed to decide whether regions of an image suggested by a set of rules contained a table. On the contrary, the main approach followed currently consists in adapting general deep learning algorithms for object detection to the problem of table detection. Namely, the main algorithm applied in this context is Faster R-CNN [31], that has been directly employed using different backbone architectures [21,23,34], and combined with deformable CNNs [36] or with image transformations [11]. Other detection algorithms such as YOLO [30] or SSD [26] have also been employed for table detection [17,21], but achieving worse results than the methods based on the Faster R-CNN algorithm. Nevertheless, training deep learning models for table detection is challenging due to the considerable amount of images that are necessary for this task — up to recently, the biggest dataset of document images containing tables was the Marmot dataset with 2,000 labelled images [18], far from the datasets employed by deep learning methods that consists of several thousands, or even millions, of images [9,33].

In order to deal with the problem of limited amount of data, one of the most successful methods applied in the literature is transfer learning [29], a technique that re-uses a model trained in a source task in a new target task. This is the approach followed in [11,34,36], where they use models trained on natural images to fine-tune their models for table detection. However, transfer learning methods are more effective when there is a close relation between the source and target domains, and, unfortunately, there is only a vague relation between natural images and document images. This issue has been faced, for instance, by applying image transformations to make document images as close as possible to natural images [11].

Another option to tackle the problem of limited data consists in acquiring and labelling more images, a task that has been undertaken for table detection in the TableBank project [23]—a dataset that consists of 417 K labelled images of documents containing tables. The TableBank dataset opens the door to apply transfer learning to not only construct models for table detection in different kinds of documents, but also to detect other objects, such as figures or formulas, that commonly appear in document images. This is the goal of the present work.

3   Materials and Methods

In this section, we explain the fine-tuning method, as well as the object detection algorithms, datasets and evaluation metrics used in this work.

3.1   Fine-Tuning

Transfer learning allows us to train models using the knowledge learned by other models instead of starting from scratch. The idea on which transfer learning techniques are based is that CNNs are designed to learn a hierarchy of features. Specifically, the lower layers of CNNs focus on generic features, while the final ones focus on specific features for the task they are working with. As explained in [29], transfer learning can be employed in different ways, and the one employed in this work is known as fine-tuning. In this technique, the weights of a network learned in a source task are employed as a basis to train a model in the destination task. In this way, the information learned in the source task is used in the destination task. This approach is especially beneficial when the source and target tasks are close to each other.

In our work, we study the effects of fine-tuning table detection algorithms from a distant domain (natural images from the Pascal VOC dataset [9]) and a close domain (images of academic documents from the TableBank dataset). To this aim, we consider the following object detection algorithms.

3.2 Object Detection Algorithms

Object detection algorithms based on deep learning can be divided into two categories [12,30]: the *two-phase algorithms*, whose first step is the generation of proposals of "interesting" regions that are classified using CNNs in a second step. And the *one-phase algorithms* that perform detection without explicitly generating region proposals. For this work, we have employed algorithms of both types. In particular, we have used the two-phase algorithm Mask R-CNN, and the one-phase algorithms RetinaNet, SSD and YOLO.

Mask R-CNN [31] is currently one of the most accurate algorithm based on a two-phase approach, and the latest version of the R-CNN family. We have used a library implemented in Keras [1] for training models with this algorithm.

RetinaNet [25] is a one-phase algorithm characterised by using the focal loss for training on a scarce set of difficult examples, and that prevents the large number of easy negatives from overwhelming the detector during training. We have used another library implemented in Keras [24] for traning models with this algorithm.

SSD [26] is a simple detection algorithm that completely eliminates proposal generation and encapsulates all computation in a single network. In this case, we have used the MXNET library [4] for training the models.

YOLO [30] frames object detection as a regression problem where a single neural network predicts bounding boxes and class probabilities directly from full images in one evaluation. Although there are several versions of YOLO, the main ideas are the same for all of them. We have used the Darknet library [2] for training models with this algorithm.

The aforementioned algorithms have been trained for detecting tables in a wide variety of document images by using the datasets presented in the following section.

3.3 Benchmarking Datasets

For this project, we have used several datasets, see Table 1. Namely, we have employed three kinds of datasets: the base datasets (which are used to train the base models), the fine-tune datasets for table detection, and the fine-tune dataset for detecting other objects in document images. The reason to consider several

table detection datasets is that there are several table layouts that are highly
dependent on the document type, and we want to prove that our approach can
be generalised to heterogeneous document images.

Table 1. Sizes of the train and test sets of the datasets

Datasets	#Train images	#Test images	Type of images
Pascal VOC	16,551	4,952	Natural images
TableBank	199,183	1,000	Academic documents
ICDAR13	178	60	Documents obtained from Google search
ICDAR17	1,200	400	Scientific papers
ICDAR19	599	198	Modern images
Invoices	515	172	Invoices
MarmotEn	744	249	Scientific papers
MarmotChi	754	252	E-books
UNLV	302	101	Technical reports, business letters, newspapers and magazines
ICDAR17FIG	1,200	400	Scientific papers
ICDAR17FOR	1,200	400	Scientific papers

Base Datasets. In this work, we have employed two datasets of considerable
size for creating the base models that are later employed for fine-tuning.

The Pascal VOC dataset [9] is a popular project designed to create and evaluate algorithms for image classification, object detection and segmentation. This
dataset consists of natural images which have been used for training different
models in the literature. Thanks to the trend of releasing models to the public,
we have employed models already trained with this dataset to apply fine-tuning
from natural images to the context of table detection.

TableBank [23] is a table detection dataset built with Word and LATEX documents that contains 417 K labeled images. For this project, we only employ
the LATEX images (199,183 images) since the Word images contain some errors
in the annotations. On the contrary to the Pascal VOC dataset, where there
were available models trained for such a dataset, we have trained models for the
TableBank dataset from scratch.

Fine-Tuning Datasets. We have used several open datasets for fine-tuning;
however, most table detection datasets only release the training set. Hence, in
this project, we have divided the training sets into two sets (75% for training
and 25% for testing) evaluating our approach. The dataset split are available
in the project webpage, and the employed datasets are listed as follows.

ICDAR13 [13] is one of the most famous datasets for table detection and structure recognition. This dataset is formed by documents extracted from Web pages and email messages. This dataset was prepared for a competition focused on the task of detecting tables, figures and mathematical equations from images. The dataset is comprised of PDF files which we converted to images to be used within our framework. The dataset contains 238 images in total, 178 were used for training and 60 for testing.

ICDAR17 [10] is a data set prepared for a competition as ICDAR13. The dataset consists of 1.600 images in total, where we can find tables, formulas and figures. The training set consists of 1,200 images, while the rest of the 400 images are used for testing. This dataset has been employed three times in our work: for the detection of tables (from now on, we will call this dataset ICDAR17), for the detection of figures (from now on, we will call this dataset ICDAR17FIG) and for the detection of formulas (from now on, we will call this dataset ICDAR17FOR).

ICDAR19 [37] is, as in the previous cases, a dataset proposed for a competition. The dataset contains two types of images: modern documents and archival ones with various formats. In this work we have only taken the modern images (797 images in total, 599 for training and 198 for testing).

Invoices is a proprietary dataset of PDF files containing invoices from several sources. The PDF files had to be converted into images. This set has 515 images in the training set and 172 in the testing set.

Marmot [18] is a dataset that shows a great variety in language type, page layout, and table styles. Over 1,500 conference and journal papers were crawled for this dataset, covering various fields, spanning from the year 1970 to latest 2011 publications. In total, 2,000 pages in PDF format were collected. The dataset is composed of Chinese (from now on MarmotChi) and English pages (from now on MarmotEn): the MarmotChi dataset was built from over 120 e-Books with diverse subject areas provided by Founder Apabi library, and no more than 15 pages were selected from each book, this dataset contains 993 images in total, 744 were used for training and 249 for testing. And the MarmotEn dataset was crawled from the Citeseer website, this dataset contains 1,006 images in total, 754 were used for training and 252 for testing.

UNLV [35] is comprised of a variety of documents which includes technical reports, business letters, newspapers and magazines. The dataset contains a total of 2,889 scanned documents where only 403 documents contain a tabular region. We only used the images containing a tabular region in our experiments: 302 for training and 101 for testing.

Using the aforementioned algorithms and datasets, we have trained several models that have been evaluated using the following metrics.

3.4 Performance Measure

In order to evaluate the constructed models for the different datasets, we employed the same metric used in the ICDAR19 competition for table detection [37]. Considering that the ground truth bounding box is represent by GTP, and that the bounding box detected by an algorithm is represented by DTP; then, the formula for finding the overlapped region between them is given by:

$$IoU(GTP, DTP) = \frac{area(GTP \cap DTP)}{area(GTP \cup DTP)}$$

IoU(GTP, DTP) represents the overlapped region between ground truth and detected bounding boxes and its value lies between zero and one.

Now, given a threshold $T \in [0, 1]$, we define the notions of True Positive at T, TP@T, False Positive at T, FP@T, and False Negative at T, FN@T. The TP@T is the number of ground truth tables that have a major overlap ($IoU \geq T$) with one of the detected tables. The FP@T indicates the number of detected tables that do not overlap ($IoU < T$) with any of the ground tables. And, FN@T indicates the number of ground truth tables that do not overlap ($IoU < T$) with any of the detected tables. From these notions, we can define the Precision at T, P@T, Recall at T, R@T, and F1-score at T, F1@T, as follows:

$$P@T = \frac{TP@T}{FP@T + TP@T}$$

$$R@T = \frac{TP@T}{FN@T + TP@T}$$

$$F1@T = \frac{2 * TP@T}{FP@T + FN@T + 2 * TP@T}$$

Finally, the final score is decided by the weighted average WAvgF1 value:

$$WAvgF1 = \frac{0.6 \times F1@0.6 + 0.7 \times F1@0.7 + 0.8 \times F1@0.8 + 0.9 \times F1@0.9}{0.6 + 0.7 + 0.8 + 0.9}$$

In the above formula, and since results with higher IoUs are more important than those with lower IoUs, we use IoU threshold as the weight of each F1 value to get a definitive performance score for convenient comparison. Using these metrics we have obtained the results presented in the following section.

4   Results

In this section, we conduct a thorough study of our approach, see Tables 2, 3, 4, and 5. Each table corresponds with the results obtained from each object detection algorithm: Table 2 contains the results that have been obtained using Mask R-CNN; Table 3, the results of RetinaNet; Table 4, the results of SSD; and, finally, Table 5 contains the results of YOLO. The tables are divided into three parts: the first row contains the results obtained for the TableBank dataset, the next 9 rows correspond with the result obtained with the models trained by fine-tuning from natural image models, and the last 9 rows correspond with the models fine-tuned from the TableBank models. All the models built in this work were trained using the default parameters in each deep learning framework, and using K80 NVIDIA GPUs.

Table 2. Results using the Mask R-CNN algorithm

	@0.6			@0.7			@0.8			@0.9			WAvgF1	Improvement
	P@0.6	R@0.6	F1@0.6	P@0.7	R@0.7	F1@0.7	P@0.8	R@0.8	F1@0.8	P@0.9	R@0.9	F1@0.9		
TableBank	0.94	0.98	0.96	0.94	0.97	0.95	0.93	0.96	0.94	0.84	0.87	0.86	0.92	
ICDAR13	0.14	0.77	0.23	0.08	0.45	0.14	0.03	0.16	0.05	0	0	0	0.09	
ICDAR17	0.32	0.85	0.46	0.28	0.75	0.41	0.17	0.47	0.25	0.04	0.1	0.06	0.27	
ICDAR17FIG	0.29	0.61	0.39	0.22	0.46	0.3	0.13	0.27	0.17	0.01	0.03	0.02	0.19	
ICDAR17FOR	0.18	0.45	0.26	0.09	0.24	0.13	0.03	0.07	0.04	0	0.01	0	0.09	
ICDAR19	0.6	0.64	0.62	0.48	0.51	0.5	0.24	0.25	0.25	0.02	0.02	0.02	0.31	
Invoices	0.38	0.56	0.45	0.28	0.42	0.34	0.16	0.23	0.19	0.02	0.03	0.02	0.22	
MarmotEn	0.37	0.75	0.49	0.28	0.58	0.38	0.08	0.17	0.11	0	0.01	0	0.21	
MarmotChi	0.52	0.83	0.64	0.48	0.76	0.59	0.32	0.51	0.39	0.07	0.11	0.08	0.39	
UNLV	0.29	0.58	0.39	0.17	0.34	0.23	0.06	0.11	0.08	0.01	0.02	0.01	0.15	
ICDAR13	0.7	0.97	0.81	0.7	0.97	0.81	0.7	0.97	0.81	0.47	0.65	0.54	0.72	0.63
ICDAR17	0.72	0.95	0.82	0.7	0.93	0.8	0.68	0.9	0.78	0.49	0.64	0.56	0.72	0.45
ICDAR17FIG	0.36	0.69	0.47	0.33	0.63	0.43	0.23	0.43	0.3	0.05	0.09	0.07	0.29	0.09
ICDAR17FOR	0.1	0.49	0.17	0.06	0.28	0.1	0.02	0.12	0.04	0	0.01	0	0.06	-0.02
ICDAR19	0.76	0.85	0.81	0.74	0.83	0.79	0.67	0.75	0.71	0.38	0.42	0.4	0.65	0.34
Invoices	0.54	0.71	0.61	0.5	0.66	0.57	0.39	0.52	0.45	0.19	0.26	0.22	0.44	0.21
MarmotEn	0.72	0.93	0.81	0.7	0.9	0.79	0.67	0.87	0.76	0.46	0.6	0.52	0.70	0.48
MarmotChi	0.82	0.98	0.89	0.82	0.98	0.89	0.81	0.96	0.88	0.62	0.73	0.67	0.82	0.42
UNLV	0.66	0.83	0.74	0.63	0.8	0.7	0.55	0.69	0.61	0.24	0.3	0.27	0.55	0.39

Table 3. Results using the RetinaNet algorithm

	@0.6			@0.7			@0.8			@0.9			WAvgF1	Improvement
	P@0.6	R@0.6	F1@0.6	P@0.7	R@0.7	F1@0.7	P@0.8	R@0.8	F1@0.8	P@0.9	R@0.9	F1@0.9		
TableBank	0.98	0.86	0.92	0.98	0.86	0.92	0.97	0.85	0.91	0.94	0.82	0.87	0.90	
ICDAR13	0.56	0.58	0.57	0.56	0.58	0.57	0.56	0.58	0.57	0.34	0.35	0.35	0.50	
ICDAR17	0.65	0.86	0.74	0.64	0.85	0.73	0.58	0.77	0.67	0.48	0.63	0.55	0.66	
ICDAR17FIG	0.57	0.61	0.59	0.7	0.76	0.73	0.73	0.79	0.76	0.74	0.8	0.77	0.72	
ICDAR17FOR	0.63	0.06	0.12	0.63	0.06	0.12	0.6	0.06	0.11	0.4	0.04	0.07	0.10	
ICDAR19	0.86	0.66	0.74	0.82	0.63	0.72	0.76	0.58	0.66	0.58	0.45	0.51	0.64	
Invoices	0.90	0.59	0.71	0.88	0.58	0.70	0.86	0.56	0.68	0.70	0.46	0.56	0.65	
MarmotEn	0.75	0.86	0.8	0.74	0.86	0.8	0.7	0.81	0.75	0.47	0.54	0.5	0.69	
MarmotChi	0.78	0.85	0.81	0.75	0.81	0.78	0.73	0.79	0.75	0.5	0.54	0.52	0.70	
UNLV	0.81	0.83	0.82	0.79	0.81	0.80	0.76	0.77	0.76	0.61	0.63	0.62	0.73	
ICDAR13	0.83	0.77	0.8	0.79	0.74	0.77	0.76	0.71	0.73	0.72	0.68	0.7	0.74	0.24
ICDAR17	0.92	0.87	0.89	0.92	0.87	0.89	0.89	0.84	0.86	0.79	0.75	0.77	0.84	0.18
ICDAR17FIG	0.76	0.79	0.77	0.74	0.77	0.76	0.72	0.75	0.74	0.64	0.66	0.65	0.72	0.001
ICDAR17FOR	0.26	0.35	0.3	0.24	0.32	0.27	0.19	0.26	0.22	0.08	0.1	0.09	0.20	0.11
ICDAR19	0.91	0.74	0.82	0.87	0.81	0.79	0.81	0.67	0.73	0.68	0.56	0.61	0.72	0.08
Invoices	0.92	0.59	0.72	0.92	0.59	0.71	0.87	0.55	0.68	0.74	0.47	0.58	0.66	0.01
MarmotEn	0.93	0.86	0.9	0.92	0.86	0.89	0.91	0.84	0.87	0.78	0.73	0.75	0.84	0.14
MarmotChi	0.87	0.87	0.87	0.85	0.85	0.85	0.83	0.83	0.83	0.69	0.7	0.69	0.80	0.10
UNLV	0.81	0.83	0.82	0.79	0.8	0.8	0.75	0.77	0.76	0.63	0.64	0.64	0.74	0.01

We start by analysing the results for the TableBank dataset, see the first row of the tables. Each model has its strenghts and weaknesses, and depending on the context we can prefer different models. The overall best model, that is the model with higher WAvgF1-score, is the Mask R-CNN model; the other three models are similar among them. If we focus on detecting as most tables as possible (R@0.6) and not detecting other artifacts as tables (P@0.6), the best model is YOLO. Finally, if we are interested in accurately detecting the regions of the tables (F1@0.9), the best model is RetinaNet. The strength of the SSD model is that it is faster than the others.

Let us focus now on the table detection datasets. In the case of models fine-tuned using natural images, the algorithms that stand out are RetinaNet and YOLO, see Tables 2 to 5 and Fig. 1. Similary, the models that achieve higher accuracies when fine-tuning from the TableBank dataset are YOLO and RetinaNet, see Fig. 2. As can be seen in Tables 2 to 5, fine-tuning from a close domain produce more accurate models that fine-tuning from an unrelated domain. However, the effects on each algorithm and dataset greatly differ. The algorithm that is more considerably boosted for table detection is Mask R-CNN, that improves up to a 60% in some cases and 42% in mean. In the case of RetinaNet, in mean it improves by 11%, YOLO a 9%, and SSD is the one with the least improvement, only a 5%.

Finally, if we consider the results for the datasets containing figures and formulas, the improvement is not as remarkable as in the detection of tables. In this case, the algorithm that takes a bigger advantage of this technique is YOLO, since in both cases it improves up to a 10%. In the case of RetinaNet, it improves the detection of formulas by 10%, while that of figures barely improves. And in the case of SSD and Mask R-CNN, they are the ones with the least improvement and even have some penalty.

As we have shown in this section, fine-tuning from the TableBank dataset can boosten table detection models. However, there is not a model that outperforms the rest, see Figs. 1 and 2. Therefore, we have released a set of tools to employ the trained models, and also employ them for constructing models using fine-tuning on custom datasets.

5   Tools for Table Detection

Using one of the generated detection model with new images is usually as simple as invoking a command with the path of the image (and, probably, some additional parameters). However, this requires the installation of several libraries and the usage of a command line interface; and, this might be challenging for non-expert users. Therefore, it is important to create simple and intuitive interfaces that might be employed by different kinds of users; otherwise, they will not be able to take advantage of the object detection models.

To disseminate our detection models, we have created a set of Jupyter notebooks, that allows users to detect tables in their images. Jupyter notebooks [22] are documents for publishing code, results and explanations in a form that is

Table 4. Results using the SSD algorithm

	@0.6			@0.7			@0.8			@0.9			WAvgF1	Improvement
	P@0.6	R@0.6	F1@0.6	P@0.7	R@0.7	F1@0.7	P@0.8	R@0.8	F1@0.8	P@0.9	R@0.9	F1@0.9		
TableBank	0.96	0.97	0.96	0.94	0.95	0.95	0.92	0.92	0.92	0.82	0.82	0.82	0.90	
ICDAR13	0.54	0.68	0.6	0.44	0.55	0.49	0.38	0.48	0.43	0.15	0.19	0.17	0.40	
ICDAR17	0.49	0.71	0.58	0.41	0.59	0.48	0.34	0.49	0.4	0.22	0.32	0.4	0.45	
ICDAR17FIG	0.7	0.8	0.75	0.68	0.77	0.72	0.61	0.69	0.65	0.34	0.38	0.36	0.59	
ICDAR17FOR	0.44	0.64	0.52	0.34	0.49	0.4	0.19	0.27	0.22	0.03	0.05	0.22	0.32	
ICDAR19	0.31	0.35	0.33	0.23	0.26	0.25	0.18	0.2	0.19	0.1	0.11	0.1	0.20	
Invoices	0.87	0.84	0.85	0.8	0.78	0.79	0.63	0.61	0.62	0.27	0.26	0.26	0.59	
MarmotEn	0.67	0.76	0.71	0.63	0.71	0.67	0.6	0.67	0.63	0.38	0.43	0.4	0.58	
MarmotChi	0.57	0.7	0.63	0.48	0.60	0.53	0.36	0.45	0.4	0.25	0.31	0.28	0.44	
UNLV	0.66	0.64	0.65	0.6	0.58	0.59	0.45	0.43	0.44	0.12	0.11	0.12	0.42	
ICDAR13	0.62	0.68	0.65	0.62	0.68	0.65	0.5	0.55	0.52	0.32	0.35	0.34	0.52	0.12
ICDAR17	0.55	0.71	0.62	0.46	0.60	0.52	0.42	0.54	0.47	0.30	0.39	0.34	0.47	0.01
ICDAR17FIG	0.27	0.77	0.41	0.26	0.72	0.38	0.21	0.58	0.31	0.10	0.28	0.15	0.29	-0.3
ICDAR17FOR	0.5	0.74	0.6	0.41	0.6	0.49	0.29	0.42	0.34	0.07	0.11	0.09	0.35	0.03
ICDAR19	0.35	0.35	0.35	0.28	0.27	0.28	0.23	0.22	0.23	0.13	0.12	0.13	0.23	0.03
Invoices	0.91	0.86	0.89	0.87	0.81	0.84	0.71	0.67	0.69	0.37	0.35	0.36	0.66	0.07
MarmotEn	0.71	0.75	0.73	0.69	0.73	0.71	0.66	0.70	0.68	0.49	0.52	0.51	0.64	0.06
MarmotChi	0.61	0.67	0.64	0.50	0.55	0.52	0.42	0.46	0.44	0.28	0.31	0.29	0.45	0.01
UNLV	0.72	0.66	0.69	0.66	0.61	0.63	0.5	0.45	0.47	0.28	0.26	0.27	0.49	0.07

Table 5. Results using the YOLO algorithm

	@0.6			@0.7			@0.8			@0.9			WAvgF1	Improvement
	P@0.6	R@0.6	F1@0.6	P@0.7	R@0.7	F1@0.7	P@0.8	R@0.8	F1@0.8	P@0.9	R@0.9	F1@0.9		
TableBank	0.98	0.99	0.98	0.98	0.99	0.98	0.96	0.97	0.96	0.74	0.75	0.75	0.90	
ICDAR13	0.92	0.58	0.6	0.61	0.61	0.61	0.57	0.55	0.56	0.31	0.32	0.32	0.50	
ICDAR17	0.9	0.94	0.92	0.88	0.93	0.9	0.78	0.82	0.8	0.39	0.41	0.4	0.72	
ICDAR17FIG	0.88	0.84	0.86	0.85	0.82	0.84	0.75	0.72	0.74	0.23	0.22	0.23	0.63	
ICDAR17FOR	0.9	0.85	0.87	0.82	0.77	0.79	0.54	0.5	0.52	0.1	0.09	0.1	0.52	
ICDAR19	0.95	0.91	0.93	0.94	0.9	0.92	0.89	0.85	0.87	0.61	0.58	0.6	0.81	
Invoices	0.89	0.87	0.88	0.84	0.82	0.83	0.7	0.68	0.69	0.26	0.26	0.26	0.63	
MarmotEn	0.9	0.96	0.93	0.87	0.93	0.9	0.76	0.81	0.78	0.32	0.34	0.33	0.70	
MarmotChi	0.95	0.96	0.96	0.93	0.94	0.94	0.88	0.89	0.89	0.61	0.62	0.61	0.83	
UNLV	0.91	0.95	0.93	0.88	0.91	0.89	0.73	0.76	0.74	0.39	0.4	0.39	0.70	
ICDAR13	1	0.65	0.78	0.95	0.61	0.75	0.9	0.58	0.71	0.6	0.39	0.47	0.66	0.15
ICDAR17	0.94	0.94	0.94	0.93	0.94	0.93	0.89	0.89	0.89	0.61	0.62	0.61	0.82	0.09
ICDAR17FIG	0.91	0.83	0.87	0.89	0.82	0.86	0.83	0.76	0.8	0.44	0.4	0.42	0.71	0.07
ICDAR17FOR	0.94	0.85	0.89	0.88	0.79	0.83	0.65	0.59	0.62	0.19	0.17	0.18	0.59	0.06
ICDAR19	0.95	0.95	0.95	0.94	0.94	0.94	0.9	0.9	0.9	0.68	0.68	0.68	0.85	0.04
Invoices	0.9	0.89	0.89	0.87	0.85	0.86	0.76	0.74	0.75	0.39	0.39	0.39	0.69	0.06
MarmotEn	0.95	0.97	0.96	0.95	0.97	0.96	0.92	0.93	0.93	0.68	0.69	0.69	0.87	0.16
MarmotChi	0.97	0.93	0.95	0.96	0.93	0.94	0.92	0.89	0.91	0.69	0.67	0.68	0.85	0.02
UNLV	0.93	0.95	0.94	0.92	0.94	0.93	0.83	0.85	0.84	0.48	0.49	0.49	0.77	0.06

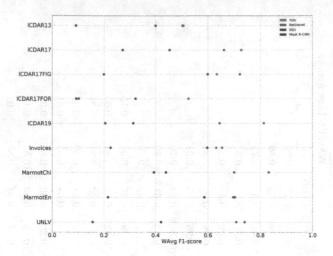

Fig. 1. Dispersion diagram using fine-tuning from natural images

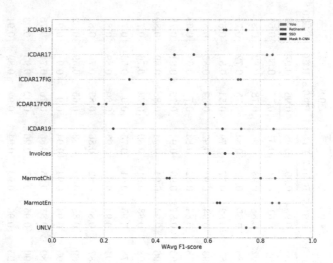

Fig. 2. Dispersion diagram using fine-tuning dataset from the TableBank

both readable and executable; and, they have been widely adopted across multiple disciplines, both for their usefulness in keeping a record of data analyses, and also for allowing reproducibility. The drawback of Jupyter notebooks is that they require the installation of several libraries. Such a problem has been overcome in our case by providing our notebooks in Google Colaboratory [5], a free Jupyter notebook environment that requires no setup and runs entirely in the cloud avoiding the installation of libraries in the local computer. The notebooks are available at https://github.com/holms-ur/fine-tuning.

In addition, in the same project webpage, due to the heterogeneity of document images containing tables, we have provided all the weights, configuration files, and necessary instructions to fine-tune any of the detection models created in this work to custom datasets containing tables.

6   Conclusion and Further Work

In this work, we have shown the benefits of using fine-tuning from a close domain in the context of table detection. In addition to the accuracy improvement, this approach avoids overfitting and solves the problem of having a small dataset. Moreover, we can highlight that apart from the Mask R-CNN algorithm, other algorithms such as YOLO and RetinaNet can achieve a good performance in the table detection task.

Since table detection is the first step towards table analysis, we plan to use this work as a basis for determining the internal structure of tables, and, eventually, extracting the semantics from table contents. Moreover, we are also interested in extending these methods to detect forms in document images.

References

1. Abdulla, W.: Mask R-CNN for object detection and instance segmentation on Keras and TensorFlow (2017). https://github.com/matterport/Mask_RCNN
2. Alexey, A.B.: YOLO darknet (2018). https://github.com/AlexeyAB/darknet
3. Cesari, F., et al.: Trainable table location in document images. In: 16th International Conference on Pattern Recognition, ICPR 2002, vol. 3, p. 30236. ACM (2002)
4. Chen, T., et al.: MXNet: A Flexible and Efficient Machine Learning Library for Heterogeneous Distributed Systems. CoRR abs/1512.01274 (2015). http://arxiv.org/abs/1512.01274
5. Colaboratory team: Google colaboratory (2017). https://colab.research.google.com
6. Costa e Silva, A.: Learning rich hidden Markov models in document analysis: table location. In: 10th International Conference on Document Analysis and Recognition, ICDAR 2010, pp. 843–847. IEEE (2009)
7. Coüasnon, B., Lemaitre, A.: Recognition of tables and forms. In: Doermann, D., Tombre, K. (eds.) Handbook of Document Image Processing and Recognition, pp. 647–677. Springer, London (2014). https://doi.org/10.1007/978-0-85729-859-1_20
8. Embley, D.W., et al.: Table-processing paradigms: a research survey. Int. J. Doc. Anal. Recogn. 8(2–3), 647–677 (2006)
9. Everingham, M., et al.: The Pascal visual object classes challenge: a retrospective. Int. J. Comput. Vision 111(1), 98–136 (2015)
10. Gao, L., Yi, X., Jiang, Z., Hao, L., Tang, Z.: ICDAR2017 competition on page object detection. In: 14th IAPR International Conference on Document Analysis and Recognition, ICDAR 2017, pp. 1417–1422 (2017)
11. Gilani, A., et al.: Table detection using deep learning. In: 14th International Conference on Document Analysis and Recognition, ICDAR 2017, pp. 771–776. IEEE (2017)

12. Girshick, R., et al.: Accurate object detection and semantic segmentation. In: 2014 IEEE Computer Society Conference on Computer Vision and Pattern Recognition, CVPR 2014, pp. 580–587. IEEE (2014)
13. Gobel, M.C., Hassan, T., Oro, E., Orsi, G.: ICDAR2013 table competition. In: 12th ICDAR Robust Reading Competition, ICDAR 2013, pp. 1449–1453. IEEE (2013)
14. Goodfellow, I., Bengio, Y., Courville, A.: Deep Learning. MIT Press (2016). http://www.deeplearningbook.org
15. Hao, L., et al.: A table detection method for PDF documents based on convolutional neural networks. In: 12th International Workshop on Document Analysis Systems, DAS 2016, pp. 287–292. IEEE (2016)
16. Hirayama, Y.: A method for table structure analysis using DP matching. In: 3rd International Conference on Document Analysis and Recognition, ICDAR 1995, pp. 583–586. IEEE (1995)
17. Huang, Y., et al.: A YOLO-based table detection method. In: 15th International Conference on Document Analysis and Recognition, ICDAR 2019 (2019)
18. Institute of Computer Science and Technology of Peking University and Institute of Digital Publishing of Founder R&D Center, China: Marmot dataset for table recognition (2011). http://www.icst.pku.edu.cn/cpdp/sjzy/index.htm
19. Jianying, H., et al.: Medium-independent table detection. In: Document Recognition and Retrieval VII. vol. 3967, pp. 583–586. International Society for Optics and Photonics (1999)
20. Kasar, T., et al.: Learning to detect tables in scanned document images using line information. In: 12th International Conference on Document Analysis and Recognition, ICDAR 2013, pp. 1185–1189. IEEE (2013)
21. Kerwat, M., George, R., Shujaee, K.: Detecting knowledge artifacts in scientific document images - comparing deep learning architectures. In: 5th International Conference on Social Networks Analysis, Management and Security, SNAMS 2018, pp. 147–152. IEEE (2018)
22. Kluyver, T., et al.: Jupyter notebooks – a publishing format for reproducible computational workflows. In: 20th International Conference on Electronic Publishing, pp. 87–90. IOS Press (2016)
23. Li, M., et al.: TableBank: Table Benchmark for Image-based Table Detection and Recognition. CoRR abs/1903.01949 (2019). http://arxiv.org/abs/1903.01949
24. Lin, T., Goyal, P., Girshick, R., He, K., Dollár., P.: Keras retinanet (2017). https://github.com/fizyr/keras-retinanet
25. Lin, T.Y., et al.: Focal loss for dense object detection. In: 16th International Conference on Computer Vision, ICCV 2017, pp. 2999–3007 (2017)
26. Liu, W., Anguelov, D., Erhan, D., Szegedy, C., Reed, S., Fu, C.-Y., Berg, A.C.: SSD: single shot multibox detector. In: Leibe, B., Matas, J., Sebe, N., Welling, M. (eds.) ECCV 2016. LNCS, vol. 9905, pp. 21–37. Springer, Cham (2016). https://doi.org/10.1007/978-3-319-46448-0_2
27. Oliveira, D.A.B., Viana, M.P.: Fast CNN-based document layout analysis. In: 14th International Conference on Computer Vision Workshops, ICCVW 2017, pp. 1173–1180. IEEE (2017)
28. Oro, E., Ruffolo, M.: PDF-TREX: an approach for recognizing and extracting tables from PDF documents. In: 10th International Conference on Document Analysis and Recognition, ICDAR 2009, pp. 906–910. IEEE (2009)
29. Razavian, A.S., et al.: CNN features off-the-shelf: an astounding baseline for recognition. In: 27th Conference on Computer Vision and Pattern Recognition Workshops, CVPRW 2014, pp. 512–519 (2014)

30. Redmon, J., Farhadi, A.: YOLOv3: an incremental improvement. CoRR abs/1804.02767 (2018). http://arxiv.org/abs/1804.02767
31. Ren, S., He, K., Girshick, R., Sun, J.: Faster R-CNN: towards real-time object detection with region proposal networks. In: Advances in Neural Information Processing Systems, vol. 28, pp. 91–99 (2015)
32. Rosebrock, A.: Deep Learning for Computer Vision with Python. PyImageSearch (2018). https://www.pyimagesearch.com/
33. Russakovsky, O., et al.: ImageNet large scale visual recognition challenge. Int. J. Comput. Vision **115**(3), 211–252 (2015). https://doi.org/10.1007/s11263-015-0816-y
34. Schreiber, S., et al.: DeepDeSRT: deep learning for detection and structure recognition of tables in document images. In: 14th International Conference on Document Analysis and Recognition, ICDAR 2017, pp. 1162–1167. IEEE (2017)
35. Shahab, A., Shafait, F., Kieninger, T., Dengel, A.: An open approach towards the benchmarking of table structure recognition systems. In: 9th IAPR International Workshop on Document Analysis Systems, DAS 2010, pp. 113–120 (2010)
36. Siddiqui, S.A., et al.: DeCNT: deep deformable CNN for table detection. IEEE Access **6**, 74151–74161 (2018)
37. Suen, C.Y., et al.: ICDAR2019 Table Competition (2019). http://icdar2019.org/
38. Zanibbi, R., Blostein, D., Cordy, J.R.: A survey of table recognition. Document Anal. Recogn. **7**(1), 1–16 (2004)

IIIT-AR-13K: A New Dataset for Graphical Object Detection in Documents

Ajoy Mondal[1(✉)], Peter Lipps[2], and C. V. Jawahar[1]

[1] Centre for Visual Information Technology,
International Institute of Information Technology, Hyderabad, India
{ajoy.mondal,jawahar}@iiit.ac.in
[2] Open Text Software GmbH, Grasbrunn/Munich, Germany
peter.lipps@opentext.com

Abstract. We introduce a new dataset for graphical object detection in business documents, more specifically annual reports. This dataset, IIIT-AR-13K, is created by manually annotating the bounding boxes of graphical or page objects in publicly available annual reports. This dataset contains a total of 13K annotated page images with objects in five different popular categories—table, figure, natural image, logo, and signature. It is the largest manually annotated dataset for graphical object detection. Annual reports created in multiple languages for several years from various companies bring high diversity into this dataset. We benchmark IIIT-AR-13K dataset with two state of the art graphical object detection techniques using Faster R-CNN [20] and Mask R-CNN [11] and establish high baselines for further research. Our dataset is highly effective as training data for developing practical solutions for graphical object detection in both business documents and technical articles. By training with IIIT-AR-13K, we demonstrate the feasibility of a single solution that can report superior performance compared to the equivalent ones trained with a much larger amount of data, for table detection. We hope that our dataset helps in advancing the research for detecting various types of graphical objects in business documents (http://cvit.iiit.ac.in/usodi/iiitar13k.php).

Keywords: Graphical object detection · Annual reports · Business documents · Faster R-CNN · Mask R-CNN

1 Introduction

Graphical objects such as tables, figures, logos, equations, natural images, signatures, play an important role in the understanding of document images. They are also important for information retrieval from document images. Each of these graphical objects contains valuable information about the document in a compact form. Localizing such graphical objects is a primary step for understanding

© Springer Nature Switzerland AG 2020
X. Bai et al. (Eds.): DAS 2020, LNCS 12116, pp. 216–230, 2020.
https://doi.org/10.1007/978-3-030-57058-3_16

documents or information extraction/retrieval from the documents. Therefore, the detection of those graphical objects from the document images has attracted a lot of attention in the research community [7,10,13–16,18,21,22,24,26,27]. Large diversity within each category of the graphical objects makes detection task challenging. Researchers have explored a variety of algorithms for detecting graphical objects in the documents. Numerous benchmark datasets are also available in this domain to evaluate the performance of newly developed algorithms. In this paper, we introduce a new dataset, IIIT-AR-13K for graphical object detection.

Before we describe the details of the new dataset, we make the following observations:

1. State of the art algorithms (such as [7,13,14,21,22,24,26]) for graphical object detection are motivated by the success of object detection in computer vision (such as Faster R-CNN and Mask R-CNN). However, the high accuracy expectations in documents (similar to the high accuracy expectations in OCR) make the graphical object detection problem, demanding and thereby different compared to that of detecting objects in natural images.
2. With documents getting digitized extensively in many business workflows, automatic processing of business documents has received significant attention in recent years. However, most of the existing datasets for graphical object detection are still created from scientific documents such as technical papers.
3. In recent years, we have started to see large automatically annotated or synthetically created datasets for graphical object detection. We hypothesize that, in high accuracy regime, carefully curated small data adds more value than automatically generated large datasets. At least, in our limited setting, we validate this hypothesis later in this paper.

Popular datasets for graphical object detection (or more specifically table detection) are ICDAR 2013 table competition dataset [8] (i.e., ICDAR-2013), ICDAR 2017 competition on page object detection dataset [5] (i.e., ICDAR-POD-2017), cTDaR [6], UNLV [23], Marmot table recognition dataset [4], DeepFigures [25], PubLayNet [30], and TableBank [15]. Most of these existing datasets are limited with respect to their size (except DeepFigures, PubLayNet, and TableBank) or category labels (except ICDAR-POD-2017, DeepFigures, and PubLayNet). Most of them (except ICDAR-2013, cTDaR, and UNLV) consist of only scientific research articles, handwritten documents, and e-books. Therefore, they are limited in variations in layouts and structural variations in appearances.

On the contrary, business documents such as annual reports, pay slips, bills, receipt copies, etc. of the various companies are more heterogeneous to their layouts and complex visual appearance of the graphical objects. In such kind of heterogeneous documents, the detection of graphical objects becomes further difficult.

In this paper, we introduce a new dataset, named IIIT-AR-13K for localizing graphical objects in the annual reports (a specific type of business documents) of various companies. For this purpose, we randomly select publicly available annual reports in English and other languages (e.g., French, Japanese, Russian, etc.) of multiple (more than ten) years of twenty-nine different companies. We manually annotate the bounding boxes of five different categories of graphical objects (i.e., table, figure, natural image, logo, and signature), which frequently appear in the annual reports. IIIT-AR-13K dataset contains 13K annotated pages with 16K tables, 3K figures, 3K natural images, 0.5K logos, and 0.6K signatures. To the best of the author's knowledge, the newly created dataset is the largest among all the existing datasets where ground truths are annotated manually for detecting graphical objects (more than one category) in documents.

We use FASTER R-CNN [20] and MASK R-CNN [11] algorithms to benchmark the newly created dataset for graphical object detection task in business documents. Experimentally, we observe that the creation of a model trained with IIIT-AR-13K dataset achieves better performance than the model trained with the larger existing datasets. From the experiments, we also observe that the model trained with the larger existing datasets achieves the best performance by fine-tuning with a small number (only 1K) of images from IIIT-AR-13K dataset.

Our major contributions/claims are summarized as follows:

- We introduce a highly variate new dataset for localizing graphical objects in documents. The newly created dataset is the largest among the existing datasets where ground truth is manually annotated for the graphical object detection task. It also has a larger label space compared to most existing datasets.
- We establish FASTER R-CNN and MASK R-CNN based benchmarks on the popular datasets. We report very high quantitative detection rates on the popular benchmarks.
- Though smaller than some of the recent datasets, this dataset is more effective as training data for the detection task due to the inherent variations in the object categories. This is empirically established by creating a unique model trained with IIIT-AR-13K dataset to detect graphical objects in all existing datasets.
- Models trained with the larger existing datasets achieve the best performance after fine-tuning with a very limited number (only 1K) of images from IIIT-AR-13K dataset.

2 Preliminaries

Graphical objects such as tables, various types of figures (e.g., bar chart, pie chart, line plot, natural image, etc.), equations, and logos in documents contain valuable information in a compact form. Understanding of document images requires localizing of such graphical objects as an initial step. Several datasets (e.g., ICDAR-2013 [8], ICDAR-POD-2017 [5], cTDaR [6], UNLV [23], Marmot [4], DeepFigures [25], PubLayNet [30], and TableBank [15]) exist in the literature, which are dedicated to localize graphical objects (more specifically tables) in the document images. Ground truth bounding boxes of ICDAR-2013, ICDAR-POD-2017, cTDaR, UNLV, and Marmot are annotated manually. Ground truth bounding boxes of DeepFigures, PubLayNet, and TableBank are generated automatically. Among these datasets, only ICDAR-POD-2017, DeepFigures, and PubLayNet are aimed to address localizing a wider class of graphical objects. Other remaining datasets are designed only for localizing tables. Some of these datasets (e.g., ICDAR-2013, cTDaR, UNLV, and TableBank) are also used for table structure recognition and table recognition tasks in the literature. Some other existing datasets such as SciTSR [2], Table2Latex [3], and PubTabNet [29] are used exclusively for table structure recognition and table recognition purpose in the literature.

2.1 Related Datasets

ICDAR-2013 [8]: This dataset is one of the most cited datasets for the task of table detection and table structure recognition. It contains 67 PDFs which corresponds to 238 page images. Among them, 40 PDFs are taken from the US Government and 27 PDFs from EU. Among 238 page images, only 135 page images contain tables, in total 150 tables. This dataset is popularly used to evaluate the algorithms for both table detection and structure recognition task.

ICDAR-POD-2017 [5]: It focuses on the detection of various graphical objects (e.g., tables, equations, and figures) in the document images. It is created by annotating 2417 document images selected from 1500 scientific papers of Cite-Seer. It includes a large variety in page layout - single-column, double-column, multi-column, and a significant variation in the object structure. This dataset is divided into (i) training set consisting of 1600 images, and (ii) test set comprising 817 images.

cTDaR [6]: This dataset consists of modern and archival documents with various formats, including document images and born-digital formats such as PDF. The images show a great variety of tables from hand-drawn accounting books to stock exchange lists and train timetable, from record books of prisoner lists, tables from printed books, production census, etc. The modern documents consist of scientific journals, forms, financial statements, etc. Annotations correspond to table regions, and cell regions are available. This dataset contains (i) training set - consists of 600 annotated archival and 600 annotated modern document images with table bounding boxes, and (ii) test set - includes 199 annotated archival and 240 annotated modern document images with table bounding boxes.

Marmot [4]: It consists of 2000 Chinese and English pages at the proportion of about 1:1. Chinese pages are selected from over 120 e-Books with diverse subject areas provided by Founder Apabi library. Not more than 15 pages are selected from each Book. English pages are selected over 1500 journal and conference papers published during 1970–2011 from the Citeseer website. The pages show a great variety in layout - one-column and two-column, language type, and table styles. This dataset is used for both table detection and structure recognition tasks.

UNLV [23]: It contains 2889 document images of various categories: technical reports, magazines, business letters, newspapers, etc. Among them, only 427 images include 558 table zones. Annotations for table regions and cell regions are available. This dataset is also used for both table detection and table structure recognition tasks.

DeepFigures [25]: Siegel *et al.* [25] create this dataset by automatically annotating pages of two large web collections of scientific documents (arXiv and PubMed). This dataset consists of 5.5 million pages with 1.4 million tables and 4.0 million figures. It can be used for detecting tables and figures in the document images.

PubLayNet [30]: *and* ***PubTabNet*** [29] PubLayNet consists of 360K page images with annotation of various layout elements such as text, title, list, figure, and table. It is created by automatically annotating 1 million PubMed Central^{TM} PDF articles. It contains 113K table regions and 126K figure regions in total. It is mainly used for layout analysis purposes. However, it can also be used for table and figure detection tasks. PubTabNet is the largest dataset for the table recognition task. It consists of 568K images of heterogeneous tables extracted from scientific articles (in PDF format). Ground truth corresponds to each table image represents structure information and text content of each cell in HTML format.

TableBank [15]: This dataset contains 417K high-quality documents with tables. This dataset is created by downloading LaTex source code from the year 2014 to the year 2018 through bulk data access in arXiv. It consists of 163K word document, 253K LaTex document and in total 417K document images with annotated table regions. Structure level annotation is available for 57K word documents, 88K LaTex documents and in total 145K document images.

2.2   Related Work in Graphical Object Detection

Localizing graphical objects (e.g., tables, figures, mathematical equations, logos, signatures, etc.) is the primary step for understanding any documents. Recent advances in object detection in natural scene images using deep learning inspire researchers [7,13,14,16,18,21,22,24,26] to develop deep learning based algorithms for detecting graphical objects in documents. In this regards, various researchers like Gilani *et al.* [7], Schreiber *et al.* [22], Siddiqui *et al.* [24] and Sun *et al.* [26] employ Faster R-CNN [20] model for detecting table in documents. In [13], the authors use YOLO [19] to detect tables in documents. Li *et al.* [16] use Generative Adversarial Networks (GAN) [9] to extract layout feature to improve table detection accuracy.

Few researchers [14,21] focus on detection of various kinds of graphical objects (not just tables). Kavasidis *et al.* [14] propose saliency based technique to detect three types of graphical objects—table, figure and mathematical equation. In [21], Mask R-CNN is explored to detect various graphical objects.

2.3   Baseline Methods for Graphical Object Detection

We use Faster R-CNN [20] and Mask R-CNN [11] as baselines for detecting graphical objects in annual reports. We use publicly available implementations of Faster

R-CNN [28] and MASK R-CNN [1] for this experiment. We train both the models with the images of training set of IIIT-AR-13K dataset for the empirical studies.

Faster R-CNN: The model is implemented using PyTorch; trained and evaluated on NVIDIA TITAN X GPU with 12 GB memory with batch size of 4. The input images are resized to a fixed size of 800 × 1024 by preserving original aspect ratio. We use the pre-trained ResNet-101 [12] backbone on MS-COCO [17] dataset. We use five different anchor scales (i.e., 32, 64, 128, 256, and 512) and five anchor ratios (i.e., 1, 2, 3, 4, and 5) so that the region proposals can cover almost every part of the image irrespective of the image size. We use stochastic gradient descent (SGD) as an optimizer with initial learning rate = 0.001 and the learning rate decays after every 5 epochs and it is equal to 0.1 times of the previous value.

Mask R-CNN: Every input image is rescaled to a fixed size of 800 × 1024 preserving the original aspect ratio. We use the pre-trained ResNet-101 [12] backbone on MS-COCO [17] dataset. We use Tensorflow/Keras for the implementation, and train and evaluate on NVIDIA TITAN X GPU that has 12 GB memory with a batch size of 1. We further use 0.5, 1, and 2 as the anchor scale values and anchor box sizes of 32, 64, 128, 256, and 512. Further, we train a total of 80 epochs. We train all FPN and subsequent layers for first 20 epochs, the next 20 epochs for training FPN + last 4 layers of ResNet-101; and last 40 epochs for training all layers of the model. During the training process, we use 0.001 as the learning rate, 0.9 as the momentum and 0.0001 as the weight decay.

Fig. 1. Sample annotated document images of IIIT-AR-13K dataset. **Dark Green:** indicates ground truth bounding box of table, **Dark Red:** indicates ground truth bounding box of figure, **Dark Blue:** indicates ground truth bounding box of natural image, **Dark Yellow:** indicates ground truth bounding box of logo and **Dark Pink:** indicates ground truth bounding box of signature. (Color figure online)

3   IIIT-AR-13K Dataset

3.1   Details of the Dataset

For detecting various types of graphical objects (e.g., table, figure, natural image, logo, and signature) in the business documents, we generate a new dataset, named IIIT-AR-13K. The newly generated dataset consists of 13K pages of publicly available annual reports. Annual reports in English and non-English languages (e.g., French, Japanese, Russian, etc.) of multiple (more than ten) years of twenty-nine different companies. Annual reports contain various types of graphical objects such as tables, various types of graphs (e.g., bar chart, pie chart, line plot, etc.), images, companies' logos, signatures, stamps, sketches, etc. However, we only consider five categories of graphical objects, e.g., table, figure (including both graphs and sketches), natural image (including images), logo, and signature.

Table 1. Statistics of our newly generated IIIT-AR-13K dataset.

Object category	IIIT-AR-13K			
	Training	Validation	Test	Total
Page	$9333 \approx 9K$	$1955 \approx 2K$	$2120 \approx 2K$	$13415 \approx 13K$
Table	$11163 \approx 11K$	$2222 \approx 2K$	$2596 \approx 3K$	$15981 \approx 16K$
Figure	$2004 \approx 2K$	$481 \approx 0.4K$	$463 \approx 0.4K$	$2948 \approx 3K$
Natural image	$1987 \approx 2K$	$438 \approx 0.4K$	$455 \approx 0.4K$	$2880 \approx 3K$
Logo	$379 \approx 0.3K$	$67 \approx 0.06K$	$135 \approx 0.1K$	$581 \approx 0.5K$
Signature	$420 \approx 0.4K$	$108 \approx 0.9K$	$92 \approx 0.09K$	$620 \approx 0.6K$

Annual reports in English and other languages of more than ten years of twenty-nine different companies are selected to increase diversity with respect to the language, layout, and each category of graphical objects. We manually annotate the bounding boxes of each type of graphical object. Figure 1 shows some sample bounding box annotated images. Finally, we have 13K annotated pages with 16K tables, 3K figures, 3K natural images, 0.5K logos, and 0.6K signatures. The dataset is divided into (i) training set consisting of 9K document images, (ii) validation set containing 2K document images, and (iii) test set consists of 2K document images. For every company, we randomly choose 70%, 15%, and remaining 15% of the total pages as training, validation, and test, respectively. Table 1 shows the statistics of the newly generated dataset.

3.2 Comparison with the Existing Datasets

Table 2 presents the comparison of our dataset with the existing datasets. From the table, it is clear that with respect to label space, all the existing datasets (except ICDAR-POD-2017, DeepFigures, and PubLayNet) containing only one object category, i.e., table, are subsets of our newly generated IIIT-AR-13K dataset (containing five object categories). Most of the existing datasets (except ICDAR-2013, cTDaR, and UNLV) consist of only scientific research articles, handwritten documents, and e-books. The newly generated dataset includes annual reports, one specific type of business document. It is the largest among the existing datasets where ground truths are annotated manually for the detection task. It consists of a large variety of pages of annual reports compared to scientific articles of existing datasets.

Table 2. Statistics of existing datasets along with newly generated dataset for graphical object detection task in document images. **T:** indicates table. **F:** indicates figure. **E:** indicates equation. **NI:** indicates natural image. **L:** indicates logo. **S:** indicates signature. **TL:** indicates title. **TT:** indicates text. **LT:** indicates list. **POD:** indicates page object detection. **TD:** indicates table detection. **TSR:** indicates table structure recognition. **TR:** indicates table recognition.

Dataset	Category label	#Images	Task				#Tables
			POD	TD	TSR	TR	
ICDAR-2013 [8]	1: T	238	✓	✓	✓	✓	150
ICDAR-POD-2017 [5]	3: T, F, E	2417	✓	✓	×	×	1020
cTDaR [6]	1: T	2K	✓	✓	✓	✓	3.6K
Marmot [4]	1: T	2K	✓	✓	×	×	958
UNLV [23]	1: T	427	✓	✓	✓	✓	558
IIIT-AR-13k	**5: T, F, NI, L, S**	**13K**	✓	✓	×	×	**16K**
DeepFigures [25][a]	2: T, F	5.5M	✓	✓	×	×	1.4M
PubLayNet [30][a]	5: T, F, TL, TT, LT	360K	✓	✓	×	×	113K
TableBank [15][a]	1: T	417K	✓	✓	×	×	417K
			×	×	✓	×	145K
SciTSR [2][b]	1: T	–	×	×	✓	✓	15K
Table2Latex [3][b]	1: T	–	×	×	✓	✓	450K
PubTabNet [29][b]	1: T	–	×	×	✓	✓	568K

[a]Ground truth bounding boxes are annotated automatically.
[b]Dataset is dedicated only for table structure recognition and table recognition.

3.3 Diversity in Object Category

We select annual reports (more than ten years) of twenty-nine different companies for creating this dataset. Due to the consideration of annual reports of several years of various companies, there is a significant variation in layouts (e.g., single-column, double-column, and triple-column) and graphical objects like tables, figures, natural images, logos, and signatures. The selected documents are heterogeneous. Heterogeneity increases the variability within each object category and also increases the similarity between object categories. A significant variation in layout structure, the similarity between object categories, and diversity within object categories make this dataset more complex for graphical detection tasks. Figure 2 illustrates the significant variations in table structures and figures in the newly generated dataset.

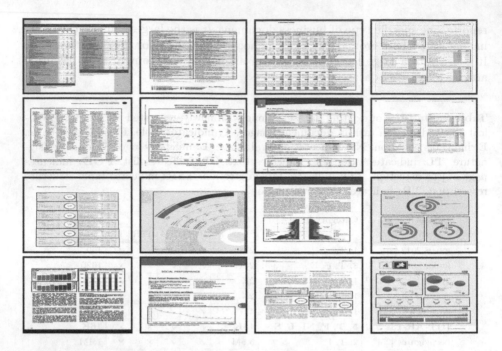

Fig. 2. Sample annotated images with large variation in tables and figures.

3.4 Performance for Detection Using Baseline Methods

Quantitative results of detection on the validation and test sets of IIIT-AR-13K dataset using baseline approaches - Faster R-CNN and Mask R-CNN are summarized in Table 3. From the table, it is observed that Mask R-CNN produces better results than Faster R-CNN. Among all the object categories, both the baselines obtain the best results for table and worst results for logo. This is because of highly imbalanced training set (11K tables and 0.3K logos).

3.5 Effectiveness of IIIT-AR-13K over Existing Larger Datasets

The newly generated dataset IIIT-AR-13K is smaller (with respect to number of document images) than some of the existing datasets (e.g., DeepFigures, TableBank, and PubLayNet) for the graphical object (i.e., table) detection task. We establish the effectiveness of the smaller IIIT-AR-13K dataset over the larger existing datasets - DeepFigures, TableBank, and PubLayNet for graphical object (i.e., table common object category of all datasets) detection task. In [15], the authors experimentally show that TableBank dataset is more effective than the largest DeepFigures dataset for table detection. In this paper, we use TableBank and PubLayNet datasets to establish the effectiveness of the newly generated dataset over the existing datasets for table detection. For this purpose, Mask R-CNN model is trained with training set of each of three datasets - TableBank,

Table 3. Quantitative results of two baseline approaches for detecting graphical objects. **R:** indicates Recall. **P:** indicates Precision. **F:** indicates F-measure. **mAP:** indicates mean average precision.

Dataset	Category	Faster R-CNN				Mask R-CNN			
		R↑	P↑	F↑	mAP↑	R↑	P↑	F↑	mAP↑
Validation	Table	0.9571	0.9260	0.9416	0.9554	**0.9824**	**0.9664**	**0.9744**	**0.9761**
	Figure	0.8607	0.7800	0.8204	0.8103	**0.8699**	**0.8326**	**0.8512**	**0.8391**
	Natural Image	0.9027	0.8607	0.8817	0.8803	**0.9461**	**0.8820**	**0.9141**	**0.9174**
	Logo	0.8566	0.4063	0.6315	0.6217	**0.8852**	**0.4122**	**0.6487**	**0.6434**
	Signature	0.9411	0.8000	0.8705	0.9135	**0.9633**	**0.8400**	**0.9016**	**0.9391**
	Average	0.9026	0.7546	0.8291	0.8362	**0.9294**	**0.7866**	**0.8580**	**0.8630**
Test	Table	0.9512	0.9234	0.9373	0.9392	**0.9711**	**0.9715**	**0.9713**	**0.9654**
	Figure	0.8530	0.7582	0.8056	0.8332	**0.8898**	**0.7872**	**0.8385**	**0.8686**
	Natural Image	0.8745	0.8631	0.8688	0.8445	**0.9179**	**0.8625**	**0.8902**	**0.8945**
	Logo	0.5933	0.3606	0.4769	0.4330	**0.6330**	**0.3920**	**0.5125**	**0.4699**
	Signature	0.8868	0.7753	0.8310	0.8981	**0.9175**	**0.7876**	**0.8525**	**0.9115**
	Average	0.8318	0.7361	0.7839	0.7896	**0.8659**	**0.7601**	**0.8130**	**0.8220**

PubLayNet, and IIIT-AR-13K. These trained models are individually evaluated on test set of existing datasets - ICDAR-2013, ICDAR-POD-2017, cTDaR, UNLV, Marmot, PubLayNet. We name these experiments as

(i) **Experiment-I:** Mask R-CNN trained with TableBank (LaTex) dataset.
(ii) **Experiment-II:** Mask R-CNN trained with TableBank (Word) dataset.
(iii) **Experiment-III:** Mask R-CNN trained with TableBank (LaTex+Word) dataset.
(iv) **Experiment-IV:** Mask R-CNN trained with PubLayNet dataset.
(v) **Experiment-V:** Mask R-CNN trained with IIIT-AR-13K dataset.

Observation-I - Without Fine-Tuning: To establish effectiveness of the newly generated IIIT-AR-13K dataset over larger existing datasets, we do experiments: Experiment-I, Experiment-II, Experiment-III, Experiment-IV, and Experiment-V. Table 4 illustrates quantitative results of these experiments. The table highlights that Experiment-V produces the best results (with respect to F-measure and mAP) for cTDaR, UNLV, and IIIT-AR-13K datasets. Though Experiment-I produces the best mAP value for Marmot dataset, the performance of Experiment-V on this dataset is very close to the performance of Experiment-I. For ICDAR-2017 dataset, the performance of Experiment-V is lower than Experiment-I and Experiment-III with respect to F-measure. The table highlights that, even though IIIT-AR-13K is significantly smaller than TableBank and PubLayNet, it is more effective for training a model for detecting tables in the existing datasets.

Figure 3 shows the predicted bounding boxes of tables in the pages of several datasets using three different models. Here, dark Green, Pink, Cyan, and Light Green colored rectangles highlight the ground truth and predicted bounding

Table 4. Performance of table detection in the existing datasets. Model is trained with only training images containing tables of the respective datasets. **TBL:** indicates TableBank (LaTex). **TBW:** indicates TableBank (Word). **TBLW:** indicates TableBank (LaTex+Word). **R:** indicates Recall. **P:** indicates Precision. **F:** indicates F-measure. **mAP:** indicates mean average precision.

Test dataset	Training dataset	Quantitative score			
		R↑	P↑	F↑	mAP↑
ICDAR-2013	TBL	0.9454	0.9341	**0.9397**	0.9264
	TBW	0.9090	**0.9433**	0.9262	0.8915
	TBLW	0.9333	0.9277	0.9305	0.9174
	PubLayNet	0.9272	0.8742	0.9007	0.9095
	IIIT-AR-13K	**0.9575**	0.8977	0.9276	**0.9393**
ICDAR-2017	TBL	0.9274	0.7016	0.8145	0.8969
	TBW	0.8107	0.5908	0.7007	0.7520
	TBLW	**0.9369**	**0.7406**	**0.8387**	**0.9035**
	PubLayNet	0.8296	0.6368	0.7332	0.7930
	IIIT-AR-13K	0.8675	0.6311	0.7493	0.7509
cTDaR	TBL	0.7006	0.7208	0.71078	0.5486
	TBW	0.6405	0.7582	0.6994	0.5628
	TBLW	0.7230	0.7481	0.7356	0.5922
	PubLayNet	0.6391	0.7231	0.6811	0.5610
	IIIT-AR-13K	**0.8097**	**0.8224**	**0.8161**	**0.7478**
UNLV	TBL	0.5806	0.6612	0.6209	0.5002
	TBW	0.5035	0.8005	0.6520	0.4491
	TBLW	0.6362	0.6787	0.6574	0.5513
	PubLayNet	0.7329	0.8329	0.7829	0.6950
	IIIT-AR-13K	**0.8602**	**0.7843**	**0.8222**	**0.7996**
Marmot	TBL	0.8860	**0.8840**	0.8850	**0.8465**
	TBW	0.8423	0.8808	0.8615	0.8026
	TBLW	**0.8919**	0.8802	**0.8860**	0.8403
	PubLayNet	0.8549	0.8214	0.8382	0.7987
	IIIT-AR-13K	0.8852	0.8075	0.8464	0.8464

boxes of tables using Experiment-I, Experiment-IV, and Experiment-V, respectively. In those images, Experiment-V correctly detects the tables. At the same time, either Experiment-I or Experiment-IV or Experiment-I and Experiment-IV fails to predict the bounding boxes of the tables accurately.

Observation-II - Fine-Tuning with Complete Training Set: From Table 5, it is also observed that the performance of Experiment-I, Experiment-II, Experiment-III, and Experiment-IV on the validation set and test set of

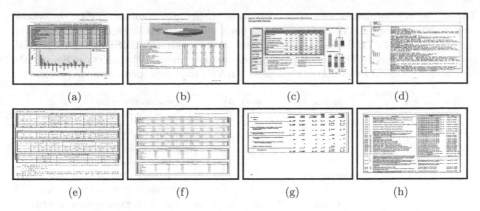

(a)   (b)   (c)   (d)

(e)   (f)   (g)   (h)

Fig. 3. Few examples of predicted bounding boxes of tables in various datasets - (a) ICDAR-2013, (b) ICADR-POD-2017, (c) CTDaR, (d) UNLV, (e) Marmot, (f) PubLayNet, (g) IIIT-AR-13K (validation), (h) IIIT-AR-13K (test) using **Experiment-I, Experiment-IV** and **Experiment-V. Experiment-I:** Mask R-CNN trained with TableBank (LaTex) dataset. **Experiment-IV:** Mask R-CNN trained with PubLayNet dataset. **Experiment-V:** Mask R-CNN trained with IIIT-AR-13K dataset. **Dark Green:** rectangle highlights the ground truth bounding boxes of tables. **Pink, Cyan and Light Green:** rectangles indicate the predicted bounding boxes of tables using **Experiment-I, Experiment-IV** and **Experiment-V**, respectively.

IIIT-AR-13K dataset is significantly worse (10% in case of F-measure and 15% in case of mAP) than the performance of Experiment-V. On the contrary, when we fine-tune with the training set of IIIT-AR-13K dataset, all the fine-tuning experiments obtain similar outputs compared to Experiment-V. In the case of the validation set of PubLayNet dataset, Experiment-IV outperforms Experiment-II and Experiment-V; and the performances of Experiment-I and Experiment-III are very close to the performance of Experiment-IV. When we fine-tune with the training set of PubLayNet dataset, the fine-tuned model achieves similar performance compared to Experiment-IV. These experiments highlight that fine-tuning with the complete training set is effective for both the larger existing datasets and as well as IIIT-AR-13K dataset.

Observation-III - Fine-Tuning with Partial Training Set: From Table 5, we observe that the performance of the trained model using TableBank and PubLayNet on validation and test sets of IIIT-AR-13K dataset is (10–20%) less than the performance of model trained with training set of IIIT-AR-13K. When we fine-tune the models with a partial training set of only 1K randomly selected images of IIIT-AR-13K, the models also achieve similar outputs (see Table 5). On the other hand, the performance of the trained model with IIIT-AR-13K dataset on validation set of PubLayNet is (10–15%) is less than models trained with TableBank and PubLayNet. When we fine-tune the model (trained with IIIT-AR-13K dataset) with the complete training set of PubLayNet, the model obtains very close output to the models trained with TableBank and PubLayNet.

Table 5. Performance of table detection using fine-tuning. **TBL:** indicates TableBank (LaTex). **TBW:** indicates TableBank (Word). **TBLW:** indicates TableBank (LaTex+Word). **F-TRS:** indicates complete training set. **P-TRS:** indicates partial training set i.e., only 1K randomly selected training images. **R:** indicates Recall. **P:** indicates Precision. **F:** indicates F-measure. **mAP:** indicates mean average precision.

Test dataset	Training dataset	Fine-tune	Quantitative score			
			R↑	P↑	F↑	mAP↑
PubLayNet (validation)	TBL		0.9859	0.6988	0.8424	0.9461
	TBW		0.9056	0.7194	0.8125	0.8276
	TBLW		0.9863	0.7258	0.8560	0.9557
	PubLayNet		**0.9886**	0.7780	0.8833	0.9776
	IIIT-AR-13K		0.9555	0.5530	0.7542	0.8696
	IIIT-AR-13K	PubLayNet(F-TRS)	0.9882	**0.7898**	**0.8890**	**0.9781**
	IIIT-AR-13K	PubLayNet(P-TRS)	0.9869	0.4653	0.7261	0.9712
IIIT-AR-13K (validation)	TBL		0.8109	0.7370	0.7739	0.7533
	TBW		0.7641	0.8214	0.7928	0.7230
	TBLW		0.8217	0.7345	0.7781	0.7453
	PubLayNet		0.8100	0.7092	0.7596	0.7382
	IIIT-AR-13K		**0.9891**	0.7998	0.8945	0.9764
	TBL	IIIT-AR-13K(F-TRS)	0.9869	0.8600	0.9234	0.9766
	TBW		0.9815	0.8065	0.8940	0.9639
	TBLW		0.9873	0.8614	0.9244	**0.9785**
	PubLayNet		0.9842	0.8344	0.9093	0.9734
	TBL	IIIT-AR-13K(P-TRS)	0.9747	0.8790	0.9269	0.9648
	TBW		0.9734	0.8767	0.9251	0.9625
	TBLW		0.9783	**0.8998**	**0.9391**	0.9717
	PubLayNet		0.9797	0.8732	0.9264	0.9687
IIIT-AR-13K (test)	TBL		0.8023	0.7428	0.7726	0.7400
	TBW		0.7704	0.8396	0.8050	0.7276
	TBLW		0.8278	0.7500	0.7889	0.7607
	PubLayNet		0.8093	0.7302	0.7697	0.7313
	IIIT-AR-13K		**0.9826**	0.8361	0.9093	0.9688
	TBL	IIIT-AR-13K(F-TRS)	0.9761	0.8774	0.9267	0.9659
	TBW		0.9753	0.8239	0.8996	0.9596
	TBLW		0.9776	0.8788	0.9282	**0.9694**
	PubLayNet		0.9753	0.8571	0.9162	0.9642
	TBL	IIIT-AR-13K(P-TRS)	0.9637	0.9022	0.9330	0.9525
	TBW		0.9699	0.8922	0.9311	0.9615
	TBLW		0.9718	**0.9091**	**0.9405**	0.9629
	PubLayNet		0.9684	0.8905	0.9294	0.9552

But when we fine-tune the model (trained with IIIT-AR-13K dataset) with the partial training set (only 1K images) of PubLayNet, the model is unable to obtain similar output (with respect to F-measure) to the models trained with TableBank and PubLayNet. These experiments also highlight that the newly created IIIT-AR-13K dataset is more effective for fine-tuning.

4   Summary and Observations

This paper presents a new dataset IIIT-AR-13K for detecting graphical objects in business documents, specifically annual reports. It consists of 13K pages of annual reports of more than ten years of twenty-nine different companies with bounding box annotation of five different object categories—table, figure, natural image, logo, and signature.

- The newly generated dataset is the largest manually annotated dataset for graphical object detection purpose, at this stage.
- This dataset has more labels and diversity compared to most of the existing datasets.
- Though IIIT-AR-13K is smaller than the existing automatic annotated datasets—DeepFigures, PubLayNet, and TableNet, the model trained with IIIT-AR-13K performs better than the model trained with larger datasets for detecting tables in document images in most cases.
- Models trained with the existing datasets also achieve better performance by fine-tuning with a limited number of training images from IIIT-AR-13K.

We believe that this dataset will aid research in detecting tables and other graphical objects in business documents.

References

1. Abdulla, W.: Mask R-CNN for object detection and instance segmentation on Keras and Tensorflow. GitHub repository (2017)
2. Chi, Z., Huang, H., Xu, H.D., Yu, H., Yin, W., Mao, X.L.: Complicated table structure recognition. arXiv (2019)
3. Deng, Y., Rosenberg, D., Mann, G.: Challenges in end-to-end neural scientific table recognition. In: ICDAR (2019)
4. Fang, J., Tao, X., Tang, Z., Qiu, R., Liu, Y.: Dataset, ground-truth and performance metrics for table detection evaluation. In: WDAS (2012)
5. Gao, L., Yi, X., Jiang, Z., Hao, L., Tang, Z.: ICDAR 2017 competition on page object detection. In: ICDAR (2017)
6. Gao, L., et al.: ICDAR 2019 competition on table detection and recognition (cTDaR). In: ICDAR (2019)
7. Gilani, A., Qasim, S.R., Malik, I., Shafait, F.: Table detection using deep learning. In: ICDAR (2017)
8. Göbel, M., Hassan, T., Oro, E., Orsi, G.: ICDAR 2013 table competition. In: ICDAR (2013)
9. Goodfellow, I., et al.: Generative adversarial nets. In: NIPS (2014)
10. Hao, L., Gao, L., Yi, X., Tang, Z.: A table detection method for PDF documents based on convolutional neural networks. In: Workshop on DAS (2016)
11. He, K., Gkioxari, G., Dollár, P., Girshick, R.: Mask R-CNN. In: ICCV (2017)
12. He, K., Zhang, X., Ren, S., Sun, J.: Deep residual learning for image recognition. In: CVPR (2016)
13. Huang, Y., et al.: A YOLO-based table detection method. In: ICDAR (2019)

14. Kavasidis, I., et al.: A saliency-based convolutional neural network for table and chart detection in digitized documents. In: Ricci, E., Rota Bulò, S., Snoek, C., Lanz, O., Messelodi, S., Sebe, N. (eds.) ICIAP 2019. LNCS, vol. 11752, pp. 292–302. Springer, Cham (2019). https://doi.org/10.1007/978-3-030-30645-8_27
15. Li, M., Cui, L., Huang, S., Wei, F., Zhou, M., Li, Z.: TableBank: table benchmark for image-based table detection and recognition. In: ICDAR (2019)
16. Li, Y., Yan, Q., Huang, Y., Gao, L., Tang, Z.: A GAN-based feature generator for table detection. In: ICDAR (2019)
17. Lin, T.-Y., et al.: Microsoft COCO: common objects in context. In: Fleet, D., Pajdla, T., Schiele, B., Tuytelaars, T. (eds.) ECCV 2014. LNCS, vol. 8693, pp. 740–755. Springer, Cham (2014). https://doi.org/10.1007/978-3-319-10602-1_48
18. Melinda, L., Bhagvati, C.: Parameter-free table detection method. In: ICDAR (2019)
19. Redmon, J., Divvala, S., Girshick, R., Farhadi, A.: You only look once: unified, real-time object detection. In: CVPR (2016)
20. Ren, S., He, K., Girshick, R., Sun, J.: Faster R-CNN: towards real-time object detection with region proposal networks. In: NIPS (2015)
21. Saha, R., Mondal, A., Jawahar, C.V.: Graphical object detection in document images. In: ICDAR (2019)
22. Schreiber, S., Agne, S., Wolf, I., Dengel, A., Ahmed, S.: Deepdesrt: deep learning for detection and structure recognition of tables in document images. In: ICDAR (2017)
23. Shahab, A., Shafait, F., Kieninger, T., Dengel, A.: An open approach towards the benchmarking of table structure recognition systems. In: DAS (2010)
24. Siddiqui, S.A., Malik, M.I., Agne, S., Dengel, A., Ahmed, S.: DeCNT: deep deformable CNN for table detection. IEEE Access 6, 74151–74161 (2018)
25. Siegel, N., Lourie, N., Power, R., Ammar, W.: Extracting scientific figures with distantly supervised neural networks. In: ACM/IEEE on Joint Conference on Digital Libraries (2018)
26. Sun, N., Zhu, Y., Hu, X.: Faster R-CNN based table detection combining corner locating. In: ICDAR (2019)
27. Tran, D.N., Tran, T.A., Oh, A., Kim, S.H., Na, I.S.: Table detection from document image using vertical arrangement of text blocks. Int. J. Contents 11, 77–85 (2015)
28. Yang, J., Lu, J., Batra, D., Parikh, D.: A faster Pytorch implementation of faster R-CNN (2017)
29. Zhong, X., ShafieiBavani, E., Yepes, A.J.: Image-based table recognition: data, model, and evaluation. arXiv (2019)
30. Zhong, X., Tang, J., Yepes, A.J.: PubLayNet: largest dataset ever for document layout analysis. In: ICDAR (2019)

Page Segmentation Using Convolutional Neural Network and Graphical Model

Xiao-Hui Li[1,2], Fei Yin[1], and Cheng-Lin Liu[1,2,3(✉)]

[1] National Laboratory of Pattern Recognition,
Institute of Automation of Chinese Academy of Sciences,
95 Zhongguancun East Road, Beijing 100190, People's Republic of China
{xiaohui.li,fyin,liucl}@nlpr.ia.ac.cn
[2] School of Artificial Intelligence, University of Chinese Academy of Sciences,
Beijing 100049, People's Republic of China
[3] CAS Center for Excellence of Brain Science and Intelligence Technology, Beijing,
People's Republic of China

Abstract. Page segmentation of document images remains a challenge due to complex layout and heterogeneous image contents. Existing deep learning based methods usually follow the general semantic segmentation or object detection frameworks, without plentiful exploration of document image characteristics. In this paper, we propose an effective method for page segmentation using convolutional neural network (CNN) and graphical model, where the CNN is powerful for extracting visual features and the graphical model explores the relationship (spatial context) between visual primitives and regions. A page image is represented as a graph whose nodes represent the primitives and edges represent the relationships between neighboring primitives. We consider two types of graphical models: graph attention network (GAT) and conditional random field (CRF). Using a convolutional feature pyramid network (FPN) for feature extraction, its parameters can be estimated jointly with the GAT. The CRF can be used for joint prediction of primitive labels, and combined with the CNN and GAT. Experimental results on the PubLayNet dataset show that our method can extract various page regions with precise boundaries. The comparison of different configurations show that GAT improves the performance when using shallow backbone CNN, but the improvement with deep backbone CNN is not evident, while CRF is always effective to improve, even when combining on top of GAT.

Keywords: Page segmentation · Graph attention network · Conditional random field · Feature pyramid network

1 Introduction

Page segmentation, or layout analysis, plays an important role in document image understanding and information extraction. It is aimed at segmenting the entire document images into small regions with homogeneous contents and high

© Springer Nature Switzerland AG 2020
X. Bai et al. (Eds.): DAS 2020, LNCS 12116, pp. 231–245, 2020.
https://doi.org/10.1007/978-3-030-57058-3_17

level semantics such as paragraphs, titles, lists, tables and figures. Page segmentation is a prerequisite for many following applications such as text line transcription, table structure analysis and figure classification, etc.

Although many efforts have been made, page segmentation remains an open problem due to complex document layout and heterogeneous image contents. Traditional methods [1] either split the document images into smaller regions progressively (top-down) or group small elements (pixels or connected components (CCs)) into larger regions (bottom-up). They depend on sophisticated handcrafted features and heuristic rules which are hard to design and prone to errors. On the contrary, deep learning based methods can automatically learn useful features from raw images, thus becoming the mainstream in the research community, recently. However, existing deep learning based methods usually follow the general semantic segmentation or object detection frameworks designed for natural scene images, leaving the intrinsic characteristics of document images ignored. As has been declared in [2], document images are very different from natural scene images in many aspects such as richness of color, organization form of objects or regions, diversity of region scales and aspect ratios, confusion of foreground and background pixels, etc. These differences limit the performance of existing deep learning based page segmentation methods on the root cause.

In this paper, we propose a conceptually simple and intuitive method for page segmentation which takes the intrinsic characteristics of document images into consideration. Specifically, we consider page regions as logical clusters of small primitives which are shared across different region categories. In our system, each document image is formulated as a graph whose nodes represent the primitives and edges represent the relationships between neighbouring primitives. The nodes and edges are then classified into different classes using their features learned with deep convolution neural networks. To improve the classification accuracy, contextual information is also integrated using two graphical models: graph attention networks (GAT) and conditional random fields (CRF). After that, the primitives are merged to generate page regions based on the classification results of primitives and relations. Our method can extract various page regions with precise boundaries and is friendly for explanation. Decent experimental results on the PubLayNet document layout analysis dataset demonstrate the effectiveness of our method. The comparison of difference network configurations reveals more detailed effects of backbone CNN models, GAT and CRF.

The rest of this paper is organized as follows. Section 2 briefly reviews related works. Section 3 gives details of the proposed method. Section 4 presents the experimental results, and Sect. 5 draws concluding remarks.

2   Related Works

2.1   Page Segmentation

Comprehensive survey of traditional page segmentation methods have been given in [1]. In this section, we focus on deep leaning based methods (falling in bottom-up methods) which are closely related to this work.

Semantic Segmentation Based Methods. These methods treat page segmentation as pixel labeling problems and use deep neural networks for image pixel classification. Meier et al. [3] use fully convolutional network (FCN) to segment Newspapers into Articles. He et al. [4] use a multi-scale multi-task FCN to segment pages into *text, figure* and *table*. To increase segmentation accuracy, Yang et al. [5] proposed a multi-modal FCN which utilizes both visual and linguistic information. To obtain instance-level page regions, Li et al. [6] improve the vanilla FCN with label pyramids and deep watershed transformation.

Object Detection Based Methods. These methods treat page segmentation as object detection problems and use general object detection frameworks designed for natural scene images for page region detection. Huang et al. [7] and Sun et al. [8] use YOLOv3 [9] or Faster R-CNN [10] to detect table regions in pages. Saha et al. [11] and Zhong et al. [12] use Faster R-CNN [10] and Mask R-CNN [13] to detect equations, texts, titles, lists, tables and figures in pages.

Bottom-Up Methods. Yi et al. [14] redesign the region proposal method with connected component analysis and use CNN and dynamic programming to classify the region proposals into different classes of regions. Li et al. [2] first extract the so-called *"line region"*s based on image processing techniques and heuristic rules, then these *"line region"*s are classified and grouped into page regions using CNN and CRF.

2.2 Graph Parsing

Another closely related work is graph parsing, which jointly modeling individual objects and relationships among them. Dai et al. [15] first use a detector to detect individual objects, then a deep relational network is used to jointly predict the labels of objects and the relations between object pairs. [16] inherits this idea and uses it for form parsing. In [17], Line-Of-Sight graphs are progressively built from connected components and symbols and CNN is used to predict the relationship between neighboring graph nodes for formula recognition.

2.3 Graphical Models

Graph Neural Networks (GNN) [18] and Probabilistic Graphical Models (PGM) [19] have been widely used in many graph analysis tasks. Generally speaking, GNN is usually used for feature-level massage fusion among neighboring nodes and edges, while PGM directly model label-level dependence and compatibility among neighboring nodes.

Graph Neural Networks. Two of the most frequently used variants of GNN are graph convolutional networks (GCN) [20] and graph attention networks (GAT) [21]. For document analysis, Ye et al. [23] propose to use GAT for online stroke classification, while Qasim et al. [24] propose to use DGCNN and GravNet for table structure analysis.

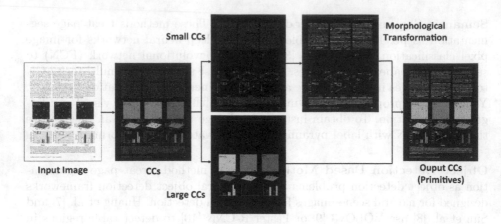

Fig. 1. Primitives Exaction. CCs are displayed with random color.

Probabilistic Graphical Models. Conditional Random Fields (CRF) [22] is one of the most popular type of PGM. For on-line stroke classification, Ye et al. [25] use CRF to integrate contextual information and optimize the parameters of CRF and backbone neural networks by joint training. Li et al. [26] use CRF for text/non-text classification and printed/handwritten text separation in complex off-line document images.

3   Proposed Method

3.1   System Overview

Our method treats page regions as clusters of small primitives of different geometric categories. For each page, the primitives are extracted using image processing techniques, then they are used to build a graph whose nodes represent the primitives and edges represent the relationships between neighbouring primitives. The nodes and edges are then classified into different classes based on features learned with deep convolution neural networks. To improve the classification accuracy, contextual information is integrated using GAT and CRF. Specifically, before classification, node and edge features can go through GAT for feature-level context fusion, then CRF and multi-layer perceptron (MLP) are used for joint node prediction and edge classification. After that, the primitives are merged to generate page regions based on predicted node and edge labels.

3.2   Primitive Extraction

The process of primitive extraction is illustrated in Fig. 1. After image binarization, we split the binary image into two separate maps with one map only contains small CCs (CCs whose height and width are both smaller than a threshold (e.g., 48 pixels), mainly belong to text) and the other only contains large

CCs (mainly belong to non-text). Then we apply mathematical morphological transformation on the first map to merge horizontally overlapping and adjacent CCs and split vertical touching CCs. By doing this, the number of CCs are greatly reduced and CCs across adjacent page regions are split. After that, we re-extract CCs in the first map and combine them with the large CCs. These CCs are used as primitives in the following steps.

The primitives are used to build a graph $G(V, E)$ for each page, where each node $v_i \in V$ represents a primitive and each edge $e_{i,j} \in E$ represents the relationship between a pair of neighboring primitives (v_i, v_j). We use a simple method to find each node's neighbors, specifically, if two primitives are horizontally or vertically overlapping and there is no other primitives exist between them, then they are neighbors of each other. Self-connections are also added.

3.3 Deep Feature Learning

We first learn a feature map with deep CNN model for the entire image. The structure of our deep CNN model can be found in Fig. 2. The basic architecture of our model is Feature Pyramid Network (FPN) [27], and the backbone of FPN can be VGG Net [28], ResNet [29], etc. After down-stream encoding and up-stream decoding, we resize and concatenate the feature maps of different scales to build the final feature map and reduce its channels to smaller number (128) using convolution with kernel size of 1×1.

To get node features F_V of the graph, RoIAlign [13] is performed on the feature map to pool the features of each node to a fixed dimension ($128 \times 5 \times 9 = 5760$) according to its bounding box, after that, a fully connected (FC) layer is used to further reduce node features to a lower dimension (512). For each edge, we first find the tight rectangular bounding box surrounding its two nodes, then RoIAlign and FC layer are used to pool and reduce the edge features to a fixed dimension (512). After that, we concatenate each edge's feature $F_{e_{i,j}}$ with features of its two nodes F_{v_i}, F_{v_j} and apply another FC layer to form the final edge features $F_{e_{i,j}}$ (dimension: 512). See Fig. 3 for more details.

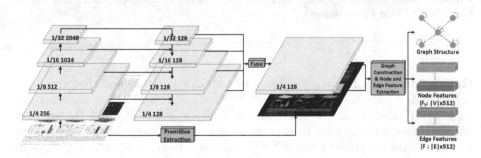

Fig. 2. Structure of our deep CNN model.

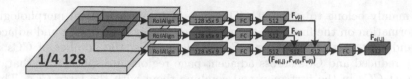

Fig. 3. Node and edge feature extraction.

3.4   Context Integration

Once we get node features F_V and edge features F_E, we can simply use multi-layer perceptron (MLP) for node and edge classification. However, local information is not enough to distinguish hard samples which belong to different classes but share similar features or belong to the same class but have totally different visual appearance, thus leading to inferior accuracy. In this paper, we investigate two widely used graph models namely GAT and CRF to integrate contextual information, aiming at improving the model performance.

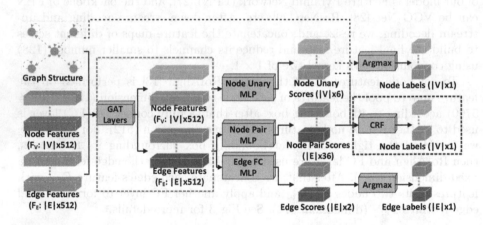

Fig. 4. Context integration with GAT and CRF. Green dotted box: GAT; blue dotted box: CRF. (Color figure online)

Graph Attention Layer. The input of our GAT layer are node features $F_V \in \mathbb{R}^{|V| \times D_V}$ and edge features $F_E \in \mathbb{R}^{|E| \times D_E}$, where V and E are node set and edge set, $|V|$ and $|E|$ are node number and edge number, D_V and D_E are node feature dimension and edge feature dimension, respectively. The layer produces new sets of node features $F'_V \in \mathbb{R}^{|V| \times D'_V}$ and edge features $F'_E \in \mathbb{R}^{|E| \times D'_E}$, as its outputs. In the layer, we first apply a shared liner transformation parametrized by a weight matrix $W_V \in \mathbb{R}^{D'_V \times D_V}$ to every node and another shared liner transformation parametrized by a weight matrix $W_E \in \mathbb{R}^{D'_E \times D_E}$ to every edge, then a shared attention mechanism $a : \mathbb{R}^{D'_V + D'_V + D'_E} \rightarrow \mathbb{R}$ is used to compute the attention coefficient $s_{i,j}$ of each neighboring node pair (v_i, v_j):

$$s_{i,j} = a(W_V F_{v_i} \| W_V F_{v_j} \| W_E F_{e_{i,j}})\tag{1}$$

where $s_{i,j} \in \mathbb{R}$ indicates the importance of node j's features to node i, $\|$ stands for concatenation, and a is a liner transformation in this work. Please be aware that except for F_{v_i} and F_{v_j}, $F_{e_{i,j}}$ is also taken into consideration for computing $s_{i,j}$ for (v_i, v_j). Then we normalize $s_{i,j}$ across all choices of j using softmax function:

$$\alpha_{i,j} = softmax(s_{i,j}) = \frac{exp(s_{i,j})}{\sum_{k \in \mathcal{N}_i} exp(s_{i,k})}\tag{2}$$

where $\mathcal{N}_i$ is the neighborhood of v_i. After obtaining the normalized attention coefficients $\alpha_{i,j}$, we can use them to compute a weighted combination of the neighborhood features, to produce the final output features for every node:

$$F'_{v_i} = \sigma\left(\sum_{j \in \mathcal{N}_i} \alpha_{i,j} W_V F_{v_j}\right)\tag{3}$$

where $\sigma(\cdot)$ is leaky ReLU activation function. Then updated node features F'_V along with the original edge features F_E are used to update the edge features:

$$F'_{e_{i,j}} = f(F'_{v_i} \| F'_{v_j} \| W_E F_{e_{i,j}})\tag{4}$$

where $f : \mathbb{R}^{D'_V + D'_V + D'_E} \rightarrow \mathbb{R}^{D'_E}$ is a liner transformation in this work.

Node features and edge features can go through multiple GAT layers to integrate contextual information in larger scope, and the output features are fused with the input ones in an element-wise way through residual connection. We don't use *multi-head attention* in this work, because in our experiments it only produce marginal improvement of accuracy at the cost of large computation and space overhead.

Conditional Random Fields. The implementation of our CRF model, including its formulation, inference and optimization, is just the same as that in [26] and [2], so we don't give its details in this paper for saving space. Here we only simply introduce its unary potentials and pairwise potentials. We formulate the unary potential function as follows:

$$U(y_i, v_i; \theta_U) = \sum_{k=1}^{K} -\lambda_k \delta(k = y_i) z_{i,k}(v_i; \theta_U),\tag{5}$$

where $\delta(\cdot)$ is the indicator function, which equals 1 if the input is true and 0 otherwise; K is the number of classes ($K = 6$ in this work); $z_{i,k}$ is the output value of *node unary MLP* that corresponds to the i-th node and the k-th class;

λ_k (=1.0 in this work) is a weight coefficient for $z_{i,k}$; θ_U is network parameters for U. Correspondingly, the pairwise potential function is formulated as follows:

$$P(y_i, y_j, e_{i,j}; \theta_P) = \sum_{k_i=1}^{K} \sum_{k_j=1}^{K} -\lambda_{k_i,k_j} \delta(k_i = y_i) \cdot \delta(k_j = y_j) z_{i,k_i,j,k_j}(e_{i,j}; \theta_P), \quad (6)$$

where z_{i,k_i,j,k_j} is the output value of *node pair MLP* corresponding to the node pair (v_i, v_j) when they are labeled with the class value (k_i, k_j); λ_{k_i,k_j} (=1.0 in this work) is a weight coefficient for z_{i,k_i,j,k_j}; θ_P is network parameters for P. The dimension of the pairwise network output for one node pair is: K^2.

3.5 Node and Edge Classification

Node Classification. For node classification, we evaluate four strategies to study the effect of GAT and CRF for context integration: neither GAT or CRF is used (MLP); only one of GAT or CRF is used (MLP+GAT and MLP+CRF); both GAT and CRF are used (MLP+GAT+CRF).

Edge Classification. For edge classification, we evaluate two strategies to study the effect of GAT for context integration: only MLP is used for classification (MLP); GAT is used for context fusion (MLP+GAT). Theoretically, CRF can also be tested for edge classification. But as the classification accuracy is already sufficiently high, we don't use CRF to reduce memory and computation cost.

3.6 Page Region Extraction

After node and edge classification, we can get page regions by merging primitives according to their labels and relationships. Here remains a problem to solve: suppose we have two primitives, node classification results tell us they belong to different classes but edge classification results tell us they belong to the same page region, which one should we trust more? According to this, we designed two strategies for page region extraction as follows:

Classifying Before Merging (CBM). Under this strategy, we merge two primitives only if their relationship is predicted to be *merge* and they share the same predicted labels. After merging, the primitives inside each page region all share the same label, which is used as the page region's label.

Merging Before Classifying (MBC). Under this strategy, we merge two primitives as long as their relationship is predicted to be *merge*, and we don't care whether they share the same label or not. After merging, the primitives inside each page region may have different labels, and the label with most number of primitives is used as the region's label.

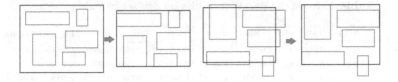

Fig. 5. Page region adjusting according to CCs inside or overlapping with it. Green: CCs; Blue: detected page regions. (Color figure online)

After page region extraction, we perform some simple post-processing operations to improve the quality of regions and filter out noise detections. These post-processing operations include adjusting page regions according to CCs inside or overlapping (more than half area) with them, see Fig. 5; merging vertically adjacent *list* regions; removing too small *text*(height < 32 and width < 32), *list*(height < 16), *table*(height < 32) and *figure*(height < 32) regions. Nothing more post-processing operations are used.

4   Experimental Results

4.1   Implementation Details

Network Structure. We test five backbone structures for our deep CNN model namely VGG-8 (each block only have one convolution layer), VGG-11, VGG-16 [28] and ResNet-50, ResNet-101 [29]. The parameters of ResNets are pre-trained on ImageNet while the parameters of VGG Nets are initialized randomly. We use only two layers of GAT, since no further accuracy improvement is observed with more GAT layers. All of our models are implemented on the PyTorch[1] platform except that the GAT layers are implemented with the DGL[2] library and the CRF is implemented with the OpenGM[3] library.

Training Settings. All models are trained end-to-end with back-propagation using cross-entropy loss function. The initial learning rate is set to 0.01, then we employ a exponential learning rate policy where the initial learning rate is multiplied by $\gamma^{-\beta \frac{iter}{max\_iter}}$ with $\gamma = 2$, $\beta = 8$ and $max\_iter = 168k$ (4 epoches) in this work. We use the momentum of 0.9 and a weight decay of 0.0005. We employ $4\times$ TITAN XP GPUs for training and batch size is 8. All images are resized to a fixed size of 512×724 (same aspect ratio as A4 paper) during model training and page segmentation, then coordinates of detected regions are rescaled to the original page size.

---

[1] https://pytorch.org/get-started/locally/.
[2] https://docs.dgl.ai/index.html.
[3] https://github.com/opengm/opengm.

Table 1. Node classification results on dev set of PubLayNet.

	MLP	GAT	CRF	Other	Text	Title	List	Table	Figure	Accuracy
VGG-8	✓	×	×	0.9259	0.9898	0.9811	0.9249	0.9886	0.9818	0.9835
	✓	✓	×	0.9284	0.9913	0.9829	0.9448	0.9935	0.9901	0.9874
	✓	×	✓	0.9315	0.9917	0.9844	0.9448	0.9924	0.9882	0.9873
	✓	✓	✓	0.9312	0.9920	0.9855	0.9514	0.9940	0.9909	0.9885
VGG-11	✓	×	×	0.9326	0.9917	0.9824	0.9440	0.9924	0.9879	0.9872
	✓	✓	×	0.9352	0.9931	0.9837	0.9619	0.9941	0.9909	0.9897
	✓	×	✓	0.9361	0.9929	0.9849	0.9579	0.9944	0.9914	0.9896
	✓	✓	✓	0.9366	0.9934	0.9859	0.9640	0.9945	0.9915	0.9902
VGG-16	✓	×	×	0.9280	0.9930	0.9831	0.9604	0.9939	0.9907	0.9892
	✓	✓	×	0.9309	0.9929	0.9844	0.9565	0.9956	0.9938	0.9899
	✓	×	✓	0.9305	0.9938	0.9858	0.9671	0.9956	0.9936	0.9909
	✓	✓	✓	0.9319	0.9935	0.9856	0.9618	0.9959	0.9942	0.9906
ResNet-50	✓	×	×	0.9341	0.9941	0.9889	0.9691	0.9962	0.9957	0.9917
	✓	✓	×	0.9354	0.9943	0.9892	0.9706	0.9959	0.9952	0.9917
	✓	×	✓	0.9357	0.9945	0.9893	0.9721	0.9967	0.9961	0.9922
	✓	✓	✓	0.9361	0.9945	0.9893	0.9726	0.9963	0.9955	0.9921
ResNet-101	✓	×	×	0.9348	0.9939	0.9883	0.9691	0.9966	0.9960	0.9917
	✓	✓	×	0.9352	0.9940	0.9882	0.9704	0.9966	0.9959	0.9918
	✓	×	✓	0.9370	0.9943	0.9891	0.9716	0.9972	0.9968	**0.9924**
	✓	✓	✓	0.9366	0.9942	0.9890	0.9707	0.9968	0.9960	0.9920

4.2 Dataset

We conducted experiments on the PubLayNet dataset [12]. PubLayNet is by far the largest dataset for document layout analysis and page segmentation. It consist of 358,353 pages in total, along with corresponding annotation files of MS COCO Object Detection task's format. It exhibits a considerable variety in both page layout styles and page region styles, including single-cloumn pages, two-column pages, multi-column pages and various kinds of *text*, *title*, *list*, *table* and *figure* regions. The dataset is split into train set, dev set and test set with 335,703 pages, 11,245 pages and 11,405 pages, respectively. Annotation files of the test set are not made public, so we evaluate our method on the dev set.

4.3 Evaluation Metric

For node and edge classification, we report F1 values (harmonic average of precision and recall) of each class and the overall accuracy or all classes. For page region extraction, we report average precision (AP) and average recall (AR) @ intersection over union (IoU) [0.50:0.95] of each class's bounding boxes and their macro average over all classes: mAP and mAR, which is used in the COCO competition[4]. We also report $AP^{IoU=0.5}$, $AP^{IoU=0.75}$ of each class to show the performance of our model at different IoU thresholds.

---

[4] http://cocodataset.org/#detection-eval

Table 2. Edge classification results on dev set of PubLayNet.

	MLP	GAT	Merge	Split	Accuracy
VGG-8	✓	✗	0.9905	0.9971	0.9955
	✓	✓	0.9912	0.9973	0.9958
VGG-11	✓	✗	0.9915	0.9974	0.9960
	✓	✓	0.9918	0.9975	0.9961
VGG-16	✓	✗	0.9922	0.9976	0.9963
	✓	✓	0.9923	0.9976	0.9964
ResNet-50	✓	✗	0.9925	0.9977	0.9964
	✓	✓	0.9921	0.9975	0.9962
ResNet-101	✓	✗	0.9930	0.9978	**0.9967**
	✓	✓	0.9925	0.9977	0.9964

4.4   Node and Edge Classification

Node and edge classification results with different backbone CNN models and context fusion methods are shown in Table 1 and Table 2, respectively.

From Table 1 we can see, all the models can produce fairly high node classification accuracy even with the shallowest backbone VGG-8 without any context fusion model. When using deeper backbone networks, the accuracy increase monotonously from 0.9835 to 0.9917. When adding GAT layers, the accuracy increases relatively more significant for shallower backbones VGG-8 and VGG-11 than deeper backbones VGG-16, ResNet-50 and ReasNet-101. The reason behind this is that convolution itself is inherently a operation of context fusion in local area, when stacking more convolution layers, contextual information in larger area is fused. GAT layers can integrate contextual information which shallower CNN models fail to learn, but become less helpful for deeper CNN models which have substantially large receptive fields. Similar observation can be made for edge classification from Table 2.

On the other hand, adding CRF can always provide a relatively larger improvement compare with its corresponding counterpart, as it can model label dependency and compatibility between neighboring nodes, which is essential for classification tasks but can not be learned using CNN or GAT.

We use the results of node and edge classification with highest accuracies for the next step of page region extraction, i.e. ResNet-101 with MLP and CRF for node classification and ResNet-101 with MLP for edge classification.

4.5   Page Region Extraction

Page region extraction results are shown in Table 3. F-RCNN and M-RCNN stand for Faster R-CNN and Mask R-CNN which are used in [12]; CMB stands for classifying before merging and MBC stands for merging before classifying.

Table 3. Page region extraction results on dev set of PubLayNet. GTA denotes ground truth adjustment (bounding box compaction).

		Text	Title	List	Table	Figure	Macro average
F-RCNN [12]	AP	0.910	0.826	0.883	0.954	0.937	0.902
M-RCNN [12]	AP	0.916	0.840	0.886	0.960	0.949	0.910
CBM	$AP^{IoU=0.5}$	0.9685	0.9729	0.9318	0.9886	0.9646	0.9653
	$AP^{IoU=0.75}$	0.9226	0.4722	0.8970	0.9831	0.8822	0.8314
	AP	0.8860	0.5270	0.8683	0.9761	0.8376	0.8190
	AR	0.8523	0.5271	0.8710	0.9691	0.8318	0.8103
MBC	$AP^{IoU=0.5}$	0.9704	0.9745	0.9451	0.9894	0.9664	0.9692
	$AP^{IoU=0.75}$	0.9248	0.4730	0.9103	0.9839	0.8845	0.8353
	AP	0.8880	0.5279	0.8811	0.9766	0.8398	0.8227
	AR	0.8528	0.5269	0.8716	0.9687	0.8318	0.8103
CBM_GTA	$AP^{IoU=0.5}$	0.9877	0.9834	0.9353	0.9899	0.9663	0.9725
	$AP^{IoU=0.75}$	0.9820	0.9810	0.9205	0.9863	0.9423	0.9624
	AP	0.9815	0.9799	0.9154	0.9827	0.9384	0.9596
	AR	0.9442	0.9799	0.9182	0.9757	0.9319	0.9500
MBC_GTA	$AP^{IoU=0.5}$	0.9896	0.9849	0.9490	0.9907	0.9680	0.9764
	$AP^{IoU=0.75}$	0.9843	0.9826	0.9339	0.9871	0.9445	0.9665
	AP	0.9837	0.9815	0.9286	0.9833	0.9406	0.9635
	AR	0.9446	0.9796	0.9185	0.9753	0.9317	0.9499

As we can see from Table 3, compared with CMB, MBC can produce higher AP for all classes, mainly because edge classification results are more reliable and MBC is robust for node classification errors in some degree since it adopt a voting strategy to obtain regions' labels.

However, compared with the baseline methods F-RCNN and M-RCNN, the APs of our method are significantly lower for all classes except *table*. From Table 3 we can see the APs of our methods for all classes are quite high when IoU = 0.5, but drop dramatically when IoU = 0.75. This usually happens when the bounding boxes of detected objects are not precise enough. So can we draw the conclusion that the proposed method is not able to detect precise bounding boxes for page regions except for *table*? While, this is not really true. After analyzing the page segmentation results, we find that page regions that our method generates have tight bounding boxes surrounding the CCs inside them, while the ground truth page regions don't. That is to say, there may exist some unnecessary white space in the ground truth bounding boxes, or some CCs belonging to certain page regions are not completely inside their bounding boxes, see the first row of Fig. 6. In other words, the ground truth bounding boxes of the PubLayNet dataset may be larger or smaller than they should be to contain all the information belonging to them. This is because they are not annotated manually but generated by automatically parsing the PDF files, thus leading to lost of errors and incompatibilities.

So what if we simply adjust the ground truth bounding boxes (the second row of Fig. 6) using the same method as for our detected regions (Fig. 5)?

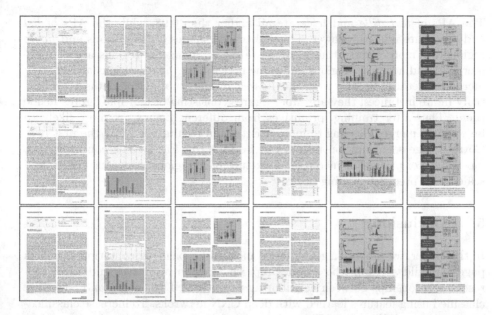

Fig. 6. Ground truth bounding box adjustment. **First row**: original ground truth bounding boxes; **second row**: adjusted ground truth bounding boxes; **third row**: page segmentation results by our method. Green: *text*; red: *title*; cyan: *list*; yellow: *table*; blue: *figure*; black: *other*. (Color figure online)

By doing this, we obtain very interesting results as shown in Table 3 denoted as CBM_GTA and MBC_GTA. As we can see, APs of MBC_GTA for all classes are much higher or comparable than previous method F-RCNN and M-RCNN.

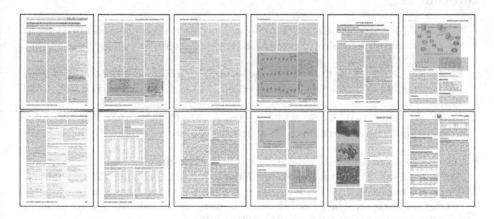

Fig. 7. Page segmentation results using the proposed method. **First row**: erroneous results; **second row**: correct results. Green: *text*; red: *title*; cyan: *list*; yellow: *table*; blue: *figure*; black: *other*. (Color figure online)

The APs of *text*, *title* and *table* are all higher than 0.98, while the APs of *list* and *figure* are relatively lower. This is because shapes of *list* regions and *figure* regions are more diverse thus harder to detect. Some *text* regions are wrongly merged to adjacent *list* regions and some neighboring *figure* regions are wrongly merged together. By the way, the *figures* in the PubLayNet dataset are not truly figure instances, they are super figures consisting of multiple figures inside them. Thus after training, primitives from different *figures* are more likely to be wrongly merged, leading to inferior detection results, see Fig. 7 the first row. Correct page segmentation results can be found in Fig. 7 the second row. For most kinds of pages, our method can extract page regions of various classes and scales with precise bounding boxes.

5   Conclusion

In this paper, we propose a graph based method for page segmentation. For each page we build a graph whose nodes represent region primitives and edges represent relationships between neighboring primitives. Then nodes and edges are classified using features learned with deep CNN netwroks. To increase classification accuracy, GAT and CRF are also used to integrate contextual information. After that, primitives are merged to generate page regions according to node and edge classification results. Our method is simple and intuitive and can generate page regions with precise bounding boxes. Experimental results on the PubLayNet dataset demonstrate the effectiveness of our system.

Acknowledgments. This work has been supported by National Natural Science Foundation of China (NSFC) Grants 61733007, 61573355 and 61721004.

References

1. Shafait, F., Keysers, D., Breuel, T.: Performance evaluation and benchmarking of six-page segmentation algorithms. IEEE Trans. Pattern Anal. Mach. Intell. **30**, 941–954 (2008)
2. Li, X.-H., Yin, F., Liu, C.-L.: Page object detection from pdf document images by deep structured prediction and supervised clustering. In: ICPR, pp. 3627–3632 (2018)
3. Meier, B., Stadelmann, T., Stampfli, J., Arnold, M., Cieliebak, M.: Fully convolutional neural networks for newspaper article segmentation. In: ICDAR, pp. 414–419 (2017)
4. He, D., Cohen, S., Price, B., Kifer, D., Lee Giles, C.: Multi-scale multi-task FCN for semantic page segmentation and table detection. In: ICDAR, pp. 254–261 (2017)
5. Yang, X., Yumer, E., Asente, P., Kraley, M., Kifer, D., Lee Giles, C.: Learning to extract semantic structure from documents using multimodal fully convolutional neural networks. In: CVPR, pp. 5315–5324 (2017)
6. Li, X.-H., Yin, F., Xue, T., Liu, L., Ogier, J.-M., Liu, C.-L.: Instance aware document image segmentation using label pyramid networks and deep watershed transformation. In: ICDAR, pp. 514–519 (2019)

7. Huang, Y., et al.: A yolo-based table detection method. In: ICDAR, pp. 813–818 (2019)
8. Sun, N., Zhu, Y., Hu, X.: Faster R-CNN based table detection combining corner locating. In: ICDAR, pp. 1314–1319 (2019)
9. Redmon, J., Farhadi, A.: YOLOv3: an incremental improvement. arXiv:1804.02767 (2018)
10. Ren, S., He, K., Girshick, R., Sun, J.: Faster R-CNN: towards real-time object detection with region proposal networks. In: NIPS, pp. 91–99 (2015)
11. Saha, R., Mondal, A., Jawahar, CV.: Graphical object detection in document images. In: ICDAR, pp. 51–58 (2019)
12. Zhong, X., Tang, J., Yepes, A.J.: PubLayNet: largest dataset ever for document layout analysis. In: ICDAR. (2019)
13. He, K., Gkioxari, G., Dollár, P., Girshick, R.: Mask R-CNN. In: ICCV, pp. 2961–2969 (2017)
14. Yi, X., Gao, L., Liao, Y., Zhang, X., Liu, R., Jiang, Z.: CNN based page object detection in document images. In: ICDAR, pp. 230–235 (2017)
15. Dai, B., Zhang, Y., Lin, D.: Detecting visual relationships with deep relational networks. In: CVPR, pp. 3076–3086 (2017)
16. Davis, B., Morse, B., Cohen, S., Price, B., Tensmeyer, C.: Deep visual template-free form parsing. In: ICDAR (2019)
17. Mahdavi, M., Condon, M., Davila, K., Zanibbi, R.: LPGA: line-of-sight parsing with graph-based attention for math formula recognition. In: ICDAR (2019)
18. Zhou, J., et al.: Graph neural networks: a review of methods and applications. arXiv:1812.08434 (2018)
19. Koller, D., Friedman, N.: Probabilistic Graphical Models: Principles and Techniques. MIT Press, Cambridge (2009)
20. Kipf, T.N., Welling, M.: Semi-supervised classification with graph convolutional networks. arXiv:1609.02907 (2016)
21. Veličković, P., Cucurull, G., Casanova, A., Romero, A., Lio, P., Bengio, Y.: Graph attention networks. arXiv:1710.10903 (2017)
22. Lafferty, J., McCallum, A., Pereira, F.C.N.: Conditional random fields: probabilistic models for segmenting and labeling sequence data (2001)
23. Ye, J.-Y., Zhang, Y.-M., Yang, Q., Liu, C.-L.: Contextual stroke classification in online handwritten documents with graph attention networks. In: ICDAR, pp. 993–998 (2019)
24. Qasim, S.R., Mahmood, H., Shafait, F.: Rethinking table recognition using graph neural networks. In: ICDAR, pp. 142–147 (2019)
25. Ye, J.-Y., Zhang, Y.-M., Liu, C.-L.: Joint training of conditional random fields and neural networks for stroke classification in online handwritten documents. In: ICPR, pp. 3264–3269 (2016)
26. Li, X.-H., Yin, F., Liu, C.-L.: Printed/handwritten texts and graphics separation in complex documents using conditional random fields. In: DAS, pp. 145–150 (2018)
27. Lin, T.-Y., Dollár, P., Girshick, R., He, K., Hariharan, B., Belongie, S.: Feature pyramid networks for object detection. In: CVPR, pp. 2117–2125 (2017)
28. Simonyan, K., Zisserman, A.: Very deep convolutional networks for large-scale image recognition. In: ICLR (2015)
29. He, K., Zhang, X., Ren, S., Sun, J.: Deep residual learning for image recognition. In: CVPR, pp. 770–778 (2016)

The Notary in the Haystack – Countering Class Imbalance in Document Processing with CNNs

Martin Leipert[1(✉)], Georg Vogeler[2], Mathias Seuret[1], Andreas Maier[1], and Vincent Christlein[1]

[1] Pattern Recogntition Lab, FAU Erlangen-Nuremberg, Erlangen, Germany
{martin.leipert,mathias.seuret,andreas.maier}@fau.de
[2] Austrian Centre for Digital Humanities, Universität Graz, Graz, Austria
georg.vogeler@uni-graz.at

Abstract. Notarial instruments are a category of documents. A notarial instrument can be distinguished from other documents by its notary sign, a prominent symbol in the certificate, which also allows to identify the document's issuer. Naturally, notarial instruments are underrepresented in regard to other documents. This makes a classification difficult because class imbalance in training data worsens the performance of Convolutional Neural Networks. In this work, we evaluate different countermeasures for this problem. They are applied to a binary classification and a segmentation task on a collection of medieval documents. In classification, notarial instruments are distinguished from other documents, while the notary sign is separated from the certificate in the segmentation task. We evaluate different techniques, such as data augmentation, under- and oversampling, as well as regularizing with focal loss. The combination of random minority oversampling and data augmentation leads to the best performance. In segmentation, we evaluate three loss-functions and their combinations, where only class-weighted dice loss was able to segment the notary sign sufficiently.

Keywords: Convolutional Neural Networks · Class imbalance · Segmentation

1 Introduction

Notarial instruments, the subject of this work, originate from medieval Italy. The ones considered in this study are recognizable for the human eye by a characteristic notary sign, see for example Fig. 1a, which served as signature to identify the issuing commissioner Around 1200 CE they appeared in today's Germany and Austria. The analysis of their spread and development is an open problem in historic research [4,17,18]. Especially as they are one particular source to reconstruct the development of monasteries. Compared to the number of medieval documents with seals, notarial instruments are underrepresented. Automatic

X. Bai et al. (Eds.): DAS 2020, LNCS 12116, pp. 246–261, 2020.
https://doi.org/10.1007/978-3-030-57058-3_18

(a) Source: Seckau, Austria, 1455, BayHStA München HU Chiemsee 23.

(b) Source: Vienna, Austria, 1460, Archiv der Erzdiözese Salzburg, AUR 2836.

Fig. 1. Examples of a notary sign (a) (https://www.monasterium.net/mom/AT-AES/Urkunden/2836/charter) and a notary document (b) (https://www.monasterium.net/mom/DE-BayHStA/HUChiemsee/023/charter).

classification and segmentation of these documents, to automatically identify the commissioner, is beneficial. Notary signs cover typically a small area of the document [16], see Fig. 1b, where the notary sign is visible in the bottom left. In classification tasks, the class notary document is severely underrepresented and the difference to other documents is only a small detail. In segmentation tasks, the area of the *notary sign* is much smaller compared to *text*. So both problems have to deal with severe class imbalance.

In this work, we conduct a comparative study for handwritten document classification and segmentation. These studies are sparse for the field of class imbalance [1], which is a major problem of the dataset at hand. Therefore, we evaluate different techniques to counter this issue: (1) augmentation, including a novel idea of data interpolation, which is easily integratable in the case of notarial instruments containing characteristic notary signs; (2) data oversampling and undersampling; (3) different losses, such as focal loss [11]; (4) lastly, we tried to enhance the performance of segmentation networks by feeding them entire regions of the layout more often to let them memorize their characteristics.

The work is structured as follows. After presenting the related work in Sect. 2, we introduce the used dataset in Sect. 3. The networks, tools and loss functions used for the study are presented in Sects. 3.1 to 3.3. Afterwards, we give an overview of the used settings in Sect. 4.1. The results are described in Sect. 4.2 for classification and Sect. 4.3 for segmentation and discussed in Sect. 4.4. Finally, the paper is concluded in Sect. 5.

2   Related Work

Countering Class Imbalance: There exist two main strategies to counter class imbalance in neural network training [9]. Data level techniques – which modify the data fed to the model – and algorithmic methods which modify the training algorithm. Usually, data level methods try to achieve class balance. From those, random minority class oversampling achieved the best performance in previous studies [1]. In contrast to traditional machine learning, oversampling does not lead to overfitting for convolutional neural networks (CNNs). Another technique is SMOTE [19], a method to interpolate new samples from training data. Furthermore, data augmentation can be used to virtually increase the amount of training data, although the performance under augmentation is always worse than with the equivalent amount of real data. Algorithmic level methods include class-weighting – to increase the weight adaptation for miss-classifications – or specific loss functions, such as focal loss [11]. The latter puts additional weight on hard to classify examples. This technique (in combination with the RetinaNet architecture) outperformed classical detectors significantly.

Document Classification and Segmentation with CNNs: Documents of the same genre have a similar visual structure and region-specific features. This allows to classify documents via CNNs. In a comparative study, Harley et al. [3] showed that pretrained CNNs (on ImageNet) that took the entire document as input, outperformed other approaches that used only document parts. Documents may be processed based on simple, textural features. For example, Oyedotun et al. [13] achieved a region identification rate (share of regions assigned to the correct class) of 84% for segmentation with a simple network taking six pixel-wise textural features as input and producing a three-class output (image, text, background). Therefore the input was fed to a single fully-connected hidden layer. Also specifically developed deep CNNs for documents exist. For example, the *dhSegment* CNN [12] consists of a contracting path and an expanding path similar to the well-known U-Net [14]. The contracting path is based on the *ResNet50* architecture [5]. The approach outperformed previous approaches in border and ornament extraction of medieval documents.

3   Materials and Methods

The used document collection originates from the *Monasterium* project. This collection of digitized medieval documents from different European archives offers descriptions and images of more than 650 000 documents [6, 15].

Subject of this study is a set of 31 836 documents, which contains 974 notarial instruments and various other types. They originate from libraries of Southern Germany and Austria. The notary signs in the documents are circumscribed by manually drawn polygons. The time range of the documents is from 800 CE to 1499 CE, where the first notarial instrument in the collection appears shortly after 1200 CE. The distributions are not entirely separable, not every notary certificate is clearly identifiable, e. g., due to a missing notary sign. The documents are 3000 px wide, but vary in spatial resolution. Their real size is between 15 cm × 20 cm and 40 cm × 50 cm with font sizes ranging from 15 px to 60 px, the majority around 30 px. The size of the notary sign ranges from about

(a) Faded font due to water damage   (b) Brown stain

Fig. 2. Examples for varying document appearance.

200 px × 200 px to 500 px × 700 px. Other varying factors are yellowish paper, low contrast, overexposure during scanning and faded ink. Stains with low contrast are frequent due to water damage, see for example Fig. 2.

3.1 Networks

The classification task is binary, i. e., distinguish notarial instruments from non-notary ones. The networks used the full image as input, rescaled and cropped to a size of 224 px × 224 px. ResNet50 [5] and DenseNet121 [7] networks are used for classification, both pretrained on ImageNet. Both networks reuse features within the network by skip-connections. While ResNet bypasses the concatenated results from previous layers, DenseNet uses the entire feature vectors of previous layers within a block. DenseNet outperforms ResNet in stability, parameter efficiency, and is less prone to overfitting [7].

The segmentation task is a four class problem: background, text, ornaments and notary signs. As in classification, images are reduced to 224 px × 224 px. This size is sufficient to identify all foreground components (text, ornament and notary sign). The patch is segmented with a U-Net [14] using 5 layers. The U-Net consists of a contracting/encoder and expanding/decoder path. Blocks on the same level are interconnected, to reconstruct the output with the input features on the specific level. The U-Net achieves large patch segmentation by an overlap strategy. Batch normalization [8] is applied in every layer after the ReLU non-linear functions to stabilize training.

3.2 Tools

The networks in the classification settings are trained with different settings, i. e., different loss functions, data augmentation and sampling methods. Adam [10] with PyTorch's default parameters ($\beta_1 = 0.9, \beta_2 = 0.999$) is used as optimizer for both the segmentation and the classification. For the classification task, training is performed for 1250 iterations and a batch size of 32. The initial learning rate is $1 \cdot 10^{-3}$, except for settings with focal loss, where it is $5 \cdot 10^{-4}$. To adapt the learning rate, a step decay learning rate scheduler with $\gamma = 0.5$ is used. The step size is set to 250. In segmentation, training is performed for 60 epochs, where the

Table 1. Used groups of augmentation effects. Italic effects only occur for moderate and/or heavy augmentation.

Group	Effects
Flip	Horizontal and vertical flip, 90°-rotation
Affine transformation	Random rescale, shift & rotation
Blur & noise	*Ordinary & gaussian noise, ordinary, gaussian motion & ordinary blur, median filter*
Distortion	Optical & grid distorsion, elastic transformation
Brightness & contrast	Random brightness & contrast adaption.
Color	Histogram equalization (CLAHE), random HSV[a]-shift, RGB-shift, channel shuffle and grey conversion
Special effect	Shadow, *snow or jpg compression*

[a]Hue Saturation Value

initial learning rate is $3 \cdot 10^{-3}$, except for settings with focal loss as part of the loss function. In the latter cases, the initial learning rate is $1 \cdot 10^{-3}$. Again a step decay learning rate scheduler is applied for learning rate adaption. The decay parameter is set to $\gamma = 0.3$ and the step size is 10. Additionally in classification, dropout with $p = 0.2$ is used for the last fully connected layer of the respective networks, which empirically improves the recognition rates.

Image Augmentation: The data augmentation strategies are derived from the previously examined dataset-intrinsic variations. Those are contrast, color, lighting, affine transformations, deformations and damages. Three dynamic image augmentation functions are built with the *Albumentations* framework [2]. The "weak" augmentation is designed to mimic the intrinsic variations of the dataset. The "moderate" augmentation uses higher parameter and probability values. The "heavy" augmentation also includes additional effects not being part of the intrinsic data variation. These functions are identical for both classification and segmentation.

They are all composed according to a certain pattern, i.e., they include a combination of up to seven groups of effects. These groups are a probabilistic combination of flip, an affine transformation, one of a group of blurs, one non-linear distortion, a special effect, a brightness-contrast variation and one of various color modifications as shown in Table 1.

The "flip" and "affine transformation" groups simulate different positions, scales and orientations of the document. The flips are used to force the network to learn the features independent of their absolute position. The "brightness & variance" and "Blur & noise" groups mimic aging paper and different acquisition qualities. The "Distortions" imitate acquisition variables, as the photographies of the documents are not always plain. The "Color" shift simulates different font and paper colors. The "special effects" group mimics either darker (shadow) or missing areas (snow).

Fig. 3. Notary certificate with the notary sign from another document inserted. Document source: BayHStA KU Rohr 327, Sign source: Archiv Erzdiözese Salzburg, D319. (https://www.monasterium.net/mom/AT-AES/Urkunden/2854/charter)

Each of these groups has an assigned occurrence probability. If the group is drawn, one effect out of the group is chosen – again according to a probability – to contribute to the augmentation. This concept is the same for all three augmentation functions. So in an edge case, exactly one effect per seven groups is applied. The occurrence probabilities of the groups and the intensity parameters of the effects are varied, resulting in different augmentation intensities. Each parameter and probability is set such that it doubles from "weak" to "heavy". As consequence for "weak" augmentation in average 1.7 effects were applied, for "moderate" 3.4 and for "heavy" 4.0. e.g., the group affine transformation occurred with a probability of 0.4 for "weak" augmentation and with a probability of 0.7 for "moderate" and "heavy". The scale limit was 0.1 for "weak", 0.15 for "moderate" and 0.2 for "heavy".

Notary Symbol Swap: This was introduced to allow meaningful interpolation between data. The system should learn the notary sign as the decisive structural element of notarial instruments. Therefore, notary signs of other documents are either inserted at the original sign's position, cf. Fig. 3 for an example, or added at an arbitrary location of the document.

Random Minority Oversampling: Used to balance the dataset, in which non-notary documents occur 32 times more frequent than notary ones. Hence, the latter ones are oversampled. This is done by random drawing, such that notary and non-notary signs occur with the same probability of 0.5. The variance within the notary set is smaller, which is combined with augmentation to compensate possible overfitting effects.

Random Majority Undersampling: In contrast to oversampling, a random subset of the non-notary documents is drawn to balance the number of samples of notary and non-notary document within the training data.

Table 2. Settings for the training of the DenseNet and the ResNet

Setting	Augmentation	Swap/add sign	Loss	Under-/oversampling
1			BCE	
2			BCE	+
3	o		BCE	+
4	+		BCE	+
5	++		BCE	+
6	+		BCE	-
7		Swap & Add	BCE	+
8	+	Swap & Add	BCE	+
9	+	Swap	BCE	+
10	+		Focal	
11	+	Swap & Add	Focal	
12	+	Swap	Focal	+

Meaningful Segment Training: In the segmentation task, background and text are largely over-represented in comparison to notary signs. To compensate this, the U-Net is trained with segments of an entire "meaningful" area of the image (text, ornaments or a notary sign) by randomly choosing one type of these segments occurring in the image, this is repeated to fill the image or resized to fill the patch. This is done with a probability of 0.5 alternating with normal notary certificates.

3.3 Loss Functions

Classification: For training, we use the binary cross entropy (BCE) loss. As alternative loss function and algorithmic imbalance compensation, focal loss [11] is used. It emphasizes examples with a high error in training. Due to this property, it achieves good performance for problems suffering from strong class imbalance. Focal loss is based on balanced cross entropy and consists of adding a weighting $(1 - p_t)^\gamma$ and (optionally) a class weighting factor α_t to the BCE function.

$$\text{FL}(p_t) = -\alpha_t (1 - p_t)^\gamma \log(p_t) \tag{1}$$

where the probability p_t is defined as:

$$p_t = \begin{cases} p & y = 1 \\ 1 - p & \text{otherwise} \end{cases} \tag{2}$$

With a high $\gamma \approx 2$, hard examples are weighted stronger and thus are focus of training. This setting is used for focal loss in classification. With a low $\gamma \approx 0.5$ the weight of samples with a small deviation is increased. With $\gamma = 0$ the function and its properties are equivalent to BCE loss. In the classification tasks $\gamma = 2$ is used.

Segmentation: We evaluate training of a U-Net using different loss functions: (1) class-weighted dice loss (weighted by inverse class frequency), (2) BCE loss, and (3) a two dimensional version of focal loss ($\gamma = 1.5$). Additionally, we combine different loss functions to evaluate if these combinations improve training behavior by compromising different optimization goals.

4   Evaluation

4.1   Evaluation Protocol

The set of 31 838 images for the classification was split into 21 227 training, 2653 validation and 7958 test images. For segmentation, the set of 887 images was split into 496 training, 98 validation and 293 test images.

For the classification task, we created twelve different evaluation settings, see Table 2. The settings test various augmentation intensities, namely "weak" (o), "moderate" (+) and "heavy" (++) augmentation, swapping of the notary sign and random adding/swapping of notary signs. Focal loss, undersampling (-) and oversampling (+) are compared under moderate augmentation. Each setting is executed and trained three times. We sum up all the results and compute common error metrics for a two class-problems: sensitivity, specificity and the F-value (a. k. a. F1-score).

The first setting serves as baseline where no imbalance counter measures are used. In the second setting, random minority oversampling as most effective counter-measure is introduced. This is further expanded in settings 3, 4, 5, where data augmentation is added to the previous setting and its intensity is steadily increased from "weak" to "heavy" to further boost the accuracy. The most effective "moderate" augmentation is used in setting 6, together with random majority undersampling as a comparison to oversampling. Settings 7 to 9 introduce notary symbol swapping/adding. This method is used as sole augmentation in setting 7. In setting 8, this technique serves as additional augmentation to "moderate" augmentation. Setting 9 uses "moderate" augmentation and symbol swapping only, to compare if it is more effective than adding. The underlying assumption is, that the correct arrangement of the sign is important for detection and the adding of a symbol at an arbitrary position could possibly reduce recognition accuracy. From setting 10 to 12, focal loss is introduced and combined with the best performing "moderate" augmentation in setting 10. In setting 11, this augmentation is further combined with swapping and adding of notary signs. Finally all successful techniques – random minority oversampling, moderate augmentation and symbol swap – are combined with focal loss, to examine, if focal loss performance could beat BCE.

For the segmentation experiments, the main variation in the settings (Table 3) is the use of segments and the combination of different loss functions. In the settings 1 to 4 with focal loss, it is tested if augmentation improves the result. In settings 5 to 10 the most effective combination of loss functions is tested. Settings 11 to 13 check if meaningful segments improve the result. The segmentation

Table 3. Settings for the training of the U-Net in the segmentation task.

Setting	Augmentation	Focal	Dice	BCE	Meaningful segments
1		1.0			
2	o	1.0			
3	+	1.0			
4	++	1.0			
5	+		1.0		
6	+			1.0	
7	+		0.5	0.5	
8	+	0.5		0.5	
9	+	0.5	0.5		
10	+	0.33	0.33	0.33	
11	+			1.0	+
12	+	1.0			+
13	+	1.0			+

results are evaluated by means of the Intersection over Union (IoU)

$$IoU = \frac{|A \cap B|}{|A \cup B|} \tag{3}$$

of the labels in the ground truth A and in the networks' prediction B.

4.2   Classification Results

The classification results for the different settings are given in Table 4a and Table 4b for ResNet and DenseNet, respectively. Ignoring class imbalance (setting 1) leads to the worst results for both networks, the ResNet model even fails to correctly classify any notarial instrument. Using oversampling without data augmentation (setting 2) gives the second worst results. When combining oversampling with data augmentation (setting 3–6, 8, 9) the results increase significantly, but start to decrease with "heavy" augmentation (setting 5). The best results are achieved using the DenseNet with a "moderate" augmentation function (setting 3, 9) and for the combination of moderate data augmentation, focal loss and oversampling (setting 12).

Swapping the notary sign (setting 7–9, 11, 12) does neither drastically improve nor worsens the results. When using only notary sign swapping & adding without data augmentation (setting 7), the results improve compared to pure oversampling (setting 2). For DenseNet, the combination of notary sign swapping with data augmentation and oversampling (setting 9) achieves the best F-measure. Notary sign swapping worsens the result slightly for ResNet when applied in combination with "moderate" data augmentation, i.e., comparing setting 8 with setting 4.

Table 4. Classification results for ResNet and DenseNet.

Setting	Sensitivity	Specificity	F-value	Setting	Sensitivity	Specificity	F-value
1	**100**	0	0	1	90.2	56.5	69.5
2	92.2	82.4	87.0	2	93.3	87.0	90.0
3	97.4	93.4	95.3	3	98.5	95.0	96.8
4	96.5	**94.8**	95.7	4	97.3	94.9	96.1
5	95.4	94.5	94.9	5	97.3	94.7	96.0
6	94.9	94.7	94.8	6	96.4	96.9	96.6
7	96.9	84.0	89.9	7	97.2	89.7	93.3
8	97.1	92.6	94.8	8	98.1	94.2	96.1
9	96.4	93.7	95.0	9	98.4	**95.3**	**96.9**
10	94.5	51.2	66.4	10	94.4	89.9	92.1
11	94.4	89.6	91.9	11	95.4	88.3	91.8
12	97.8	94.0	**95.9**	12	**99.0**	94.8	**96.9**
(a) ResNet				(b) DenseNet			

Comparing undersampling (setting 6) with oversampling (setting 4), the results are inconclusive, oversampling works slightly better for ResNet and the other way around for DenseNet. Focal loss (setting 10–12) is also able to compensate data imbalance, but does not achieve the same recognition rates when applied without oversampling (10, 11). It performs significantly worse than oversampling on moderately augmented data (setting 10 vs. setting 4). When adding swapping to oversampling, the accuracy increases for ResNet but does not improve for DenseNet.

In nearly all settings (apart from setting 11), DenseNet outperforms the corresponding ResNet setting. In particular, all DenseNet settings with over- or undersampling in combination with augmentation perform better than the best ResNet setting.

4.3 Results of the Segmentation

Table 5 gives the results for all settings which were tried in this study. See Table 3 for individual details about them.

Focal Loss with Different Augmentations (Setting 1–4): Figure 4 shows the results of focal loss for different augmentation levels as colored areas. At all augmentation levels, the network trained with focal loss is able to distinguish well text and background. However, it fails entirely to segment ornaments or notary signs, see also Table 5 setting 1–4. Additionally, the IoU declines slightly with stronger augmentation as the best result is achieved without augmentation.

Moderate Augmentation and Different Loss Functions (Setting 3, 5, 6): A qualitative comparison of the different losses in combination with moderate augmentation can be found in Fig. 5, the color scheme is similar to the focal loss example. Dice loss is well able to segment notary signs, although it segments a larger area

Table 5. Intersection over Union (IoU) for different area types.

Setting	Background	Text	Ornament	Notary Sign	mean IoU
1	0.8820	0.8848	0.0000	0.0022	0.5896
2	0.8639	0.8711	0.0000	0.0000	0.5783
3	**0.8931**	0.8797	0.0000	0.0016	0.5915
4	0.8694	0.8544	0.0000	0.0041	0.5760
5	0.8791	0.8741	0.0000	0.0603	0.6045
6	0.8507	0.8503	**0.0049**	0.2902	0.6637
7	0.8857	**0.8833**	0.0000	**0.4101**	**0.7264**
8	0.8833	0.8622	0.0000	0.0000	0.5818
9	0.8818	0.8747	0.0000	0.0003	0.5856
10	0.8860	0.8705	0.0000	0.0001	0.5855
11	0.8255	0.7995	0.0000	0.0007	0.5419
12	0.7358	0.7977	0.0027	0.1942	0.5759
13	0.8494	0.8237	0.0000	0.0045	0.5592
Area [%]	**63.95**	**34.28**	**0.07**	**1.70**	

around the sign. Also the text is segmented mainly correctly, with comparable results to BCE and focal loss. Additionally, Dice loss delivers the most confident result. The only drawback are artifacts, such as the lower area of the notary sign in Fig. 4b that is classified as "text". Also non-existing "notary signs" appear several times in the lower left corner of documents. BCE segments text and background relatively accurately (setting 5), even better than Dice (setting 6), cf. Table 5. The segmentation for the text is especially tight. However it fails, similar to focal loss, to segment notary signs. Focal loss behaves comparable to BCE, with the only difference that BCE segments the notary signs slightly better, however, still insufficiently.

Combined Loss Functions (Setting 7–10): Figure 6 shows the results of the combined loss functions. Dice and BCE (setting 7) are a good combination providing overall accurate boundaries and segmented the notary sign clearly and overall achieve the best results in Table 5. All combinations with focal loss perform significantly worse.

Training with Meaningful Region Snippets (Setting 11–13): The region snippets deteriorate the result's quality in terms of confidence and accuracy. This is valid for both repeating and scaling, where the first performs better in experiments than the latter. This is also demonstrated in Fig. 7. It leads to artifacts and area miss-classifications. For any used loss function, the prediction is worse than with the corresponding network trained without this training scheme.

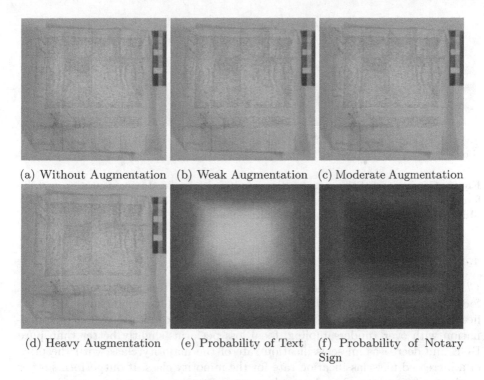

(a) Without Augmentation   (b) Weak Augmentation   (c) Moderate Augmentation

(d) Heavy Augmentation   (e) Probability of Text   (f) Probability of Notary Sign

Fig. 4. Comparison of the segmentation results with focal loss and different augmentations. In the plots *a*, *b*, *c* & *d* red color indicates class "background", green "text" and blue "notary sign". Source: BayHStA München KU Niederaltaich 951. (Color figure online)

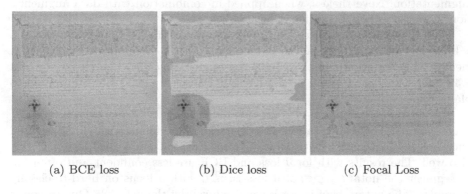

(a) BCE loss   (b) Dice loss   (c) Focal Loss

Fig. 5. Comparison of the segmentation results with moderate augmentation and different loss functions (color scheme is identical to Fig. 4). Source: Archiv der Erzdiözese Salzburg, D467. (Color figure online)

(a) BCE and Dice (b) BCE and Focal (c) Dice and Focal (d) BCE, Dice and
Loss   Loss   Loss   Focal Loss

Fig. 6. Comparison of the segmentation results with moderate augmentation and different loss function combinations (color scheme is identical to Fig. 4). Source: BayHStA München KU Aldersbach 877 (Color figure online)

4.4   Discussion

Classification: Countering class imbalance is important, data augmentation, focal loss, under- and oversampling improve the results significantly in classification. However, focal loss performs worse as single countermeasure. In combination with over-/undersampling, focal loss performs slightly better than pure BCE and decreases missclassification rate of the majority class, with the price of a increased missclassification rate for the minority class. It outperforms other settings, when it is combined with data augmentation, notary sign swapping and oversampling.

Data interpolation with swapping and adding notary signs is a valid counter strategy for class imbalance. The boost is however weak in comparison to data augmentation. Nevertheless, when applied in combination with data augmentation, it works as best boost for accuracy.

Data augmentation clearly improves the prediction. Despite this, there is a limit for that effect when the data quality starts to diminish (shown with "heavy" augmentation). Improvements by augmenting data contribute up to a 6.3% increase (ResNet) and 6.7% increase (DenseNet) in F-value compared to plain oversampling.

Segmentation: For segmentation, the best results are found with class-weighted Dice loss in combination with Binary Cross Entropy. Pure Dice loss performs also well. The results with focal loss and BCE are less confident and often fail to segment the notary sign. Focal loss is used with a focus on hard-to-classify examples. The notary sign is not recognized well by the network. One reason is probably that the manual segmentation lead to non-distinct boundaries, causing areas assigned to the wrong class to be treated as hard examples. The BCE loss function does not add class-weighting, in contrast to the used class-weighted Dice loss. We assume that with accurately estimated examples and clearly separable distributions, focal loss would perform better in segmentation since focal loss would compensate the missing class weight. Another factor, which reduces the

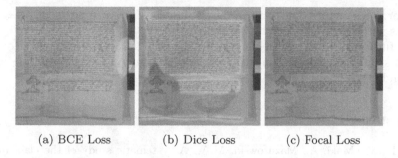

(a) BCE Loss   (b) Dice Loss   (c) Focal Loss

Fig. 7. Results of segmentation trained with snippets (color scheme is identical to Fig. 4). Source: BayHStA München KU Niederaltaich 2526 (Color figure online)

training success with focal loss, is that some documents have a very bad data quality, with a strongly faded text. The desired effect of an improved behavior for combined loss functions occurs with Dice and BCE loss. All other combinations – all including focal loss – deteriorate the results quality. The reasons are probably the same as for plain focal loss.

Feeding segments during training led to no improvements. When using scaled areas scaling to fill the image, the size of fed areas varies largely, as notary signs are usually much smaller than text area. This probably leads to inconsistency in the kernels, to decreased prediction quality and confidence, and to frequent artifacts. Repeating them probably conflicts with the overlap strategy of UNet and deteriorates region border recognition. An implementation where the areas would be of normal scale, but notary signs are fed more frequently without repeating, would probably lead to better results.

5   Conclusion

An optimal counter strategy for class-imbalance in two class classification problems is found with oversampling, notary sign swapping and moderate augmentation. Focal loss also works as imbalance counter. The results are comparable as with BCE when we use the same strategies as the best method, otherwise it performs worse. We identify the interpolation of data by swapping notary signs as a valid technique for data augmentation. However note that normal data augmentation is more efficient.

In segmentation, Dice loss with class weighting works well for data with high class imbalance. In combination with BCE its result even improves. This delivers clear segmentation results and sharp borders. Additionally, a good notary sign segmentation is achieved. Augmentation in case of class imbalance leads to an improvement for the focal loss function. Focal loss fails to segment the notary sign and is inappropriate for this application and data quality. Apparently, focal loss has problems with the deviations within the data, which are not always that clearly separable. Also feeding meaningful segments does not lead to improved behavior, presumably due to too large differences between segment sizes and conflicts with the UNet's overlap strategy.

For future work, we would like to vary the imbalance to see the effect on the current strategies and also improve the notary sign swapping technique using noise and other transformations.

References

1. Buda, M., Maki, A., Mazurowski, M.A.: A systematic study of the class imbalance problem in convolutional neural networks. Neural Netw. **106**, 249–259 (2018). https://doi.org/10.1016/j.neunet.2018.07.011
2. Buslaev, A.V., Parinov, A., Khvedchenya, E., Iglovikov, V.I., Kalinin, A.A.: Albumentations: fast and flexible image augmentations. arXiv (2018). https://doi.org/10.3390/info11020125
3. Harley, A.W., Ufkes, A., Derpanis, K.G.: Evaluation of deep convolutional nets for document image classification and retrieval. In: 13th International Conference on Document Analysis and Recognition (ICDAR), pp. 991–995 (2015)
4. Härtel, R.: Notarielle und kirchliche Urkunden im frühen und hohen Mittelalter, Historische Hilfswissenschaften, vol. 4. Böhlau (2011). https://doi.org/10.1524/hzhz.2012.0384
5. He, K., Zhang, X., Ren, S., Sun, J.: Deep residual learning for image recognition. In: 2016 IEEE Conference on Computer Vision and Pattern Recognition (CVPR), pp. 770–778 (2016). https://doi.org/10.1109/CVPR.2016.90
6. Heinz, K.: Monasterium.net: Auf dem Weg zu einem mitteleuropäischen Urkundenportal. Digitale Diplomatik. Neue Technologien in der historischen Arbeit mit Urkunden, pp. 70–77 (2009)
7. Huang, G., Liu, Z., Van Der Maaten, L., Weinberger, K.Q.: Densely connected convolutional networks. In: IEEE Conference on Computer Vision and Pattern Recognition (CVPR), pp. 4700–4708 (2017). https://doi.org/10.1109/CVPR.2017.243
8. Ioffe, S., Szegedy, C.: Batch normalization: accelerating deep network training by reducing internal covariate shift. In: International Conference on International Conference on Machine Learning (ICML), pp. 448–456 (2015)
9. Johnson, J.M., Khoshgoftaar, T.M.: Survey on deep learning with class imbalance. J. Big Data **6**(1), 1–54 (2019). https://doi.org/10.1186/s40537-019-0192-5
10. Kingma, D.P., Ba, J.: Adam: a method for stochastic optimization. In: International Conference on Learning Representations (ICLR) (2015)
11. Lin, T., Goyal, P., Girshick, R., He, K., Dollár, P.: Focal loss for dense object detection. IEEE Trans. Pattern Anal. Mach. Intell. **42**(2), 318–327 (2020). https://doi.org/10.1109/TPAMI.2018.2858826
12. Oliveira, S.A., Seguin, B., Kaplan, F.: dhSegment: a generic deep-learning approach for document segmentation. In: 2018 16th International Conference on Frontiers in Handwriting Recognition (ICFHR), pp. 7–12 (2018). https://doi.org/10.1109/ICFHR-2018.2018.00011
13. Oyedotun, O.K., Khashman, A.: Document segmentation using textural features summarization and feedforward neural network. Appl. Intell. **45**(1), 198–212 (2016). https://doi.org/10.1007/s10489-015-0753-z

14. Ronneberger, O., Fischer, P., Brox, T.: U-Net: convolutional networks for biomedical image segmentation. In: Navab, N., Hornegger, J., Wells, W.M., Frangi, A.F. (eds.) MICCAI 2015. LNCS, vol. 9351, pp. 234–241. Springer, Cham (2015). https://doi.org/10.1007/978-3-319-24574-4_28

15. Vogeler, G.: 'Monasterium.net' - eine Infrastruktur für diplomatische Forschung. Das Mittelalter **24**(1), 247–252 (2019). https://doi.org/10.1515/mial-2019-0022

16. Weileder, M.: Das virtuelle deutsche Urkundennetzwerk : Ein Kooperationsprojekt zur Online-Bereitstellung von Urkunden im Kontext der Erschließung. Digitale Urkundenpräsentationen **6**, 83–94 (2011)

17. Weileder, M.: Spätmittelalterliche Notarsurkunden aus virtuellen Archiven. In: Sonderveröffentlichung der Staatlichen Archive Bayerns, vol. 10, pp. 50–57. Bayerische Staatsbibliothek (2014)

18. Weileder, M.: Spätmittelalterliche Notarsurkunden: Prokuratorien, beglaubigte Abschriften und Delegatenurkunden aus bayerischen und österreichischen Beständen, Archiv für Diplomatik. Beiheft, vol. 18. Böhlau (2019)

19. Wong, S.C., Gatt, A., Stamatescu, V., McDonnell, M.D.: Understanding data augmentation for classification: when to warp? In: 2016 International Conference on Digital Image Computing: Techniques and Applications (DICTA), pp. 1–6 (2016). https://doi.org/10.1109/DICTA.2016.7797091

Evaluation of Neural Network Classification Systems on Document Stream

Joris Voerman[1,2]([envelope]), Aurélie Joseph[2], Mickael Coustaty[1],
Vincent Poulain d'Andecy[2], and Jean-Marc Ogier[1]

[1] La Rochelle Université, L3i Avenue Michel Crépeau, 17042 La Rochelle, France
{joris.voerman,mickael.coustaty,jean-marc.ogier}@univ-lr.fr
[2] Yooz, 1 Rue Fleming, 17000 La Rochelle, France
{joris.voerman,aurelie.joseph,vincent.poulaindandecy}@getyooz.com

Abstract. One major drawback of state of the art Neural Networks (NN)-based approaches for document classification purposes is the large number of training samples required to obtain an efficient classification. The minimum required number is around one thousand annotated documents for each class. In many cases it is very difficult, if not impossible, to gather this number of samples in real industrial processes. In this paper, we analyse the efficiency of NN-based document classification systems in a sub-optimal training case, based on the situation of a company's document stream. We evaluated three different approaches, one based on image content and two on textual content. The evaluation was divided into four parts: a reference case, to assess the performance of the system in the lab; two cases that each simulate a specific difficulty linked to document stream processing; and a realistic case that combined all of these difficulties. The realistic case highlighted the fact that there is a significant drop in the efficiency of NN-Based document classification systems. Although they remain efficient for well represented classes (with an over-fitting of the system for those classes), it is impossible for them to handle appropriately less well represented classes. NN-Based document classification systems need to be adapted to resolve these two problems before they can be considered for use in a company's document stream.

Keywords: Document classification · Image processing · Language processing

1 Introduction

Companies generate a large amount of documents everyday, internally or external entities. Processing all of these documents required a lot of resources. For this reason, many companies use automatic system like Digital Mailroom [29] to reduce the workload of those documents processing. Companies like ABBYY, KOFAX, PARASCRIPT, YOOZ, etc. propose performing solutions that can process electronic and paper documents. For the classification task, most of them

© Springer Nature Switzerland AG 2020
X. Bai et al. (Eds.): DAS 2020, LNCS 12116, pp. 262–276, 2020.
https://doi.org/10.1007/978-3-030-57058-3_19

use a combination of several classifiers specialized for one or some type of documents, with at least one based on Machine Learning techniques. However, all these systems have the same issue: they need to be retrained each time new classes are added. This maintenance is costly in time and in workload because it needs to build a learning dataset. An option to overcome this problem is to use an incremental learning system like [29] or [6], but they are not able to compete with state of the art deep learning systems in performances.

One main objective of this paper is to explore the adaptation of neural network systems to Digital Mailroom context for classification and extraction of main information inside all companies documents. Thereby, in such industrial context, an error (i.e. a misclassified document) has more impact than a rejection (i.e. a document rejected by the system and tagged with no class) because when a document is rejected, an operator will be warned and will correct it. On the opposite, an error will not be highlighted as a rejection and the error will be propagated into the next steps of the system.

One of the best possible modelization of Digital Mailroom entries is a document stream model. A document stream is a sequence of documents that appear irregularly in time. It can be very heterogeneous and composed of numerous classes unequally represented inside the stream. Indeed a document stream is generally composed, in a first time, of a core group of some classes highly represented, that is the majority of the stream. In a second time, remaining documents are unequally distributed between the majority of other classes, less represented than previous. Many classes are composed of only few documents that do not allow an efficient training. In addition, the class composition of a document stream is in constant growing, spontaneous new category or new sub-category variation appear time-to-time. And finally, the number of documents per class increases as the content of a document stream evolves.

This definition highlights two main constraints. First, in a real document stream application, classes representation are unbalanced. When a training set is generated from a document stream, it will be by nature unbalanced. This could reduce neural network methods performances if low represented classes are insufficiently trained and impact higher represented classes performances like noises. Then, no train set generated from a document stream could represent all real cases. The training phase will then be incomplete because the domain changes endlessly. These unexpected/incomplete classes, at first glance, cannot be managed by a neural network system.

The objectives of this experimentation is to evaluate the adaptation of neural network classification methods to these two constraints. Also, to determine and quantify impacts on network training phases and possible modifications that can be applied on neural network systems to counter or lessen this impact.

The next section will overview related works and methods that will be compared in the experimentation from Sect. 4. Then, we will describe the testing protocol used by this evaluation in the Sect. 3 and we will conclude and open perspectives in the Sect. 5.

2 Related Work

2.1 Overview

Solutions for document classification by machine learning methods can be divided in two approaches: Static Learning and Incremental Learning. Static Learning is based on a stable training corpus, supposed representative of the domain. Mostly, neural network methods follows this approach and can be themselves distinguished in two categories, regarding the fact they are whether based on pixels information or texts.

The first step to process texts with a neural network is the word representation. The most widely used techniques currently rely on the word embedding approach, like Word2Vec [21], GloVe [24] and more recently BERT [7]. These methods enhance previous textual feature like bag-of-words and word occurrences matrix. To limit potential noises, a word selection is commonly apply with basically a stop-word suppression ("the", "and", etc.). More advanced strategies used information gain, mutual information [5], L1 Regularization [22] or TF-IDF [14] to select useful feature. The second step is the classification itself with multiple model, mainly Recurrent Neural Network (RNN) [9] and Convolutional Neural Network (CNN) [15,32], and more recently a combination of these two structures, named RCNN, like in [18]. The RNN approaches are generally reinforced by specific recurrent cell architecture like Long Short-Term Memory LSTM [12] and Gated Recurrent Unit GRU [4]. Some industrial application use this type of network like CloudScan [23] for invoice classification and information retrieval.

For the image classification, the principal category of neural network method used is currently the pixel-based CNN with a high diversity of structures: ResNets [11], InceptionResNet [30], DenseNets [13], etc. But these approaches are not only restrictive to image and can be extended to documents classification without any text interpretation, as the RVL-CDIP dataset challenge [10].

All these methods are highly efficient for document classification when the Static Learning condition is met, but as explained in the introduction, document stream does not encounter this condition. The second approach has been designed to solve this issue with an Incremental Learning process. The main idea is to reinforce the system gradually at each data encounter and no more in one training phase. The incremental approach for neural network is relatively recent because their current structure were not designed for this task. However, two potential approaches have emerged recently: Structural Adaptation [25] or Training Adaptation [17,28]. Out of neural network, several classifiers based on classical algorithm can be find like incremental SVM [20], K-means [1] and Growing Neural Gas (IGNG) [3] or industrial applications like INTELLIX [29] or InDUS [6].

In addition of previous methods, two trends appeared in the last decade and are closed to our situation for classification of low represented classes: Zero-shot learning and One-Shot/Few-Shot Learning.

The first, Zero-Shot learning is an image processing challenge where some classes are need to be classified without previous training samples [31]. This description seems to be perfect for document stream classification, but all methods are based on transfer learning from a complete domain, mainly textual, where all classes are represented. This cannot be applied because no complete domain exists for our case.

The second, One-Shot Learning, is also an image processing challenge but where some classes are represented by only few, at least one, samples. This is the case for some low represented class from document stream. Two main groups of approaches exist: first group relies on the use of Bayesian-based techniques like in [19] and [8]; the second group relies on modified neural network architecture like the Neural Turing Machine (NTM), the Memory Augmented Neural Network (MANN) [27] and the Siamese Network [16].

In order to assess the performances of those categories of approaches on the dedicated case of document stream classification, we propose in the next section to provide a more detailed description of the evaluated methods. These ones have been chosen for their specificities that will be presented hereafter.

2.2 Compared Methods

Active Adaptive Incremental Neural Gas (A2ING) [3]

A2ING is a document stream classification method by Neural Gas that use an active semi-supervised sequential training.

The classification by Neural Gas is a machine learning method inspired by human brain. All classes are represented by centroids, called neurons. These centroids are put in a space composed in accordance with features chosen to describe the data, here documents. A document is then associated to the closest centroid in this feature space. The training is done on the position of centroids in the space and the classification range. This range is associate to each centroid and limit the distance allowed to express themselves. This range limitation is equally used to detect new classes.

This version is inspired by the Adaptive Incremental Neural Gas (AING) method, which was designed to relies on an incremental learning with an adaptive training phases. A2ING is a version with a active semi-supervised sequential training. Active mean that the training phases need an operator to correct algorithm answers because the learning dataset contains labelled and unlabelled data, called semi-supervised.

INDUS [6], developed by the french company YOOZ, is design to classify document stream. This system relies on the A2ING classification algorithm, and uses the textual information as feature to characterize documents. More precisely, it used the TF-IDF. INDUS has also a souvenir system allow remembering classes with few documents in order to assess performance improvement during the training phase. This study compare it with neural network on document stream classification.

Holistic Convolutional Neural Network (HCNN) [10]

HCNN is an image classifier using a CNN based on the pixels information. Its weights are initialized by a model trained on ImageNet 2012 challenge database [26]. According to authors, this fine-tuned model offers better performances and this method has the best results of their study on the RVL-CDIP dataset proposed in the same article. This method take as input a fixed sized image, resized if necessary, and train pixel-level feature by multiple CNN layers to distinguish classes. The classification itself is done by three successive fully-connected layers. We chose the HCNN approach as a baseline for RVL-CDIP classification task using only the image information.

Recurrent Convolutional Neural Network (RCNN) [18]

RCNN is a bidirectional recurrent CNN for text classification. It uses a word embedding like word2vec skip-gram model for text representation and is divided in two steps. The first step is a bidirectional RNN, our implementation used LSTM cell in the RNN layers. This network compute completely the two sides context of each word to enhance the word embedding. Contexts are the right and left sides of the word and are computed from the beginning to the end of the text. The second step is a CNN feed by the result of the bi-RNN (Fig. 1). This network end in a fully connected layer to classify input text.

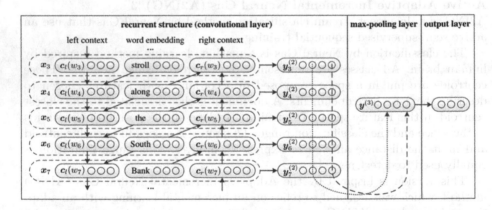

Fig. 1. The structure of the recurrent convolutional network scheme by [18] with sentence sample.

Textual Convolutional Neural Network (TCNN) [15,32]

This system is a combination inspired by two CNN designed for text classification. The combination relies on a character-level CNN [32] where each layer is replaced by a multi-channel CNN [15]. The result in a strong text classifier by vocabulary that has more basics feature than RCNN. This could have an importance in this evaluation with a non-optimal training set.

3 Testing Protocol

3.1 Dataset

The data-set used for this evaluation is RVL-CDIP [10], which is a subset of the IIT-CDIP [2] data-set. It is composed of 400 000 documents equally distributed in 16 classes. Classes correspond to a variety of industrial and administrative documents from the tobacco's companies. More specifically, the 16 classes are: letter, memo, email, file-folder, form, handwritten, invoice, advertisement, budget, news, article, presentation, scientific publication, questionnaire, resume, scientific report and specification. Some classes does not contain any usable text, like the file-folder one, or very few text like presentation. On the contrary, the scientific reports are mainly composed of text. Moreover, this advertisement class have an high structural diversity unlike the resume. For each class, 20 000 documents as used for the training set, 2500 for the validation set and 2500 for the test set. Originally, images have variable sizes, a resolution of 200 or 300 dpi and could be composed of one or multiple pages. For this work, we standardized all images to 754 * 1000 pixels, 72 dpi and one page.

In order to evaluate the language processing based methods, we applied a recent OCR software on the IIT-CDIP equivalent images for a better quality. The creator of RVL-CDIP data-set originally recommends to use their IIT-CDIP OCR text file but they were not organized in the same way than the RVL-CDIP images. Text files are computed on multiple page documents that was not in RVL-CDIP with an old OCR (2006). The new text files solve all those problems as they are computed on only one page with ABBYY-FineReader (version 14.0.107.232).

In addition, a private dataset provided by the Yooz company was used to challenge its own state of the art method [3]. This dataset is a subset of a real document stream from Yooz's customers. It is composed of 23 577 documents unequally distributed in 47 classes, 15 491 documents (65.71%) of them are used for training, 2203 (9.34%) for validation and 5883 (24,95%) for test. Each class contains between 1 to 4075 documents. The distribution of documents between classes is illustrated by Fig. 2. The main drawback of this data-set is its unbalanced and incomplete test set. With it, it is impossible to avoid the possibility of statistical aberration, in particular for classes represented by only one or zero documents in the test set. In addition, classes with a weak representation does not really impact global performances even if its specified classification accuracy is null. In fact, only a small number of over-represented classes correctly classified is enough to get a high accuracy score.

3.2 Evaluation Method

This section proposes to introduce the evaluation process apply in the next section. To evaluate each cases, we have to modify the training set consequently and only the training set. Classes impacted by the modification and documents used by the generated set was chosen randomly. For each run a new random

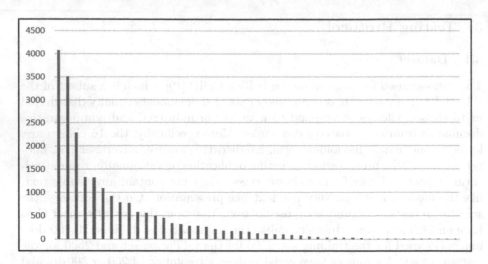

Fig. 2. Documents distribution between classes in Yooz private dataset. This introduce the number of document associated to each of 47 dataset classes

set was generated according to the process described to the corresponding case. Obviously comparison results between methods are computed on the same randomly generated set.

Que les résultats de chaque méthodes sont envoyé à un système de rejet ce qui permet d'évaluer le niveau de définition de chaque classe par le système et à quel point ces définition lui permettent bien les différencier entre elle.

In addition, an identical rejection system process each method results to evaluate the confidence level according to each class and the system capacity to distinguish classes between them even if the training is sub-optimal. The rejection system is an experimentally computed threshold applied on confidences scores obtained from the network. The last layer have one cell per class, and a sigmoid function provides the probability that the input image belong to each class. If the highest value is lower than the threshold, we reject the input in order to not miss-classify it. This is a deliberately simple rejection process because we wanted to evaluate only method performances and not the rejection system himself.

As explained in introduction, a rejected document is considered as less important than a error. To include this in the evaluation, we choose to do not used classic F1-Score but a F0.5-score. It improves the importance of precision beside recall with a modification of β variable of F-score equation (with $\beta = 0.5$):

$$(1 + \beta^2)\frac{Precision \cdot Recall}{\beta^2 \cdot Precision + Recall} \tag{1}$$

The other measures used are the Recall, the Precision and the Rejection Rate. The last measure is the global system accuracy that not includes the rejection system for its calculation, with the objective to display raw performances of neural network.

4   Experiments

The proposed evaluation tends to compare different methods in four cases: ideal one with same number of document in training, validation and test; unbalanced and incomplete cases where all classes are unbalanced or incomplete; a realistic case which mimics a real document stream. This comparison aims to determine the efficiency and the adaptation of neural network for document stream classification. The first one is used as reference, the two next methods are corresponding to two specific stream difficulties, and the last one simulate a realistic document stream. The four next subsections are devoted to each of those cases, with a presentation of the case, of the modifications applied on the training dataset, an overall analysis of the results, a methods adaptation analysis and a table with results.

4.1   Ideal Case

In Ideal case, all the training set of RVL-CDIP is used for the training phase of each method. It corresponds to the traditional evaluation conditions of neural network, and we will use it as a benchmark for methods efficiency comparison.

Results display in Table 1 show that the HCNN method got the best results. This can be explained by the fact that the dataset is more favorable to deep networks designed for image content recognition than the one dedicated to the textual content. Indeed, RVL-CDIP vocabulary is generally poor and many classes have a low number of words or in the worst case no words like the "file folders" class. In addition, an entire class and many other documents contain handwritten text that cannot be processed by the conventional OCR used here for the text extraction. Indeed, two classes are very complicated for text processing because they contain too few words and two others are also complicated but with slightly more words.

TCNN have the best performances for text content methods, it does not need too many words to work but the four complicated classes reduce drastically its accuracy. A low number of words is even more detrimental for RCNN that uses also as feature the word order inside the document text that need more text resources to be efficient. A2ING method suffer, in addition, of the huge number of documents because it was not designed to manage this.

4.2   Unbalanced Case

This second case simulates the unbalanced representation distribution between classes in a document stream. The objectives are to see if this unbalanced distribution impacts the neural network performances and how they are altered. The training set was modified with a reduction of some class distribution as follows:

Table 1. Result for ideal case

Value/Method	A2ING	HCNN	RCNN	TCNN
Accuracy	31.06%	**88.36%**	68.15%	79.67%
Precision	**95.94%**	95.42%	86.10%	93.72%
Recall	23.53%	**84.64%**	55.34%	69.91%
F0.5-score	59.39%	**93.05%**	77.48%	87.74%

all classes are divided in four groups of four classes. Each group is linked to a percentage of their original number of documents and organize in tier. These tier are respectively tier 5%, 10%, 50% and 100%. This distribution is a modelization based on real document streams.

All systems are affected by the unbalanced training set and lose between 9% and 11% of their accuracy as introduced by Table 2. The effect of this unbalanced training process highlights the over-fitting problem of the most represented classes (tier 100%, 50%) and an under-fitting (due to an insufficient training) for lesser represented classes (tier 10%, 5%). This unbalanced case results in a recall value higher than the precision for the first tier and the reverse for the second. The system seems to create trash classes with lesser well defined of tier 100% classes. These classes have a very low precision (less than 50%) in comparison of other (around 75%). However, our proposed rejection system allows stabilizing the precision of high tier classes but to the detriment of low tier recall.

RCNN keep the worst result and it is the second in accuracy loss. It is the only method where the rejection system has entirely eliminate last tier classes (tier 5%). The network has not been trained enough to manage these classes and their confidence scores were too low. HCNN is the most impacted method for accuracy reduction (-11.10%) but it keeps the highest global performances. Last tier seems to be the biggest problem instead of tier 10% that keep generally acceptable performances. TCNN is less impacted than RCNN and can handle the last tier unlike it.

4.3   Incomplete Case

As explained in introduction of this article, it is impossible to generate a training set that contains all classes from a real case document stream. Any system used for document stream classification have to handle new/unexpected classes or at least reject them as noise. This test is probably the most difficult for neural network because they are absolutely not designed for this type of situation. The objectives is then to analyse in particular the rejection result for reduced classes, the effect of noises during training on complete class performances. For this case the training set is split in two equal groups. The first one corresponds to complete classes and uses all the documents for the training phase. The second group is the noisy classes. We considered those classes in a similar way to the one proposed

Table 2. Result for unbalanced case and rejection rate means (RRM) for each classes tier

Value/Method	HCNN	RCNN	TCNN
Accuracy	**78.26%**	57.08%	71.41%
Precision	89.14%	78.76%	**91.63%**
Recall	**76.25%**	51.68%	56.73%
F0.5-score	**86.22%**	71.29%	81.59%
RRM tier 5%	41.74%	**84.85%**	55.98%
RRM tier 10%	30.31%	**46.64%**	42.99%
RRM tier 50%	**14.37%**	35.43%	58.53%
RRM tier 100%	**8.60%**	26.35%	15.58%

in the one-shot learning challenge [8,16,19,27]. So only one document is used to train them.

As expected, the obtained results are bad in accordance to Table 3. No method has well classified even only one document from the noisy classes, and the rejection process did not balanced enough the impact of noise on complete class performances. Only some classes with a specific vocabulary and highly formatted structure keep high performances, the other become trash classes and gather all noisy classes documents. The rejection rate is higher for the noisy classes than the complete ones, with in average 46.35% of documents rejected. But it is far from enough.

Again, the confidence scores of the RCNN approach do not allow differentiating enough the document classes, and the rejection system does not work well. The rejection rate is high for all classes and the noisy classes are not really more rejected than the others. Only one class obtained a good result as it is strongly defined by its vocabulary. TCNN and HCNN has got much better performances, in particular for specific vocabulary classes and for formatted structure classes. Moreover, they have an higher rejection rate for noisy classes.

4.4 Realistic Case

This scenario is a combination of the two previous cases, and was designed to be as close as possible to real documents stream. All classes are divided in five groups with in first time an incomplete group of four classes. In a second time, four groups of three classes with the same system than unbalanced case, with tier 5%, 10%, 50% and 100%. With this distribution we can finally simulate results of neural network methods on a document stream with an ideal test set and evaluate the global impact of document streams on neural network method efficiency.

We can observe in Table 4 that the obtained results are better than the ones from the incomplete case. This can be explained by the fact that fewer classes were incomplete. On the contrary, the obtained results are lower that

Table 3. Result for incomplete case and rejection rate means (RRM) for each classes groups

Value/Method	HCNN	RCNN	TCNN
Accuracy	**45.99%**	40.86%	43.97%
Precision	60.75%	47.40%	**61.38%**
Recall	**71.66%**	68.89%	60.10%
F0.5-score	**62.66%**	50.55%	61.12%
RRM noise	47.22%	36.03%	**55.84%**
RRM complete	**9.46%**	26.19%	23.97%

the ones from the unbalanced case with the addition of noisy classes. There is no prominent differences between each unbalanced tier results and those from Sect. 4.2, expect for the precision of 100% tier classes that gather in addition noisy classes documents. On another hand, the rejection rate for incomplete classes is higher here, around 66.22% on average. Like for Sect. 4.3, no documents of incomplete classes have been correctly classified. On the whole, all methods have lost between −23% and −28% of accuracy, so around one third of their original performances. Conclusions of individual result of each method is similar to the two previous cases.

The first tests with a artificial reduction of unbalanced seems to slightly improve performances of methods (with the rejection system). This modification result in a reduction of train samples for the 100% tier, to equal the 50% tier (so 6 classes with 50% of their original training samples). This seems to support the importance of balancing classes in a neural network train set because at first glance, this reduction of training sample numbers should decrease the method performances.

5   Conclusion

5.1   Results Summary

In a general analysis based on Table 5, we can say that neural network classification systems are unreliable in the situation of industrial document stream. They cannot handle very low represented or unexpected classes. They can deal, with difficulty, slightly more represented classes and they have high result for over-represented classes even if the precision was reduced by sub-represented classes that they gathered. The impact of incomplete classes is the main problem with the impossibility to add new unexpected class encountered to the classification system. This seems to be unsolvable without an important structural adaptation.

The unbalanced number of document per class inside the training set reduces the performances. The neural network based systems then tends to over-fit their model for the highest represented classes, to the detriment of the lesser well defined ones. The less represented classes are affected by sub-optimal training.

Table 4. Result for realistic case and rejection rate means (RRM) for each classes groups/tier

Value/Method	A2ING	HCNN	RCNN	TCNN
Accuracy	16.27%	**62.46%**	44.70%	51.93%
Precision	**84.63%**	77.55%	68.45%	78.50%
Recall	14.00%	**71.82%**	41.98%	46.98%
F0.5-score	42.13%	**76.33%**	60.78%	69.21%
RRM noise	**97.20%**	55.49%	71.02%	72.16%
RRM tier 5%	87.85%	30.01%	**89.29%**	66.79%
RRM tier 10%	**85.20%**	29.13%	59.16%	63.87%
RRM tier 50%	77.81%	**10.35%**	48.12%	33.08%
RRM tier 100%	78.19%	**6.81%**	18.16%	22.83%

In another hand, the unbalanced distribution of classes problem can be subdued by an adaptation of the training set. For instance, reducing the gap between the highest represented classes and the lowest represented classes, has a positive effect on the thresholded performances (*i.e.* the ones obtained with the rejection process) by reducing the over-fitting problem and getting a better modelization between them. The next step for this ways could be to improve the representation of low classes by the generation of samples like in [19].

5.2   Neural Networks Robustness Conclusion

On the whole, neural network classification systems are highly preforming in case where training data was not a problem. But for document stream classification, they cannot manage low represented and unexpected classes with same performances. Although, this type of classes represents more than a half of all stream classes.

Results introduce by Table 6 computed on the private Yooz's dataset do not lead to the same conclusion, all methods performances are high because the test set composition hide low classes impact on global result. Indeed, if the evaluation set is incomplete and unbalanced like the training set, low represented classes are impossible to evaluate, with sometimes only one are two documents. Individual class performance are not reliable for this dataset because low classes have not enough test samples. In addition, this too few samples does not impact global performances. In fact, its is possible to achieved a 70% accuracy with only the classification of three main classes (Fig. 2).

In an other hand, this dataset was build for the evaluation of text approach method and HCNN offer a surprisingly high score on it despite is composition mostly verbose. The RVL-CDIP dataset are probably not the best choice to compare text and image approaches, it was clearly designed for image classification method.

Table 5. Summary table RVL-CDIP

Case/Method	Acc	Pre	Rec	Acc	Pre	Rec
RVL-CDIP	A2ING			HCNN		
Ideal	31.06%	**95.94%**	23.53%	**88.36%**	95.42%	**84.64%**
Unbalanced	–	–	–	**78.26%**	89.14%	**76.25%**
Incomplete	–	–	–	**45.99%**	60.75%	**71.66%**
Realistic	16.27%	**84.63%**	14.00%	**62.46%**	77.55%	**71.82%**
RVL-CDIP	RCNN			TCNN		
Ideal	68.15%	86.10%	55.34%	79.67%	93.72%	69.91%
Unbalanced	57.08%	78.76%	51.68%	71.41%	**91.63%**	56.73%
Incomplete	40.86%	47.40%	68.89%	43.97%	**61.38%**	60.10%
Realistic	44.70%	68.45%	41.98%	51.93%	78.50%	46.98%

Table 6. Summary table private data-set

Case/Method	Acc	Pre	Rec	Acc	Pre	Rec
Private D-S	A2ING			HCNN		
True	85.37%	97.88%	80.79%	89.26%	98.13%	88.39%
Private D-S	RCNN			TCNN		
True	88.63%	95.67%	78.83%	**93.44%**	**98.97%**	**89.28%**

5.3 Perspectives

We need more test to determine the limit of training samples required by each method to stay reliable. Equally, it can be interesting to determine a formula that describe the behaviour of network efficiency as function of unbalance rate. One-shot learning methods like NTM and MANN [27] could be a option to balance impact of incomplete or very low represented classes on performances. The best solution seems to be the apart classification of this type of classes.

Finally, text or image approaches are both highly performing but not on same classes and the weakness of one could be balanced by the strength of the other on this dataset. A good multi-modal system could give better result as same as an improvement of rejection system.

References

1. Aaron, B., Tamir, D.E., Rishe, N.D., Kandel, A.: Dynamic incremental k-means clustering. In: 2014 International Conference on Computational Science and Computational Intelligence, vol. 1, pp. 308–313. IEEE (2014)
2. Baron, J.R., Lewis, D.D., Oard, D.W.: TREC 2006 legal track overview. In: TREC. Citeseer (2006)

3. Bouguelia, M.-R., Belaïd, Y., Belaïd, A.: A stream-based semi-supervised active learning approach for document classification. In: 2013 12th International Conference on Document Analysis and Recognition, pp. 611–615. IEEE (2013)
4. Cho, K., et al.: Learning phrase representations using RNN encoder-decoder for statistical machine translation. arXiv preprint arXiv:1406.1078 (2014)
5. Cover, T.M., Thomas, J.A.: Elements of Information Theory. Wiley, New York (2012)
6. d'Andecy, V.P., Joseph, A., Ogier, J.-M.: Indus: incremental document understanding system focus on document classification. In: 2018 13th IAPR International Workshop on Document Analysis Systems (DAS), pp. 239–244. IEEE (2018)
7. Devlin, J., Chang, M.-W., Lee, K., Toutanova, K.: Bert: pre-training of deep bidirectional transformers for language understanding. arXiv preprint arXiv:1810.04805 (2018)
8. Fei-Fei, L., Fergus, R., Perona, P.: One-shot learning of object categories. IEEE Trans. Pattern Anal. Mach. Intell. **28**(4), 594–611 (2006)
9. Graves, A., Mohamed, A.-R., Hinton, G.: Speech recognition with deep recurrent neural networks. In: 2013 IEEE International Conference on Acoustics, Speech and Signal Processing, pp. 6645–6649. IEEE (2013)
10. Harley, A.W., Ufkes, A., Derpanis, K.G.: Evaluation of deep convolutional nets for document image classification and retrieval. In: 2015 13th International Conference on Document Analysis and Recognition (ICDAR), pp. 991–995. IEEE (2015)
11. He, K., Zhang, X., Ren, S., Sun, J.: Deep residual learning for image recognition. In: Proceedings of the IEEE Conference on Computer Vision and Pattern Recognition, pp. 770–778 (2016)
12. Hochreiter, S., Schmidhuber, J.: Long short-term memory. Neural Comput. **9**(8), 1735–1780 (1997)
13. Huang, G., Liu, Z., Van Der Maaten, L., Weinberger, K.Q.: Densely connected convolutional networks. In: Proceedings of the IEEE Conference on Computer Vision and Pattern Recognition, pp. 4700–4708 (2017)
14. Jones, K.S.: A statistical interpretation of term specificity and its application in retrieval. J. Doc. **28**, 11–21 (1972)
15. Kim, Y.: Convolutional neural networks for sentence classification. arXiv preprint arXiv:1408.5882 (2014)
16. Koch, G., Zemel, R., Salakhutdinov, R.: Siamese neural networks for one-shot image recognition. In: ICML Deep Learning Workshop, Lille, vol. 2 (2015)
17. Kochurov, M., Garipov, T., Podoprikhin, D., Molchanov, D., Ashukha, A., Vetrov, D.: Bayesian incremental learning for deep neural networks. arXiv preprint arXiv:1802.07329 (2018)
18. Lai, S., Xu, L., Liu, K., Zhao, J.: Recurrent convolutional neural networks for text classification. In: Twenty-Ninth AAAI Conference on Artificial Intelligence (2015)
19. Lake, B.M., Salakhutdinov, R., Tenenbaum, J.B.: Human-level concept learning through probabilistic program induction. Science **350**(6266), 1332–1338 (2015)
20. Laskov, P., Gehl, C., Krüger, S., Müller, K.-R.: Incremental support vector learning: analysis, implementation and applications. J. Mach. Learn. Res. **7**(Sep), 1909–1936 (2006)
21. Mikolov, T., Chen, K., Corrado, G., Dean, J.: Efficient estimation of word representations in vector space. arXiv preprint arXiv:1301.3781 (2013)
22. Ng, A.Y.: Feature selection, L1 vs. L2 regularization, and rotational invariance. In: Proceedings of the Twenty-First International Conference on Machine Learning, p. 78 (2004)

23. Palm, R.B., Winther, O., Laws, F.: Cloudscan-a configuration-free invoice analysis system using recurrent neural networks. In: 2017 14th IAPR International Conference on Document Analysis and Recognition (ICDAR), vol. 1, pp. 406–413. IEEE (2017)

24. Pennington, J., Socher, R., Manning, C.D.: Glove: global vectors for word representation. In: Proceedings of the 2014 Conference on Empirical Methods in Natural Language Processing (EMNLP), pp. 1532–1543 (2014)

25. Rosenfeld, A., Tsotsos, J.K.: Incremental learning through deep adaptation. IEEE Trans. Pattern Anal. Mach. Intell. (2018)

26. Russakovsky, O., et al.: Imagenet large scale visual recognition challenge. Int. J. Comput. Vis. **115**(3), 211–252 (2015)

27. Santoro, A., Bartunov, S., Botvinick, M., Wierstra, D., Lillicrap, T.: One-shot learning with memory-augmented neural networks. arXiv preprint arXiv:1605.06065 (2016)

28. Sarwar, S.S., Ankit, A., Roy, K.: Incremental learning in deep convolutional neural networks using partial network sharing. IEEE Access **8**, 4615–4628 (2019)

29. Schuster, D., et al.: Intellix-end-user trained information extraction for document archiving. In: 2013 12th International Conference on Document Analysis and Recognition, pp. 101–105. IEEE (2013)

30. Szegedy, C., Ioffe, S., Vanhoucke, V., Alemi, A.A.: Inception-v4, inception-ResNet and the impact of residual connections on learning. In: Thirty-First AAAI Conference on Artificial Intelligence (2017)

31. Xian, Y., Schiele, B., Akata, Z.: Zero-shot learning-the good, the bad and the ugly. In: Proceedings of the IEEE Conference on Computer Vision and Pattern Recognition, pp. 4582–4591 (2017)

32. Zhang, X., Zhao, J., LeCun, Y.: Character-level convolutional networks for text classification. In: Advances in Neural Information Processing Systems, pp. 649–657 (2015)

Computerized Counting of Individuals in Ottoman Population Registers with Deep Learning

Yekta Said Can$^{(\boxtimes)}$ and Mustafa Erdem Kabadayı

Koc University, Rumelifeneri Yolu, 34450 Sarıyer, Istanbul, Turkey
ycan@ku.edu.tr

Abstract. The digitalization of historical documents continues to gain pace for further processing and extract meanings from these documents. Page segmentation and layout analysis are crucial for historical document analysis systems. Errors in these steps will create difficulties in the information retrieval processes. Degradation of documents, digitization errors and varying layout styles complicate the segmentation of historical documents. The properties of Arabic scripts such as connected letters, ligatures, diacritics and different writing styles make it even more challenging to process Arabic historical documents. In this study, we developed an automatic system for counting registered individuals and assigning them to populated places by using a CNN-based architecture. To evaluate the performance of our system, we created a labeled dataset of registers obtained from the first wave of population registers of the Ottoman Empire held between the 1840s–1860s. We achieved promising results for classifying different types of objects and counting the individuals and assigning them to populated places.

Keywords: Page segmentation · Historical document analysis · Convolutional Neural Networks · Arabic layout analysis

1 Introduction

Historical documents are precious cultural resources that provide the examination of historical, social and economic aspects of the past [1]. The digitization of them also provides immediate access for researchers and the public to these archives. However, for maintenance reasons, access to them might not be possible or could be limited. Furthermore, we can analyze and infer new information from these documents after the digitalization processes. For digitalizing the historical documents, page segmentation of different areas is a critical process for

This work has been supported by European Research Council (ERC) Project: "Industrialisation and Urban Growth from the mid-nineteenth century Ottoman Empire to Contemporary Turkey in a Comparative Perspective, 1850–2000" under the European Union's Horizon 2020 research and innovation programme grant agreement No. 679097.

X. Bai et al. (Eds.): DAS 2020, LNCS 12116, pp. 277–290, 2020.
https://doi.org/10.1007/978-3-030-57058-3_20

further document analysis and information retrieval [2]. Page segmentation techniques analyze the document by dividing the image into different regions such as backgrounds, texts, graphics, decorations [3]. Historical document segmentation is more challenging because of the degradation of document images, digitization errors and variable layout types. Therefore, it is difficult to segment them by applying projection-based or rule-based methods [3].

Page segmentation errors have a direct impact on the output of the OCR which converts handwritten or printed text into digitized characters [2]. Therefore, page segmentation techniques for the historical documents become important for the correct digitization. We can examine the literature on page segmentation under three subcategories [3]. The first category is the granular based techniques which combine the pixels and fundamental elements into large components [4,5] and [6]. The second category is the block-based techniques that divide the pages into little regions and then combine into large homogenous areas [7] and [8]. The last one is the texture-based methods which extracts textual features classify objects with different labels [9,10] and [11]. Except for the block-based techniques, these methods work in a bottom-up manner [3]. The bottom-up mechanisms have better performance with documents in variable layout formats [3]. However, they are expensive in terms of computational power because there are plenty of pixels or small elements to classify and connect [3]. Still, the advancement of technology of CPUs and GPUs alleviates this burden. Feature extraction and classifier algorithm design are very crucial for the performance of page segmentation methods. Although document image analysis has started with more traditional machine learning classifiers, with the emergence of Convolutional Neural Networks (CNNs), most studies use them because of their better performance.

Arabic script is used in writing different languages, e.g., Ottoman, Arabic, Urdu, Kurdish, Persian[12]. It could be written in different manners which complicate the page segmentation procedure. It is a cursive script in which connected letters create ligatures [12]. Arabic words could further include dots and diacritics which causes even more difficulties in the page segmentation [12].

In this study, we developed a software that automatically segments pages and recognizes objects for counting the Ottoman population registered in populated places. Our data comes from the first population registers of the Ottoman Empire that is realized in the 1840s. These registers are the results of an unprecedented administrative operation, which aimed to register each and every male subject of the empire, irrespective of age, ethnic or religious affiliation, military or financial status. Therefore, they aimed to have universal coverage for the male populace and thus these registers can be called (proto-) censuses. The geographical coverage of these registers is the entire Ottoman Empire in the mid-nineteenth century, which encompassed the territories of around two dozen successor states of today in Southeast Europe and the Middle East. For this paper, we are focusing on two locations: Nicaea in western Anatolia in Turkey, and Svishtov a Danubian town in Bulgaria.

In these censuses, officers prepared manuscripts without using handwritten or printed tables. Furthermore, there is not any pre-determined page structure. Page layouts can differ in different districts. There were also structural changes depending on the officer. We created a labeled dataset to give as an input to the supervised learning algorithms. In this dataset, different regions and objects are marked with different colors. We then classified all pixels and connected the regions comprising of the same type of pixels. We recognized the populated place starting points and person objects on these unstructured handwritten pages and counted the number of people in all populated places and pages. Our system successfully counts the population in different populated places.

The structure of the remaining parts of the paper is as follows. In Sect. 2, the related work in historical document analysis will be reviewed. We described the structure of the created database in Sect. 3. Our method for page segmentation and object recognition is described in Sect. 4. Experimental results and discussion are presented in Sect. 5. We present the conclusion and future works of the study in Sect. 6.

2 Related Works

Document image analysis studies have started in the early 1980's [13]. Laven et al. [14] developed a statistical learning based page segmentation system. They created a dataset that includes 932 page images of academic journals and labeled physical layout information manually. By using a logistic regression classifier, they achieved approximately 99% accuracy with 25 labels. The algorithm for segmentation was a variation of the XY-cut algorithm [15]. Arabic document layout analysis has also been studied with traditional algorithms in the literature. Hesham et al. [12] developed an automatic layout detection system for Arabic documents. They also added line segmentation support. After applying Sauvola binarization [16], noise filtering and skewness correction algorithms, they classified text and non-text regions with the Support Vector Machine (SVM) algorithm. They further segmented lines and words.

In some cases, the historical documents might have a tabular structure which makes it easier to analyze the layout. Zhang et al. [17] developed a system for analyzing Japanese Personnel Record 1956 (PR1956) documents which includes company information in a tabular structure. They segmented the document by using the tables and applied Japanese OCR techniques to segmented images. Richarz et al. [18] also implemented a semi-supervised OCR system on historical weather documents with printed tables. They scanned 58 pages and applied segmentation by using the printed tables. Afterward, they recognized digits and seven letters in the document.

After the emergence of neural networks, they are also tested on Arabic document analysis systems. Bukhari et al. [6] developed an automatic layout detection system. They classified the main body and the side text by using the MultiLayer Perceptron (MLP) algorithm. They created a dataset consisting of 38 historical document images from a private library in the old city of Jerusalem.

They achieved 95% classification accuracy. Convolutional Neural Network is also a type of deep neural network that can be used for most of the image processing applications [19]. CNN and Long Short Term Memory (LSTM) used for document layout analysis of scientific journal papers written in English in [20] and [21]. Amer et al. proposed a CNN-based document layout analysis system for Arabic newspapers and Arabic printed texts. They achieved approximately 90% accuracy in finding text and non-text regions.

CNNs are also used for segmenting historical documents. As mentioned previously, historical document analysis has new challenges when compared to the modern printed text layout analysis, such as degraded images, variable layouts and digitization errors. The Arabic language also creates more difficulties for document segmentation due to its cursive nature where letters are connected by forming ligatures. Words may also contain dots and diacritics which could be problematic for segmentation algorithms. Although, there are studies applying CNNs to historical documents [22], [3] and [2], to the best of our knowledge, this study is the first to apply CNN-based segmentation and object recognition in historical handwritten Arabic document analysis literature.

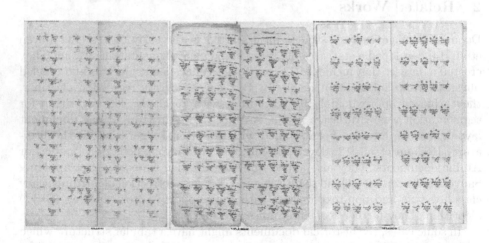

Fig. 1. Three sample pages of the registers belong to three different districts. The layout of pages can change between districts.

3   Structure of the Registers

Our case study is focusing on the registers of Nicaea and Svistov district registers, NFS.d. 1411, 1452, and NFS.d. 6314, respectively, available at the Turkish Presidency State Archives of the Republic of Turkey – Department of Ottoman Archives in jpeg format upon request. We aim to develop a methodology to be implemented for an efficient distant reading of similar registers from various regions of the empire prepared between the 1840s and the 1860s. As mentioned above, these registers provide detailed demographic information on male members of the households, i.e., names, family relations, ages, and occupations. Females in the households were not registered. The registers became available for research at the Ottoman state archives in Turkey, as recently as 2011. Their total number is around 11,000. Until now, they have not been subject to any systematic study. Only individual registers were transliterated in a piecemeal manner. The digital images of the recordings are usually around the size of 2100×3000 pixels.

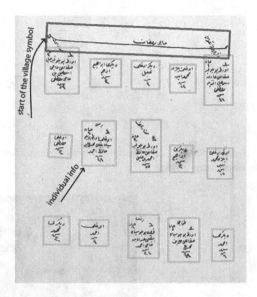

Fig. 2. Start of the populated place symbol and an individual cluster are demonstrated.

As mentioned previously, the layout of these registers can change from district to district (see Fig. 1) which makes our task more complicated. In this study, we work with the generic properties of these documents. The first property is the populated place start symbol. This symbol is used in most of the districts and can mark the end of the previous populated place and start of the new one (see Fig. 2). The remaining clusters are in the registers are individuals counted in the census and they include demographic information about them. There are also updates in these registers which marks the individuals when they go to the military service or decease. The officers generally draw a line on the individual and sometimes mistakenly connect the individual with an adjacent one which can cause some errors in the segmentation algorithm (see Fig. 3).

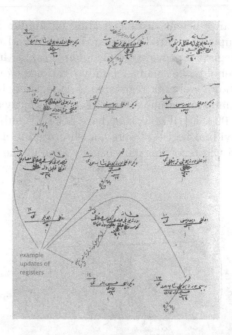

Fig. 3. Example updates of registers are shown. Some of them can connect two individuals and can cause clustering errors.

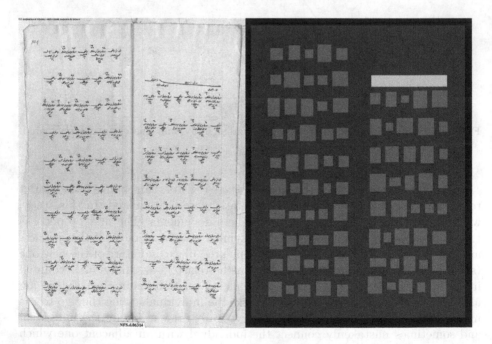

Fig. 4. A sample register page and its labeled version are demonstrated. Different colors represent different object types. (Color figure online)

4 Automatic Page Segmentation and Object Recognition System for Counting Ottoman Population

4.1 Creating a Dataset

To be able to use the dhSegment toolbox [22], we created a dataset with labels. We created four different classes. The first one is the background which is the region between the page borders and document borders. We marked this region as black. The second class is the page region and it is marked with blue. The third one is the start of a populated place object and we colored it with green. The last one is the individual registers and we marked them with red. We marked 173 pages with the described labels. 51 of them belong to the Svistov district and 122 of them belong to the Nicaea district. An example original image and labeled version are shown in Fig. 4.

4.2 Training the CNN Architecture

In order to train a CNN for our system, we used dhSegment [22] toolbox. This toolbox trained a system using the deep residual pretrained Resnet-50 architecture [23]. The toolbox has both a contracting path (follows the deep residual network in Resnet-50 [23]) and an expanding path which maps low resolution features to the original high resolution features (see terminology for expanding and contracting paths in [24]) [22]. The expanding path consists of five blocks and a convolutional layer for pixel classification and each deconvolutional step consists of upscaling of an image, concatenation of feature map to a contracting one, 3×3 convolutional and one Relu layer blocks.

In order to train the model, the toolbox used L2 regularization with 10^{-6} weight decay [22]. Xavier initialization [25] and Adam optimizer [26] are applied. Batch renormalization [27] is employed for refraining from a lack of diversity problem. The toolbox further downsized pictures and divided them into 300 $\times$ 300 patches for better fitting into the memory and providing support for training with batches. With the addition of margins, border effects are prevented. Because of the usage of pre-measured weights in the network, the training time is decreased substantially [22]. The training process exploits a variety of on-the-fly data augmentation techniques like rotation, scaling and mirroring. The system outputs the probabilities of each pixel belonging to one of the trained object types. Detailed metrics of one of the trained models by the integration of Tensorboard is shown in Fig. 5.

4.3 Preparing the Dataset for Evaluation

We trained three different models for evaluating the performance of our system. The first two models were trained with a register of one district and tested them with a completely different district's register. For the last model, we further combined our two registers and trained a combined model. This model is tested with 10-fold cross-validation.

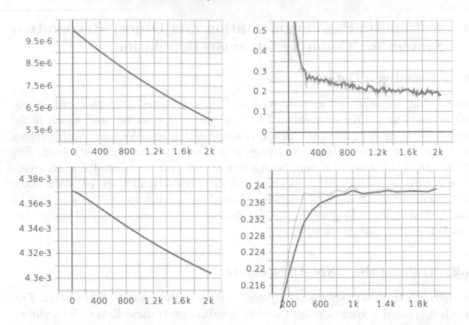

Fig. 5. Training metrics are demonstrated. In the top left, learning rate, in top right loss function, in bottom left regularized loss and in the bottom right global steps per second metrics are demonstrated. The subfigures are created with Tensorboard.

4.4   Post-processing

In our problem, we have four different classes: a background, page, an individual and the start of a populated place, namely. Therefore, we evaluated the probabilities of pixels that belong to one of the classes. For each class, there is a binarized matrix showing the probabilities that a pixel belongs to them. By using these matrices, pixels should be connected and components should be created. Connected component analysis tool [22] is used for creating objects. After the objects are constructed for all classes, the performance of our system could be measured.

4.5   Assigning Individuals to the Populated Places

This toolbox [22] finds the objects in all pages by supporting batch processing. However, for our purposes, we need the number of people in any populated place. To this end, we designed an algorithm for counting people and assigning them to the populated places. The flowchart of our algorithm can be seen in Fig. 6.

Firstly, we recorded the x and y coordinates of the rectangles of the found objects. The object could be of populated place start or individual type. Furthermore, they divided each page into two blocks and we have to consider this structure also. We defined a center of gravity for each object. It is computed by averaging all four coordinates of the rectangle surrounds the object. Due to the

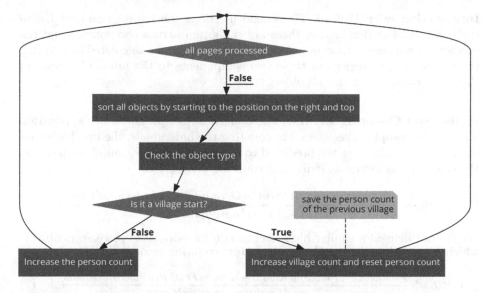

Fig. 6. Flowchart of our populated-place assigning algorithm

structure of the Arabic language, if an object is closer to the top of the page and right of the page than any other object, it comes before. However, if the object is in the left block of a page, without looking at the distance to top, it comes after any object in the right block of the page. We first sorted populated place start objects. For all individual objects, we compared their position on page and page number with all populated place start objects. If the individual object is after a populated place start object N and before populated place start object N + 1, we assigned the individual to populated place N.

5 Experimental Results and Discussion

In this section, we first define the metrics used for evaluating our system. We then present our results and discuss them.

5.1 Metrics

To evaluate our system performance, we used four different metrics. The first two metrics are low-level evaluators and they are widely used in object detection problems. We defined the third and fourth high-level metrics to evaluate the accuracy of our system.

Pixel-Wise Classification Accuracy: The first metric is the pixel-wise accuracy. It can be calculated by dividing the accurately classified pixels in all documents to the number of all pixels in all documents (for all object types).

Intersection over Union: The second metric is the Intersection over Union (IoU) metric. For this metric, there are the ground-truth components and the predicted components from our model. This metric can be calculated by dividing the intersection of regions of these two components to the union of regions of these two components (for all object types).

High-Level Counting Errors: These metrics are specific to our application for counting people in registers. For counting the individuals, the first high-level metric can be defined as the predicted count errors over the ground truth count. We can call this metric as individual counting error (ICE).

$$ICE = \left\| \frac{PredictedIndividualCount - GroundTruthIndividualCount}{GroundTruthIndividualCount} \right\| \quad (1)$$

We further defined a similar high-level metric for populated-place start objects which is named as the populated-place start counting error (VSCE).

$$VSCE = \left\| \frac{PredictedPopPlaceStartCount - GroundTruthPopPlaceStartCount}{GroundTruthPopPlaceStartCount} \right\| \quad (2)$$

5.2   Results and Discussion

We have two registers from the Nicaea district and one register from the Svistov district. In model 1, we trained with Nicaea registers and tested with the Svistov registers. In model 2, we trained a model with the Svistov district register and tested with the Nicaea registers. We further tested 10-fold cross validation on registers in the same district. In model 3, we trained and tested the model on the Svistov registers and in model 4, we trained and test on Nicaea registers with 10-fold cross-validation. In model 5, we combined the whole dataset and evaluated the model with 10-fold cross-validation. The pixel-wise accuracy, IoU and counting error results are provided in Table 1. Note that the first three metrics are provided for a different number of object classification. The last two metrics are the error of finding the number of individuals and the populated-place start objects. We further provided correctly predicted and mistakenly predicted raw binarized images in Fig. 7 and 8, respectively. The best ICE results are obtained when the Svistov registers are used for the training. The worst accuracy is obtained when the system is trained with Nicaea registers and tested with the Svistov register. Furthermore, the populated-place start counting error is 0% for all models which means that our system can recognize populated-place start objects perfectly.

As mentioned before, the layout of registers depends on the districts and the officer. For our registers, individuals in Nicaea are widely separated, whereas the distance between registers is less in Svistov registers. The average number of registered individuals in a Nicaea register page is approximately 40 and 80 in a Svistov register which confirms the above statement. Therefore, when the system is trained with loosely put Nicaea registers and tested in closely written clusters in Svistov, the counting error increases and the number of mistakes for counting multiple registers as one start to occur (see Fig. 8). Whereas, if we change training and test parts, the system error for counting objects approaches to 100% as we expected. If we mix the dataset and apply 10-fold cross-validation, we achieved counting errors in between. For our purposes, although high-level metrics are more crucial, low-level metrics showed the general performance of our system. They are also beneficial for comparing the performances of different models. Furthermore, even though IoU metric results are low, our classification errors are close to 0%. It could be inferred that the structure of registers is suitable for automatic object classification systems. The documents do not have printed tables, but their tabular-like structures make it easier to cluster and classify them.

Table 1. Results with different metrics are presented for five different models.

Trained on	Tested with	Pixel-wise Acc. (%)	IoU (%)	ICE (%)	VSCE (%)
Nicaea	Nicaea	94.25	80.91	1.65	0
Nicaea	Svistov	92.72	73.6	11.57	0
Svistov	Svistov	93.04	72.29	0.76	0
Svistov	Nicaea	85.36	47.95	0.27	0
Mixed	Mixed	87.26	48.54	2.26	0

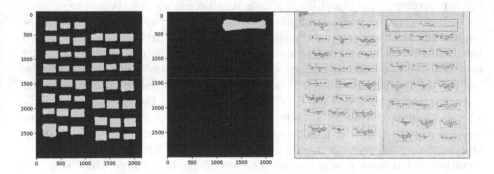

Fig. 7. A sample prediction made by our system. In the left, a binarized prediction image for counting individuals, in the middle a binarized image for counting populated-place start and in the right, the objects are enclosed with rectangular boxes. Green boxes for individual register counting and the red box for counting the populated-place start object. (color figure online)

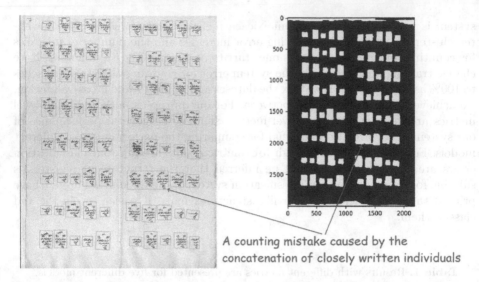

A counting mistake caused by the
concatenation of closely written individuals

Fig. 8. A sample counting mistake. All three individual registers are counted as 1. This results in two missing records in our automatic counting system.

6   Conclusion and Future Works

In this study, we developed an automatic individual counting system for the registers recorded in the first censuses of the Ottoman Empire which are held between 1840–1860. The registers are written in Arabic script and their layouts highly depend on the district and the officer in charge. We created a labeled dataset for three registers and evaluated our system on this dataset. We further developed an algorithm for assigning people to populated-places after detecting individual people and populated-place start symbols. For counting the populated-place start symbols, we achieved 0% error. Furthermore, we achieved the maximum individual counting error of 0.27%. We inferred from these results that the models should be trained with closely placed and noisy registers (Svistov register in our case study). When these models are tested with a clean and a loosely placed one (Nicaea register in this case study), the system counts individuals accurately. However, if a model is trained with a loosely placed register and tested with closely placed one, the number of counting errors is increasing. Our aim is to develop a generic system that can be implemented for efficient counting and distant reading of all registers prepared between the 1840s and the 1860s. Since it is impossible to label all registers, we will strategically label the closely placed and noisy ones to develop such a system. As future works, we plan to develop an automatic handwriting recognition system for the segmented individual register objects.

References

1. Kim, M.S., Cho, K.T., Kwag, H.K., Kim, J.H.: Segmentation of handwritten characters for digitalizing Korean historical documents. In: Marinai, S., Dengel, A.R. (eds.) DAS 2004. LNCS, vol. 3163, pp. 114–124. Springer, Heidelberg (2004). https://doi.org/10.1007/978-3-540-28640-0_11
2. Wick, C., Puppe, F.: Fully convolutional neural networks for page segmentation of historical document images. In: 2018 13th IAPR International Workshop on Document Analysis Systems (DAS), pp. 287–292. IEEE (2018)
3. Xu, Y., He, W., Yin, F., Liu, C.-L.: Page segmentation for historical handwritten documents using fully convolutional networks. In: 2017 14th IAPR International Conference on Document Analysis and Recognition (ICDAR), vol. 1, pp. 541–546. IEEE (2017)
4. Baechler, M., Ingold, R.: Multi resolution layout analysis of medieval manuscripts using dynamic MLP. In: 2011 International Conference on Document Analysis and Recognition, pp. 1185–1189. IEEE (2011)
5. Garz, A., Sablatnig, R., Diem, M.: Layout analysis for historical manuscripts using sift features. In: 2011 International Conference on Document Analysis and Recognition, pp. 508–512. IEEE (2011)
6. Bukhari, S.S., Breuel, T.M., Asi, A., El-Sana, J.: Layout analysis for Arabic historical document images using machine learning. In: 2012 International Conference on Frontiers in Handwriting Recognition, pp. 639–644. IEEE (2012)
7. Uttama, S., Ogier, J.-M., Loonis, P.: Top-down segmentation of ancient graphical drop caps: lettrines. In: Proceedings of 6th IAPR International Workshop on Graphics Recognition, Hong Kong, pp. 87–96 (2005)
8. Ouwayed, N., Belaïd, A.: Multi-oriented text line extraction from handwritten Arabic documents (2008)
9. Cohen, R., Asi, A., Kedem, K., El-Sana, J., Dinstein, I.: Robust text and drawing segmentation algorithm for historical documents. In: Proceedings of the 2nd International Workshop on Historical Document Imaging and Processing, pp. 110–117. ACM (2013)
10. Asi, A., Cohen, R., Kedem, K., El-Sana, J., Dinstein, I.: A coarse-to-fine approach for layout analysis of ancient manuscripts. In: 2014 14th International Conference on Frontiers in Handwriting Recognition, pp. 140–145. IEEE (2014)
11. Chen, K., Wei, H., Hennebert, J., Ingold, R., Liwicki, M.: Page segmentation for historical handwritten document images using color and texture features. In: 2014 14th International Conference on Frontiers in Handwriting Recognition, pp. 488–493. IEEE (2014)
12. Hesham, A.M., Rashwan, M.A.A., Al-Barhamtoshy, H.M., Abdou, S.M., Badr, A.A., Farag, I.: Arabic document layout analysis. Pattern Anal. Appl. 20(4), 1275–1287 (2017). https://doi.org/10.1007/s10044-017-0595-x
13. Nagy, G.: Twenty years of document image analysis in PAMI. IEEE Trans. Pattern Anal. Mach. Intell. 1, 38–62 (2000)
14. Laven, K., Leishman, S., Roweis, S.: A statistical learning approach to document image analysis. In: Eighth International Conference on Document Analysis and Recognition (ICDAR 2005), pp. 357–361. IEEE (2005)
15. Ha, J., Haralick, R.M., Phillips, I.T.: Recursive X-Y cut using bounding boxes of connected components. In: Proceedings of 3rd International Conference on Document Analysis and Recognition, vol. 2, pp. 952–955, August 1995

16. Sauvola, J., Seppanen, T., Haapakoski, S., Pietikainen, M.: Adaptive document binarization. In: Proceedings of the Fourth International Conference on Document Analysis and Recognition, vol. 1, pp. 147–152. IEEE (1997)
17. Zhang, K., Shen, Z., Zhou, J., Dell, M.: Information extraction from text regions with complex tabular structure (2019)
18. Richarz, J., Fink, G.A.: Towards semi-supervised transcription of handwritten historical weather reports. In: 10th IAPR International Workshop on Document Analysis Systems, pp. 180–184. IEEE (2012)
19. Matsumoto, T., et al.: Several image processing examples by CNN. In: IEEE International Workshop on Cellular Neural Networks and their Applications, pp. 100–111, December 1990
20. Breuel, T.M.: Robust, simple page segmentation using hybrid convolutional MDL-STM networks. In: 2017 14th IAPR International Conference on Document Analysis and Recognition (ICDAR), vol. 01, pp. 733–740, November 2017
21. Augusto Borges Oliveira, D., Palhares Viana, M.: Fast CNN-based document layout analysis. In: Proceedings of the IEEE International Conference on Computer Vision, pp. 1173–1180 (2017)
22. Ares Oliveira, S., Seguin, B., Kaplan, F.: dhSegment: a generic deep-learning approach for document segmentation. In: 2018 16th International Conference on Frontiers in Handwriting Recognition (ICFHR), pp. 7–12, August 2018
23. He, K., Zhang, X., Ren, S., Sun, J.: Deep residual learning for image recognition. In: Proceedings of the IEEE Conference on Computer Vision and Pattern Recognition, pp. 770–778 (2016)
24. Ronneberger, O., Fischer, P., Brox, T.: U-Net: convolutional networks for biomedical image segmentation. In: Navab, N., Hornegger, J., Wells, W.M., Frangi, A.F. (eds.) MICCAI 2015. LNCS, vol. 9351, pp. 234–241. Springer, Cham (2015). https://doi.org/10.1007/978-3-319-24574-4_28
25. Glorot, X., Bengio, Y.: Understanding the difficulty of training deep feedforward neural networks. In: Proceedings of the Thirteenth International Conference on Artificial Intelligence and Statistics, pp. 249–256 (2010)
26. Kingma, D.P., Ba, J.: Adam: a method for stochastic optimization. arXiv preprint arXiv:1412.6980 (2014)
27. Ioffe, S.: Batch renormalization: towards reducing minibatch dependence in batch-normalized models. In: Advances in Neural Information Processing Systems, pp. 1945–1953 (2017)

Word Embedding and Spotting

Annotation-Free Learning of Deep Representations for Word Spotting Using Synthetic Data and Self Labeling

Fabian Wolf$^{(\boxtimes)}$ ⓘ and Gernot A. Fink ⓘ

Department of Computer Science, TU Dortmund University,
44227 Dortmund, Germany
{fabian.wolf,gernot.fink}@cs.tu-dortmund.de

Abstract. Word spotting is a popular tool for supporting the first exploration of historic, handwritten document collections. Today, the best performing methods rely on machine learning techniques, which require a high amount of annotated training material. As training data is usually not available in the application scenario, *annotation-free* methods aim at solving the retrieval task without representative training samples. In this work, we present an *annotation-free* method that still employs machine learning techniques and therefore outperforms other *annotation-free* approaches. The weakly supervised training scheme relies on a lexicon, that does not need to precisely fit the dataset. In combination with a confidence based selection of pseudo-labeled training samples, we achieve state-of-the-art *query-by-example* performances. Furthermore, our method allows to perform *query-by-string*, which is usually not the case for other *annotation-free* methods

Keywords: Word spotting · Annotation-free · Weakly supervised

1 Introduction

The digitization of documents sparked the creation of huge digital document collections that are a massive source of knowledge. Especially, historic and handwritten documents are of high interest for historians. Nonetheless, information retrieval from huge document collections is still cumbersome. Basic functionalities such as an automatic search for word occurrences are extremely challenging, due to the high visual variability of handwriting and degradation effects. Traditional approaches like *optical character recognition* often struggle when it comes to historic collections. In these cases, word spotting methods that do not aim at transcribing the entire document offer a viable alternative [8]. Word spotting describes the retrieval task of finding the most probable occurrences of a word of interest in a document collection. As the system provides a ranked list of alternatives, it is up to the expert and his domain knowledge to decide which entities are finally relevant.

X. Bai et al. (Eds.): DAS 2020, LNCS 12116, pp. 293–308, 2020.
https://doi.org/10.1007/978-3-030-57058-3_21

Considering document analysis research, machine learning strongly influenced word spotting methods and a multitude of systems emerged [8]. Common taxonomies distinguish methods based on the type of query representation, a previously or simultaneously performed segmentation step and the necessity of a training procedure. Most systems represent the query either by an exemplar image (*query-by-example*) [19–21] or a string representation (*query-by-string*) [13, 27, 32]. In order to localize a query, the documents need to be segmented into word images. *Segmentation-based* methods [13, 20, 27] assume that this segmentation step is performed independently beforehand. In contrast, *segmentation-free* methods such as [22, 33] aim at solving the retrieval and segmentation problem jointly. Another distinction commonly made concerns the use of machine learning methods. So called *learning-free* techniques rely on expert designed feature representations [20, 30] and they are usually directly applicable as they do not rely on a learning phase. Motivated by the success in other computer vision tasks, machine *learning-based* techniques and especially convolutional neural networks dominate the field of word spotting today [27, 28, 33].

The distinction between *learning-free* and *learning-based* word spotting methods suggests that applying machine learning methods is a disadvantage in itself. This is only true for supervised learning approaches that require huge amounts of annotated training material in order to be successful. In cases where learning can be applied without such a requirement, we can not see any disadvantage of leveraging the power of machine learning for estimating models of handwriting for word-spotting purposes. We therefore suggest to distinguish methods based on the requirement of training data. Methods that do not rely on any manually labeled samples will be termed *annotation-free* as opposed to *annotation-based* techniques relying on supervised learning, as most current word spotting approaches based on deep learning do [27, 28]. Today, almost all *annotation-free* methods are also *learning-free* as it is not straight forward to devise a successful learning method that can be applied if manual annotations are not available. These *learning-free* methods provide a feature embedding that encodes the visual appearance of a word [20, 30]. As no model for the appearance of handwriting is learned, *query-by-string* is usually out of scope for these approaches.

In this work, we propose an *annotation-free* method for *segmentation-based* word spotting that overcomes this drawback by performing learning without requiring any manually labeled data. The proposed method uses a synthetic dataset to train an initial model. Due to the supervised training on the synthetic dataset, the model is capable to perform *query-by-string* word spotting. This initial model is then transferred to the target domain iteratively in a semi-supervised manner. Our method exploits the use of a lexicon which is used to perform word recognition to generate pseudo-labels for the target domain. The selection of pseudo-labels used to train the network is based on a confidence measure. We show that a confidence based selection is superior to randomly selecting training samples and already a rough estimate of the lexicon is sufficient to outperform other *annotation-free* methods. The proposed training scheme is summarized in Fig. 1.

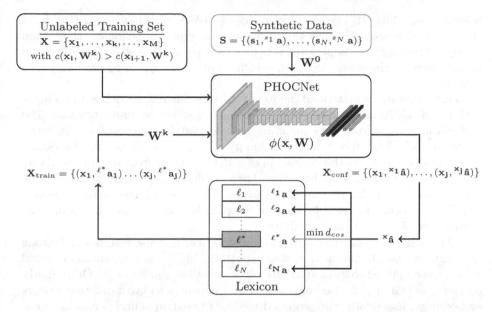

Fig. 1. Semi-supervised training scheme: First an initial model is trained on synthetic data. The model is then iteratively transferred to the target domain by training on confidently estimated samples which are pseudo-labeled with lexicon based recognition.

2   Related Work

In order to solve the retrieval task of word spotting, a system evaluates the similarity between document image regions and a query. Similar to popular recognition models, many methods exploit the sequential structure of handwriting. In an early work on *segmentation-based* word spotting, Rath and Manmatha proposed to use dynamic time warping to quantify the similarity of two word images based on the optimal alignment of their sequences [19]. Other sequential models such as *hidden Markov models* (HMM) [21,31] and *recurrent neural networks* [17] were also successfully used for word spotting and they are still popular today [28].

Traditional feature extraction methods also saw high popularity, due to their success in other computer vision tasks. In this case, the general approach is to embed the visual appearance of a word image in a feature vector. This *holistic* representation can then be easily compared to other document regions or image queries with a simple distance measure. Traditional descriptors based on gradient orientations such as HOG, LBP and SIFT have been shown to be suitable to capture the characteristics of handwriting [2,10,20]. Usually the descriptors of an image patch are accumulated in a histogram following the idea of a *Bag of Visual Words (BoVW)*. Since such a histogram vector neglects any spatial relations between descriptors, it is necessary to combine the approach with an additional model. For example, [1] and [24] use a pyramidal scheme to add spatial information, while [21] encodes a sequence of BoVW vectors with an HMM. As these feature-based approaches only embed the visual appearance, they often struggle when spotting is performed across multiple writing styles. Queries need to be given in the image domain, which only allows *query-by-example* word spotting.

These limitations motivated the use of *learning-based* approaches. In an influential work [3], Almazan et al. proposed the use of attribute representations. The method aims at learning the mapping between word images and a *Pyramidal Histogram of Characters*, which is a binary vector encoding the spatial occurrences of characters. As the derivation of a PHOC vector from a string is trivial, word images and strings can be mapped in a common embedding space, allowing *query-by-string*. In [3], the visual appearance of a word image is first encoded in a Fisher Vector, followed by a set of Support Vector Machines predicting the presence or absence of each attribute.

The application of neural networks and deep learning also resulted in a strong performance gain in the field of word spotting. In [26], a convolutional neural network is employed to learn attribute representations similar to [3]. Other methods such as [12] or [32] also employed neural networks to learn different feature embeddings. Essentially, the proposed methods based on neural networks significantly outperformed all previous approaches for *query-by-example* and *query-by-string*. While the discussed networks are usually trained on segmented word images, it has been shown that the approaches can be effectively adapted to the *segmentation-free* scenario by using word hypotheses [22] or region proposal networks [33].

Although *learning-based* methods showed exceptional performances on almost all benchmarks, they rely on a tremendous amount of training data. Considering the application of a word spotting tool, which is the exploration of a so far unknown document collection, the assumption that annotated documents are available rarely holds. This problem is far from being exclusive to word spotting and it has been of interest to the computer vision and machine learning communities in general. *Semi-supervised learning* describes the concept of using unlabeled data in combination with only a limited amount of annotated training samples [5]. In [6], this approach was successfully applied for a document analysis tasks. The authors used a word spotting system on a unlabeled dataset to generate additional labeled samples for a handwriting recognition system. A special type of *semi-supervised* methods employ so called self-labeling techniques [29], which have been also studied for neural networks [14]. In general, an initial model is trained on an annotated dataset and later used to generate labels for unlabeled data of the target domain. These pseudo-labeled data samples are integrated in the training scheme to further adapt the model.

Transfer-learning describes another approach to reduce the need of training data. It has been shown that data from another domain can be used to efficiently pre-train a model. In [13], a synthetic dataset for training word spotting models is proposed. Annotated training samples are rendered from computer fonts that resemble handwriting. The resulting dataset is used for pre-training a network that is then fine tuned on samples from the target domain. As shown in [9], training a model exclusively on synthetic data does not allow for state of the art performances. Anyways, the amount of training data necessary to achieve competitive results can be reduced significantly.

The lack of training data is a crucial problem for word spotting on historic datasets. While being clearly outperformed by fully-supervised methods and missing the possibility to perform *query-by-string* word spotting, *annotation-free*, feature-based methods are still receiving attention from the research community [20, 30].

3 Method

Our method evolves around a basic word spotting system based on an attribute CNN. We use the TPP-PHOCNet architecture proposed in [27] to estimate the attribute representation of an input word image. A 4-level PHOC representation of partitions 1, 2, 4 and 8 serves as a word string embedding. In all experiments, the assumed alphabet is the Latin alphabet plus digits, which results in an attribute vector $\mathbf{a} \in (0,1)^D$ with $D = 540$. Given a trained network the system allows to perform word spotting and lexicon-based word recognition, as described in Sect. 3.1.

The proposed training scheme presented in Sect. 3.2 does not require any manually annotated training material. Starting with an initial model, the system exploits the use of automatically generated pseudo-labels for the target domain. In order to enhance the accuracy of the generated labels, only a subset of the predicted pseudo-labels is used during the next training cycle. The selection of samples is based on an estimate of how confident the network is in its predictions. In this work, we compare the three confidence measures described in Sect. 3.3.

3.1 Word Spotting and Recognition

Given a trained PHOCNet with weights $\mathbf{W}$, the network constitutes a function ϕ that estimates the desired attribute representation $\hat{\mathbf{a}} = \phi(\mathbf{x}, \mathbf{W})$ for an input word image $\mathbf{x}$. In the segmentation-based scenario, word spotting is then performed by ranking all word images of the document according to their similarity to a query. In this work, similarity is measured by the cosine dissimilarity d_{cos} between the estimated attribute vector and the query vector. Depending on the query paradigm, the query vector $^q\mathbf{a}$ is either directly derived from a query string or estimated from a query word image $\mathbf{q}$ with $^q\hat{\mathbf{a}} = \phi(\mathbf{q}, \mathbf{W})$.

Word recognition is performed in a similar manner. Let $\mathbb{L}$ be a lexicon of size N with a set of corresponding attribute representations $^{\ell_i}\mathbf{a}$, $i \in 1, \ldots, N$. Based on the estimated attribute vector $^x\hat{\mathbf{a}}$ for the input image $\mathbf{x}$, word recognition reduces to a nearest neighbour search over the lexicon. Therefore, the recognition result ℓ^* is given by:

$$\ell^* = \underset{\ell \in \mathbb{L}}{\arg\min}\, d_{cos}\left(^\ell\mathbf{a}, {}^x\hat{\mathbf{a}}\right). \tag{1}$$

Algorithm 1: Semi-supervised training procedure

Input : Synthetic data $\mathbf{S} = \{(\mathbf{s}_1, {}^{s_1}\mathbf{a}), \ldots, (\mathbf{s}_N, {}^{s_N}\mathbf{a})\}$; unlabeled training
images $\mathbf{X} = \{\mathbf{x}_1, \ldots, \mathbf{x}_M\}$; number of training cycles K;
PHOCNet $\phi(\cdot, \mathbf{W})$; confidence measure $c(\cdot, \mathbf{W})$;

1   Train initial model $\phi(\cdot, \mathbf{W}^0)$ on $\mathbf{S}$;
2   **for** $k \leftarrow 0$ **to** K **do**
3   Estimate attribute representation ${}^{\mathbf{x}}\hat{\mathbf{a}} = \phi(\mathbf{x}, \mathbf{W}^k)$ for each element $\mathbf{x}$ in $\mathbf{X}$;
4   Sort $\mathbf{X}$ w.r.t. confidence: $\mathbf{X} = \{\mathbf{x}_1, \ldots, \mathbf{x}_j, \ldots, \mathbf{x}_M\}$ with
     $c(\mathbf{x}_i, \mathbf{W}^k) > c(\mathbf{x}_{i+1}, \mathbf{W}^k)$;
5   Select j most confident samples $\mathbf{X}_{conf} = \{\mathbf{x}_1, \ldots, \mathbf{x}_j\}$;
6   Generate pseudo labels with word recognition
     $\mathbf{X}_{train} = \{(\mathbf{x}_1, {}^{\ell_1^*}\mathbf{a}), \ldots, (\mathbf{x}_j, {}^{\ell_j^*}\mathbf{a})\}$;
7      $\mathbf{W}^{k+1} \leftarrow$ train $\phi(\cdot, \mathbf{W}^k)$ on $\mathbf{X}_{train}$

3.2 Training Scheme

The proposed training scheme is summarized in Algorithm 1. Initially, we train
the model $\phi(\cdot, \mathbf{W})$ on a purely synthetically generated dataset. Its images $\mathbf{s}_i$ are
generated based on computer fonts that resemble handwriting. The correspond-
ing attribute representations ${}^{s_i}\mathbf{a}$ are known and their creation does not cause
any manual annotation effort.

In order to further improve the initial model $\phi(\cdot, \mathbf{W}^0)$, we exploit the unlabeled target dataset. First, an estimate of the attribute representation ${}^{\mathbf{x}}\hat{\mathbf{a}}$ is
computed for each word image $\mathbf{x}$ in the target dataset $\mathbf{X}$. As shown in [9], a
model only trained on synthetic data does not yield a very high performance
and it is likely that the estimated attribute vectors are highly inaccurate. In
order to still derive a reliable pseudo-label, unreliable samples are removed and
a lexicon serves as an additional source of domain information.

Inaccurate estimates of attribute vectors are identified by the use of a confidence measure. Essentially, a confidence measure constitutes a function $c(\cdot, \mathbf{W})$
that quantifies the quality of the network outputs based on its current weights
$\mathbf{W}$. In this work, we investigate three different approaches to measure the network's confidence, see Sect. 3.3.

In each cycle a fixed percentage of confidently estimated samples is selected.
For each sample in this confident part of the unlabeled dataset $\mathbf{X}_{conf} = \{\mathbf{x}_1, \ldots, \mathbf{x}_j\}$ a pseudo-label is generated. The label l^* is derived by performing word recognition with the lexicon $\mathbb{L}$, as described in Sect. 3.1. Therefore,
the attribute vector representation used as a training target is derived from the
lexicon entry with minimal cosine dissimilarity to the estimated attribute representation ${}^{\mathbf{x}}\hat{\mathbf{a}}$. The resulting dataset with pseudo-labels is then used for training
in order to further adapt the model.

The process of estimating attribute representations based on the current
state of the model, selecting confident samples and generating pseudo-labels is
performed repeatedly for K cycles. As the training set of pseudo-labeled data

$\mathbf{X}_{train}$ is comparably small and the model therefore prone to overfitting the same regularization techniques as proposed in [27] are used. The set of training samples is augmented using random affine transformations. Based on the predicted labels, word classes are balanced in the resulting augmented training set. As an additional regularization measure, the PHOCNet architecture employs dropout in its fully connected layers.

3.3 Confidence Measures

As shown in [34], confidence measures allow to quantify the quality of an attribute vector prediction. Furthermore, recognition accuracies are higher on more confident parts of a dataset, making it a suitable tool for pseudo-label selection. Following the approach of [23], we model each Attribute A_i as a binary random variable following a Bernoulli distribution. The output of the attribute CNN is given by $^{\times}\hat{\mathbf{a}} = \phi(\mathbf{x}, \mathbf{W})$. Each element of the output vector is considered an estimate $^{\times}\hat{a}_i \approx p(A_i = 1|\mathbf{x})$ for the probability of the i-th attribute being present in the word image $\mathbf{x}$.

Sigmoid Activation. Based on the work in [34], we derive a confidence measure directly from the network outputs. Each sigmoid activation provides a pseudo probability for the estimate $\hat{a}_i$. To estimate the confidence of an entire attribute vector we follow the approach of [34] and we sum over the estimates of all active attributes. Neglecting inactive attributes, i.e., attributes with a pseudo probability of $\hat{a}_i < 0.5$, resulted in a slightly better performance in our experiments. We believe that this is due to the estimated attribute vector being almost binary with only a few attributes close to one. It seems that the confidence estimation for the high number of absent attributes disturbs the overall assessment of the attribute vector. The resulting confidence measure $c(\mathbf{x}, \mathbf{W})$ is then given by

$$c(\mathbf{x}, \mathbf{W}) = \sum_{^{\times}\hat{a}_i > 0.5} \phi(\mathbf{x}, \mathbf{W})_i \approx \sum_{^{\times}\hat{a}_i > 0.5} p(A_i = 1|\mathbf{x}). \tag{2}$$

Test Dropout. Another approach to estimate uncertainty is to use dropout as an approximation [7]. A confidence measure can be derived by applying dropout layers at test time. By performing multiple forward passes a variance can be observed for each attribute estimate $\hat{a}_i$. In this case the assumption is that for a confident prediction the estimate remains constant although neurons are dropped in the dropout layers. The approach is directly applicable to the PHOCNet as shown in [34]. All fully connected layers except the last one are applying dropout with a probability of 0.5. We calculate the mean over all attribute variances over 100 forward passes. A high confidence corresponds to a low mean attribute variance.

Entropy. A well known concept from information theory is to use entropy to measure the amount of information received by observing a random variable

[4]. The observation of a random variable with minimal entropy does not hold any information. Therefore, there is no uncertainty about the realization of the random variable. In this case, a low entropy corresponds to a high confidence in the network's predictions. Following the interpretation of an attribute as a Bernoulli distributed random variable A_i, its entropy is given by

$$H(A_i) = -\hat{a}_i \log \hat{a}_i - (1 - \hat{a}_i) \log(1 - \hat{a}_i). \tag{3}$$

To model the confidence of an entire attribute vector, we compute the negative joint entropy over all attributes. As in [23], we assume conditional independence among attributes. The joint entropy over all attributes is then computed by the sum over the entropies of the individual random variables.

$$c(\mathbf{x}, \mathbf{W}) = -H(A_1, \ldots, A_D) = -\sum_{i=1}^{D} H(A_i)$$
$$= \sum_{i=1}^{D} \hat{a}_i \log \hat{a}_i + (1 - \hat{a}_i) \log(1 - \hat{a}_i). \tag{4}$$

4 Experiments

We evaluate our method on four benchmark datasets for *segmentation-based* word spotting. In those cases where an annotated training set is available, we do not make use of any labels. For details on the datasets and the specific evaluation protocols see Sect. 4.1. We use *mean average precision* (mAP) in all our experiments to measure performance and to allow for a direct comparison to other methods [8]. As the provision of an exact lexicon can be quite a limitation in an application scenario, Sect. 4.3 presents experiments on different choices of lexicons. Sect. 4.4 provides an evaluation of different confidence measures and investigates the question whether a confidence based selection of samples is superior to random sampling. In Sect. 4.5, we compare our method to the state-of-the-art and especially to *annotation-free* methods.

4.1 Datasets

George Washington. The George Washington (GW) dataset has been one of the first datasets used to evaluate *segmentation-based* word spotting [19]. The documents were published by the Library of Congress, Washington DC, USA and they contain letters written by George Washington and his secretaries. In general, the writing style of the historic dataset is rather homogeneous. The benchmark contains 4860 segmented and annotated word images. As no distinctive separation in training and test partition exist, we follow the four-fold cross validation protocol presented in [3]. Although we train our network on the training splits, we do not make use of the annotations. Images from the test split are considered to represent unknown data and are therefore only used for evaluation.

IAM. The IAM database was created to train and to evaluate handwriting recognition models [15]. 657 different writers contributed to the creation of the benchmark, leading to a huge variety of writing styles. In total over 115000 annotated word images are split into a training, validation and test partition. No writer contributed to more than one partition. Due to its size and the strong variations in writing styles the IAM database became another widespread benchmark for word spotting [8]. The common approach for word spotting is to use each word image (*query-by-example*) or each unique transcription (*query-by-string*) of the test set as a query once. Stop words are not used as queries but still kept in the test set as distractors.

Bentham. The Bentham datasets originated from the project *Transcribe Bentham* and were used for the keyword spotting competitions at the *International Conference on Frontiers in Handwriting Recognition 2014* (BT14) [16] and at the *International Conference on Document Analysis and Recognition 2015* (BT15) [18]. The historic datasets contain documents written by the English philosopher Jeremy Bentham and show some considerable variations in writing styles. Both competitions define a *segmentation-based, query-by-example* benchmark. The BT14 set consists of 10370 segmented word images and a set of 320 designated queries. For BT15 the dataset was extended to 13657 word images and a significantly larger number of queries of 1421.

IIIT-HWS. We use a synthetically generated dataset to train an initial model without any manual annotation effort. The IIIT-HWS dataset, proposed in [13], was created from computer fonts that resemble handwriting. Based on a dictionary containing 90000 words, a total number of 1 million word images were created and successfully used to train a word spotting model. We use the published dataset to train our models and did not make any changes to the generation process.

4.2 Training Details

As the proposed method is based on the TPP-PHOCNet architecture, we mainly stick to the hyperparameters that have been proven successful in [27]. We train the network in an end to end fashion with Binary Cross Entropy and the ADAM optimizer. All input word images are inverted, such that the actual writing, presumably dark pixels, is represented by a value of one. In all experiments, we use a batch size of 10, weight decay of $5 \cdot 10^{-5}$ and we employ a momentum with mean 0.9 and variance 0.999.

Our model is initially trained on the IIIT-HWS dataset for 70000 iterations with a learning rate of 10^{-4}, followed by another 10000 training iterations with a learning rate of 10^{-5}. We follow the approach of [9] and randomly resize the synthetic word images during training to cope with differently sized images in the benchmark datasets. Each synthetic word image is scaled by a random factor within the interval $[1, 2)$.

Table 1. Evaluation of different lexicons. Pseudo-labels are selected randomly in all cases. Results reported as mAP [%].

Lexicon	GW		IAM		BT14		BT15	
	QbE	QbS	QbE	QbS	QbE	QbS	QbE	QbS
None (*)	46.6	57.9	16.0	39.5	18.1	–	16.4	–
Language Based	73.1	64.0	56.9	77.1	79.2	–	65.2	–
Closed	**87.8**	**87.8**	**63.7**	**83.6**	–	–	–	–
Bentham	–	–	–	–	**84.3**	–	**69.1**	–

*Initial model, no weakly supervised training.

The following training phase, which is only weakly supervised by a lexicon, is performed in multiple cycles. After each cycle, old pseudo-labels are neglected and a new set is generated with word recognition and the selection scheme. On all datasets except IAM we create a total number of 10000 samples using the augmentation method presented in [26]. Word classes are balanced based on the pseudo-labels. Due to the bigger size of the IAM database, we augment the pseudo-labeled samples to 30000 images. For each cycle, the network is trained for one epoch with respect to the augmented training set and a learning rate of 10^{-5}. In our experiments, we train the network for $K = 20$ cycles. After selecting 10% of the pseudo-labels as training samples during the first 10 cycles we increase the percentage to 60%.

4.3 Lexicon

In a first set of experiments, we investigate how crucial the prior knowledge on the lexicon is. All experiments do not make use of a confidence measure but perform the selection of pseudo-labeled samples randomly. We experiment with three different types of lexicons and compare the results against the performance of the network after training only on synthetic data. First we assume that only the language of the document collection is known. To derive a lexicon for all our datasets, we use the 10000 most common English words. Note that this results in 13.5% out-of-vocabulary words on GW, and 10.4% on the IAM database. Due to the lack of transcriptions, we cannot report out of vocabulary percentages for the Bentham datasets. As all samples in the GW and IAM datasets are labeled, we can create a closed lexicon containing all training and test transcriptions of the respective datasets. Even though, this is the most precise lexicon resulting in no out-of-vocabulary words, we argue that in an application scenario an exact lexicon is usually not available. In case of the Bentham datasets, we investigate another lexicon that is based on the manual line-level annotations published in [18]. This resembles the case that some related texts, potentially written by the same author, are available and provide a more precise lexicon.

Table 1 presents the resulting spotting performances with respect to the different lexicons. In general, performances increase substantially by training on the handwritten samples from the target domain under weak supervision.

Table 2. Evaluation of confidence measures. All experiments use a language based lexicon. Results reported as mAP [%]. Best *annotation-free* results are marked in bold.

Confidence	GW		IAM		BT14		BT15	
	QbE	QbS	QbE	QbS	QbE	QbS	QbE	QbS
Random	73.1	64.0	56.9	77.1	79.2	–	65.2	–
Entropy	79.5	82.1	**62.6**	**81.3**	84.2	–	75.2	–
Sigmoid	**83.2**	**82.3**	62.6	81.0	**87.2**	–	**76.3**	–
Test dropout	50.4	39.7	19.5	34.3	23.4	–	18.6	–
$d_{\cos}(^{\times}\hat{a},{}^{t}\mathbf{a})$	93.8	94.3	75.0	87.7	–	–	–	–

Already the approximate lexicon based on the modern English language results in high performance gains also for the historic benchmarks. For the closed as well as the related Bentham lexicon, it can be seen that performances increase with a more precise lexicon. Nonetheless, we would argue that in the considered scenario only a language based lexicon, which does not require any additional information on the texts besides their language, is a reasonable option.

4.4   Confidence Measures

As discussed in Sect. 3.3, a confidence measure can be used to identify parts of a dataset that have higher recognition accuracies. In our experiments, we use the three approaches described in Sect. 3.3 to quantify confidence. We only select the most confident pseudo-labeled samples to continue training. Furthermore, we conducted another experiment that uses the cosine dissimilarity between the estimated attribute representation $^{\times}\hat{a}$ and the actual transcription $^{t}\hat{a}$ as a confidence measure. This is motivated by the idea that a confidence measure essentially quantifies the quality of the attribute estimation, which corresponds to the similarity between estimation and annotation. Although in practice the cosine dissimilarity cannot be computed without a given annotation, it gives us an upper bound on how well the method would perform with a perfect confidence estimation.

Table 2 presents the results of the experiments, which are conducted with the different confidence measures. For entropy and sigmoid activations, we observe a performance gain on all benchmarks compared to a random sample selection. Despite the clear probabilistic interpretation, using entropy performs only on par with sigmoid activations and it is slightly outperformed on the presumably simpler datasets of GW and BT14. The use of test dropout does not yield any satisfactory results and even performs worse than a random approach. We observed that test dropout only gives high confidences for rather short words, which makes the selected pseudo-labeled samples not very suitable as training samples. A longer word potentially provides a bigger set of correct annotation on the attribute level, even in cases where the pseudo-label is wrong.

Table 3. Comparison on GW and IAM. Results reported as mAP [%]. Best *annotation-free* results are marked in bold, best overall in italic.

Method	Annotations [n]	GW		IAM	
		QbE	QbS	QbE	QbS
Languaged Based & Sigmoid	0	**83.2**	**82.3**	**62.6**	**81.0**
Almazan et al. [2]	0	49.4	–	–	–
Sfikas et al. [25]	0	58.3	–	13.2	–
DTW [3]	0	60.6	–	12.3	–
FV [3]	0	62.7	–	15.6	–
Retsinas et al. [20]	0	77.1	–	28.1	–
Gurjar et al. [9]	0	39.8	48.9	26.2	36.5
Gurjar et al. [9]	1000	95.7	96.5	55.3	74.0
AttributeSVM [3]	Complete	93.0	91.2	55.7	73.7
TPP-PHOCNet [27]	Complete	97.9	96.7	84.8	92.9
STPP-PHOCNet [23]	Complete	97.7	96.8	89.2	*95.4*
Deep Embed [11]	Complete	*98.0*	*98.8*	*90.3*	94.0
Triplet-CNN [32]	Complete	98.0	93.6	81.5	89.4

Considering the use of cosine dissimilarity, it can be seen that a more accurate confidence estimation can still improve performance. The proposed method in combination with cosine dissimilarity outperforms all other confidence measures, suggesting that the proposed confidence measures are providing suboptimal estimates only.

4.5 Comparison

In order to allow for a fair comparison to the state-of-the-art we only consider the performance of our method with respect to sigmoid activation as a confidence measure and a language based lexicon. Compared to other *annotation-free* methods, the only additional prior knowledge which we exploit, is the language of the considered documents. Table 3 reports the performance of the proposed method and other *annotation-free* and *annotation-based* approaches on the GW and IAM dataset. The best results so far that do not require training material are reported in [20]. Our method achieves higher mean average precisions on both datasets. Note that the difference is substantially higher in case of the IAM database. The work in [20] is heavily based on a specific feature design to incorporate visual appearance, which is quite suitable for the homogeneous appearance of the GW dataset. Nonetheless, our method outperforms all other *annotation-free* methods, while the difference is more substantial on datasets as the IAM database where writing styles and visual appearance vary strongly.

In [9], experiments were presented that show how performance increases, when a limited number of annotated samples is used to fine tune a network

Table 4. Comparison for the *annotation-free, query-by-example* benchmark on the Bentham datasets. Results reported as mAP [%]. Best results are marked in bold.

Method	BT14	BT15
	QbE	QbE
Languaged Based & Sigmoid	**87.2**	**76.3**
Aldavert et al. [1]	46.5	–
Almazan et al. [3]	51.3	–
Kovalchuk et al. [10]	52.4	–
CVC [18]	–	30.0
PRG [18]	–	42.4
Sfikas et al. [25]	53.6	41.5
Zagoris et al. [35]	60.0	50.1
Retsinas et al. [20]	71.1	58.4

similar to ours. While a number of 1 000 annotated samples are sufficient to outperform our semi-supervised approach on GW, we still achieve better performances on IAM. This suggests that our model is able to learn characteristics across different writing styles without relying on any annotations. Due to the lack of annotated training data from the target domain, our method performs worse compared to fully supervised approaches.

The experiments on both Bentham datasets reported in Table 4 support our observations. As the benchmarks are considered *annotation-free*, no word image labels are provided. Therefore, we cannot report any quantitative evaluation of *query-by-string* word spotting. While outperforming all other methods in the *query-by-example* case, our method additionally offers the possibility to perform *query-by-string*, which is not the case for all other *annotation-free* approaches.

5   Conclusions

In this work, we show that an *annotation-free* method for *segmentation-based* word spotting, which does not use any manually annotated training material, can still successfully employ machine learning techniques. Compared to other methods that do not include a learning phase, this leads to significant improvements in performance. The proposed method relies on a lexicon that provides additional domain information. Our experiments show that already a language based lexicon, which does not necessarily precisely correspond to the considered documents, is sufficient to achieve state-of-the-art performances. We successfully make use of a confidence measure to select pseudo-labeled samples during training to boost overall performance. Additionally, our method provides the capability to perform *query-by-string* word spotting, which is usually not the case for other *annotation-free* approaches. Therefore, our method is highly suitable for the exploration of heterogeneous datasets where no training material is available.

References

1. Aldavert, D., Rusiñol, M., Toledo, R., Lladós, J.: A study of bag-of-visual-words representations for handwritten keyword spotting. Int. J. Doc. Anal. Recogn. **18**(3), 223–234 (2015)
2. Almazán, J., Gordo, A., Fornés, A., Valveny, E.: Efficient exemplar word spotting. In: British Machine Vision Conference Surrey, UK (2012)
3. Almazán, J., Gordo, A., Fornés, A., Valveny, E.: Word spotting and recognition with embedded attributes. IEEE Trans. Pattern Anal. Mach. Intell. **36**(12), 2552–2566 (2014)
4. Bishop, C.M.: Pattern Recognition and Machine Learning. Information Science and Statistics. Springer-Verlag New York, Inc., Secaucus (2006)
5. van Engelen, J.E., Hoos, H.H.: A survey on semi-supervised learning. Mach. Learn. **109**(2), 373–440 (2019). https://doi.org/10.1007/s10994-019-05855-6
6. Frinken, V., Baumgartner, M., Fischer, A., Bunke, H.: Semi-supervised learning for cursive handwriting recognition using keyword spotting. In: Proceedings of International Conference on Frontiers in Handwriting Recognition, Bari, Italy, pp. 49–54 (2012)
7. Gal, Y., Ghahramani, Z.: Dropout as a Bayesian approximation: Representing model uncertainty in deep learning, New York City, NY, USA (2016)
8. Giotis, A.P., Sfikas, G., Gatos, B., Nikou, C.: A survey of document image word spotting techniques. Pattern Recogn. **68**, 310–332 (2017)
9. Gurjar, N., Sudholt, S., Fink, G.A.: Learning deep representations for word spotting under weak supervision. In: Proceedings of International Workshop on Document Analysis Systems, Vienna, Austria, pp. 7–12 (2018)
10. Kovalchuk, A., Wolf, L., Dershowitz, N.: A simple and fast word spotting method. In: Proceedings of International Conference on Frontiers in Handwriting Recognition, Crete, Greece, pp. 3–8 (2014)
11. Krishnan, P., Dutta, K., Jawahar, C.V.: Word spotting and recognition using deep embedding. In: Proceedings of International Workshop on Document Analysis Systems, Vienna, Austria, pp. 1–6 (2018)
12. Krishnan, P., Dutta, K., Jawahar, C.: Deep feature embedding for accurate recognition and retrieval of handwritten text. In: Proceedings of International Conference on Frontiers in Handwriting Recognition, Shenzhen, China,pp. 289–294 (2016)
13. Krishnan, P., Jawahar, C.V.: HWNet v2: an efficient word image representation for handwritten documents. Int. J. Doc. Anal. Recogn. **22**(4), 387–405 (2019)
14. Lee, D.H.: Pseudo-label: the simple and efficient semi-supervised learning method for deep neural networks. In: ICML 2013 Workshop: Challenges in Representation Learning (WREPL) (2013)
15. Marti, U., Bunke, H.: The IAM-database: an English sentence database for offline handwriting recognition. Int. J. Doc. Anal. Recogn. **5**(1), 39–46 (2002)
16. Pratikakis, I., Zagoris, K., Gatos, B., Louloudis, G., Stamatopoulos, N.: ICFHR 2014 competition on handwritten keyword spotting (H-KWS 2014). In: Proceedings of International Conference on Frontiers in Handwriting Recognition, Crete, Greece, pp. 814–819 (2014)
17. Puigcerver, J.: A probabilistic formulation of keyword spotting. Dissertation, Universitat Politècnica de València, València, Spain (2018)
18. Puigcerver, J., Toselli, A., Vidal, E.: ICDAR 2015 competition on keyword spotting for handwritten documents. In: Proceedings of International Conference on Document Analysis and Recognition, Nancy, France, pp. 1176–1180 (2015)

19. Rath, T.M., Manmatha, R.: Word spotting for historical documents. Int. J. Doc. Anal. Recogn. **9**(2–4), 139–152 (2007)
20. Retsinas, G., Louloudis, G., Stamatopoulos, N., Gatos, B.: Efficient learning-free keyword spotting. IEEE Trans. Pattern Anal. Mach. Intell. **41**(7), 1587–1600 (2019)
21. Rothacker, L.: Segmentation-free word spotting with bag-of-features hidden Markov models. Dissertation, TU Dortmund University, Dortmund, Germany (2019)
22. Rothacker, L., Sudholt, S., Rusakov, E., Kasperidus, M., Fink, G.A.: Word hypotheses for segmentation-free word spotting in historic document images. In: Proceedings of International Conference on Document Analysis and Recognition, Kyoto, Japan (2017)
23. Rusakov, E., Rothacker, L., Mo, H., Fink, G.A.: A probabilistic retrieval model for word spotting based on direct attribute prediction. In: Proceedings of International Conference on Frontiers in Handwriting Recognition, Niagara Falls, NY, USA, pp. 38–43 (2018)
24. Rusiñol, M., Aldavert, D., Toledo, R., Lladós, J.: Efficient segmentation-free keyword spotting in historical document collections. Pattern Recogn. **48**(2), 545–555 (2015)
25. Sfikas, G., Retsinas, G., Gatos, B.: Zoning aggregated hypercolumns for keyword spotting. In: Proceedings of International Conference on Frontiers in Handwriting Recognition, pp. 283–288 (2016)
26. Sudholt, S., Fink, G.A.: PHOCNet: a deep convolutional neural network for word spotting in handwritten documents. In: Proceedings of International Conference on Frontiers in Handwriting Recognition, Shenzhen, China, pp. 277–282 (2016)
27. Sudholt, S., Fink, G.A.: Attribute CNNs for word spotting in handwritten documents. Int. J. Doc. Anal. Recogn. (IJDAR) **21**(3), 199–218 (2018). https://doi.org/10.1007/s10032-018-0295-0
28. Toselli, A.H., Romero, V., Vidal, E., Sánchez, J.A.: Making two vast historical manuscript collections searchable and extracting meaningful textual features through large-scale probabilistic indexing. In: Proceedings of International Conference on Document Analysis and Recognition, Sydney, NSW, Australia, pp. 108–113 (2019)
29. Triguero, I., García, S., Herrera, F.: Self-labeled techniques for semi-supervised learning: taxonomy, software and empirical study. Knowl. Inf. Syst. **42**(2), 245–284 (2013). https://doi.org/10.1007/s10115-013-0706-y
30. Vats, E., Hast, A., Fornés, A.: Training-free and segmentation-free word spotting using feature matching and query expansion. In: Proceedings of International Conference on Document Analysis and Recognition, Sydney, NSW, Australia, pp. 1294–1299 (2019)
31. Vidal, E., Toselli, A.H., Puigcerver, J.: High performance query-by-example keyword spotting using query-by-string techniques. In: Proceedings of International Conference on Document Analysis and Recognition, Nancy, France,pp. 741–745 (2015)
32. Wilkinson, T., Brun, A.: Semantic and verbatim word spotting using deep neural networks. In: Proceedings of International Conference on Frontiers in Handwriting Recognition, pp. 307–312 (2016)
33. Wilkinson, T., Lindström, J., Brun, A.: Neural CTRL-F: segmentation-free query-by-string word spotting in handwritten manuscript collections. In: Proceedings of International Conference on Computer Vision, Venice, Italy,pp. 4443–4452 (2017)

34. Wolf, F., Oberdiek, P., Fink, G.A.: Exploring confidence measures for word spotting in heterogeneous datasets. In: Proceedings of International Conference on
Document Analysis and Recognition, Sydney, NSW, Australia, pp. 583–588 (2019)
35. Zagoris, K., Pratikakis, I., Gatos, B.: Unsupervised word spotting in historical
handwritten document images using document-oriented local features. IEEE Trans.
Image Process. **26**(8), 4032–4041 (2017)

Fused Text Recogniser and Deep Embeddings Improve Word Recognition and Retrieval

Siddhant Bansal$^{(\boxtimes)}$, Praveen Krishnan, and C. V. Jawahar

Center for Visual Information Technology, IIIT, Hyderabad, India
siddhant.bansal@students.iiit.ac.in, praveen.krishnan@research.iiit.ac.in,
jawahar@iiit.ac.in

Abstract. Recognition and retrieval of textual content from the large document collections have been a powerful use case for the document image analysis community. Often the word is the basic unit for recognition as well as retrieval. Systems that rely only on the text recogniser's (OCR) output are not robust enough in many situations, especially when the word recognition rates are poor, as in the case of historic documents or digital libraries. An alternative has been word spotting based methods that retrieve/match words based on a holistic representation of the word. In this paper, we fuse the noisy output of text recogniser with a deep embeddings representation derived out of the entire word. We use average and max fusion for improving the ranked results in the case of retrieval. We validate our methods on a collection of Hindi documents. We improve word recognition rate by 1.4% and retrieval by 11.13% in the mAP.

Keywords: Word recognition · Word retrieval · Deep embeddings · Text recogniser · Word spotting

1 Introduction

Presence of large document collections like the Project Gutenberg [2] and the Digital Library of India (DLI) [5] has provided access to many books in English and Indian languages. Such document collections cover a broad range of disciplines like history, languages, art and science, thereby, providing free access to a vast amount of information. For the creation of such libraries, the books are converted to machine-readable text by using Optical Character Recognition (OCR) solutions.

Use of OCR has also enabled successful retrieval of relevant content in document collections. In this work, we aim at improving the word recognition and retrieval performance for the Hindi language. This is challenging since the data in these libraries consists of degraded images and complex printing styles. Current methods providing content-level access to a large corpus can be divided into two classes: (a) text recognition (OCR) and (b) word spotting. Text recognition-based

© Springer Nature Switzerland AG 2020
X. Bai et al. (Eds.): DAS 2020, LNCS 12116, pp. 309–323, 2020.
https://doi.org/10.1007/978-3-030-57058-3_22

	Input Image	Baseline Prediction	Prediction using Confidence Score	Lexicon based Prediction
(a)	आलोचनाएं	आलोचनाए	आलोचनाएं	आलोचनाएं
(b)	बेचैनी	बेचनी	बेचैनी	बेचैनी
(c)	बर्हिमुखी	बरहिमुखी	बरहिमुखी	बर्हिमुखी
(d)	टेलीफ़ोन	टेलीफोन	टेलीफोन	टेलीफ़ोन

Fig. 1. In this figure we show the word recognition results. In (a) and (b), the correct word is selected using both the confidence score and lexicon based methods proposed in this work. In (c) and (d), the correct output is predicted only by the lexicon based prediction. Both the methods are successful in capturing very low level details that are missed by the baseline text recognition methods.

approaches have shown to perform well in many situations [9,19,23]. Using this technique, scanned documents are converted to machine-readable text and then the search is carried out for the queries on the generated text. On the other hand, word spotting is a recognition-free approach for text recognition. Here, embeddings for the words in the documents are extracted and the nearest neighbour search is performed to get a ranked list. Various attempts [7,8,18,22] have been made for creating systems using recognition-free approaches also. However, all of these methods fall short in taking advantage of both, i.e., effectively fusing recognition-based and recognition-free methods.

We aim at exploring the complementary behavior of recognition-based and recognition-free approaches. For that, we generate the predictions by a CRNN [23] style text recognition network, and get the deep embeddings using the End2End network [16]. Figure 1 shows the qualitative results obtained for word recognition using methods proposed in this work. As it can be seen in the Fig. 1, baseline word recognition system fails to recognise some of the minute details. For example, in Fig. 1(a) and (d) the baseline word recognition system fails to predict the '.' present in the words. Whereas, in Fig. 1 (b) and (c) the baseline recognition system fails to identify a couple of *matras*. Using the methods proposed in this work, we are able to capture the minute details missed by the baseline work recognition system and contribute towards improving the word recognition system.

All the methods proposed in this work are analysed on the Hindi language. Though our experimental validation is limited to printed Hindi books, we believe, the methods are generic, and independent to underlying recognition and embedding networks. Word retrieval systems using word spotting are known to have a higher recall, whereas, text recognition-based systems provide higher precision [17]. Exploring the aforementioned fact, in this work, we propose various ways of fusing text recognition and word spotting based systems for improving word recognition and retrieval.

1.1 Related Works

We build upon the recent work from text recognition and word spotting. We propose a new set of methods which improves these existing methods by using their complementary properties. In this section we explore the work done in (a) text recognition and (b) word spotting.

Modern text recognition solutions are typically modelled as a Seq2Seq problem using Recurrent Neural Networks (RNNs). In such a setting, convolutional layers are used for extracting features, leading to Convolutional Recurrent Neural Network (CRNN) based solutions. Transcriptions for the same are generated using different forms of recurrent networks. For example, Garain et al. [11] use BLSTMs along with CTC loss [12] and Adak et al. [3] use CNNs as features extractors coupled with RNN for sequence classification. Similarly, Sun et al. [27] use a convolutional network as the feature extractor and multi-directional (MDir) LSTMs as the recurrent units. Pham et al. [19] also use the convolutional network as feature extractor whereas they use multi-dimensional long short-term memory (MDLSTM) as the recurrent units. Chen et al. [9] use a variation of the LSTM unit, which they call SeqMDLSTM. All these methods heavily rely on recognition-based approach for word recognition and do not explore recognition-free approaches for improving word recognition and retrieval. Whereas, we use a CNN-RNN hybrid architecture first proposed in [23] for generating the textual transcriptions. To add to it we use the End2End network [16] for generating deep embeddings of the transcriptions to further improve the word recognition using methods defined in Sect. 3.

In word spotting the central idea is to compute the holistic representation for the word image. Initial methods can be traced back to [21] where Rath et al. uses profile features to represent the word images and then compared them using a distance metric. Many deep learning approaches like [13,14] have been proposed in the domain of scene text recognition for improved text spotting. Poznanski et al. [20] adopted VGGNet [24] by using multiple fully connected layers, for recognising attributes of Pyramidal Histogram of Characters (PHOC). Different CNN architectures [15,25,26,28] were proposed which uses PHOC defined embedding spaces for embedding features. On the other hand, Sudholt et al. [25] suggested an architecture which embeds image features to PHOC attributes by using sigmoid activation in the last layer. It uses the final layer to get a holistic representation of the images for word spotting and is referred to as PHOCNet. Methods prior to deep learning use handcrafted features [6,18], and Bag of Visual Words approaches [22] for comparing the query image with all other images in the database. Various attempts of retrieving Indian texts using word spotting methods have been illustrated in [7,8,18,22]. All these methods greatly explore the recognition-free approach for word recognition. They do not study the fusion of recognition-free approach with recognition-based approaches to use the complementary information from both the systems.

In this paper, we explore various methods using which we are able to fuse both recognition-based and recognition-free approaches. We convert the text transcriptions generated by recognition-based methods to deep embeddings using recognition-free approach and improve word recognition and retrieval. Our major contributions are:

1. We propose techniques for improving word recognition by incorporating multiple hypotheses and corresponding deep embeddings. Using this technique, we report an average improvement in the word accuracy by 1.4%.
2. Similarly for improving word retrieval we propose techniques using both recognition-based and recognition-free approaches. Using approaches introduced in this paper, we report an average improvement in the mAP score by 11.12%. We release an implementation at our webpage[1].

2   Baseline Methods

We first explain the two baseline methods that we are using in this work for our task.

2.1   Text Recognition

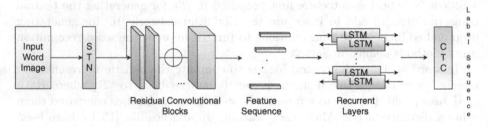

Fig. 2. CRNN architecture takes in a word image passes it through the Spatial Transform (STN) layer (for correcting the affine transformation applicable to the word images), followed by residual convolutional blocks which learns the feature maps. These feature maps are further given as an input to the BLSTM layer.

The problem of text recognition involves converting the content of an image to textual transcriptions. For this purpose, we use CNN-RNN hybrid architecture which was first proposed by [23]. Figure 2 shows the CRNN with the STN layer proposed in [10]. Here, from the last convolutional layer we obtain a feature map $F_l \in \mathbb{R}^{\alpha \times \beta \times \gamma}$ which is passed as an input to the BLSTM layers as a sequence of γ feature vectors, each represented as $F_{l+1} \in \mathbb{R}^{\alpha \times \beta}$. Here l is the layer ID. CTC loss [12] is used to train the network with best path decoding.

---

[1] http://cvit.iiit.ac.in/research/projects/cvit-projects/fused-text-recogniser-and-de ep-embeddings-improve-word-recognition-and-retrieval.

2.2   Word Spotting

We use the End2End network proposed in [16] to learn the textual and word image embeddings. Figure 3 shows this architecture. Feature extraction and embedding are the major components of the network. The network consists of two input streams - Real Stream and Label Stream as shown in Fig. 3. The real stream takes in real-word images as input and feeds it into a deep residual network which computes the features. The label stream gets divided into two different streams - the PHOC [4] feature extractor and a convolutional network. A synthetic image of the current label is given as an input to the convolutional network which in turn calculates its feature representation. This feature representation is in turn concatenated with the vectorial representation which is calculated using PHOC. Features generated from both the streams are fed to the label embedding layer which is responsible for projecting the embeddings in a common feature space where both the embeddings are in close proximity.

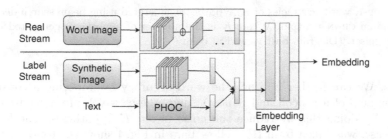

Fig. 3. End2End[16] network for learning both textual embedding using a multi-task loss function.

3   Fusing Word Recognition and Word Spotting

We propose a system that leverages the best traits of both the recognition-based and recognition-free approaches. Given the document images, word images are cropped out using the word bounding box information available as part of the annotation. The word images are fed into a pre-trained text recognition network as described in the Sect. 2.1 to get the textual transcriptions. Similarly, the word images are also fed into the End2End network [16] for getting word images' deep embeddings.

3.1   Word Recognition Using Multiple Hypotheses

To improve word recognition, we use the beam search decoding algorithm for generating K hypotheses from the text recognition system. Figure 4 shows a graph of top-k word accuracy vs. K, where K is the number of hypotheses generated by the text recogniser. Bars in dark blue show the top-k accuracy. Here we consider the output to be correct if the correct word occurs once in K

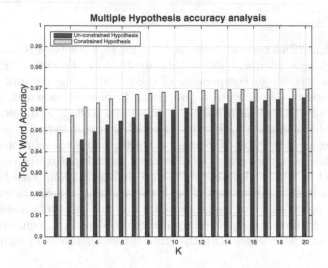

Fig. 4. Top-K word accuracies of K hypotheses generated using beam search decoding algorithm on CRNN's output. Where K is the number of hypotheses generated by the text recogniser. (Diagram best viewed in color)

outputs. We can further improve the word accuracy by imposing a constraint with the help of a lexicon and removing hypotheses which are not part of the lexicon. By doing this we reduce the noise in our K hypotheses. The lexicon, in our case, was taken from [1]. Yellow bars in Fig. 4 show the top-k accuracy after filtering predictions using the lexicon. In this work we devise methods using which we can select the best hypothesis from these K hypotheses. We use deep embeddings generated by the End2End network. All the K hypotheses are passed through the End2End network [16] and their deep embeddings are denoted by $E_{m_j} \ \forall j \in \{1, ..., K\}$.

Notation. Given a dataset of n word images their embeddings are denoted by $E_{w_i} \ \forall i \in \{1, ..., n\}$. E_{img} represents the embedding of the word image we want to recognise. All the text inputs are passed through the label stream shown in the Fig. 3. Input query text, text recogniser's noisy output and text recogniser's multiple hypotheses when converted to embeddings (using the label stream) are denoted by E_t, $E_{n_i} \ \forall i \in \{1, ..., n\}$, and $E_{m_j} \ \forall j \in \{1, ..., K\}$ respectively.

Baseline Word Recognition System. Figure 5(a) shows the baseline method that generates a single output. The baseline results for constrained and unconstrained approaches can be seen at $K = 1$ in Fig. 4. It can be concluded from Fig. 4 that, correct word is more likely to be present as K increases. In this section we present methods which allows us to improve the word recogniser's output by selecting the correct word out of K predictions.

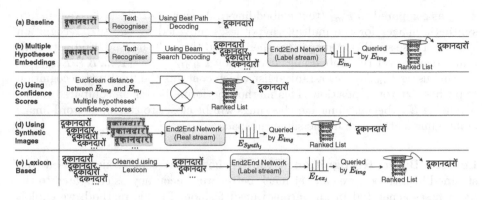

Fig. 5. (a) A single output text is generated using the best path decoding algorithm, (b) multiple hypotheses are generated using the beam search decoding algorithm, which are passed through the End2End network. The embedding generated (E_{m_j}) is queried by E_{img} to get a ranked list of predictions. Prediction at the first position is the new output. (c) Distance between E_{img} and E_{m_j} is summed with the confidence score to generate the ranked list. (d) Synthetic images are generated corresponding to the multiple hypotheses which are then converted to embedding (E_{Synth_j}) using the End2End network. E_{Synth_j} is queried by E_{img} to get a ranked list of predictions. (e) Here multiple hypotheses are limited using a lexicon, which is converted to embedding (E_{Lex_j}) using the End2End network. E_{Lex_j} is queried by E_{img} to get a ranked list. (Diagram best viewed in color)

Using Multiple Hypotheses' Embeddings. As it can be seen in Fig. 5(b), an input image is given to the text recogniser. The text recogniser uses beam search decoding algorithm for generating multiple hypotheses for the input image. These hypotheses are then passed through the label stream of the End2End network to generate E_{m_j}. E_{m_j} is queried by E_{img} to get a ranked list. The string at the top-1 position of the ranked list is considered to be the text recogniser's new output.

Using Confidence Scores. A confidence score is associated with E_{m_j} generated by the beam search decoding algorithm. This method uses that confidence score to select a better prediction. As seen in Fig. 5(c) the Euclidean distance between E_{img} and E_{m_j} is summed with the confidence score. This new summed score is used to re-rank the ranked list obtained by querying E_{img} on E_{m_j}. The string at the top-1 position of the re-ranked list is considered to be the text recogniser's new output.

Using Synthetic Images. In this method we bring E_{m_j} closer to E_{img} by converting multiple hypotheses to synthetic images[2]. Here, we exploit the fact that in the same subspace, the synthetic image's embeddings will lie closer to

---

[2] Generated using https://pango.gnome.org.

E_{img} as compared to E_{m_j} (text embeddings). As shown in Fig. 5(d), we generate synthetic images for the multiple hypotheses provided by the text recognition system. These synthetic images are then passed through the End2End network's real stream to get the embeddings denoted by $E_{synth_j} \; \forall j \in \{1, ..., K\}$. E_{synth_j} is queried using E_{img} to get a ranked list which contains the final text recognition hypothesis at top-1 position. This method performs better as compared to the one using E_{m_j} because image embeddings will be closer to the input word-image embedding as they are in the same subspace.

Lexicon Based Recognition. Figure 4 shows that using lexicon-based constrained hypotheses, we can get much better word accuracy as compared to the hypotheses generated in an unconstrained fashion. In this method, we exploit this fact and limit the multiple hypotheses generated from the text recognition system. As shown in Fig. 5(e), we limit the multiple hypotheses after they are generated. These filtered hypotheses are then passed through the label stream of the End2End network. The embeddings generated are denoted by $E_{Lex_j} \; \forall j \in \{1, ..., K\}$. E_{m_j} is now replaced by E_{Lex_j}. This method outperforms the previous methods as it decreases the noise in the hypotheses by constraining the number of words to choose from.

3.2 Word Retrieval Using Fusion and Re-ranking

For creating a retrieval system, we get the deep embeddings for the input query text (E_t) and also for the word images (E_{w_i}) from the documents on which we wish to query the input text. To use the best traits of the text recognition system, we also convert the text generated by the text recognition system to deep embeddings (E_{n_i}). In this section, we propose various techniques for improving word retrieval.

Baseline Word Retrieval System. There are three major ways to perform baseline word retrieval experiment. The first method focuses on demonstrating the word retrieval capabilities of embeddings in a "query by string" setting. Shown in Fig. 6(a), E_t is queried on E_{w_i} to get a ranked list. The second method focuses on demonstrating the word retrieval capabilities of embeddings in a "query by example" setting. Here input is a word image which is converted to embedding E_{img} using the real stream of the End2End network. E_{img} is used to query E_{w_i} to get a ranked list. The third method focuses on demonstrating the word retrieval when done on E_{n_i} in a "query by string" setting. Here, E_t is queried on E_{n_i} to get a ranked list. This method is a measure of how well the text recogniser is performing without the help of E_{w_i}. The performance of this method is directly proportional to the performance of the text recogniser.

Another way in which we can measure the performance of text recogniser's noisy output is by using the edit distance. Edit distance measures the number of operations needed to transform one string into another. We calculate the edit distance between input query text and text recogniser's noisy output. The

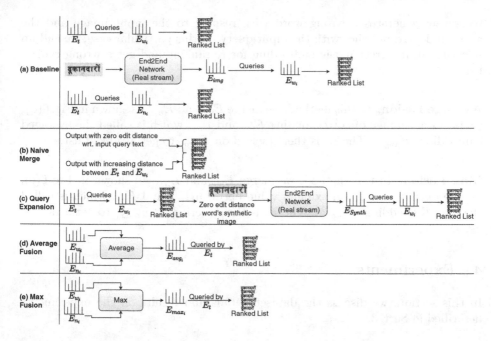

Fig. 6. (a) Baseline for word retrieval queries E_t on E_{w_i} and E_{n_i}. (b) Naive Merge attempts to exploit the best traits of text recognition and word spotting. (c) In Query Expansion, we merge the "query by string" and "query by example" setting. (d), (e) Average and Max Fusion methods fuse E_{w_i} and E_{n_i} by performing average and max operation respectively.

ranked list is created in an increasing order of edit distance. This experiment gauges the contribution of the text recogniser in the word retrieval process.

Naive Merge. As shown in the Fig. 6(b), we get the initial ranked list by calculating the edit distance between input text and noisy text recogniser's output. The remaining words are arranged in the order of increasing Euclidean distance between E_t and E_{w_i}. This method exploits the best traits of both the methods - text recognition and word spotting for creating a ranked list.

Query Expansion. In this method, initially we use the "query by string" setting and in the second stage "query by example" setting is used. The "query by example" setting works much better than the "query by string" setting, as image embedding is used to query on E_{w_i} and image embeddings lie closer in the subspace. As shown in Fig. 6(c), from the initial ranked list, we get a word with zero edit distance. A synthetic image corresponding to this word is generated. It is then fed to the real stream of the End2End network. We then get that word's synthetic image embedding (E_{Synth}). It is then queried on E_{w_i} to get a re-ranked list. The only case in which this method can fail is, when the text recogniser generates a wrong word with respect to the input image and the generated word matches with the input query; in this particular case, we end up selecting an incorrect image embedding for re-ranking and get a wrong ranked list.

Average Fusion. In this method, we merge E_{w_i} and E_{n_i}. As shown in Fig. 6(d), we perform average of corresponding E_{w_i} and E_{n_i}, which is called the averaged embedding (E_{avg_i}). The E_t is then queried on E_{avg_i} to get a ranked list.

Max Fusion. In this method, we merge E_{w_i} and E_{n_i}. As shown in Fig. 6(e), we the output having a maximum value between E_{w_i} and E_{n_i}, which is called the max Embedding (E_{max_i}). The E_t is then queried on E_{max_i} to get a ranked list.

4   Experiments

In this section, we discuss the dataset details and results on the experiments described in Sect. 3.

4.1   Dataset and Evaluation Metrics

Table 1. Dataset details

Dataset name	Annotated	#Pages	#Words	Usage
Dataset1	Yes	1389	396087	Training
Dataset2.1	Yes	402	105475	Word recognition
Dataset2.2	Yes	500	120000	Word retrieval

We use two type of data collections for implementing and evaluating various strategies for word retrieval and recognition. In the first collection (Dataset1), the books were scanned and annotated internally. In the second collection (Dateset2) the books were randomly sampled from the DLI [5] collection. Books in this collection range from different time periods. They consists of variety of font sizes and are highly degraded. Dataset2 is further divided into Dataset2.1 and Dataset2.2 for word recognition and word retrieval respectively. The word retrieval experiments are performed using $17,337$ queries, which is a set of unique words from 500 pages selected. Dataset 2.1 and 2.2 have some pages common among them. Table 1 summarises the datasets used in this work. To get models which generalize well and are unbiased towards any particular dataset, we train our End2End network and word recogniser on Dataset1 and test the models on Dataset2. Both of these datasets contain annotated books.

We evaluate our word recognition system in terms of word accuracy which is $1 - WER$ (Word Error Rate), where WER is defined as $\frac{S+D+I}{S+D+C}$. Here S is the count of substitutions, D is the count of deletions, I is the count of insertions and C is the count of correct words. All our word retrieval methods are evaluated using the mAP score, which is defined as $mAP = \frac{\sum_{q=1}^{Q} AvgP(q)}{Q}$. Here Q is the number of queries, $AvgP(q)$ is the average precision for each query.

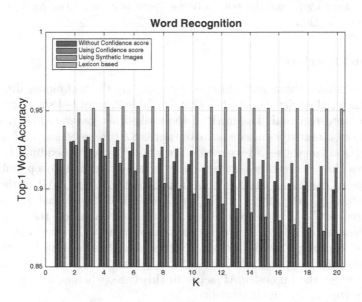

Fig. 7. Word recognition results. Here K denotes the number of hypotheses generated by the text recogniser. (Diagram best viewed in color)

4.2 Word Recognition

This section shows the results obtained by applying the techniques discussed in Sect. 3.1 for word recognition on Dataset2.1. There is no pre-processing done on these words and contain a high amount of noise, like presence of skewing, marks and cuts. Figure 7 shows the results of word recognition experiments. The bars in dark blue colour show the variation in the word accuracies as K is increased for the experiment in which we use the multiple hypotheses' embeddings as shown in Fig. 5(b). It is evident as the K is increased, at first, the word accuracy increases and then, starts to decrease; the reason is as K increases the noise starts to increase and the algorithm tends to choose incorrect predictions. In this method for $K = 3$, we report the highest word accuracy that is 93.12%, which is 1.2% more than accuracy at $K = 1$.

The bars in light blue show the results for the method where we incorporate the confidence information. It shows a higher gain in the accuracy score as K is increased. At $K = 3$, we report the highest word accuracy, which is 1.4% more than the accuracy at $K = 1$. Green bars in Fig. 7 show the result for experiment using synthetic images. At $K = 2$ we report the highest word accuracy which is 0.65% more than the accuracy at $K = 1$. The lexicon-based method shows the best result of all with the highest accuracy reaching to 95.26%, which is 1.25% more as compared to $K = 1$ (shown using yellow bars in Fig. 7). And it can be observed that the results are more consistent in this case, and don't drop as they do for the other methods; the reason being there is a very little amount of noise in the top predictions.

4.3   Word Retrieval

This section shows the results obtained by applying the techniques discussed in Sect. 3.2 for word retrieval on $Dataset2.2$. In Table 2, rows 1–4 show the baseline results for word retrieval. The mAP of the ranked list generated by calculating edit distance on text recogniser's noisy output is 82.15%, whereas mAP for query by string (QbS) method on text recogniser's noisy word embeddings is 90.18%. This proves that converting text recogniser's noisy output to deep embeddings (E_{n_i}) helps in capturing the most useful information. The mAP of the ranked list generated by querying E_t on E_{w_i} is 92.80%; this shows that E_{n_i} is noisier as compared to E_{w_i}. These results lead us to the conclusion that we can use the complementary information from the text recogniser's noisy text and word images for improving word retrieval. Highest mAP is observed in the query by example (QbE) setting which is 96.52%. As we have used image input for querying we get the highest mAP score in this case, whereas in other cases we are using string as the input modality.

Table 2. This table summarises the results for all the word retrieval experiments. It can be observed that methods using "query by example" setting work best in the baseline as well as the re-ranking case. In the case of fusion, average fusion proves to be the one showing the most improvement.

Experiment type	Input modality	Experiment	mAP
Baseline	String	Edit Distance on Text Recogniser's Outputs	82.15
	String	QbS on Text Recogniser's Embeddings	90.18
	String	**QbS Word Image Embeddings**	**92.80**
	Image	**QbE Word Image Embeddings**	**96.52**
Re-ranking	String	Naive merge	92.10
	String	**Query Expansion**	**93.18**
Fusion	**String**	**Average Fusion**	**93.07**
	String	Max. Fusion	92.79

Row 5–6 in Table 2 shows the results obtained by re-ranking the ranked lists. The mAP of the ranked list obtained by naive merge is 92.10%. Naive

merge shows an improvement over baselines by using the best output of the text recogniser for the initial ranked list. And arranging the remaining word images in increasing order of Euclidean distance between E_t and E_{w_i}. The mAP of the ranked list obtained by query expansion is 93.18%. Query expansion shows improvement over all the previous methods as here we get a second-ranked list by using the text recogniser's best output's word image. It is intuitive that, if we use a word-image embedding for querying E_{w_i} the mAP will improve.

Rows 7–8 in Table 2 shows the results obtained by fusing E_{n_i} and E_{w_i}. The ranked list obtained by E_{avg_i} gives us an mAP score of 93.07%, and the one obtained by E_{max_i} is 92.79%. The fusion methods show an improvement over the baselines in QbS as they contain information of both E_{w_i} and E_{n_i}.

4.4   Failure Cases

Figure 8 shows instances where both word recognition and retrieval fail to achieve the desired results.

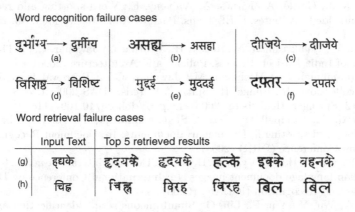

Fig. 8. Failure cases

In case (a) and (c) the image is degraded due to which we are not able to capture the image information well and end up generating the wrong prediction. In case (b), (d), (e), and (f) the characters resemble closely to other characters which leads to a wrong output. To add to this, in the case of (b), (d), and (e) the characters are rare which increases the confusion. Cases (g) and (h) show the example where word retrieval was not able to perform well. For case (g) and (h), the input query has a rare character which resembles other characters, due to which the words are incorrectly retrieved. Though, for (h), using query expansion technique, we were successful in retrieving one out of two instances of the query.

5   Conclusion

To summarise, we improve the word retrieval process by using the deep embeddings generated by the End2End network [16]. We have shown that by using the complementary information of text recogniser and word spotting methods, we can create word recognition and retrieval system capable of performing better than both of the individual systems. We plan to explore various other fusion techniques apart from average and max fusion for improving word retrieval.

References

1. Hindi word frequency list. http://cse.iitkgp.ac.in/resgrp/cnerg/qa/fire13translit/index.html. Accessed 26 Dec 2019
2. Project Gutenberg. https://www.gutenberg.org/wiki/Main_Page. Accessed 26 Dec 2019
3. Adak, C., Chaudhuri, B.B., Blumenstein, M.: Offline cursive Bengali word recognition using CNNs with a recurrent model. In: International Conference on Frontiers in Handwriting Recognition (ICHFR) (2016)
4. Almazán, J., Gordo, A., Fornés, A., Valveny, E.: Word spotting and recognition with embedded attributes. IEEE Trans. Pattern Anal. Mach. Intell. **36**, 2552–2566 (2014)
5. Ambati, V., Balakrishnan, N., Reddy, R., Pratha, L., Jawahar, C.V.: The Digital Library of India Project: Process, Policies and Architecture (2006)
6. Balasubramanian, A., Meshesha, M., Jawahar, C.V.: Retrieval from document image collections. In: Bunke, H., Spitz, A.L. (eds.) DAS 2006. LNCS, vol. 3872, pp. 1–12. Springer, Heidelberg (2006). https://doi.org/10.1007/11669487_1
7. Bhardwaj, A., Kompalli, S., Setlur, S., Govindaraju, V.: An OCR based approach for word spotting in Devanagari documents. In: Document Recognition and Retrieval Conference (DRR) (2008)
8. Chaudhury, S., Sethi, G., Vyas, A., Harit, G.: Devising interactive access techniques for Indian language document images. In: International Conference on Document Analysis and Recognition (ICDAR) (2003)
9. Chen, Z., Wu, Y., Yin, F., Liu, C.: Simultaneous script identification and handwriting recognition via multi-task learning of recurrent neural networks. In: International Conference on Document Analysis and Recognition (ICDAR) (2017)
10. Dutta, K., Krishnan, P., Mathew, M., Jawahar, C.V.: Improving CNN-RNN hybrid networks for handwriting recognition. In: International Conference on Frontiers in Handwriting Recognition (ICHFR) (2018)
11. Garain, U., Mioulet, L., Chaudhuri, B.B., Chatelain, C., Paquet, T.: Unconstrained Bengali handwriting recognition with recurrent models. In: International Conference on Document Analysis and Recognition (ICDAR) (2015)
12. Graves, A., Fernández, S., Gomez, F., Schmidhuber, J.: Connectionist temporal classification: labelling unsegmented sequence data with recurrent neural networks. In: International Conference on Machine Learning (ICML) (2006)
13. Jaderberg, M., Simonyan, K., Vedaldi, A., Zisserman, A.: Synthetic data and artificial neural networks for natural scene text recognition. CoRR (2014)
14. Jaderberg, M., Vedaldi, A., Zisserman, A.: Deep features for text spotting. In: Fleet, D., Pajdla, T., Schiele, B., Tuytelaars, T. (eds.) ECCV 2014. LNCS, vol. 8692, pp. 512–528. Springer, Cham (2014). https://doi.org/10.1007/978-3-319-10593-2_34

15. Krishnan, P., Dutta, K., Jawahar, C.V.: Deep feature embedding for accurate recognition and retrieval of handwritten text. In: International Conference on Frontiers in Handwriting Recognition (ICHFR) (2016)
16. Krishnan, P., Dutta, K., Jawahar, C.V.: Word spotting and recognition using deep embedding. In: Document Analysis Systems (DAS) (2018)
17. Krishnan, P., Shekhar, R., Jawahar, C.: Content level access to Digital Library of India pages. In: ACM International Conference Proceeding Series (ICPS) (2012)
18. Meshesha, M., Jawahar, C.V.: Matching word images for content-based retrieval from printed document images. Int. J. Doc. Anal. Recogn. (IJDAR) **11**, 29–38 (2008). https://doi.org/10.1007/s10032-008-0067-3
19. Pham, V., Bluche, T., Kermorvant, C., Louradour, J.: Dropout improves recurrent neural networks for handwriting recognition. In: International Conference on Frontiers in Handwriting Recognition (ICHFR) (2014)
20. Poznanski, A., Wolf, L.: CNN-N-Gram for handwriting word recognition. In: Computer Vision and Pattern Recognition (CVPR) (2016)
21. Rath, T., Manmatha, R.: Word spotting for historical documents. Int. J. Doc. Anal. Recogn. (IJDAR) **9**, 299 (2007). https://doi.org/10.1007/s10032-006-0035-8
22. Shekhar, R., Jawahar, C.V.: Word image retrieval using bag of visual words. In: Document Analysis Systems (DAS) (2012)
23. Shi, B., Bai, X., Yao, C.: An end-to-end trainable neural network for image-based sequence recognition and its application to scene text recognition. CoRR
24. Simonyan, K., Zisserman, A.: Very deep convolutional networks for large-scale image recognition. arXiv 1409.1556 (2014)
25. Sudholt, S., Fink, G.A.: PHOCNet: a deep convolutional neural network for word spotting in handwritten documents. CoRR (2016)
26. Sudholt, S., Fink, G.A.: Attribute CNNs for word spotting in handwritten documents. CoRR (2017)
27. Sun, Z., Jin, L., Xie, Z., Feng, Z., Zhang, S.: Convolutional multi-directional recurrent network for offline handwritten text recognition. In: Conference on Frontiers in Handwriting Recognition (ICHFR) (2016)
28. Wilkinson, T., Brun, A.: Semantic and verbatim word spotting using deep neural networks. In: International Conference on Frontiers in Handwriting Recognition (ICHFR) (2016)

A Named Entity Extraction System for Historical Financial Data

Wassim Swaileh[2](✉), Thierry Paquet[1], Sébastien Adam[1],
and Andres Rojas Camacho[1]

[1] LITIS EA4108, University of Rouen Normandie, Rouen, France
{thierry.paquet,sebastien.adam,andres.camacho}@univ-rouen.fr
[2] ETIS, UMR 8051, CY Cergy Paris Université, ENSEA, CNRS, Cergy, France
wassim.swaileh@cyu.fr

Abstract. Access to long-run historical data in the field of social sciences, economics and political sciences has been identified as one necessary condition to understand the dynamics of the past and the way those dynamics structure our present and future. Financial yearbooks are historical records reporting on information about the companies of stock exchanges. This paper concentrates on the description of the key components that implement a financial information extraction system from financial yearbooks. The proposed system consists in three steps: OCR, linked named entities extraction, active learning. The core of the system is related to linked named entities extraction (LNE). LNE are coherent n-tuple of named entities describing high level semantic information. In this respect we developed, tested and compared a CRF and a hybrid RNN/CRF based system. Active learning allows to cope with the lack of annotated data for training the system. Promising performance results are reported on two yearbooks (the French Desfossé yearbook (1962) and the German Handbuch (1914–15)) and for two LNE extraction tasks: capital information of companies and constitution information of companies.

Keywords: Linked named entities extraction · Active learning · CRF · String embedding

1 Introduction

Access to long-run historical data in the field of social sciences, economics and political sciences has been identified as one necessary condition to understand the dynamics of the past and the way those dynamics structure our present and future. In this context, the EURHISFIRM project has been founded by the EU to develop a Research Infrastructure with a focal point on the integration of financial and corporate governance historical information of firms. During its design phase, one of the objectives of EURHISFIRM is to design and develop an intelligent and collaborative system for the extraction and enrichment of data from historical paper sources such as yearbooks and price lists that were published over years in the many European stock exchanges.

© Springer Nature Switzerland AG 2020
X. Bai et al. (Eds.): DAS 2020, LNCS 12116, pp. 324–340, 2020.
https://doi.org/10.1007/978-3-030-57058-3_23

In this paper, we focus on the processing of yearbooks. Figure 1 shows two examples of such documents, one German (GR) and one French (FR). These yearly publications were intended to provide updated information about the companies of the stock exchanges, including their name, date of creation, financial status, governing board members, headquarters address, branch address, financial information such as capital amount, date and amount of capital increase, balance sheet of the year including assets and liabilities, etc. As shown on these two examples, Yearbooks have mostly textual contents organised in specific sections, on the contrary of prices lists that contain tabular data. As a consequence, extracting information from yearbooks requires the design of a general named entity extraction system from OCRed yearbooks. This represents a real complex challenge related to document image segmentation, optical character recognition (OCR) and linked named entity recognition (NER).

This paper concentrates on the description of a key component of the EURHISFIRM platform that implements a financial information extraction system from Yearbooks. The rest of this paper is organised as follows: Sect. 2 gives a brief overview of related works. The system architecture is then described in Sect. 3. Section 4 emphasises on the key component of the system: the Linked Financial Named Entities (LFNE) extraction. In this purpose we study a Conditional Random Field (CRF) based approach and a hybrid recurrent Neural Networks/CRF (RNN-CRF) based approach. Then, in Sect. 5, we report on the system performance on two specific sections of the two Yearbooks under study. We also analyse the system's performance when introducing an active learning strategy in order to cope with the lack of annotated training data.

2   Related Work

Information Extraction (IE) from born digital texts has been extensively studied in the field of Natural Language Processing (NLP), notably through the well known Message Understanding Conferences (MUC) that were organised from 1987 to 1997. In 1995, the first competition on Named Entity Recognition (NER) was introduced during MUC-6[1]. Since 1999, the yearly conference on Natural Language Learning (CoNLL) covers a large framework of topics about NLP, mostly through machine learning approaches. Information extraction from scanned documents has been by fare much less studied. The contributions in this respect have concentrated on analysing the performance degradation caused by OCR errors [9]. While some studies have been carried out on synthetic data by introducing character errors generated randomly, most recent studies have been motivated by digitisation projects of historical documents so as to enhance search performance and offering high level semantic indexing and search. In general, OCR quality is erratic on historical documents and there is a tendency for unusual non-alphabetic characters to appear. In addition, the OCR system has problems with layout and is frequently unable to distinguish marginal notes from the main body of the text, giving rise to discontinuous sequences of words

---

[1] https://cs.nyu.edu/faculty/grishman/muc6.html.

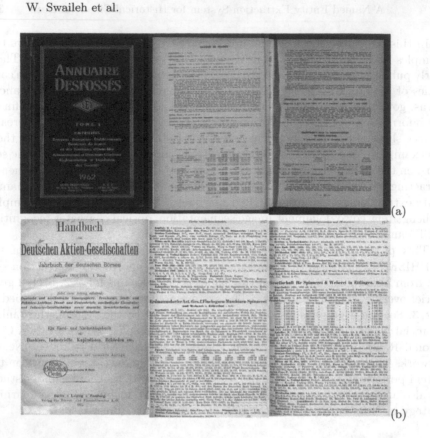

Fig. 1. Samples of (a) French yearbook "Annuaire Desfossés, 1962", (b) German yearbook "handbuch, 1914/1915".

which can confound a NER system. In [2] the authors investigate person and place names recognition in digitised records of proceedings of the British parliament. The experiments show better recognition for person names (F = 0.75) than for location names (F = 0.66). This may be explained by the fact that person name recognition is more dependent on finding patterns in the text while location name recognition is more dependent on gazetteer resources. In [10] the authors focus on full name extraction on a corpus composed of 12 titles spanning a diverse range of printed historical documents with relevance to genealogy and family history research. The results show a certain correlation between WER and F-measure of the NER systems. From the analysis of errors the authors conclude that word order errors play a bigger role in extraction errors than do character recognition errors, which seems reasonable as extraction systems intensively exploit contextual word neighbours information. [12] evaluate the efficacy of some available tools for accurately extracting semantic entities that could be used for automatically tagging and classifying documents by using uncorrected OCR outputs from OCRopus and tesseract. The test data came from the Wiener

Library, London, and King's College London's Serving Soldier archive. Performance of NER were on overall lower than those reported in the literature due to the difference of the data sets generally used to train and test NER systems. The authors suggest that automatically extracted entities should be validated using controlled vocabularies or other kinds of domain knowledge in order to circumvent the variability in spelling these entities in general. In [4] the authors analyse the performance of the stanford NER [7] on a snapshot of the Trove[2] newspaper collection at the National Library of Australia (NLA). The results show that the pre-trained Stanford NER performs slightly better (F1-score = 0.71) that the trained Stanford NER (F1-score = 0.67) considering location, person and organisation names. All of these studies have considered Named Entity recognition as a post-processing stage to OCR and they show that NER performance is affected by the OCR errors, mostly word errors. Toledo et al. [14] proposed a standalone architecture, that integrates image analysis through a CNN and a BLSTM neural network for extracting family names and other entities from Spanish medieval handwritten birth records. This architecture is not affected by the accumulation of errors of the traditional methods. This approach shows excellent performance on named entity extraction but requires a preliminary word segmentation stage to operate, which remains a limitation.

We have seen in this literature review that most NER tasks applied on document images have considered OCR prior to NER. We now need to give a brief but specific overview of the NER literature. Named entity recognition approaches can be categorised into three main groups of approaches; 1) rule based approaches [15] 2) statistical approaches [1,6] 3) mixed rule-based/statistical approaches [14]. The rule-based approaches were used in early NER systems where entity features were specified in advance by domain experts. These approaches are not flexible towards inter-entity and intra-entity ambiguities (confusions). They are known to provide a good precision but generally with a low recall. Besides they require a lot of human efforts to be developed as each rule is designed manually [15].

Several statistical models have been applied for NER. We can classify them in two main groups; 1) traditional statistical models 2) neural networks models. Among the various statistical models proposed in the literature such as Support Vector Machines or Hidden Markov Models, CRF [8] have emerged as the state of the art approaches. For any of these models, one needs to define handcrafted features that describe the syntactic and semantic nature of the entities to be extracted in the text, before training the model on a sufficient amount of annotated data. More recently, Recurrent Neural Networks (RNN) coupled with CRF have emerged as the new state of art architecture for NER [6]. Indeed, Recurrent Neural Networks allow training word embeddings with unannotated data, which then serve as efficient features for the NER task. Word embeddings however have some limitations as they are average representations of word's context in the training dataset, and they do not capture the specific context in which a word occur in a specific sentence. Besides, out of vocabulary words cannot be associated to relevant embedding as they do not pertain to the training dataset.

---

[2] http://trove.nla.gov.au/.

To circumvent these limitations Akbik et al. [1] introduced the concept of string embedding which capture the left and right context of characters in texts, making the system free of any dictionary. Moreover, string embedding allow the design of dynamic word embeddings, where a word receives an embedding depending on its specific context within the sentence under study. A word may gets different embeddings when occurring in different contexts. In 2018, this approach achieved new state of art results on the CoNLL2003 NER data set.

3 System Architecture

Financial Yearbooks contain structured information about companies which are organised in sections. Sections are paragraphs or groups of paragraphs that report on some specific financial information related to a company. A section sometimes brings the actual composition of the governing board through a list of persons, while sometimes it brings some values of some financial indicators such as the capital of the company with the value of shares and the number of shares that compose the capital. Moreover, some sections also report on the evolution of these indicators over years. These historical records on these financial indicators are of particular interest for experts as they would bring access to long run financial time series to be analysed. Figure 2 gives one example page of the French Desfossé 1962 yearbook (left) and of the German Handbuch 1914–15 (right).

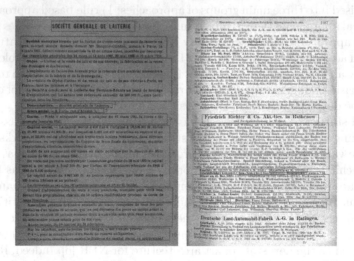

Fig. 2. Section structure of the two yearbooks, the French Desfossé 1962 (left) and the German handbuch 1914–15 (right).

We can see that for any of these sources, sections (highlighted in color rectangles) start with a bold title that identifies the type of the section, then the text of the section is running over one single or multiple paragraphs. In these examples we can notice that some sections of the German yearbook may run one after the other within a single paragraph, whereas sections of the French yearbook always start with a new paragraph. In the same way, a description of a company always start with a new page of the French Yearbooks, whereas multiple companies can occur within a single page of the German yearbook. It is to be noticed that every sections do not contain information of similar interest for scientific financial research, and that each yearbook is reporting financial information in different ways. However there are some stable information which any yearbook is reporting. Among the most important ones we can mention the following section types: *Founded, Purpose, Governing Board, Capital, Balance Sheets, Fiscal Year.* Through these examples we can also notice that sections contain different types of information. Some of them are lists of persons, others contain addresses, dates, amounts etc. There is a need to develop a general named entity extraction system that can be adapted specifically to each section type.

From the presentation of these examples, we can now introduce the general architecture of the extraction system that was designed for multilingual financial information extraction in yearbooks which is depicted on Fig. 3. Considering the generic system that we aim to develop, we choose to organise the processing chain in a sequential manner by first running the OCR and then the extraction system. Both sources have been processed with Omnipage[3]. We observed a very good OCR quality on the French yearbook while performance on the German yearbook appear to be slightly lower. In both cases the quality was considered sufficiently good to carry out the extraction process. No OCR correction was applied prior to the extraction process.

The OCR produces a set of text blocks which are ordered from top to bottom. Then a standard but language specific pre-processing stage allows to detect each textual component as a token. Depending on the layout of the document, section detection is performed using a rule-based approach by exploiting on a set of predefined keywords. For example, the keyword CAPITAL (French) or Kapital (German) identifies the sections that contain essential information about the company financial status.

Then information extraction is performed on each identified section and using a specific extraction algorithm. Following the literature review we choose to develop a machine learning approach for the following reasons. It offers a generic framework that can be optimised efficiently for every section to deal with. The approach is language agnostic provided human experts can provide training data. Tagging conventions of texts (words and numerical sequences) can be defined with experts (Historians) so that both the historians and the IT specialists arrive to a better understanding of the real final needs. Notice that there is no unique approach here for adopting tagging conventions, and we may follow an iterative process through trials and errors before arriving to the final tag set for a specific section. More details will be given about tagging conventions in Sect. 4.1.

Once the tagging conventions have been defined the system follows the steps depicted on Fig. Sreffig:system. Human tagging of a training data set is required before training the system. Depending on the difficulty of the extraction task of the section considered, the size of the data set has to be adjusted conveniently but

---

[3] https://www.kofax.com/Products/omnipage.

we have no reliable estimate of it. This process may be iterated to get acceptable results on the test data set. Another strategy to circumvent from the lack of annotated data is to iterate training and annotation of new examples through active learning [13]. In this manner we may ask the user to annotate only the most relevant examples for which the system is less confident. We implemented an active learning scheme that is presented in Sect. 4.4.

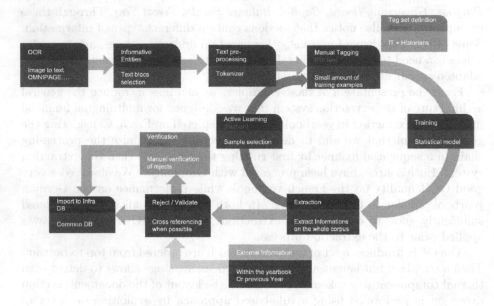

Fig. 3. The flow of processes of the system.

4   Linked Named Entity Extraction

As highlighted in Sect. 3, financial yearbooks contain many information organised in sections. Each section is reporting about specific financial information and requires a dedicated extraction module. Some sections are simply lists of items such as lists of persons for which simple rules encoded with regular expressions may suffice to get the information extracted. Some other sections are much more difficult to analyse as shown on Fig. 4 which gives two examples of the section "Capital" for both French and German yearbooks.

4.1   Tagging Conventions

The two sections on Fig. 4 highlight the typical content of financial yearbooks whatever the stock exchange and the language used. The section "Capital" reports the current capital amount of the company, its currency, the number and amount of shares that build the total capital amount. Moreover, the section "capital" is also reporting the history of the capital by showing the dates when

Fig. 4. Information to be extracted from the section Capital for both the French and German yearbooks with the tagging conventions.

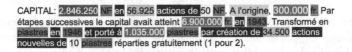

Fig. 5. Tagging the linked named entities, with tag "Link" in blue. (Color figure online)

the capital changed from the creation date of the company. Each capital change should occur with the date of change, the new capital value, the currency, the new number of shares with their amount. As these named entities are reported through a textual description and not placed into a table, a certain variability was introduced in phrasing the text at the time of publication. In addition, some information are sometimes partially missing. Figure 4 shows a case where the same tag set can be used for French and German. Notice that irrelevant words in the text are labelled with the tag "Other", as is the standard convention adopted for information extraction. One other important aspect is related to how these various information should be linked together to provide timely coherent n-tuples of information in a tabular form as follows: date - capital amount - currency - number of shares - amount of share. Such a 5-tuple is made of linked named entities and we wish the extraction process to extract not only each individual entities but also their linking attributes with the other entities they relate to. In this purpose we have introduced a specific "Link" tag that serves for tagging every non informative words within a single n-tuple, so that a n-tuple is any sequence of tags between two "Other" tags, see Fig. 5.

4.2   CRF Based Model

A Conditional Random Field models (CRF) [5] allows to compute the conditional probability of a sequence of labels $Y = \{y_1, y_2, \ldots, y_T\}$ given a sequence of input features $X = \{x_1, x_2, \ldots, x_T\}$ with the following equation:

$$P(Y|X) = \frac{1}{Z_0} exp\Big(\sum_{t=1}^{T} \sum_{k} w_k \times \phi_k(y_{t-1}, y_t, x, t)\Big) \tag{1}$$

where $\phi(y_{t-1}, y_t, x, t)$ is a feature function that maps the entire input sequence of features X paired with the entire output sequence of labels Y to some d-dimensional feature vector. Weight parameters w_k are optimised during training. The normalisation factor Z_0 is introduced to make sure that the sum of all conditional probabilities is equal to 1. Once the optimal weights $\hat{w}$ are estimated, the most likely sequence of output labels $\hat{Y}$ for a given sequence of input features X is estimated as follows:

$$\hat{Y} = arg \max_{Y} P(Y|X) \tag{2}$$

CRF have been introduced for Natural Language Processing by considering binary features. $\phi(y_{t-1}, y_t, x, t)$ is a binary feature function that is set to 1 when labels and input tokens match a certain property. In this study, we use a 5 tokens width sliding window and a set of 33×5 template features which produce a total of thousands of binary features depending on the section considered.

4.3   String Embedding (BLSTM-CRF) Model

In the literature, Recurrent Neural Network (RNN) architectures have been introduced so as to learn tokens embeddings. These embeddings are then used in place of the handcrafted features in a CRF model. Most of the state of the art NER systems use pre-trained word embeddings with a standard BLSTM-CRF setup [1,3,6,11]. In addition to pre-trained word embeddings Lample et al. [6] have introduced character-level word embeddings so as to circumvent from possible out of Vocabulary words. Similarly Peter et al. [11] introduced contextual word embeddings extracted from a multi-layer bidirectional language model of tokens (biLM). Recently, Akbik et al. [1] have used the internal states of two LSTM character language models to build contextual word embeddings, namely contextual string embeddings. Compared to other state of the art systems, this model is able to provide embeddings to any word and not only the known vocabulary words of the training set. Each language model consists of a single layer of 2048 Long Short Term Memory (LSTM) cells. A language model estimates the probability $P(x_{0:T})$ of a sequence of characters $(x_0, \ldots, x_T \Leftrightarrow x_{0:T})$ with the following equation.

$$P(x_{0:T}) = \prod_{t=0}^{T} P(x_t|x_{0:t-1}) \tag{3}$$

where $P(x_t|x_{0:t-1})$ is the probability of observing a character given its past. A forward language model $(\overrightarrow{LM})$ computes the conditional probability using the LSTM hidden states as follows:

$$P(x_t|x_{0:t-1}) \approx \prod_{t=0}^{T} P(x_t|\overrightarrow{h_t}; \theta) \tag{4}$$

where $\overrightarrow{h_t}$ represents a view of the LSTM of the past sequence of characters of character x_t while θ represents the model parameters. Similarly, a backward language model $(\overleftarrow{LM})$ computes the probability in the reverse direction as follows:

$$P(x_t|x_{t+1:T}) \approx \prod_{t=0}^{T} P(x_t|\overleftarrow{h_t}; \theta) \tag{5}$$

The word embedding w_i of word i that starts at character x_b and ends at character x_e in the sentence is obtained by the concatenation of the hidden states of the forward and the backward LM as follows:

$$w_i = \left[\overleftarrow{h}_{b-1}, \overrightarrow{h}_{e+1}\right] \tag{6}$$

Notice that the two character language models can be trained on un-annotated large corpora as they are trained to predict the next/previous character. Then, following the architecture proposed in [1], we use a hybrid BILSTM/CRF model for named entity recognition. A word level BLSTM captures word context in the sentence and its internal state feeds a CRF in place of handcrafted features. The word BLSTM is fed by the string embedding representation. This BILSTM/CRF architecture is trained on for each specific Named Entity Recognition task, while it is fed by the string embedding representation that is pre-trained on a large corpus of the language choosen. In the following experiments we used pre-trained string embeddings proposed by the authors for French and German.

4.4   Active Learning Scheme

Due to the lack of annotated data, we have introduced an active learning scheme. First, we start by training the extraction model with a few annotated examples. The trained model is then used to predict the annotation of all the unseen examples of the test data set. These automatically annotated examples are sorted according to their labelling score. The examples with a labelling score higher that 0.9 are used as additional training examples to the first training data set for a new training iteration. The examples with labelling score less than 0.5 are considered as bad examples. Thus a small set of those examples are annotated manually for enhancing the capacity of the extraction model towards this kind of bad examples.

5 Experiments

We report the evaluation results for the extraction of the linked named entities on two sections of two yearbooks: the CAPITAL and CONSTITUTION sections of the French Desfossé 1962, and the Kapital section of the German Handbuch 1914–1915. The French Desfossé 1962 Yearbook consists of 2376 pages, among which 1544 CAPITAL sections and 1547 CONSTITUTION sections have been detected. The German Handbuch 1914–1915 yearbook consists of 5174 pages in which we detected 3971 Kapital sections. Among the detected sections, 181 CAPITAL section, 91 CONSTITUTION section and 195 Kapital sections have been manually annotated to form the training and test data sets.

CRF-Model Configuration: We used the crf++ toolkit[4] for training the CRF-based extraction models. We used the default training parameters, the cut-off threshold for features selection is $f = 3$, and the C hyper-parameter that prevents over fitting has been set to $C = 1.5$.

BLSTM-CRF Model Configuration: We followed the training scheme introduced in [1] with a slight modification. In our configuration, we decreased the number of LSTM units from 256×2 to 64×2; we also set the learning rate to 0.2 with a mini-batch size of 8.

5.1 Extraction Tasks

CAPITAL Section Named Entity Extraction: The information to be extracted from this section are every capital amounts, currencies and change dates. The tag set was derived from the examples in Fig. 5.

Kapital Section Named Entity Extraction: Kapital section contains the same set of named entities to be extracted as for the CAPITAL section. In addition, two new named entities have been considered; *Cap-decr* and *Cap-incr*. These two labels refer to a increase or a decrease of the capital. Table 1 shows the extraction results on the CAPITAL and Kapital sections and using the CRF and the BLSTM-CRF extraction models. We observe very good performance on the CAPITAL section with both the CRF and the BLSTM-CRF model with small differences. For every entities we obtain precision and recall higher than 95% while the average F1-score is higher than 96%. We also observe similar excellent performance on the Kapital section. However, the BLSTM-CRF model performs better than the CRF model. Both the CRF and the BLSTM models can not deal with the Capital decrease entity due to lack of example in the training set.

---

[4] https://taku910.github.io/crfpp/.

Table 1. Performance obtained by the CRF and BLSTM models on the CAPITAL and Kapital sections.

Entity tags	CAPITAL section						Kapital section					
	CRF model			BLSTM-CRF model			CRF model			BLSTM-CRF model		
	Precision %	Recall %	F1-score	Precision %	Recall %	F1-score	Precision %	Recall %	F1-score	Precision %	Recall %	F1-score
ini-amount	95.97	95.94	95.94	96,78	95,96	96,35	89.81	77.97	83.37	97.15	89.68	93.16
ini-date	100	100	100	98,75	98,75	98,75	100	85.24	91.84	98.75	90.06	92.69
chg-amount	97.32	96.95	97.13	97,05	97,98	97,51	93.28	74.10	82.30	90.37	85.23	87.69
chg-date	97.72	94.55	96	98,04	96,05	97,01	90.20	81.73	85.73	91.64	89.01	90.29
last-amount	96.74	91.22	93.72	87,28	93,56	89,78	96.92	96.38	96.63	99.48	98.46	98.95
currency	97.25	94.12	95.63	96,09	96,46	96,26	96.91	89.79	93.19	97.78	95.85	96.79
link	98.07	95.12	96.54	97,68	95,63	96,63	91.71	84.89	88.07	90.94	90.24	90.49
Cap-decr	–	–	–	–	–	–	0	0	0	0	0	0
Cap-incr	–	–	–	–	–	–	91.26	89.30	90.23	91.48	94.42	92.88
Overall	97.57	95.27	96.39	96,69	96,43	96,55	93.65	85.84	89.55	94.16	92.17	93.13

Table 2. Performance of the CRF and the BLSTM models for extracting the named entities of the CONSTITUTION section

CONSTITUTION section	CRF model			BLSTM-CRF model		
Entity tags	Precision %	Recall %	F1-score	Precision %	Recall %	F1-score
ini-status	91.30	95.45	93.33	94.51	96.59	95.50
ini-startdate	86.43	82.61	84.48	86.33	94.52	90.18
ini-enddate	50	33.25	39.94	59.17	42.92	44.03
ini-period	100	100	100	95.83	100	97.81
chg-status	50	50	50	50	33.33	40
chg-startdate	0	0	0	37.50	25	29
chg-enddate	0	0	0	0	0	0
chg-period	100	66.67	80	100	66.67	80
link	91.11	87.23	89.13	90.54	88.83	89.61
Overall	89.38	81.45	85.23	87.72	85.26	86.42

CONSTITUTION Section Named Entity Extraction Task. From this section, we want to extract information about the company legal status, the date of creation, the period of activity and expiration date if applicable. We have introduced nine tags for this section, defined as follows: 1) *ini-status*: initial legal status of the company once created. 2) *ini-startdate*: the company creation date. 3) *ini-enddate*: the company expiration date. 4) *ini-period*: the company activity period. 5) *chg-status*: the changed legal status of the company. 6) *chg-startdate*: the start date of the changed legal status of the company. 7) *chg-enddate*: the end date of the changed legal status of the company. 9) *link*: the linking tag. In Table 2, we report the results on the CONSTITUTION section using the CRF and BLSTM-CRF models. Due to the small size of the training data set, the results show lower performance compared to those reported on the CAPITAL and Kapital sections.

5.2   Entity Linking

Once the entities have been extracted, we link them into tuples called chunks. We consider three different chunks on the CAPTITAL and Kapital sections; 1) the *ini-chunk* consists of the *ini-date*, *ini-amount* and *currency* labelled tokens. 2) the *chg-chunk* includes the *chg-date*, *chg-amount* and *currency* labelled tokens. 3) the *last-chunk* enclose *last-amount* and *currency* labelled tokens. Notice that the date associated with the *last-amount* entity is the date of the yearbook (1962), and for this reason we don't consider extracting this information.

Table 3. Chunk extraction performance with the minimal distance entity linking method and using the link tag learning method

Linking method	Minimal distance			Link tag		
Chuncks	Precision %	Recall %	F1-score	Precision %	Recall %	F1-score
ini-chunk	96.80	94.46	95.62	95.21	95.94	95.57
chg-chunk	89.72	87.89	88.80	91.58	91.42	91.50
last-chunk	91.73	90.03	90.87	97.18	91.23	94.11
Overall	90.93	89.08	89.99	92.84	92.01	92.43

Table 4. System performance results obtained by the CRF and BLSTM models for extracting and linking the named entities of the CAPITAL and Kapital sections

Linking chunks	CAPITAL section						Kapital section					
	CRF model			BLSTM-CRF model			CRF model			BLSTM-CRF model		
	Precision %	Recall %	F1-score	Precision %	Recall %	F1-score	Precision %	Recall %	F1-score	Precision %	Recall %	F1-score
ini-chunk	95.21	95.94	95.57	94.61	94.78	94.66	80.20	73.84	76.78	87.56	83.43	85.43
chg-chunk	91.58	91.42	91.50	93.96	94.74	94.35	70.18	63.64	66.68	75.66	71.74	73.58
last-chunk	97.18	91.23	94.11	89.98	94.76	92.26	95.82	96.82	96.30	95.42	98.42	98.41
Overall	92.84	92.01	92.43	93.39	94.75	94.06	78.94	73.91	76.31	83.72	80.50	82.05

Table 5. Results obtained by the CRF and the BLSTM-CRF models for extracting the linked named entities on the CONSTITUTION section

CONSTITUTION section	CRF model			BLSTM-CRF model		
Linking chuncks	Precision %	Recall %	F1-score	Precision %	Recall %	F1-score
ini-chunk	72.44	81.23	76.51	74.88	85.57	79.84
chg-chunk	37.50	17.71	22.42	30.38	27.98	28.26
Overall	67.55	65.63	66.47	64.57	71.49	67.79

We experimented two methods for linking the entities into chunks. The minimum distance method regroups entities with their closest neighbour entity. Using the *link* tag introduced in Sect. 4, we are able to learn how to link the entities.

We then consider a sequence of linked entities, as entities of the same chunk. The two methods have been evaluated on the CAPITAL section using the CRF model, see Table 3. The results show better performance when using the learned tag *link*. We analyse the performance of the Linked Named Entities on the CAPITAL and Kapital sections given on Table 4. The results show very good performance for extracting and linking the named entities of the CAPITAL section ≥90%, with the best results obtained by the BLSTM-CRF model. Performance on the German Kapital section is 10 points lower those for the French CAPITAL section. This may come from the effect of introducing two more tags (Cap-incr and Cap-decr) but this may also come from the lower regularity of the German style of writing, compared to the French yearbook. We consider two different chunks on the CONSTITUTION section. The *ini-chunk* is a group of the *ini-status*, *ini-startdate*, *ini-enddate* and *ini-period* labels. The *chg-chunk* consists of the *chg-status*, *chg-startdate*, *chg-enddate* and *chg-period* labels. The low performance in extracting the named entities on the CONSTITUTION section also affect the linking task as illustrated in Table 5.

5.3   Active Learning

To show the effectiveness of the active learning scheme, we conducted three experiments on the CAPITAL section of the French Defossee 1926 Yearbook. During these experiments, we used 200 manually annotated examples for evaluating the performance of the trained models.

Table 6. Performance with manually annotated examples with respect to the size of the training data set

Size of training data set	Precision	Recall	F1-score	Nb of gen. examples (score ≥ 90%)
M10	92.74	60.04	72.89	118
M20	93.81	83.28	88.23	170
M30	93.14	84.36	88.53	216
M40	94.97	90.67	92.77	248
M50	95.51	90.74	93.06	258

Table 7. Performance with 10 annotated examples in addition to automatically generated examples whose labelling score ≥ 90%

Size of training data set	Precision	Recall	F1-score	Nb of gen. examples (score ≥ 90%)
M10	92.74	60.04	72.89	118
M10+A118	91.97	58.55	71.55	188
M10+A188	93.42	56.51	70.42	332
M10+A332	94.64	54.23	68.95	406
M10+406	94.6	50.04	65.45	438

Table 8. System performance when training the extraction model on 10 manually annotated examples augmented by the generated examples whose labelling score ≥ 90% in addition to 10 corrected examples

Size of training data set	Precision	Recall	F1-score	Nb. of gen. examples (score ≥ 90%)
M10	92.74	60.04	72.89	118
M10+A118+C10	93.66	87.24	90.34	216
M20+A216+C10	93.04	92.95	92.99	300
M10+A300+C10	93.99	94.53	94.26	382
M10+A382+C10	94.36	95.15	94.76	476

The first experiment shows the effect of increasing the size of the training data set on the performance. Training process starts with ten manually annotated examples (M10), then by adding 10 more examples at each iteration we observe an improvement of the extraction performance in term of precision, recall and F1-score form 92.74, 60.04, 72.89 to 95.51, 90.74, 93.06 as illustrated in Table 6. At the same time, we applied the trained models on the whole set of unlabelled examples found in the yearbook.

In the second experiment, we studied the effect of increasing the training data set with the examples labelled by the model it-self whose labelling score is higher than 0.9. At the beginning of the training process, we used only ten manually annotated examples for training the initial CRF model. Then, we apply the initial model on the whole set of unlabelled examples to obtain their labels with their labelling scores. For the next training iteration, the training data set consists of the ten manually annotated examples (M10) plus 118 automatically labelled examples that have received a labelling score ≥ 0.9. Repeating this process five times we can get 418 automatically annotated training examples. We observe a precision increase from 92.74% up to 95.51% but the overall recall and F1-score degraded at every training iteration, as illustrated in Table 7.

From the second experiment, we can say that the model learns better the same examples by specialising to almost similar examples. To tackle this problem, the training data set must contain more heterogeneous examples. We introduced this notion in the third experiment during which we not only inject labelled data with high scores but also some poorly labelled examples with a labelling score < 0.5 (C10) which are manually corrected and then introduced in a new training data set for the next training iteration. After five active learning iterations, we observe a quick increase of recall and F1-score with a slight degradation of precision (see Table 8.). In comparison with the results obtained from the first experiment, we observe that with only 30 automatically selected and manually annotated examples and three training iterations, the performance (precision: 93.04; recall: 92.95; F1-score: 92.99) reach the performance obtained during the first experiment (precision: 95.51; recall: 90.74; F1-score: 93.06) for which we used 50 training examples and five training iterations.

6   Conclusion

Financial yearbooks are historical records reporting on information about the companies of stock exchanges, including their name, date of creation, financial status, governing board members, headquarters and branches addresses, financial information such as capital amount, date and amount of capital increase, balance sheet of the year including assets and liabilities, etc. In this paper we have presented the key components that implement a financial information extraction system from financial yearbooks. The proposed system consists in three steps: OCR, linked named entities extraction, active learning. The core of the system is related to linked named entities extraction (LNE) with CRF and BLSTM-CRF. The experiments have been conducted on two yearbooks with two languages (French and German). Very promising results have been obtained on three different extraction tasks on three different sections showing very good performance with state of the art named entity extraction models that are specialised to each entities with a limited amount of annotated data through the introduction of active learning.

Acknowledgement. This work has received funding from the European Union's Horizon 2020 research and innovation programme under grant agreement N 777489.

References

1. Akbik, A., Blythe, D., Vollgraf, R.: Contextual string embeddings for sequence labeling. In: Proceedings of the 27th International Conference on Computational Linguistics, pp. 1638–1649 (2018)
2. Grover, C., Givon, S., Tobin, R., Ball, J.: Named entity recognition for digitised historical texts. In: LREC (2008)
3. Huang, Z., Xu, W., Yu, K.: Bidirectional LSTM-CRF models for sequence tagging. arXiv preprint arXiv:1508.01991 (2015)
4. Kim, S.M., Cassidy, S.: Finding names in trove: named entity recognition for Australian historical newspapers. In: ALTA (2015)
5. Lafferty, J., McCallum, A., Pereira, F.C.: Conditional random fields: Probabilistic models for segmenting and labeling sequence data (2001)
6. Lample, G., Ballesteros, M., Subramanian, S., Kawakami, K., Dyer, C.: Neural architectures for named entity recognition. arXiv preprint arXiv:1603.01360 (2016)
7. Manning, C., Surdeanu, M., Bauer, J., Finkel, J., Bethard, S., McClosky, D.: The Stanford CoreNLP natural language processing toolkit. In: Proceedings of 52nd Annual Meeting of the Association for Computational Linguistics: System Demonstrations, pp. 55–60, June 2014
8. McCallum, A., Li, W.: Early results for named entity recognition with conditional random fields, feature induction and web-enhanced lexicons. In: Proceedings of the seventh conference on Natural language learning at HLT-NAACL 2003 (2003)
9. Miller, D., Boisen, S., Schwartz, R., Stone, R., Weischedel, R.: Named entity extraction from noisy input: speech and OCR. In: Proceedings of the Sixth Conference on Applied Natural Language Processing, ANLC 2000, pp. 316–324. Association for Computational Linguistics (2000)

10. Packer, T., et al.: Extracting person names from diverse and noisy OCR text. In: Proceedings of the Fourth Workshop on Analytics for Noisy Unstructured Text Data (2010)

11. Peters, M.E., et al.: Deep contextualized word representations. arXiv preprint arXiv:1802.05365 (2018)

12. Rodriquez, K.J., Bryant, M., Blanke, T., Luszczynska, M.: Comparison of named entity recognition tools for raw OCR text. In: KONVENS (2012)

13. Shen, Y., Yun, H., Lipton, Z.C., Kronrod, Y., Anandkumar, A.: Deep active learning for named entity recognition. arXiv preprint arXiv:1707.05928 (2017)

14. Toledo, J.I., Carbonell, M., Fornés, A., Lladós, J.: Information extraction from historical handwritten document images with a context-aware neural model. Pattern Recogn. **86**, 27–36 (2019)

15. Wang, S., Xu, R., Liu, B., Gui, L., Zhou, Y.: Financial named entity recognition based on conditional random fields and information entropy. In: 2014 International Conference on Machine Learning and Cybernetics, vol. 2, pp. 838–843. IEEE (2014)

Effect of Text Color on Word Embeddings

Masaya Ikoma, Brian Kenji Iwana$^{(\boxtimes)}$ (iD), and Seiichi Uchida (iD)

Kyushu University, Fukuoka, Japan
{masaya.ikoma,brian}@human.ait.kyushu-u.ac.jp, uchida@ait.kyushu-u.ac.jp

Abstract. In natural scenes and documents, we can find a correlation between text and its color. For instance, the word, "hot," is often printed in red, while "cold" is often in blue. This correlation can be thought of as a feature that represents the semantic difference between the words. Based on this observation, we propose the idea of using text color for word embeddings. While text-only word embeddings (e.g. word2vec) have been extremely successful, they often represent antonyms as similar since they are often interchangeable in sentences. In this paper, we try two tasks to verify the usefulness of text color in understanding the meanings of words, especially in identifying synonyms and antonyms. First, we quantify the color distribution of words from the book cover images and analyze the correlation between the color and meaning of the word. Second, we try to retrain word embeddings with the color distribution of words as a constraint. By observing the changes in the word embeddings of synonyms and antonyms before and after re-training, we aim to understand the kind of words that have positive or negative effects in their word embeddings when incorporating text color information.

Keywords: Word embedding · Text color

1 Introduction

In natural scenes and documents, the color of text can correlate to the meaning of the text. For example, as shown in Fig. 1, the word, "hot," is often printed in red and the word, "cold," is often blue. This correlation is caused by various reasons. The color of the object described by a text can be a reason; for example, words relating to plants are often printed in green. Visual saliency (i.e., visual prominence) is another reason; for example, cautions and warnings are often printed in red or yellow for a higher visual saliency. Another important reason is the *impression* of color; for example, red gives an excitement impression and pink gives a feminine impression, according to color psychology [2].

Based on the above observations, one might imagine that the correlation between a word and its color is useful for *word embeddings*. Figure 2 demonstrates this concept. Word embedding is the technique to convert a word into a vector that represents the meaning of the word. Word2vec [9,10] is a standard and popular technique to create word embeddings. In addition, recently,

© Springer Nature Switzerland AG 2020
X. Bai et al. (Eds.): DAS 2020, LNCS 12116, pp. 341–355, 2020.
https://doi.org/10.1007/978-3-030-57058-3_24

Fig. 1. Example of antonyms printed in different colors.

BERT [1] and its variants have become popular. Although those word embedding methods have made large progress in natural language processing (NLP), further improvement is still necessary. For example, word2vec often gives similar vectors for antonyms (such as hot and cold) since they are often grammatically interchangeable in sentences. We, therefore, expect that color usage would be helpful to have better vector representations for such cases, as shown in Fig. 2.

However, we must note that the correlation between a word and its color is not always strong. For example, the word, "food," will be printed in various colors depending on the type of food. Moreover, the color of the word, "the," will not have any specific trend, except for the general trend that words are often printed in achromatic color [3].

The purpose of this paper is to understand what kinds of words have a positive or negative or no effect in their vector representation by the incorporation of color information. For this purpose, we first determine the color distribution (i.e., color histogram) of each word by using word images collected from book titles of about 210,000 book cover images. Since book cover images (especially titles) are carefully created by professional designers, we can expect that the word images from them are more helpful to find the meaningful correlation between a word and its color than word images with thoughtless color usage. We then analyze the color histograms to find the words whose color histogram is very different from the standard one.

After those observations, we develop a new word-embedding method that can reflect the color usage of each word. On one hand, we expect positive effects using the colors; for example, antonyms with contrasting color usages, like "hot" and "cold," will have more discriminative semantic vectors (i.e., vectors that represent the meanings of the words) than the standard word embeddings. On the other hand, we expect negative effects, since there are cases with no strong word-color correlation. Accordingly, we observe the improved or degraded cases from the standard word2vec embeddings.

The main contributions of this paper are summarized as follows:

- To the authors' best knowledge, this is the first trial of an objective analysis of the correlation between color and meaning of words, by using a large number of real-word images.

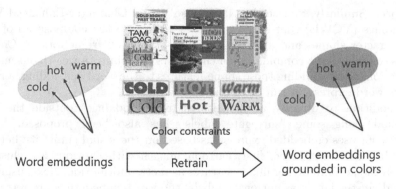

Fig. 2. The concept of this paper. Correlation between colors and meanings of individual words *might* be helpful to give better semantic vectors.

- We discover words and word classes whose color usage is very particular.
- We proposed a novel word-embedding method that reflects the color usage of the word. If a pair of words have similar color usages, their semantic vectors become more similar through the proposed method.
- Since synonyms and antonyms do not always have similar and dissimilar color usages, the effect of word-embedding was not always positive. However, we also revealed that many synonyms and antonyms have positive effects from the proposed word-embedding method.

2   Related Work

Word2vec is one of the most famous modern word embedding methods [9,10]. For word2vec, two types of models, Continuous-Bag-of-Words (CBoW) and Skip-gram, have been proposed. In both models, it is assumed that words with similar meanings in sentences are used in similar contexts. In other words, interchangeable words in a sentence can be said to have similar meanings. Word2vec uses this assumption to derive a semantic vector of the word by using a neural network.

Although word2vec gives reasonable semantic vectors for most words, it still has a drawback caused by the above assumption. Considering two sentences, "I drink cold water" and "I drink hot water", the antonym words, "cold" and "hot" are interchangeable and will have similar semantic vectors despite their opposite meanings. Similar situations often happen for (especially, adjective and adverb) antonyms and their semantic difference is underestimated.

To make semantic vectors more discriminative, a lot multimodal word embedding methods have been proposed to capture word semantics from multiple modalities. For example, Multimodal Skip-gram (MMSG) [7]. In normal Skip-gram, the co-occurrence of words in sentences is used to predict the context of a word. However, in MMSG, in addition to the co-occurrence of words, image features related to the word are simultaneously used. This makes it possible to generate word embeddings by taking into account the visual information of the object. Similarly, Sun et al. [14] propose Visual Character-Enhanced Word Embeddings (VCWE) that extract features related to the composition of Chinese character shapes and applies them to a Skip-gram framework. A Chinese word is composed of a combination of characters containing a wealth of semantic information, and by adding those shape-features to the word embedding, a richer Chinese word expression can be obtained.

In addition to directly using features from modalities outside the text, a method using some "surrogate labels" has also been proposed. Visual word2vec [6] uses embedded expressions based on the visual similarity between words. In Visual word2vec, first, cartoon images with captions are clustered based on the accompanying image features as meta information. Next, using the obtained cluster labels as surrogate labels, the word-embeddings are retrained. During this, the label of the cluster to which the image belongs is predicted from the words in the caption of the image using a CBoW framework. As a result, words belonging to the same cluster have their similarity increased. For example, words with high visual relevance, such as "eat" and "stare," have their embedded expressions changed to more similar values. A similar technique can be used to learn word-embeddings based on any modality. In Sound-word2vec [15], surrogate labels obtained by clustering the sounds are predicted from word tags, thereby acquiring word embedding that takes into account relevance in sounds.

In addition, there are methods to acquire multimodal word embeddings based on Autoencoders. In Visually Enhanced Word Embeddings (ViEW) [4], an autoencoder is used to encode linguistic representations into an image-like visual representation encoding. Next, a hidden layer vector is used as a multimodal representation of the word.

The purpose of this paper is to investigate the relationship between words observed in real environments and their colors and to analyze how the color information can affect the recognition of word semantics. Although there were several data-driven trials on font color usage [3,13], there is no trial to correlate the font color usage to the meaning of words, to the authors' best knowledge. Throughout this trial, we first observe the color usages on words and word categories, then propose a word-embedding method with color information, and finally reveal how the font color usage is useful and not useful for word embedding.

3   Collecting Word Images from Book Covers

The book covers sourced for the experiment are from the Book Cover Dataset[1] [5]. This dataset consists of 207,572 book cover images from Amazon Inc. Figure 1 shows some example book cover images.

In order to extract the words from the book cover images, the Efficient and Accurate Scene Text detector (EAST) [16] is used. EAST is a popular text detector that implements a Fully-Convolutional Network (FCN) [8] combined with locality-aware Non-Maximum-Suppression (NMS) to create multi-oriented

---

[1] https://github.com/uchidalab/book-dataset.

text boxes. After detecting the words using EAST, a Convolutional Recurrent Neural Network (CRNN) [12] is used to recognize the text boxes into text strings. Note that book title information is given as the meta-data of the Book Cover Dataset and thus it is possible to remove the words that are not contained in the book title. Consequently, we collected 358,712 images of words. Their vocabulary size (i.e., the number of different words) was 18,758. Each word has 19 images on average, 4,492 at maximum, and 1 at minimum. Note that stop words, proper nouns, digits, compounded words (e.g., "don't" and "all-goals") are removed from the collection[2]. In addition, we identify word variants (e.g., "meets" = "meet"), by using a lemmatizer[3].

4 Color Usage of a Word as a Histogram

In this paper, we represent the color of a word in the CIELAB color space. Each color is represented by a three-dimensional vector (L^*, a^*, b^*), where L^* is lightness ($\sim$intensity) and a^* and b^* are color channels. The advantage of using CIELAB is that the Euclidean distance in the CIELAB color space is approximately relative to human's perceptual difference.

CIELAB color values of each word in the detected bounding box are determined as follows. The first step is the separation of the character pixels from the background pixels. Fortunately, most word images on book covers have a high contrast between the characters and the background (possibly due to better legibility). Therefore, a simple Otsu binarization [11] technique was enough for the separation. As the second step, the CIELAB color vector (L^*, a^*, b^*) is determined for a word by taking the average of all character pixels.

In order to analyze the distribution of colors, we implement a color histogram. Specifically, all colors are quantized into one of 13 basic colors (pink, red, orange, brown, yellow, olive, yellow-green, green, blue, purple, white, gray, and black), which are defined by as basic colors in the ISCC-NBS system. The color quantization is performed using the nearest neighbor rule with Euclidean distance in CIELAB space. If a word occurs P times, in the dataset, we initially have a color histogram with P votes for $K = 13$ bins. We then normalize it to be its total votes equal to one, in order to remove the difference of P.

In the following analysis, we introduce the following two conditions. First, we do not use the word images whose word color is achromatic (i.e., white or black or gray) *and* background color is chromatic (i.e., one of the 10 chromatic colors). A survey of color usages of scene texts and their background [3] proved that if the background color is chromatic and the foreground text is achromatic then the designer typically uses the background color to portray meaning. We, therefore, extend this idea to book cover title colors and remove the word images in this case. Note that if both title and background colors are achromatic, we

---

[2] From 579,240 images of words given by EAST, 492,363 images remain after removing non-title words and misrecognized words. Finally, 358,712 images remain after removing stop words, etc.

[3] www.nltk.org/_modules/nltk/stem/wordnet.html.

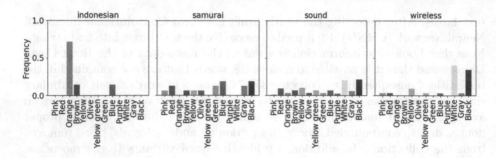

Fig. 3. Color histogram examples. (Color figure online)

do not remove the image. For the second condition, we removed the histogram for words which occur less than five times to limit the analysis to reliable color usages.

Consequently, we have color histograms of 6,762 words. Figure 3 shows several examples, where each bar of the color histogram is painted by the corresponding basic color. The leftmost histogram, "indonesian," shows a peak on a specific color, "Orange," whereas the second and the third show rather flat usages. As shown in the rightmost histogram, we sometimes find histograms with frequent use of achromatic colors.

5   Words with Particular Color Usages

5.1   How to Find Words with Particular Color Usage

Our interest is to find the words with particular color usage rather than words with standard color usage because particular color usage might show specific correlations between the meanings and the color of the word. Figure 4 shows a scatter plot for finding such words. Each dot corresponds to a word and the horizontal axis is the color variance $\sum_k (h_w^k - \mu)^2 / K$, where $h_w = (h_w^1, \ldots, h_w^K)$ is the (normalized) color histogram of the word $w \in V$ and $\mu = \sum_k h_w^k / K = 1/K$. When this variance is large for a word w, the histogram is not flat but has some peaks at specific colors; that is, the word w has a more particular color usage. The vertical axis is the distance from the average color histogram $\bar{h}$, i.e., $\|h_w - \bar{h}\|$, where $\bar{h} = \sum_w h_w / |V|$, where $|V|$ is the vocabulary size. When this distance is large for word w, the histogram h_w is largely deviated from the average color histogram $\bar{h}$ and word w has a more particular color usage. Note again, $K = 13$.

By splitting each axis by its average, we have four regions in Fig. 4. Among them, we call the region with red color *the first quadrant*. Similarly, orange, green, and blue regions as the second, third, and fourth quadrants, respectively. Each word belonging to the first quadrant will have a particular color usage. In contrast, the words in the third quadrant will have a flat and standard color usage. The first, second, third, and fourth quadrants contain 1,973,

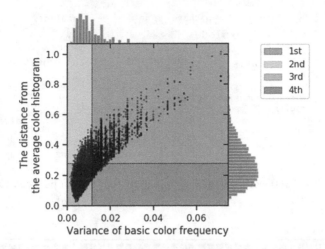

Fig. 4. The distribution of 6,762 words in a two-dimensional plane evaluating the particularity of the color usage of the word. (Color figure online)

1,081, 3,381, and 327 words, respectively. This suggests that words with particular color usages are not dominant but exist to a certain amount. Specifically, although $3,381/6,762 \sim 50\%$ words have an average and flat color usage and their color usage is not reflecting their meanings, the remaining 50% words have a particular color usage which might reflect their meanings. In addition, the first quadrant with $1,973/6,762 \sim 30\%$ words have a very particular color usage.

5.2 Words with Particular Color Usage

Figure 5 shows a word-cloud of the words in the first quadrant of Fig. 4, colored by each word's most frequent basic color. Different from typical word-clouds, the size of the word is proportional to the particularity of the word, which is calculated as the product of "the distance from the average color histogram" and "variance of basic color frequency" in Fig. 4. As a general tendency, achromatic colors are the most frequent color for most words and it can be seen that the meaning and color of words are not necessarily related. On the other hand, some relationships between the color and the meaning of words can be found. For example, "premier" is reminiscent of position and victory and is yellow while "amour" means love and frequently appears in pink, etc.

5.3 Word Category with Particular Color Usages

Words can be categorized into hierarchical groups, thus, in this section, we analyze the relationship between word category and color usage. Figure 6 shows the ratio of the four quadrants for 45 word categories, which are defined as *lex-names*[4] in WordNet. In this figure, the categories are ranked by the ratio words

---

[4] wordnet.princeton.edu/documentation/lexnames5wn.

Fig. 5. Examples of 100 words in the first quadrant that have a particular color usage. (Color figure online)

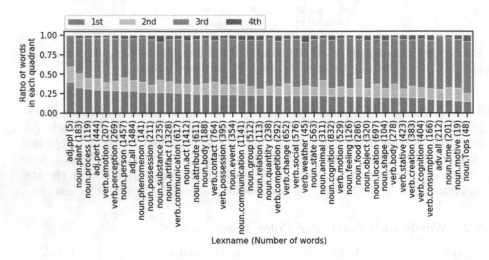

Fig. 6. The ratio of words in each quadrant by word category (i.e., lexnames).

in the first quadrant. Note that a word can belong to multiple categories by the hierarchical structure of WordNet.

The category "noun.plant" is high-ranked; namely, words relating to plants have particular color usages. This is because words related to plants tend to appear in green or their fruit color. Figure 7 shows the color histograms of "lemon" and "strawberry" and sample books of the dominant colors. In particular, "lemon" often appears as yellow or olive and "strawberry" frequently is in red. It is noteworthy that the rank of "noun.plant" is much higher than the rank of "noun.food." This is because of the large variations of foods which result in the large variations of color usages.

The category "verb.emotion" is also high-ranked. As shown in Fig. 7, the color histograms of "anger" shows a high frequency of warm reddish colors. Compared to "anger", the word "sorrow" is not printed with red but more with pale colors. This may be because the impression of colors affects the word color usages.

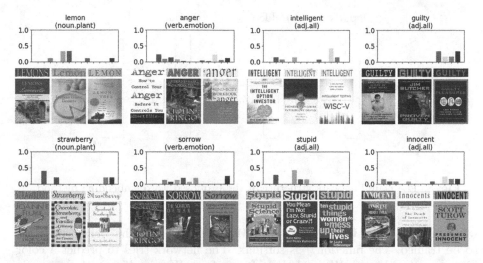

Fig. 7. Words with their color histogram and image examples. The parenthesized term (e.g., "noun.plant") is the category of the word. (Color figure online)

It should be emphasized that categories of adjectives such as "adj.all" (all adjective clusters) and "adj.pert" (relational adjectives, i.e., so-called pertainyms) are also high-ranked. These categories contain many antonym adjective pairs, like "hot" and "cold," as well as synonym adjective pairs. Figure 7 also shows the color histograms of pairs of antonyms, "intelligent" and "stupid", and "guilty" and "innocent". The word, "intelligent," tends to be printed in paler color than "stupid." The word, "guilty," was printed in purple more than "innocent" and did not show up as a warm color.

6   Word Embeddings Grounded in Color

6.1   Methodology

Based on the analysis results of Sect. 5, we attempt to create a word embedding grounded on word color usage using a neural network. Figure 8 shows the network structure of the proposed word embedding method. This network is inspired by Sound-Word2vec [15]. Specifically, this network is similar to an autoencoder or the skip-gram of word2vec. Its input x_w is the V-dimensional one-hot representation of the input word w and its output should be close to the K-dimensional vector h_w representing the color histogram of w. The value $K = 13$ is the number of bins of the color histogram. Accordingly, its output is not a V-dimensional one-hot vector but a K-dimensional vector representing a color histogram with K bins. The N-dimensional output vector from the hidden layer is the semantic vector representation of the input word w.

The neural network is trained to minimize the following loss function L:

$$L = \sum_w \|h_w - f(W_O W_I x_w)\|, \tag{1}$$

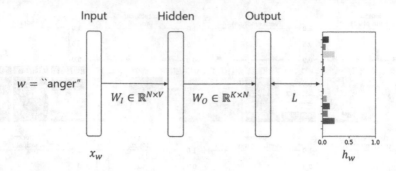

Fig. 8. The neural network structure of the proposed word embedding method.

where the matrices $W_I \in \mathbb{R}^{N \times V}$ and $W_O \in \mathbb{R}^{K \times N}$ are the weight matrices to be trained. The function f is softmax. The network tries to predict a color histogram h_w through W_O and the softmax from $W_I x_w$, which is the semantic vector representation of the word w. The matrix W_I is initialized as $\bar{W}_I$ which is a word-embedding matrix trained by a word-embedding method, such as word2vec. The matrix W_O is initialized with random values.

By updating W_I and W_O along with the minimization of L, semantic vectors become more different (similar) between words with different (similar) color usages. In other words, the proposed method retrains W_I from $\bar{W}_I$ for giving more different (similar) semantic vectors for a pair of words, if they have different (similar) color usages. As noted above, typical word embedding methods, such as word2vec, often give similar semantic vectors even for a pair of antonyms. We, therefore, can expect that they have more different vectors if they have different color usages, as shown in Fig. 2. More specifically, for a pair of antonyms, x_{w_1} and x_{w_2}, we can expect $s_{w_1,w_2} - \bar{s}_{w_1,w_2} < 0$, where $\bar{s}_{w_1,w_2} = \langle \bar{W}_I x_{w_1}, \bar{W}_I x_{w_2} \rangle / \|\bar{W}_I x_{w_1}\| \|\bar{W}_I x_{w_2}\|$ is the cosine similarity of the semantic vectors for w_1 and w_2 *without* color usage information, and $s_{w_1,w_2} = \langle W_I x_{w_1}, W_I x_{w_2} \rangle / \|W_I x_{w_1}\| \|W_I x_{w_2}\|$ is *with* color usage information.

6.2   Evaluation

To analyze the positive or negative effect of using the color usage information of words, we focused pairs of synonyms and antonyms and evaluated their similarity before and after integrating the color usage information. For the initial matrix $\bar{W}_I$, GoogleNews-vectors (word2vec trained with Google News corpus) were used[5]. We used synonym pairs and antonym pairs defined in Wordnet.

---

[5] The vocabulary assumed in GoogleNews-vectors is slightly different from that of our book title words. Specifically, 198 words in the book titles do not appear in GoogleNews-vectors. We, therefore, used 6,762-198 = 6,514 words in the experiment in this section. The data of GoogleNews-vectors is provided by https://code.google.com/archive/p/word2vec/.

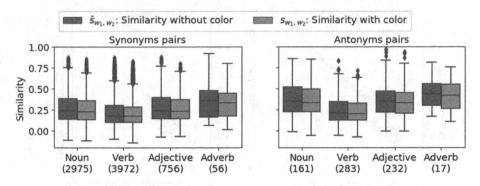

Fig. 9. Distributions of the similarity for synonym pairs (left) and antonym pairs (right).

Effect of Word Color on Similarity in Synonyms and Antonyms. As a quantitative evaluation, we observed the distributions of $\bar{s}_{w_1,w_2}$ and s_{w_1,w_2} for all of the synonym pairs and antonym pairs. Figure 9 shows the similarity distributions with and without color. The highlight of this result is that the similarity between antonym pairs is *successfully* decreased when using color. This suggests that antonym pairs tend to have different color usages. However, this decrease in similarity also happens for the synonym pairs. Thus, we cannot assume that synonyms are always printed in similar colors. Figure 9 also shows that the range of the similarity values for synonym pairs is similar to antonym pairs, even without color usage, that is, even by the original word2vec. (Ideally, synonym pairs should have higher similarities than antonym pairs.) This clearly shows the difficulty of the word embedding task.

A more detailed analysis is shown in Fig. 10. This plot shows $(\bar{s}_{w_1,w_2}, s_{w_1,w_2})$ for each synonym and antonym pair. Each point corresponds to a word pair. If a point is on the diagonal line, the corresponding word pair is not affected by the use of the color information in their semantic vector representation. The color of each point represents the similarity of the color information (i.e., the similarity of the color histograms of the paired words). It is observed that the proposed neural network works appropriately. In fact, pairs with different color usages (blue dots) are trained to have more different semantic vectors, that is, $\bar{s}_{w_1,w_2} > s_{w_1,w_2}$. In contrast, pairs with similar color usages (red dots) are trained to have more similar vectors, that is, $\bar{s}_{w_1,w_2} < s_{w_1,w_2}$.

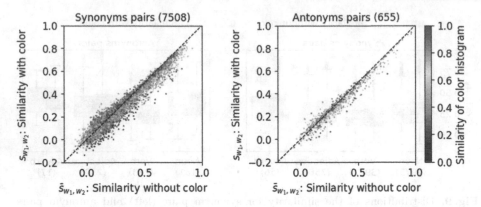

Fig. 10. Similarity of the semantic vectors with or without color information. Each dot corresponds to a pair of synonyms (left) or antonyms (right). (Color figure online)

Another observation of Fig. 10 reveals that there are many blue points for synonym pairs and red points for antonym pairs. Namely, synonym pairs often have different color usages and antonym pairs have similar usages. As we anticipated, the correlations between color usages and meanings for words are not always consistent and not always strongly correlated. This fact coincides with the quantitative evaluation result of Fig. 9.

Examples of synonyms and antonyms with an improved similarity between two words are shown in the Tables 1 and 2. In Table 1, improvement is defined by a case where the similarity of the word embedding is increased for synonyms. Conversely, in Table 2, improvement is defined by a case where the similarity is reduced for antonyms. The lexnames in these tables indicate the category in which the two words are treated as synonyms or antonyms. From these, it can be seen that the word pairs used in verbs occupy the top of synonyms and word pairs of adjectives are common in antonyms. Next, Fig. 11 shows examples of color histograms of synonym and antonym word pairs. For the synonym pairs, the colors tend to be achromatic. This is because words, such as verbs, do not necessarily associate with color impressions and the visual characteristics of verbs tend to achromatic. As a result, it can be seen that the word embedding with achromatic colors tended to be updated as more similar under the proposed method. In contrast, for antonyms, frequent chromatic differences between the word pairs can be confirmed. The correlation between the psychological impression of the color and the meaning of the word is strong. Thus, it can be said that for these words, namely adjectives, it is a good example of compensating for the differences in opposite words.

Table 1. Top 20 synonym pairs with similarity increase by color information.

Word1	Word2	$s - \bar{s}$	Lexnames
direct	taken	0.185	verb.motion verb.competition
broadcasting	spread	0.176	Verb.communication
hire	taken	0.160	Verb.possession
rolling	vagabond	0.157	verb.motion
floating	vagabond	0.156	adj.all
breed	covering	0.149	verb.contact
fade	slice	0.149	noun.act
gain	hitting	0.146	verb.motion
lease	taken	0.141	verb.possession
beam	broadcasting	0.137	verb.communication
engage	taken	0.133	verb.possession
shoot	taken	0.130	verb.communication
conducting	direct	0.130	verb.motion verb.creation
press	squeeze	0.129	verb.contact
later	recent	0.128	adj.all
affect	touching	0.128	verb.change
demand	taken	0.125	verb.stative
capture	charming	0.124	verb.emotion
drop	spent	0.124	verb.possession
pose	position	0.122	verb.contact

Table 2. Top 20 antonym pairs with similarity decrease by color information.

Word1	Word2	$s - \bar{s}$	Lexnames
certain	uncertain	−0.192	adj.all
feminine	masculine	−0.184	adj.all
sit	standing	−0.158	verb.contact
nonviolent	violent	−0.137	adj.all
enduring	enjoy	−0.136	verb.perception
sit	stand	−0.135	verb.contact
correct	wrong	−0.133	adj.all verb.social
even	odd	−0.132	adj.all
spoken	written	−0.130	adj.all
certainty	doubt	−0.128	noun.cognition
laugh	weep	−0.127	verb.body
indoor	outdoor	−0.125	adj.all
lay	sit	−0.120	verb.contact
noisy	quiet	−0.119	adj.all
cry	laugh	−0.119	verb.body
host	parasite	−0.118	noun.animal
immortal	mortal	−0.118	adj.all
shrink	stretch	−0.118	verb.change
confident	shy	−0.117	adj.all
better	worse	−0.117	adj.all

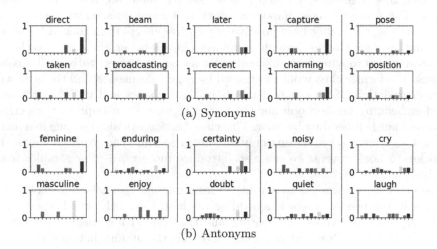

(a) Synonyms

(b) Antonyms

Fig. 11. Example of (a) synonym pairs and (b) antonym pairs with improved similarity. Upper and lower histograms indicate pairs.

7   Conclusion

To the authors' best knowledge, this paper is the first trial to analyze the relationship between the meaning and color of words, by using a data-driven (i.e., totally non-subjective) approach. For this analysis, 358,712 word images were collected from the titles of 207,572 book cover images. First, we observed the words and the word categories with their color usages. Our analysis revealed the existence of words with *particular*, i.e. specific, color usage. For example, words relating to plants have more particular color usages than words for foods. On the other hand, we also found there are many words whose color usage is not particular.

We then observed the effect of the color usage information has on the word embedding, i.e. semantic vector representation of words. To do this, we developed a novel word-embedding method, which is inspired by Sound-Word2vec [15]. The method modifies the word-embedding matrix given by word2vec to reflect color usages of individual words. Specifically, the similarity of the words with different color usages become more different and vise versa. We confirmed that the proposed method can provide a reasonable semantic vector representation. By using antonym pairs and synonym pairs, we also confirmed that color information has "positive" or "negative" effects for the semantic vector representation. This is because there are synonym pairs with similar color usages but also synonyms with different color usages. The same situation happens to antonym pairs.

Since this is the first trial on a new research topic, we have many future works. The first and most important work is a more detailed analysis of the "class" of the words that have positive or negative effects from their color usage information. As we found through the analysis in this paper, there are two classes of antonyms; antonyms with different color usages and those with similar color usages. The former class will be improved by color information and the later will have degradation. If we can find trends in these classes and use the proposed word-embedding method only for the former class, we can improve the performance of word embedding by color. This classification can also be done in a more general way, instead of focusing only on antonyms and synonyms. Of course, in addition to book covers, we can also introduce more word-color samples from the internet.

Another future work is to consider a loss function specialized for color histogram in the proposed word-embedding method. Currently, our loss function is the L^2 distance between the output and the actual color usage. This means we treat 13 basic colors independently and thus the affinity between the basic colors is not considered. For example, red and orange are treated independently but a word often with red can be printed with orange. Moreover, the current results depend too much on the choice of 13 colors and their color space (e.g. CIELAB). Therefore, it will be meaningful to find a more suitable way for color usage representation.

Acknowledgment. This work was supported by JSPS KAKENHI Grant Number JP17H06100.

References

1. Devlin, J., Chang, M.W., Lee, K., Toutanova, K.: Bert: pre-training of deep bidirectional transformers for language understanding. arXiv preprint arXiv:1810.04805 (2018)
2. Elliot, A.J., Maier, M.A.: Color psychology: effects of perceiving color on psychological functioning in humans. Annu. Rev. Psychol. **65**(1), 95–120 (2014)
3. Gao, R., Eguchi, S., Uchida, S.: True color distributions of scene text and background. In: International Conference on Document Analysis and Recognition, pp. 506–510 (2015)
4. Hasegawa, M., Kobayashi, T., Hayashi, Y.: Incorporating visual features into word embeddings: a bimodal autoencoder-based approach. In: International Conference on Computational Semantics-Short Papers (2017)
5. Iwana, B.K., Rizvi, S.T.R., Ahmed, S., Dengel, A., Uchida, S.: Judging a book by its cover. arXiv preprint arXiv:1610.09204 (2017)
6. Kottur, S., Vedantam, R., Moura, J.M., Parikh, D.: Visual Word2Vec (vis-w2v): learning visually grounded word embeddings using abstract scenes. In: IEEE Conference on Computer Vision and Pattern Recognition, pp. 4985–4994 (2016)
7. Lazaridou, A., Pham, N.T., Baroni, M.: Combining language and vision with a multimodal skip-gram model. In: Conference of the North American Chapter of the Association for Computational Linguistics: Human Language Technologies (2015)
8. Long, J., Shelhamer, E., Darrell, T.: Fully convolutional networks for semantic segmentation. In: IEEE Conference on Computer Vision and Pattern Recognition, pp. 3431–3440 (2015)
9. Mikolov, T., Chen, K., Corrado, G., Dean, J.: Efficient estimation of word representations in vector space. arXiv preprint arXiv:1301.3781 (2013)
10. Mikolov, T., Sutskever, I., Chen, K., Corrado, G.S., Dean, J.: Distributed representations of words and phrases and their compositionality. In: Advances in Neural Information Processing Systems, pp. 3111–3119 (2013)
11. Otsu, N.: A threshold selection method from gray-level histograms. IEEE Trans. Syst. Man Cybern. **9**(1), 62–66 (1979)
12. Shi, B., Bai, X., Yao, C.: An end-to-end trainable neural network for image-based sequence recognition and its application to scene text recognition. IEEE Trans. Pattern Anal. Mach. Intell. **39**(11), 2298–2304 (2016)
13. Shinahara, Y., Karamatsu, T., Harada, D., Yamaguchi, K., Uchida, S.: Serif or Sans: visual font analytics on book covers and online advertisements. In: International Conference on Document Analysis and Recognition (2019)
14. Sun, C., Qiu, X., Huang, X.: VCWE: visual character-enhanced word embeddings. arXiv preprint arXiv:1902.08795 (2019)
15. Vijayakumar, A., Vedantam, R., Parikh, D.: Sound-word2vec: learning word representations grounded in sounds. In: Conference on Empirical Methods in Natural Language Processing (2017)
16. Zhou, X., et al.: East: an efficient and accurate scene text detector. In: IEEE Conference on Computer Vision and Pattern Recognition, pp. 5551–5560 (2017)

Document Data Extraction System Based on Visual Words Codebook

Vasily Loginov[1,2] , Aleksandr Valiukov[1] , Stanislav Semenov[1] ,
and Ivan Zagaynov[1,2(✉)]

[1] R&D Department, ABBYY Production LLC, Moscow, Russia
{vasily.loginov,aleksandr.valiukov,stanislav.semenov,
ivan.zagaynov}@abbyy.com
[2] Moscow Institute of Physics and Technology (National Research University),
Moscow, Russia

Abstract. We propose a new approach to extraction important fields from business documents such as invoices, receipts, identity cards, etc. In our approach, field detection is based on image data only and does not require large labeled datasets for learning. Method can be used on its own or as an assisting technique in approaches based on text recognition results. The main idea is to generate a codebook of visual words from such documents similar to the Bag-of-Words method. The codebook is then used to calculate statistical predicates for document fields positions based on the spatial appearance of visual words in the document. Inspired by Locally Likely Arrangement Hashing algorithm, we use the centers of connected components extracted from a set of preprocessed document images as our keypoints, but we use a different type of compound local descriptors. Target field positions are predicted using conditional histograms collected at the fixed positions of particular visual words. The integrated prediction is calculated as a linear combination of the predictions from all the detected visual words. Predictions for cells are calculated from a 16×16 spatial grid. The proposed method was tested on various different datasets. On our private invoice dataset, the proposed method achieved an average top-10 accuracy of 0.918 for predicting the occurrence of a field center in a document area constituting $\sim 3.8\%$ of the entire document, using only 5 labeled invoices with previously unseen layout for training.

Keywords: Data extraction · Few-shot learning · Layout analysis · Visual words

1 Introduction

Automatic information extraction from business documents is still a challenging task due to the semi-structured nature of such documents [3]. While an instance of a specific document type contains a predefined set of document fields to be

X. Bai et al. (Eds.): DAS 2020, LNCS 12116, pp. 356–370, 2020.
https://doi.org/10.1007/978-3-030-57058-3_25

extracted (e.g. date, currency, or total amount), the positioning and representation of these fields is not constrained in any way. Documents issued by a certain company, however, usually have a specific layout.

Popular word classification approaches to information extraction, e.g. [7], require huge datasets of labeled images, which is not feasible for many real-life information extraction tasks.

Convolutional neural networks (CNN) have been used extensively for document segmentation and document classification and, more broadly, for detecting text in natural scenes (e.g. [2]). In the case of CNNs, nets are also trained on explicitly labeled datasets with information about the targets (e.g. pixel level labels, bounding boxes, etc.).

The latest deep neural network architectures can be trained directly end-to-end to extract relevant information. Method proposed in [8] takes the spatial structure into account by using convolutional operations on concatenated document text and image modalities, with the text of the document extracted using an Optical Character Recognition (OCR) engine.

Our main goal has been to develop a system capable of predicting the positions of document fields on documents with new layouts (or even on documents of new types) that were previously never seen by our system, with learning performed on a small number of documents labeled by the user. To achieve this goal, we propose a method that relies exclusively on the modality of document images, as the complex spatial structure of business documents is clearly reflected in their image modality. The proposed method takes into account the spatial structure of documents, using as a basis a very popular approach in computer vision, Bag-of-Words (BoW) model [10].

The BoW model has been widely used for natural image retrieval tasks and is based on a variety of keypoint detectors/descriptors, SIFT and SURF features being especially popular. State-of-the-art key-region detectors and local descriptors have also been successfully used for document representation in document classification and retrieval scenarios [4], in logo spotting [9], and in document matching [1].

However, document images are distinctly different from natural scenes, as document images have an explicit structure and high contrast, resulting in the detection of numerous standard key regions. Classically detected keypoints do not carry any particular semantic or structural meaning for the documents. Methods specifically designed for document images make explicit use of document characteristics in their feature representations. In [12], it was proposed to use as keypoints the centers of connected components detected through blurring and subsequent thresholding, and a new affine invariant descriptor Locally Likely Arrangement Hashing (LLAH) that encodes the relative positions of key regions. In [5], key regions are detected by applying the MSER algorithm [6] to morphologically preprocessed document images.

Inspired by the results of [5,12] and following the BoW approach, we generate a document-oriented codebook of visual words based on key regions detected by MSER and several types of compound local descriptors, containing both photometric and geometric information about the region. The visual codebook is then used to calculate statistical predicates for document field positions based on correlations between visual words and document fields.

2   Method

The main idea of our method is to build a codebook of visual words from a bank of documents (this is similar to the BoW approach) and apply the visual codebook to calculate statistical predicates for document field positions based on the spatial appearance of the visual words on the document. We use connected components extracted by the MSER algorithm from a set of morphologically preprocessed document images as our key regions. Next, local descriptors can be calculated in such key regions using various different techniques. The codebook consists of the centers of clusters obtained for the local descriptors (such centers are also known as "visual words"). We use the mutual information (MI) of two random variables, the position of a document field position, and the position of a particular visual word as a measure of quality for that visual word. The integrated quality of the visual codebook can be estimated as the average value of MI over all visual words. We predict target document field positions via conditional histograms collected at the fixed positions of the individual visual words. The integrated prediction of field position is calculated as a linear combination of the predictions from all the individual visual words detected on the document.

2.1   Keypoints Regions Extraction

To extract a keypoint region from a document image, we apply a MSER detector after morphological preprocessing. More specifically, we combine all the MSER regions detected on the original document image and on its copies obtained by a sequential application of an erosion operation. Examples of extracted rectangles of MSER regions of different sizes are shown in Fig. 1.

MSER regions are roughly equivalent to the connected components of a document image produced over all the possible thresholdings of the image. Such key regions correspond to the structural elements of the document (i.e. characters, words, lines, etc.). Combined with iterative erosion preprocessing, the MSER algorithm provides an efficient multi-scale analysis framework. It has been shown that MSER regions perform well in matching tasks of document analysis [11].

2.2   Calculation of Local Descriptors

Various local descriptors have been used in document image processing, both photometric (e.g. SIFT) and geometric (e.g. LLAH [12]). In our work we considered the following photometric descriptors of extracted MSER regions: popular SIFT, SURF and two descriptors composed using DFT or DWT coefficients (all were calculated for a grayscale image). Additionally, we concatenate the photometric descriptor with the geometric descriptor. The last one can consist of

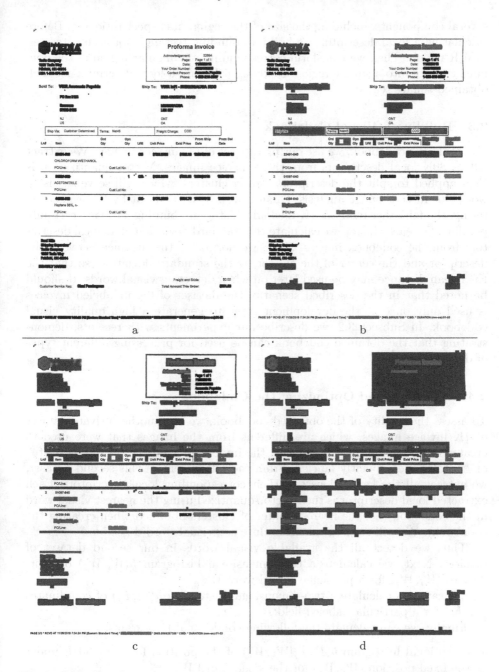

Fig. 1. An original invoice image (a) and bounded rectangles of MSER regions of different sizes extracted from the image. The area of the extracted regions is less than 0.005 (b), 0.01 (c), 0.05 (d) of the image area. The color of a region represents the size of the area (the smallest region is shown in red and the largest region is shown in blue). (Color figure online)

several components, including the size of the region, its aspect ratio, etc. Before calculating a local descriptor, we build a bounding rectangle for each extracted MSER region. Then we transform corresponding rectangular region of the document image into a square region. Next, we calculate a local descriptor for each obtained square region.

2.3   Building a Visual Codebook

To obtain the codebook, we use our private set of 6K invoice images. We extract 50K – 80K local descriptors from these invoice images. Vector quantization is then applied to split the descriptors into N clusters, which will serve as visual words for further image analysis. Quantization is carried out by K-means clustering, though other methods (K-medoids, histogram binning, etc.) are certainly possible. For each cluster, we calculate the standard deviation of its local descriptors from the codebook images. Next, we normalize the distance between the descriptor and the center of the cluster by the standard deviation, so that the Euclidean distance may be used later on when detecting visual words. It should be noted that in the described scenario, the dataset of 6K unlabeled invoices is used only once at the development stage to generate a high quality visual codebook. In Subsect. 3.2, we describe our experiments with receipts, demonstrating that the obtained codebook can be used for processing different types of documents.

2.4   Assessing and Optimizing the Codebook

To assess the quality of the obtained codebook, we use another private dataset of 1K invoice images, which are different from the images that were used to create the codebook. In this dataset, the important fields (e.g. "Invoice Date" or "Total") are explicitly labeled. From each document in this second dataset, we extract all the key regions and their corresponding local descriptors. Each extracted local descriptor is then vector-quantized using the nearest visual word in the codebook (i.e. the nearest center of clusters obtained when creating the codebook). We will refer to this procedure as "visual word detection."

Thus we detect all the available visual words in our second dataset of invoices. Next, we calculate a two-dimensional histogram $h(W_i, W_j)$ of coordinates (W_i, W_j) for a particular visual word W.

We can also calculate a two-dimensional histogram $h(F_i, F_j)$ of coordinates (F_i, F_j) for a particular labeled field F.

Finally, we can calculate the following conditional histograms:

- conditional histogram $h(F_i, F_j | W_k, W_l)$ of the position for the field F under the fixed position (W_k, W_l) for the visual word W,
- conditional histogram $h(W_i, W_j | F_k, F_l)$ of the position for the word W under the fixed position (F_k, F_l) for the invoice field F.

Bin values of two dimensional histograms are calculated for the cells from a spatial grid of $M \times N$ elements. We set $M = N = 16$ for invoice images.

If we have all of the above histograms, we can calculate the mutual information $MI(W, F)$ of two random variables, the position of the document field F, and the position of the visual word W as

$$MI(W, F) = H(F) - H(F|W) = H(W) - H(W|F), \qquad (1)$$

where $H(F)$, $H(W)$ are the marginal entropies of random positions F and W, calculated using the histograms $h(F_i, F_j)$ and $h(W_i, W_j)$;

$H(F/W)$ is the conditional entropy of F given that the value of W is known, calculated using the conditional histogram $h(F_i, F_j/W_k, W_l)$ and subsequent averaging of the result over all possible positions (W_k, W_l).

A similar approach is used for $H(W/F)$.

The mutual information $MI(W, F)$ of two random variables, the position of the document field F, and the position of the word W is a measure of the mutual dependence between the two variables. Hence, if we average MI over all the visual words in the codebook, we may use MI as an integrated quality measure of the codebook for a particular document filed F (e.g. "Total" field in the case of invoices).

We determined that the best values of MI corresponded to the following values of the main codebook parameters:

- the photometric local descriptor is composed using DFT coefficients;
- for DFT calculation, the bounding rectangle of an extracted MSER region on a grayscale image is transformed into a square area of 16×16 pixels;
- the geometric descriptor consists of only two components – the size of the MSER region and its aspect ratio;
- both descriptors are concatenated into a compound local descriptor as components, and the weight of the geometric descriptor is equal to $1/10$;
- the size of the codebook $N = 600$.

Again, it should be noted that this optimization procedure, using a large number of labeled invoices, is performed only once at the development stage.

2.5   Calculating Statistical Predicates

Once we have a visual codebook built on 6K unlabeled invoice images and optimized on 1K labeled invoice images, we can calculate a statistical predicate $P(F_j)$ for the position of a field F_j on any invoice document.

As in the previous section, for each visual word from the codebook we can calculate a conditional histogram $h(F_i, F_j|W_k, W_l)$ of the position for the particular field F under the fixed position (W_k, W_l) for the visual word W over the labeled dataset. However, when calculating this histogram, we use a shift S of the field F position relative to the fixed position of the word W for spatial coordinates.

Then we can calculate the integral two-dimensional histogram $h(S(F, W))$ of the shift S of the position of the field F that will incorporate the shifts relative to all the possible positions of visual word W in the labeled dataset. The set of

N shift histograms $h(S(F, W_j))$ for all the visual words W_j from the codebook, together with the codebook itself, are the complete data which is sufficient to calculate statistical predicates of invoice fields positions in our method.

If we are presented with a completely new document from which fields must be extracted, we first detect all the visual words. Then, for each instance of the codebook visual word W_k, we calculate the predicate $P_{ik}(F)$ of the possible position of the field F using the appropriate shift histogram $h(S(F, W_k))$, stored together with the codebook. The integral predicate $P_k(F)$ of the possible position of the filed F based on all the instances of the visual word W_k is calculated as the sum of the individual predicates $P_{ik}(F)$ for all the instances of the visual word W_k in the document.

Note that for an instance of the visual word W_k on the document, a portion of the shift histogram $h(S(F, W_k))$ may not contribute to the calculation of the predicate for this visual word. This is because big shifts may result to a field F position estimation, which lies outside of the area of the document image.

The integral predicate $P(F)$ of the possible position of the field F based on the appearance of all the visual words on the document may be calculated as a linear combination of the individual predicates $P_k(F)$ from the various visual words W_k detected in the document. Figure 2 demonstrates the statistical predicates of the "Total" field on an invoice image. Note that individual predicates based on individual visual words may poorly predict the position of a field, but an integral predicate, calculated over all the instances of all the visual words detected on a document, performs sufficiently well (typically, we detected 70–120 visual words per invoice).

Fig. 2. From left: the original image, the integral predicate for the "Total" field, an individual predicate for the "Total" field based on an instance of an individual visual word with index 411 from our codebook of 600 words. The "Total" field is marked by a blue rectangle. The instance of the individual visual word is marked by a green rectangle. The color palette shows the colors used for different predicate values (from 0 at bottom to the maximum value at the top of the palette). The size of the grid is 16×16 elements. (Color figure online)

It can be seen that our statistical predicate $P(i, j|F)$ is a two-dimensional array of the probabilities of the document field F appearing in different cells (i, j) of the spatial grid $M \times N$ imposed on the image. When calculating histograms using the dataset of labeled invoice images, we assume that a particular cell contains the field F (or the word W) if the center of the field's (or word's) rectangle is located inside the cell.

The prediction of the position of a field F may be determined by the position of the elements of the predicate array with top n values. We refer to the grid cells containing n maximum values of the predicate $P(i, j|F)$ as "top-n cells." In our experiments, we used the following metrics to measure the accuracy of the proposed method:

- top-1 accuracy, which is the percentage of correct predictions based on the grid cell with the top value of the statistical predicate $P(i, j/F)$;
- top-3 accuracy, which is the percentage of correct predictions based on the grid cells with top 3 values of the statistical predicate $P(i, j/F)$;
- top-5 accuracy, which is the percentage of correct predictions based on the grid cells with top 3 values of the statistical predicate $P(i, j/F)$.

3 Experiments

In all our experiments, we used the visual codebook created using a dataset of 6k unlabeled invoice images with approximately 72K local descriptors. This is a random subset of our private database of 60K invoices, which includes a variety of documents from different countries and vendors. The local descriptor for the codebook was composed using the DFT brightness coefficients and the size and aspect ratio of grayscale image regions detected by the MSER algorithm. The codebook contained 600 visual words.

3.1 Invoices

For invoices, we used our private dataset described above and focus on the detection of three fields: "Total," "Currency," and "Invoice Date." We describe the detection of only three fields here, but the proposed method imposes no limits on the number of extracted fields.

To calculate two-dimensional histograms, we apply a grid of 16×16 cells. A subset of 34 images sharing the same layout and originating from the same vendor was arbitrarily split into 15 training images and 19 test images. Examples of images from this subset are shown in Fig. 3.

For accuracy measurements in all our experiments we apply cross-validation. In this experiment, the integrated predicate was calculated as a simple sum of the predicates of individual visual words. The results are shown in Table 1. Note that accuracy differs for the three fields. This is because their positions fluctuated within different ranges even within the same layout. Note also that we obtained a top-10 accuracy of 0.918 (averaged over 3 fields) using only 5 labeled images for

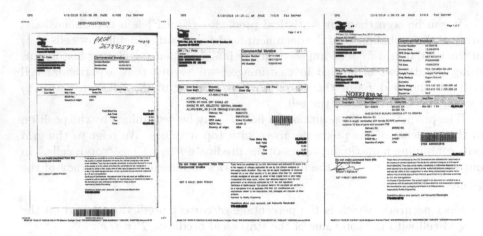

Fig. 3. Examples of invoice images from the experimental subset. All the invoices have the same layout and were issued by the same vendor.

training. In doing so, we measured the accuracy of the prediction that the field can appear in the area of ~3.8% of the entire image. This information will prove valuable when the proposed method is used as an assisting tool in approaches based on text recognition. Due to certain technical limitations, Table 1 does not contain detailed data for this accuracy assessment.

Table 1. Accuracy for invoices sharing the single layout.

Number of labeled training images	1	3	5	10	15
Top-1 accuracy					
Total	0.105	0.190	0.235	0.215	0.225
Currency	0.411	0.621	0.632	0.658	0.679
Invoice Date	0.414	0.573	0.709	0.695	0.705
Average over fields	0.310	0.461	0.525	0.523	0.536
Top-3 accuracy					
Total	0.290	0.335	0.400	0.495	0.525
Currency	0.616	0.937	0.968	0.995	1.000
Invoice Date	0.595	0.936	0.955	0.964	0.977
Average over fields	0.500	0.736	0.774	0.818	0.834
Top-5 accuracy					
Total	0.390	0.440	0.510	0.680	0.730
Currency	0.753	0.958	0.984	1.000	1.000
Invoice Date	0.691	0.973	0.991	1.000	1.000
Average over fields	0.611	0.790	0.828	0.893	0.910

3.2   Receipts

A public dataset of labeled receipts from the ICDAR 2019 Robust Reading Challenge on Scanned Receipts OCR and Information Extraction was used in our experiments. The dataset is available for download at https://rrc.cvc.uab.es/?ch=13.

We focused on three fields: "Total," "Company," and "Date." To calculate two-dimensional histograms, we used a grid of 26×10 cells. This grid size was chosen to reflect the geometry of the average receipt.

Experiment 1. For the first experiment, we chose 360 images from the dataset where receipts occupied the entire image. This was done by simply filtering the images by their dimensions. The resulting subset contained a lot of different layouts from different companies.

Images were arbitrarily split into 300 training images and 60 testing images. We calculated the statistical predicates of the field positions on the training images and measured the accuracy of our predicates on the testing images. In this experiment, the integrated predicate was calculated as a simple sum of the predicates of the individual visual words. The results are shown in Table 2.

Table 2. Accuracy for a mixture of receipts with varying layouts. 300 labeled images were used for training. The prediction is the sum of the individual predicates.

Field	Top-1 accuracy	Top-3 accuracy	Top-5 accuracy
Total	0.367	0.567	0.672
Date	0.372	0.537	0.544
Company	0.465	0.751	0.826
Average over fields	0.401	0.618	0.681

Table 3. Accuracy for a mixture of receipts with varying layouts. 300 labeled images were used for training. The prediction is the linear combination of the individual predicates. The weights are equal to the mutual information values for the individual visual words in degree 2.

Field	Top-1 accuracy	Top-3 accuracy	Top-5 accuracy
Total	0.406	0.628	0.722
Date	0.330	0.494	0.546
Company	0.497	0.766	0.816
Average over fields	0.411	0.629	0.695

Table 3 shows the results of the same experiment, but the integrated predicate for the field F is calculated as a linear combination of the predicates from the individual visual words, with their weights equal to the mutual information values $MI(F, W_k)$ for the individual visual words W_k in degree 2, see (1). The average accuracy here is about 1% higher than in Table 2 where we used a simple sum to calculate the integrated predicate. From this we conclude that calculation of integrated predictions can be further optimized and may become the subject of future research.

Experiment 2. For the second experiment, we first chose a subset of 34 receipts sharing the same layout and originating from the same company (Subset A). Examples of images from Subset A are shown in Fig. 4. Subset A was arbitrarily split into 10 training images and 24 test images. In this experiment, the integrated predicate was calculated as a simple sum of the predicates of the individual visual words. The results obtained on Subset A are shown in Table 4. Note that we obtained a top-10 accuracy of 0.769 using only 5 labeled images for training.

Fig. 4. Examples of receipts from Subset A. Notice the large variations in the field positions within the same receipt layout.

For this experiment, we had a limited number of receipts sharing the same layout. We assume, however, that we can achieve greater accuracy by simply using more training images.

Next, we chose another subset, this time containing 18 receipts (Subset B). Subset B was arbitrarily split into 5 images for training and 13 images for testing.

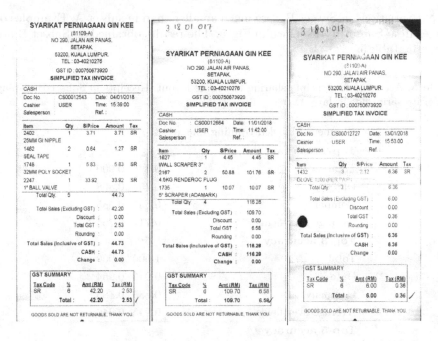

Fig. 5. Examples of receipts from Subset B. Notice the moderate variations in the field positions within the same receipt layout.

Table 4. Accuracy for receipts from sharing the same layout. Receipts in Subset A display large variations in layout.

Number of labeled training images	3	5	10
Top-1 accuracy			
Total	0.282	0.328	0.362
Date	0.242	0.264	0.307
Company	0.432	0.457	0.539
Average over fields	0.319	0.350	0.403
Top-3 accuracy			
Total	0.424	0.548	0.657
Date	0.385	0.454	0.537
Company	0.634	0.677	0.785
Average over fields	0.481	0.560	0.660
Top-5 accuracy			
Total	0.506	0.623	0.713
Date	0.441	0.541	0.628
Company	0.742	0.754	0.858
Average over fields	0.563	0.639	0.733

Layout variations in Subset B were much smaller than in subset A. Examples of images from Subset B are shown in Fig. 5. The results obtained on Subset B are shown in Table 5.

Table 5. Accuracy for receipts sharing the same layout. Receipts in Subset B display moderate variations in layout.

Number of labeled training images	3	5
Top-1 accuracy		
Total	0.846	0.900
Date	0.754	0.808
Company	0.531	0.585
Average over fields	0.710	0.764
Top-3 accuracy		
Total	1.000	1.000
Date	0.946	0.954
Company	0.885	0.885
Average over fields	0.945	0.946
Top-5 accuracy		
Total	1.000	1.000
Date	0.962	0.962
Company	0.954	0.946
Average over fields	0.972	0.969

The experiments with receipts demonstrate that the average top-3 accuracy was greater for documents sharing the same layout than for documents with different layouts. Moreover, same-layout receipts were successfully trained on only 10 labeled documents, while it took 300 labeled documents to achieve similar levels of accuracy for receipts with different layouts.

Note that the number of test images in our experiments was relatively small (13..24), as our datasets contained only a limited number of images sharing the same layout.

4   Conclusion and Future Work

In this paper, we presented a system of document field extraction based on a visual codebook. The proposed system is intended for a processing scenario where only a small number of labeled documents is available to the user for training purposes. Our experiments with a publicly available dataset of receipts demonstrated that the system performs reasonably well on documents sharing the same layout and displaying moderate variations in the field positions.

We achieved the following values of average top-5 accuracy with only 5 labeled receipt images used for training:
- 0.639 for experimental Subset A with large layout variations;
- 0.969 for experimental Subset B with moderate layout variations.

The experiments on receipts were conducted using a visual codebook built on 6 K dataset of unlabeled documents of a different type (invoices).

At the same time, an average top-5 accuracy of 0.828 was achieved for invoice images sharing the same layout with only 5 labeled documents used for training. The top-10 accuracy was 0.918. Top-10 accuracy in our experiments is the accuracy with which the system can predict that the center of a field will occur in an area constituting $\sim 3.8\%$ of the entire image.

We may conclude that the proposed method can be used on its own when the user can or is willing to label only a few training documents, which is a frequent situation in many real-life tasks. The proposed method may also be used as an assisting technique in approaches based on text recognition or to facilitate the training of neural networks.

Future research may involve adapting the proposed method to processing ID documents. Preliminary experiments with ID documents have shown that a visual codebook built on grayscale invoice images can perform reasonably well on some ID fields. Better performance may be achieved if an invoice-based codebook is enriched with important visual words corresponding to elements of colored and textured backgrounds of ID documents.

Correlations between performance and the parameters of our codebook should be further investigated.

Another possible approach to improving the performance of the proposed system is to use modern automatic optimization algorithms, especially differential evolution. This should allow us to determine more accurate values for two dozen parameters. Additional local descriptors can also be easily added.

The system's accuracy may also be improved by optimizing the calculation of integrated predictions from individual visual words contributions.

References

1. Augereau, O., Journet, N., Domenger, J.P.: Semi-structured document image matching and recognition. In: Zanibbi, R., Coüasnon, B. (eds.) Document Recognition and Retrieval XX, vol. 8658, pp. 13–24. International Society for Optics and Photonics, SPIE (2013). https://doi.org/10.1117/12.2003911
2. Borisyuk, F., Gordo, A., Sivakumar, V.: Rosetta: large scale system for text detection and recognition in images (2019)
3. Cristani, M., Bertolaso, A., Scannapieco, S., Tomazzoli, C.: Future paradigms of automated processing of business documents. Int. J. Inf. Manag. **40**, 67–75 (2018). https://doi.org/10.1016/j.ijinfomgt.2018.01.010
4. Daher, H., Bouguelia, M.R., Belaïd, A., D'Andecy, V.P.: Multipage administrative document stream segmentation. In: 2014 22nd International Conference on Pattern Recognition, pp. 966–971 (2014)

5. Gao, H., et al.: Key-region detection for document images - application to administrative document retrieval. In: 2013 12th International Conference on Document Analysis and Recognition, pp. 230–234 (2013)
6. Matas, J., Chum, O., Urban, M., Pajdla, T.: Robust wide baseline stereo from maximally stable extremal regions. Image Vis. Comput. **22**, 761–767 (2004). https://doi.org/10.1016/j.imavis.2004.02.006
7. Palm, R., Winther, O., Laws, F.: CloudScan - a configuration-free invoice analysis system using recurrent neural networks, November 2017. https://doi.org/10.1109/ICDAR.2017.74
8. Palm, R.B., Laws, F., Winther, O.: Attend, copy, parse end-to-end information extraction from documents. In: 2019 International Conference on Document Analysis and Recognition (ICDAR), pp. 329–336 (2019)
9. Rusiñol, M., Lladós, J.: Logo spotting by a bag-of-words approach for document categorization. In: 2009 10th International Conference on Document Analysis and Recognition, pp. 111–115, January 2009. https://doi.org/10.1109/ICDAR.2009.103
10. Sivic, J., Zisserman, A.: Video Google: a text retrieval approach to object matching in videos. In: IEEE International Conference on Computer Vision, vol. 2, pp. 1470–1477 (2003)
11. Sun, W., Kise, K.: Similar manga retrieval using visual vocabulary based on regions of interest. In: 2011 International Conference on Document Analysis and Recognition, pp. 1075–1079 (2011)
12. Takeda, K., Kise, K., Iwamura, M.: Real-time document image retrieval for a 10 million pages database with a memory efficient and stability improved LLAH. In: 2011 International Conference on Document Analysis and Recognition (ICDAR 2011), Beijing, China, 18–21 September 2011, pp. 1054–1058 (2011). https://doi.org/10.1109/ICDAR.2011.213

Automated Transcription for Pre-modern Japanese Kuzushiji Documents by Random Lines Erasure and Curriculum Training

Anh Duc Le[✉]

Center for Open Data in the Humanities, Tokyo, Japan
anh@ism.ac.jp

Abstract. Recognizing the full-page of Japanese historical documents is a challenging problem due to the complex layout/background and difficulty of writing styles, such as cursive and connected characters. Most of the previous methods divided the recognition process into character segmentation and recognition. However, those methods just provide character bounding boxes and classes without text transcription. In this paper, we enlarge our previous human-inspired recognition system from multiple lines to the full-page of Kuzushiji documents. The human-inspired recognition system simulates the human eye movement during the reading process. For the lack of training data, we propose a random text line erasure approach that randomly erases text lines and distorts documents. For the convergence problem of the recognition system for full-page documents, we employ curriculum training that trains the recognition system step by step from the easy level (several text lines of documents) to the difficult level (full-page documents). We tested the step training approach and random text line erasure approach on the dataset of the Kuzushiji recognition competition on Kaggle. The results of the experiments demonstrate the effectiveness of our proposed approaches. These results are competitive with other participants of the Kuzushiji recognition competition.

Keywords: Kuzushiji recognition · Full-page document recognition · Random text line erasure approach

1 Introduction

Japan had been using Kuzushiji or cursive writing style shortly after Chinese characters got into the country in the 8th century. Kuzushiji writing system is constructed from three types of characters, which are Kanji (Chinese character in the Japanese language), Hentaigana (Hiragana), and Katakana, like the current Japanese writing system. One characteristic of classical Japanese, which is very different from the modern one, is that Hentaigana has more than one form of writing. For simplifying and unifying the writing system, the Japanese government standardized Japanese language textbooks in 1900 [1]. This makes the Kuzushiji writing system is incompatible with modern printing systems. Therefore, most Japanese natives cannot read books written by Kuzushiji. We have a lot of digitized documents collected in libraries and museums throughout the country. However, it takes a lot of time to transcribe them into modern

© Springer Nature Switzerland AG 2020
X. Bai et al. (Eds.): DAS 2020, LNCS 12116, pp. 371–382, 2020.
https://doi.org/10.1007/978-3-030-57058-3_26

Japanese characters since they are difficult to read even for Kuzushiji's experts. As a result, a majority of these books have not yet been transcribed into modern Japanese characters, and most of the knowledge, history, and culture contained within these texts are inaccessible for people.

In 2017, our center (CODH) provided the Kuzushiji dataset to organize the 21st Pattern Recognition and Media Understanding Algorithm Contest for Kuzushiji recognition [2]. The competition focused on recognizing isolated characters or multiple lines of hentaigana, which are easier than whole documents, like in Fig. 1. Nguyen et al. won the competition by developing recognition systems based on convolutional neural network and Bidirectional Long Short-Term Memory [3]. In our previous work, we proposed a human-inspired reading system to recognize Kuzushi characters on PRMU Algorithm Contest [4]. The recognition system based on an attention-based encoder-decoder approach to simulate human reading behavior. We achieved better accuracy than the winner of the competition. Tarin et al. presented Kuzushi-MNIST, which contains ten classes of hiragana, Kuzushi-49, which contains 49 classes of hiragana, and Kuzishi-Kanji, which contains 3832 classes of Kanji [5]. The datasets are benchmarks to engage the machine learning community into the world of classical Japanese literature.

Fig. 1. An example of Kuzushi document which contains cursive and connected characters.

The above works for isolated characters and multiple lines of Hetaigana are not suitable for making transcriptions of full-page Kuzushiji documents. Recently, our center organized the Kuzushiji Recognition competition on Kaggle, which requires participants to make predictions of character class and location [6]. The competition provides the ground truth bounding boxes of characters in documents, and the task for participants is to predict character location and class. Here, We briefly summarize the top method in the competition.

Tascj employed ensembling models of two Cascade R-CNN for character detector and recognition [7]. The final model achieved 95% accuracy on the private test set and was ranked the first place in the competition.

Konstantin Lopuhin got second place in the competition [8]. He employed the Faster-RCNN model with ResNet backbone for character detection and recognition. For improving the accuracy, he also utilized pseudo labels for the testing set in the training process and ensembling of six models. The final model achieved 95% accuracy on the private test set but was ranked second place in the competition due to the higher number of submissions.

Kenji employed a two-stage approach and False Positive Predictor [9]. He utilized Faster RCNN for character detection and five classification models for character recogntion. The False Positive Predictor is used for removing false detection. The final model achieved 94.4% accuracy on the private test set and was ranked third place in the competition.

We observed that there are 13 teams achieved accuracy higher than 90%. Most of them employed an object detection based approach which requires bounding boxes of characters, ensembling models, and data augmentation. However, to make transcription, a process of making transcription from the character classes and locations is needed. The above systems need an additional post-process to generate text lines. In this paper, we enlarge our previous human-inspired recognition system from multiple lines of Hentaigana to the full-page of Kuzushiji documents. The system is able to generate transcription from an input image without any post-processing. For the convergence problem of the recognition system for full-page documents, we propose a step training approach that trains the recognition system step by step from the easy level (several text lines of documents) to the difficult level (full-page documents). For the lack of training data, we propose a random text line erasure approach that randomly erases text lines and distorts documents.

The following of this paper is organized as follows. Section 2 briefly describes the overview of the human-inspired recognition system. Section 3 presents the step learning approach for training the recognition system on the full-page Kuzushiji document dataset. Section 4 presents the random text line erasure approach for data generation. Finally, Sect. 5 and 6 draw the experiments and a conclusion of the paper.

2   Overview of Human Inspired Recognition System

Our recognition system is based on the attention-based encoder-decoder approach. The architecture of our recognition system is shown in Fig. 2. It contains two modules: a DenseNet for feature extraction and an LSTM Decoder with an attention model for generating the target characters. We employed a similar setting for the system as our previous works [4]. The advantage of our model is that it requires images and corresponding transcriptions without bounding boxes of characters.

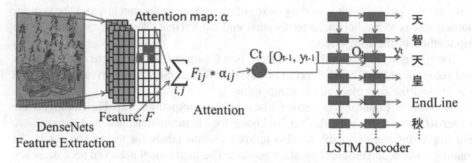

Fig. 2. The architecture of the human inspired recognition system.

At each time step t, the decoder predicts symbol y_t based on the embedding vector of the previous decoded symbol E_{yt-1}, the current hidden state of the decoder h_t, and the current context vector c_t as the following equation:

$$p(y_t|y_1,\ldots y_{t-1},F) = softmax\left(W\left(E_{hy_{t-1}} + W_h * h_t + W_c * c_t\right)\right)$$

The hidden state is calculated by an LSTM. The context vector c_t is computed by the attention mechanism.

3 Data Generation

3.1 Data Preprocessing

As mentioned earlier, the Center for Open Data in Humanities, the National Institute of Japanese Literature, and the National Institute of Informatics hosted the Kuzushiji recognition on Kaggle. The training dataset provides complete bounding boxes and character codes for all characters on a page. Since our recognition system requires the output as text lines, we have to preprocess the data provided by the competition. We need to preprocess the data to make text lines. We concatenate bounding boxes of characters into vertical lines. Then, we sorted vertical lines from right to left to make the ground truth for the input image. Figure 3 shows an example of the preprocessing process. Note that the competition does not provide bounding boxes for annotation characters (small characters), so the annotation characters should be ignored during the recognition process.

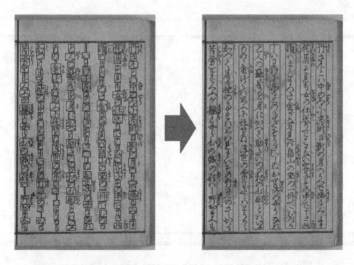

Fig. 3. An example of the preprocessing to make text lines from bounding boxes of characters.

3.2 Random Text Line Erasure

To train a big deep learning system like Human Inspired Recognition System, we need a huge number of labeled data. However, the dataset for the Kuzushiji document is very small (around 4000 images for training). Here, we integrate a random text line erasure and perspective skewing to generate more data from the available data. The random text line erasure forces the recognition system to learn to recognize line by line. The perspective skewing transforms the angle of looking images. Here, we employ the left-right skew to preserve the features of vertical text lines. This helps us to improve the performance of the recognition system.

The process of data generation is shown in Fig. 4. First, we randomly select k text lines from the input image. Then, we remove the selected text lines from the ground truth and also erase the corresponding text lines in the input image. For erasing text lines, we replace the color of the text lines by the background color of the image. Finally, we employ elastic distortion to create the final image. Figure 5 shows an example of the random text line erasure, while Fig. 6 shows that of elastic distortion. The red line in Fig. 6 is the mask of the perspective skewing. We ensure the content of images is not lose when we use the perspective skewing. In each image, we randomly erasure 40% of text lines.

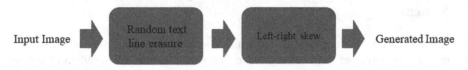

Fig. 4. The process of the random text line erasure.

Input image   Select k text lines   Erase the selected text lines

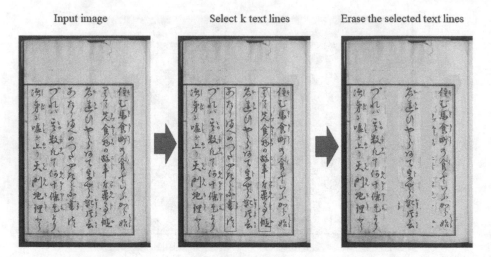

Fig. 5. The process of the random text line erasure.

Random left-right skew

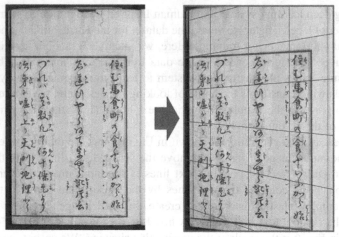

Fig. 6. The process of the random skew left-right. (Color figure online)

4 Curriculum Learning for Human-Inspired Recognition System

We can not train the human-inspired recognition system on the full page of documents because the system does not converge. The first reason is that the problem of recognizing the full page of Kuzushiji documents is very hard. The second reason is the limitation of memory. Therefore, we just train the system with small batch size. Therefore, it affects to speed of learning, the stability of the network during training.

To solve this problem, we employ curriculum learning which had proposed by Elman et al. in 1993 [10] and then shown to improve the performance of deep neural networks in several tasks by Bengio et al. in 2009 [11]. Curriculum learning is a type of learning which starts with easy examples of a task and then gradually increases the task difficulty. Humans have been learning according to this principle for decades. We employ this idea to train the recognition system.

First, we construct an easy dataset by splitting full-page documents into single or multiple lines. Figure 7 shows the process of generating multiple lines dataset. In the previous work [4], we were able to train the recognition on multiple lines. Recognition of multiple lines is easier than that of a full-page of documents. Therefore, we first train the recognition with images of multiple lines. Then, we add full-page images to the training set. Finally, we add the generated dataset to the training set. Figure 8 shows the three stages of curriculum learning with different datasets and learning rates.

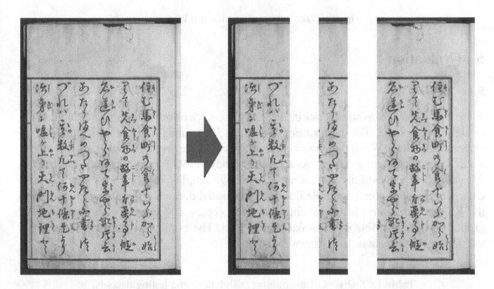

Fig. 7. An example of multiple lines generation

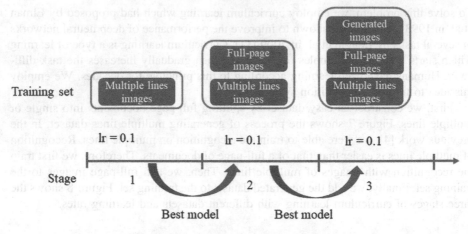

Fig. 8. Three stages of curriculum learning.

5   Evaluation

5.1   Dataset

We employ the Kuzushiji dataset in the Kaggle competition to train and evaluate our recognition system. The competition provides 3881 images for training and 4150 images for testing. We divide the training images into the origin training and validation set as the ratio 9:1, respectively. As a result, we have 3493 for training and 388 images for validation. For additional training datasets, we employ random lines splitting and data generation to create multiple lines and generated datasets. For testing, we employ the Kaggle public and private testing sets and prepare 400 images with transcription ground truth from Mina de honkoku project [12]. The number of images for training, validation, testing datasets is shown in Table 1.

Table 1. Statistics of the training, validation, and testing datasets.

Dataset	# of images
Origin training	3,493
Multiple lines	9,499
Random text line erasure	3,493
Validation	388
Kaggle public testing	1,369
Kaggle private testing	2,781
Transcription testing	400

5.2 Evaluation Metrics

In order to measure the performance of our recognition system, we use two metrics: Character Recognition Rate (CRR) for evaluating transcription generation and F1 for predicting character class and location. Character Recognition Rate is shown in the following equations:

$$CRR(T, h(T)) = 100 - \frac{1}{Z} \sum_{(I,s) \in T} ED(s, h(I))$$

Where T is a testing set which contains input-target pairs (I, s), $h(I)$ is the output of a recognition system, Z is the total number of target character in T, and $ED(s, h(I))$ is the edit distance function which computes the Levenshtein distance between two strings s and $h(I)$.

F1 metrics is generally employed for evaluating character detection task. The detail of the metrics are shown in the following equations:

$$Precision = \frac{Number\ of\ correct\ predicted\ characters}{Number\ of\ predicted\ characters}$$

$$Recall = \frac{Number\ of\ correct\ predicted\ characters}{Number\ of\ groundtruth\ characters}$$

$$F1 = 2 * \frac{Precision * Recall}{Precision + Recall}$$

5.3 Training

We train the recognition system in 3 stages as shown in Fig. 8. We call the best system in each stage as S_1, S_2, S_3, respectively. Three systems have the same architecture. The system for the previous stage is used to pre-train the system for the current stage. For each stage, we used the AdaDelta algorithm with gradient clipping to learn the parameters. The learning rate was set to 0.1. The training process was stopped when the CRR on the validation set did not improve after ten epochs.

5.4 Experimental Results

In the first experiment, We evaluate S_1, S_2, S_3 on the transcription testing set. Table 2 shows the CRRs of different recognition systems. By employing curriculum learning, we improve the CRR from 86.28% to 88.19%, and by data generation, we improve the CRR from 88.19% to 89.51%. If we do not use curriculum learning, the recognition systems do not converge during training. This result verified that curriculum learning and data generation is useful and effective for full-page document recognition.

Table 2. Performance of the recognition systems on transcription testing set.

System	Transcription testing
	CRR (%)
S_1	86.28
S_2	88.19
S_3	89.51

Since our recognition systems predict only transcriptions without locations of characters, we do not have location information for completing the task in the Kaggle competition. However, as the description in Sect. 2, we use the attention mechanism for predicting a character. Therefore, the attention mechanism may track the location of the character in the input image. We employ the following heuristic to get the location of a character. When the decoder predicts a character, the location of the character is set as the maximum probability of the attention map. We create the submission file containing character class and location information for the systems S_1, S_2, S_3 on the second experiment. Table 3 shows the F1 score on the Kaggle public and private testing sets. We achieved 81.8% and 81.9% of the F1 score on the F1 score on the Kaggle public and private testing sets, respectively. Although our result is lower than the top participants who achieved 95%, our systems do not use any location information of characters during training. The top participants had used many techniques to improve the accuracy that we have not done yet. For example, they used the ensembling of many models and data augmentation. They used object detection to detect locations of characters while we do not use the locations of characters during training.

Table 3. Performance of the recognition systems on Kaggle public and private testing sets.

System	Kaggle public testing	Kaggle private testing
	F1 (%)	F1 (%)
S_1	74.4	74.5
S_2	81.3	81.4
S_3	81.8	81.9

Since the F1 score on Kaggle competition is low, we visualize the location of character by attention mechanism on several samples to know more about false prediction. Figure 9 shows the recognition result for a part of a document. Blue characters and yellow dots are predicted characters and locations by the recognition system. The bounding boxes are from the ground truth. The correct characters are in red bounding boxes, while incorrect characters are in purple and green bounding boxes. From our observation, there are two types of frequently incorrect predictions. The first type is that the system predicts an incorrect character but a correct location as characters in purple bounding boxes in Fig. 9. The second type is that the system predicts a correct character but an incorrect location as characters in green bounding boxes in Fig. 9. The second type makes the F1 score in Kaggle testing sets low.

Fig. 9. An example of incorrectly predicted characters. (Color figure online)

Figure 10 shows an example of the recognition result in the transcription testing set. Characters with red marks are incorrect recognition. Some of them are able to revise by a language model.

Based on the above frequently incorrect prediction, we suggest future works to improve accuracy. For the first type of error, we should make more variations of characters, such as applying distortion on every character in documents. For the second type of error, we may use location information of characters to supervise the attention model during the training process.

Fig. 10. An example of recognition results on the transcription testing set. (Color figure online)

6 Conclusion

In this paper, we have proposed the random text line erasure for data generation and training the human-inspired recognition system for the full-page of Japanese historical documents by curriculum learning. The efficiency of the proposed system was demonstrated through experiments. We achieved 89.51% of CRR and 81.9% of the F1 score on transcription testing and Kaggle testing sets, respectively. We plan to improve the detection system by post-processing in the future.

References

1. Takahiro, K.: Syllabary seen in the textbook of the Meiji first year. The Bulletin of Jissen Women's Junior College, pp. 109–119 (2013)
2. Kuzushiji challenge. http://codh.rois.ac.jp/kuzushiji-challenge/
3. Nguyen, H.T., Ly, N.T., Nguyen, C.K., Nguyen, C.T., Nakagawa, M.: Attempts to recognize anomalously deformed Kana in Japanese historical documents. In: Proceedings of the 2017 Workshop on Historical Document and Processing, Kyoto, Japan, pp. 31–36, November 2017
4. Le, A.D., Clanuwat, T., Kitamoto, A.: A human-inspired recognition system for pre-modern Japanese historical documents. IEEE Access 7, 84163–84169 (2019)
5. Clanuwat, T., Bober-Irizar, M., Kitamoto, A., Lamb, A., Yamamoto, K., Ha, D.: Deep learning for classical Japanese literature. arXiv:1812.01718
6. Kuzushiji Recognition. https://www.kaggle.com/c/kuzushiji-recognition/
7. The first place solution. https://www.kaggle.com/c/kuzushiji-recognition/discussion/112788
8. The second place solution. https://www.kaggle.com/c/kuzushiji-recognition/discussion/112712
9. The third place solution. https://www.kaggle.com/c/kuzushiji-recognition/discussion/113049
10. Elman, J.L.: Learning and development in neural networks: the importance of starting small. Cognition 48(1), 71–99 (1993)
11. Bengio, Y., et al.: Curriculum learning. In: Proceedings of the 26th Annual International Conference on Machine Learning, pp. 41–48 (2009)
12. Mina de honkoku project. https://honkoku.org/

Representative Image Selection for Data Efficient Word Spotting

Florian Westphal[1]([✉]) [iD], Håkan Grahn[1] [iD], and Niklas Lavesson[1,2] [iD]

[1] Blekinge Institute of Technology, Karlskrona, Sweden
{florian.westphal,hakan.grahn}@bth.se
[2] Jönköping University, Jönköping, Sweden
niklas.lavesson@ju.se

Abstract. This paper compares three different word image representations as base for label free sample selection for word spotting in historical handwritten documents. These representations are a temporal pyramid representation based on pixel counts, a graph based representation, and a pyramidal histogram of characters (PHOC) representation predicted by a PHOCNet trained on synthetic data. We show that the PHOC representation can help to reduce the amount of required training samples by up to 69% depending on the dataset, if it is learned iteratively in an active learning like fashion. While this works for larger datasets containing about 1 700 images, for smaller datasets with 100 images, we find that the temporal pyramid and the graph representation perform better.

Keywords: Word spotting · Sample selection · Graph representation · PHOCNet · Active learning

1 Introduction

Being able to search a collection of historical handwritten document images for occurrences of a particular word, for example a name, can be a great help to historians and genealogists in their work. Furthermore, it can create a more simplified access to historical documents for a broader public. This task of searching for words in images either by string or by example is called word spotting [7]. In recent years, the performance of word spotting systems has greatly improved through the use of learning based approaches using convolutional neural networks (CNNs), for example in the work by Sudholt and Fink [25] or by Krishnan et al. [14]. However, one drawback of these approaches is the large amount of labeled data required to train these approaches.

In this paper, we explore sample selection as one possible way to improve the word spotting performance of a learning based system without increasing the number of labeled samples. This will make it possible to reduce the amount

---

This work is part of the research project "Scalable resource-efficient systems for big data analytics" funded by the Knowledge Foundation (grant: 20140032) in Sweden.

X. Bai et al. (Eds.): DAS 2020, LNCS 12116, pp. 383–397, 2020.
https://doi.org/10.1007/978-3-030-57058-3_27

of data to label without compromising performance. The main idea is to select word images from an unlabeled set of images, which represent different aspects of the target collection as well as possible, given the limitation in the number of samples to select. This diversity in selected samples should lead to a better word spotting performance, given the same number of images, compared to randomly selecting images to label. In order to achieve diversity in the aspects relevant for word spotting, it is important to represent the unlabeled word images in a suitable way to facilitate the selection of such a diverse training set.

The main contribution of this paper is the proposal and evaluation of three different word image representations for sample selection. The evaluated representations are a temporal pyramid representation based on foreground pixel counts, a graph representation based on word graphs and a Pyramidal Histogram of Characters (PHOC) representation based on predicted PHOCs.

We show that it is possible, depending on the dataset, to reduce the required amount of training data by up to 69% without leading to a statically significant reduction in word spotting performance by learning to extract a suitable PHOC representation of word images using an active learning [20] approach. While this approach works well when around 1 700 word images can be labeled, we show that the temporal pyramid and the graph representation can perform as good or better than random sampling for small dataset sizes of 100 images.

2   Related Work

While sample selection has not been applied to word spotting, it has been used for other document analysis tasks, such as character recognition and binarization. For character recognition, Rayar et al. [19] show that by selecting training samples based on the bridge vectors in a relative neighborhood graph it is possible to reduce the size of the original training set by almost 63% without reducing the recognition accuracy. However, one drawback of this approach is that it relies on label information in the original training set. Krantz and Westphal [13] follow a similar strategy for selecting training samples for image binarization using a relative neighborhood graph. In contrast to Rayar et al., their approach avoids the need for a labeled set to choose from by creating pseudo labels through clustering. In this way, they reduce the training set on average by 49.5%, which leads to a decrease in binarization performance by 0.55%.

Other methods, which have been used to increase the performance of word spotting approaches given only a limited number of training samples are data augmentation and pre-training on synthetic data. One possible augmentation strategy for word spotting is to apply different image transformations, such as shear, rotation and translation to the image, as has been proposed by Sudholt and Fink [25]. Gurjar et al. [11] have shown that pre-training a CNN based word spotting approach with the synthetic dataset by Krishnan and Jawahar [15] can achieve a reasonable word spotting performance, even with only few training samples. Since the achieved improvements for both of these methods are independent from the particular training set, it is likely that improvements achieved through the application of augmentation, pre-training and sample selection will add up. Thus, the sample selection approaches evaluated in this paper will add an additional way to reduce the amount of training data required for training a word spotting model on top of augmentation and pre-training.

3   Background

3.1   Word Spotting with PHOCNet

One CNN based approach to word spotting, which allows query by example, as well as query by string scenarios, is PHOCNet [24]. This neural network based on the VGG16 architecture [21] predicts the pyramidal histogram of characters (PHOC) [2] from a given input image. The PHOC is a vector of binary values, which indicate the presence or absence of characters from a predefined character set in different parts of a word. Since the PHOC can be derived from a string, it is possible to search for specific word images by predicting their PHOCs and then identifying those word images whose PHOCs are most similar to the generated search PHOC. Similarly, in a query by example scenario, the PHOC of the query image can be predicted and can then be used to find word images with similar PHOCs.

3.2   Graph Matching

Graph matching is a commonly used approach in pattern recognition [6,10] and has been applied also to word spotting [23]. The main idea here is to represent a word image as a graph, as shown in Fig. 1. By computing the graph edit distance between this search graph and other word graphs, it is possible to search a dataset of word images for the most similar instances.

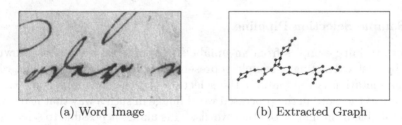

(a) Word Image   (b) Extracted Graph

Fig. 1. Word graph for one image from the Alvermann Konzilsprotokolle dataset [18].

For extracting graphs from word images, we use the projection based approach by Stauffer et al. [22]. This approach identifies first the nodes of a graph by splitting the binarized word image into segments using horizontal and vertical projection profiles, identifying gaps between characters. These segments are further subdivided horizontally and vertically in predefined intervals D_h and

D_v. The center of mass of each of these sub-segments is chosen as one node of the graph. Undirected edges are added between two nodes if their respective segments are connected by the lines of the skeletonized word.

Since computing the exact graph edit distance is an NP-hard problem, efficient graph matching requires suitable approximation algorithms. One of these algorithms is the Hausdorff edit distance (HED) [9]. The HED represents a lower bound to the exact graph edit distance and its computation has a quadratic time complexity. This is achieved by matching nodes and edges between the graphs independently, without considering other node or edge assignments.

3.3 Sample Selection with Iterative Projection and Matching (IPM)

While there are many different sampling strategies, such as K-medoids sampling, dissimilarity-based sparse subset selection [8], structured sparse dictionary selection [26] or representative selection on a hypersphere [27], in this paper, we focus on the iterative projection and matching (IPM) approach by Zaeemzadeh et al. [29]. In this approach, sample selection is achieved by reducing an $M \times N$ matrix $\mathbf{A}$ of M training sample representations $\mathbf{a} \in \mathbb{R}^N$ to a $K \times N$ matrix containing K selected samples. All vectors $\mathbf{a}$ representing the training samples are normalized, denoted as $\tilde{\mathbf{a}}$, to lie on the unit sphere. The resulting matrix of unit vectors is denoted as $\tilde{\mathbf{A}}$. The algorithm picks iteratively samples by first finding the most representative direction for the current matrix $\tilde{\mathbf{A}}$ in the unit sphere. Second, the unit vector $\tilde{\mathbf{a}}$ with the smallest angle to the chosen direction is selected and all unit vectors in $\tilde{\mathbf{A}}$ are projected onto the null space of the selected sample, effectively creating a lower dimensional subspace from which the following sample will be chosen. This process is repeated until K samples were selected.

4   Sample Selection for Word Spotting

4.1   Sample Selection Pipeline

We select training samples from an unlabeled set of word images, as shown in Fig. 2, by first extracting a sample representation for each word image. Since this representation is the base for the selection process, it is important that it characterizes the word in the processed word image in such a way that it becomes distinguishable from images of other words. This makes it possible to select a set of images of different words, which is the dimension of diversity relevant for the learning task, i.e., PHOC prediction. In this paper, we evaluate three different word image representations, *viz.* a temporal pyramid representation (Sect. 4.2), a graph representation (Sect. 4.3), and a PHOC representation (Sect. 4.4).

After extracting the respective representations, we compute the pairwise distances between all word image representations using a suitable distance metric. In this way, a word image x_i is represented as a vector $\mathbf{x_i} \in \mathbb{R}_+^N$ containing its distance to all N word images. The $N \times N$ distance matrix $\mathbf{D}$ is then used

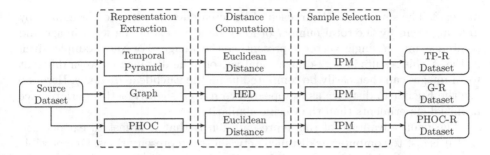

Fig. 2. Sample selection pipeline for the three evaluated word image representations.

as input to IPM [29] to select the K most representative samples. One advantage of representing the word images as vector of distances is that it leads to an N dimensional vector space with $N > K$. This is beneficial for IPM, since the IPM algorithm effectively reduces the dimensionality of its input matrix through projection for every chosen sample. The IPM algorithm then yields the most representative K samples given the provided distance matrix **D**.

4.2 Temporal Pyramid Representation (TP-R)

One simplistic approach to represent a word image is the proposed temporal pyramid representation (TP-R). This approach represents a word image as vector of pixel counts in different segments of the word image.

In order to ensure that the pixel counts represent the actual word in the image, we remove noise by first binarizing the image. Furthermore, we remove possible word fragments from adjacent words in the image by finding at most four appropriate seams to cut away fragments close to the four borders of the image using the seam carving algorithm [3]. Lastly, we resize the word image to remove the created white space around the word, as illustrated in Fig. 3.

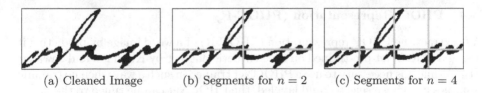

(a) Cleaned Image   (b) Segments for $n = 2$ (c) Segments for $n = 4$

Fig. 3. Word graph for one image from the Alvermann Konzilsprotokolle dataset [18].

Similar to temporal pyramid pooling [25], we split the input vertically into n segments for different values of n. However, we also split the input image horizontally at the vertical center of mass to make it possible to distinguish ascenders and descenders. The resulting segments for $n = 2$ and $n = 4$ are shown in Fig. 3. The sums of pixels for each of the created segments are normalized by dividing them by the total number of foreground pixels in the word image and combined into a single vector representing the image. For the example given in Fig. 3, this results in a 12-dimensional vector. The distance between different word images can then easily be computed using the Euclidean distance. Here, we use the Euclidean distance as it appeared to have a slightly better performance in initial experiments than the Cosine similarity.

The main advantage of this representation is that it is fast to extract and that it is fast to compute the distance between two representations. One possible disadvantage is that this approach is susceptible to variations in writing style, such as slant, skew or spacing variations within a word.

4.3 Graph Representation (G-R)

Another possible word image representation is its graph representation (G-R). We extract the graph by first pre-processing the word image as described in Sect. 4.2 and then deriving the word graph from the cleaned image using the projection based approach by Stauffer et al. [22].

Since the graph representations cannot serve as input to the used sample selection algorithm, we compute the pairwise distances between the word graphs using the Hausdorff edit distance (HED) [9]. For the HED computation we use the cost model proposed by Stauffer et al. [23]. This model defines a node insertion and deletion cost τ_v, an edge insertion and deletion cost τ_e, as well as a parameter α to trade-off node edit costs with edge edit costs and a parameter β to trade-off shifts in node location in horizontal direction with shifts in vertical direction when computing node substitution costs.

The main advantage of this representation is that it captures the shape of each character in the word and is thus more resistant to variations in writing style than TP-R. However, while it is fast to extract the graph representation from a word image, computing the HED requires considerably longer computation times than computing the Euclidean distance.

4.4 PHOC Representation (PHOC-R)

We extract the PHOC representation (PHOC-R) of a word image by predicting it using a PHOCNet pre-trained on the synthetic dataset by Krishnan and Jawahar [14]. While it is necessary to use a PHOCNet trained in such a way when no training samples have been selected and labeled, the PHOCNet can be tuned to the target dataset, as soon as a small set of word images has been selected. In this way, the PHOC representation can get more suitable for the target dataset the more samples have been labeled. Thus, this representation facilitates active learning [20]. After extracting the PHOCs, the pairwise Euclidean distances are computed, as shown in Fig. 2. Here, the Euclidean distance is used instead of the Cosine similarity, which is common in PHOC based retrieval, since initial tests have shown a better sample selection performance for the Euclidean distance.

The main advantage of this representation is that it is closely aligned with the learning task and that it can improve based on the number of training samples available. Furthermore, the distance computation between two representations is fast. One possible disadvantage of this approach is that it requires the PHOCNet to be retrained whenever more training data becomes available. Since it is reasonable to select only few samples initially, when the PHOCNet's performance can be expected to be poor, this can lead to several time intensive training runs until the desired training set size is reached.

5 Experiment Design

5.1 Experiment Setup

We conduct all experiments in this paper using the PHOCNet implementation[1] by Sudholt and Fink [24]. For the sample selection, we use the IPM implementation[2] by Zaeemzadeh et al. [29] and we compute the HED between graphs using the HED implementation in GEDLIB[3] by Blumenthal et al. [4,5], which we modified to support the cost model by Stauffer et al. [23]. The code used to conduct all experiments as well as the raw data is available online[4]. All experiments were conducted on a computer with an Intel i9-7900X CPU @ 3.3 GHz, 32 GB DDR4 RAM and two Nvidia GeForce GTX 1080 Ti.

We evaluate the different word image representations using the Alvermann Konzilsprotokolle (AK) dataset and the Botany dataset from the ICFHR2016 Handwritten Keyword Spotting Competition (H-KWS) [18], as well as the George Washington (GW) dataset by Lavrenko et al. [16]. For the two H-KWS datasets, we use their respective Train II partition as source dataset to select samples from and we use the size of their respective Train I partition as the maximum number of samples to select. This means for the Botany dataset, for example, that the largest selected training dataset with 1 684 images is chosen from a dataset with 3 611 images. Since there is no official partitioning of the George Washington dataset into train, validation and test set, we use the partitioning for the first of the four folds used by Almazán et al. [2]. We combine the train and validation set into one source dataset and use the size of the train dataset as the maximum number of samples to select. For all datasets, we select training sets with 100, 500, and the dataset specific maximum number of samples.

Since the extraction of TP-R and G-R requires binarized input images, we binarize all word images using Howe's binarization algorithm [12]. We predict the required binarization parameters using the approach by Westphal et al. [28], which has been trained on the dataset of the 2013 document image binarization

---

[1] https://github.com/ssudholt/phocnet.
[2] https://github.com/zaeemzadeh/IPM.
[3] https://github.com/dbblumenthal/gedlib.
[4] https://github.com/FlorianWestphal/DAS2020.

contest [17]. After visual inspection of one image, we adjust one of the predicted parameters for the Alvermann Konzilsprotokolle dataset and the George Washington dataset to increase the recall.

Apart from the pre-processing, the graph extraction and HED computation requires a number of dataset specific parameters, such as D_v and D_h for the graph extraction and τ_v, τ_e, α, and β for the HED computation. In our experiments, we use the parameters suggested by Stauffer et al. [23] for each of the respective datasets. Since these parameters were obtained through parameter tuning on labeled samples of the respective datasets, the obtained results have to be seen as an upper bound for the performance. This is the case, since we assume that no labeled data is available before performing the sample selection.

In order to make use of the ability of PHOC-R to adapt to a given dataset by training the used PHOCNet, we initially select only 50 images using the PHOCs predicted by a PHOCNet pre-trained on the HW-SYNTH dataset by Krishnan and Jawahar [15]. Then this PHOCNet is trained further using the selected 50 images. Based on the PHOCs predicted by this network, another 50 images are selected to obtain the first training set containing 100 images. This is continued by training the first PHOCNet, which was trained on the HW-SYNTH dataset, on the selected 100 images. Once the network is trained, its predicted PHOCs are used to select 400 additional images to obtain the next largest training set containing 500 images. In a similar fashion, the samples for the dataset specific maximum number of samples are chosen.

Apart from using the three evaluated word image representations for sample selection, we also use random sample selection as baseline for comparison. In order to obtain a more reliable result, we randomly select three training sets for each training set size and each of the used datasets. For this selection approach, we report the average results over these three training sets. Thus, we evaluate in total four sample selection approaches, *viz.* TP-R, G-R, PHOC-R and Random.

5.2   Training Setup

In this paper, we use a PHOCNet with temporal pyramid pooling (TPP) layer, as described by Sudholt and Fink [25]. As described by Gurjar et al. [11], we train the PHOCNet using stochastic gradient descend and use binary cross entropy as loss function when training to predict PHOCs.

For the selected training sets containing 100 and 500 images, we pre-train the network using the HW-SYNTH dataset [15] for 80 000 weight update steps with a batch size of 10, a momentum of 0.9, a weight decay of $5 \cdot 10^{-5}$ and an initial learning rate of 10^{-4}, which is decreased after 70 000 steps to 10^{-5}. The training is continued with the same momentum vector using the selected training samples at a constant learning rate of 10^{-5}. For the training sets containing 100 images, this training is continued for 40 000 steps, while for those sets containing 500 images training is continued for 70 000 steps. These training durations have been chosen to ensure convergence on all datasets. When using the dataset specific maximum number of samples, no pre-training is used and the PHOCNet is trained from scratch for 80 000 weight update steps with the same parameters as when training on the HW-SYNTH dataset.

We extract the character set, which is used to define the dataset specific PHOC vector, for each dataset from all transcriptions in its source dataset. Here, we make use of all available transcriptions, since it is reasonable to assume that a suitable character set can be derived given knowledge about the dataset for which the PHOCNet should be trained. In this paper, we perform case insensitive word spotting. Therefore, we convert all ground truth labels to upper case. Since the PHOC descriptors are dataset specific, the HW-SYNTH pre-training has to be performed for each dataset.

In order to increase the generalization performance of the trained PHOCNet, we use data augmentation. While we augment the images of the HW-SYNTH dataset by varying their size, as described by Gurjar et al. [11], we use the augmentation strategy proposed by Sudholt and Fink [25] for the evaluated datasets. Depending on the number of weight update steps, we generate 400 000 or 500 000 augmented samples for each training set. When the original training set size is only 100 images, we generate only 400 000 images, since this is the maximum number of images the network can process in 40 000 update steps.

5.3   Evaluation Procedure

We evaluate the word spotting performance of the trained PHOCNets using the official evaluation dataset in case of the Alvermann Konzilsprotokolle dataset and the Botany dataset. For the George Washington dataset, we follow the evaluation procedure by Almazán et al. [2]. This means that we use the transcriptions of all word images in the test dataset as query strings for the query by string (QbS) case. For the query by example (QbE) case, we use all test images as query image, as long as at least one other image of the same word is present in the test dataset. In both cases, the images are retrieved from the test. In the QbE case, the respective query image is removed from the retrieval result.

We evaluate the retrieval performance using the mean average precision (mAP), which is the average over the average precisions (APs) for all evaluation queries. The average precision AP_q for a query q is defined as follows:

$$AP_q = \frac{\sum_{i=1}^{n} P(R_{q,i}, E_q) \cdot I[r_i \in E_q]}{|E_q|} \quad (1)$$

Here, $R_{q,i} \subseteq R_{q,n}$ denotes the set of the first i, out of n, retrieved elements for the query q. The set of relevant or expected elements for q is denoted by E_q and $I[\cdot]$ is the indicator function, which is 1 if the ith retrieved element $r_i \in R_{q,i}$ is relevant with respect to q and 0 otherwise. The function $P(R, E)$ returns the precision given the set of retrieved elements R and the set of relevant or expected elements E. It is defined as follows:

$$P(R, E) = \frac{|R \cap E|}{|R|} \quad (2)$$

6   Results and Analysis

Tables 1, 2, and 3 show the word spotting results in mAP for the QbE and QbS case for all three training set sizes for the four evaluated sample selection approaches TP-R, G-R, PHOC-R and Random. Furthermore, the last two rows of those tables show the QbE and QbS results for the case that the PHOCNet is trained on all images in the source dataset. Figure 4 shows the relative number of unique words within the selected training sets with respect to the total number of unique words in the respective source set in relation to the relative mAP results. These results are relative to the mAP achieved by training on the whole source dataset. In this way, it is possible to show the mAP results for all datasets, selection approaches, query types and training set sizes in one scatter plot.

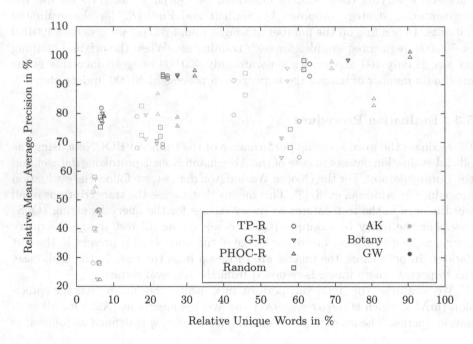

Fig. 4. Relationship between the relative number of unique selected words and the achieved relative mean average precision for the four evaluated sample selection approaches and the three evaluated datasets.

In order to compare the results for the four selection approaches for each training set size and each of the two query types, i.e., QbE and QbS, we perform a one-way repeated measures analysis of variances (ANOVA) on the average precisions (APs) obtained from each evaluation query. We use this statistic test, since the same query, i.e., the subject or independent variable, is answered by PHOCNets trained using differently selected training sets, i.e., the treatment, resulting in AP results, i.e., the dependent variable. This setup corresponds to the one-way repeated measures design. We avoid making assumptions about the distribution of the obtained AP results by applying a rank transform to the APs of the evaluated approaches before applying the ANOVA test.

If the applied ANOVA test indicates a statistically significant difference between the APs obtained for the four sample selection approaches, we perform a pairwise Wilcoxon rank sum test with Holm correction to identify between which approaches exist statistically significant differences. The application of the Holm correction to the pairwise tests is important to avoid the multiple comparisons problem. We show the results of these pairwise comparisons in the tables in form of score values indicated in brackets behind each of the mAP results. These score values are computed by adding one to the score of a method if it is statistically significantly better than another method at the $p < 0.05$ level and we reduce the score by one if a method is statistically significantly worse than another method. Therefore, the score values within one comparison group add up to zero. If the ANOVA test does not indicate any statistically significant differences between approaches at the $p < 0.05$ level, the score value is left blank.

Table 1. Word spotting results for the Alvermann Konzilsprotokolle dataset in mAP (%) and the corresponding score values indicating statistically significantly better (+1) or worse (−1) results at the $p < 0.05$ level.

		TP-R	G-R	PHOC-R	Random	All
100	QbE	51.08 (−1)	55.45 (±0)	52.73 (±0)	56.07 (+1)	—
	QbS	26.88 (−2)	38.45 (+1)	37.55 (±0)	38.36 (+1)	—
500	QbE	76.41 (−1)	76.95 (−1)	83.96 (+3)	79.91 (−1)	—
	QbS	65.13 (−1)	68.13 (±0)	76.27 (+1)	69.36 (±0)	—
1 849	QbE	88.45 (−2)	92.47 (+2)	94.64 (+2)	90.80 (−2)	—
	QbS	76.12 (−1)	87.35 (+1)	91.41 (+2)	82.67 (−2)	—
5 968	QbE	—	—	—	—	96.71
	QbS	—	—	—	—	95.71

Based on the results for the Alvermann Konzilsprotokolle (AK) dataset in Table 1, one can see that TP-R performs statistically significantly worse than at least one other method for all query types and training set sizes. One possible reason for this could be variations in writing style, such as varying spacing between characters or differences in size for certain characters in a word, which lead to different TP-Rs for the same word, making it more likely that images of the same word are selected. Figure 4 confirms this by showing that for the AK dataset TP-R selects a lower ratio of unique words per selected training set than any other selection approach.

For G-R and PHOC-R, one can see from Table 1 that these approaches perform better compared to the other two approaches, the more training samples are selected. While this is to be expected for PHOC-R, since its word image representation improves with the number of available training samples, it is surprising for G-R. One explanation for this result may be that it is not G-R, which improves, but rather that random selection fails to make use of the possibility to select a more diverse sample when the training set is large. This can be seen in Fig. 4, which shows that G-R selects nearly the same ratio of unique words as TP-R and Random for the two smaller training sets, but selects a higher ratio of unique words than both of them for the largest training set.

Lastly, we test for statistically significant differences between the word spotting results obtained by training on the largest training set selected using G-R and PHOC-R and the results obtained using the whole source dataset for training. We find that there is no statistically significant difference between the training set selected using PHOC-R and the source dataset. This indicates that by using PHOC-R, it is possible to reduce the training set for the AK dataset by 69% without statistically significant reduction in word spotting performance.

Table 2. Word spotting results for the Botany dataset in mAP (%) and the corresponding score values indicating statistically significantly better $(+1)$ or worse (-1) results at the $p < 0.05$ level.

		TP-R	G-R	PHOC-R	Random	All
100	QbE	36.73 (—)	36.24 (—)	35.06 (—)	36.52 (—)	—
	QbS	23.83 (±0)	18.94 (−1)	18.73 (−1)	23.50 (+2)	—
500	QbE	54.62 (±0)	54.86 (±0)	59.59 (±0)	56.80 (±0)	—
	QbS	57.31 (±0)	59.27 (±0)	66.04 (±0)	63.12 (±0)	—
1684	QbE	60.89 (±0)	56.33 (−1)	65.16 (+2)	58.70 (−1)	—
	QbS	54.40 (−1)	54.31 (−1)	71.41 (+3)	57.21 (−1)	—
3611	QbE	—	—	—	—	78.75
	QbS	—	—	—	—	83.90

Similar to the AK dataset, one can see for the results of the Botany dataset in Table 2 that the word spotting performance of PHOC-R improves compared to the other three approaches, the more training samples are available. However, we do not see a similar improvement for G-R. One reason for this may be the larger variation in writing styles in the Botany dataset leading to less unique words to be selected, as shown in Fig. 4. However, this cannot be the only explanation, since such variations should affect TP-R even more than G-R. Another possible aspect could be binarization errors, which affect G-R more than TP-R, since they can make the graph representations of the same word appear to be different. TP-R is presumably more robust towards binarization issues, due to the normalization of the word image's pixel counts.

When comparing the word spotting performance of a PHOCNet trained on the complete source dataset with the performance of a model trained on the largest training set selected using PHOC-R, we find the former to perform statistically significant better than the latter. This may be due to the large writing style variations in the Botany dataset, which lead to the necessity for larger training datasets.

Table 3. Word spotting results for the George Washington dataset in mAP (%) and the corresponding score values indicating statistically significantly better ($+1$) or worse (-1) results at the $p < 0.05$ level.

		TP-R	G-R	PHOC-R	Random	All
100	QbE	77.05 ($+3$)	72.13 (-2)	74.47 ($+1$)	74.07 (-2)	—
	QbS	71.68 ($+1$)	73.81 ($+1$)	72.73 ($+1$)	69.38 (-3)	—
500	QbE	86.77 (±0)	87.88 ($+1$)	87.45 ($+1$)	87.39 (-2)	—
	QbS	85.62 (±0)	85.60 (±0)	87.32 ($+1$)	86.00 (-1)	—
1823	QbE	91.39 (±0)	92.72 (±0)	95.22 ($+3$)	92.50 (-3)	—
	QbS	85.55 (-2)	90.62 ($+2$)	91.53 ($+2$)	87.67 (-2)	—
3645	QbE	—	—	—	—	94.09
	QbS	—	—	—	—	92.08

For the George Washington (GW) dataset, the relative word spotting performance for G-R and PHOC-R is comparable to the AK dataset with respect to the fact that their performance improves compared to the other two approaches, the more training samples are selected (cf. Table 3). However, TP-R performs better on this dataset than on the AK dataset, in particular when only 100 training samples are selected. This may be the case, since there is less variation in writing style in the GW dataset, which makes it easier for TP-R to select word images with more unique words. We compare the word spotting performance achieved by training on the source dataset with the performance achieved by training on the largest selected training sets. We find that there is no statistically significant difference between the performance of G-R, PHOC-R and the source dataset in the QbS case and that there is no statistically significant different between the performance of PHOC-R and the source dataset in the QbE case. Therefore, we conclude that using PHOC-R can help to reduce the amount of training data required for the GW dataset by almost 50%.

7 Conclusion

In this paper, we have evaluated three different word image representations for unlabeled sample selection for word spotting. The evaluated representations are a temporal pyramid representation (TP-R), a graph based representation (G-R) and a pyramidal histogram of characters representation (PHOC-R). We have shown that, depending on the dataset, the use of PHOC-R can lead to a reduction in the number of required training samples of up to 69% without reducing the word spotting performance in a statistically significant way. We have argued that the main reason for PHOC-R's performance is that it learns to adjust its word image representation in an active learning like fashion. Thus, it performs best when a larger amount of training samples can be retrieved iteratively. For selecting smaller training sets, we have shown that G-R and TP-R can perform as well or better than random selection. We have pointed out that current limitations of these two word image representations may be due to binarization issues, as well as their susceptibility to variations in writing style. Possible future work would be to improve the used binarization method and to increase TP-R's robustness by applying skew and slant correction to the word images. Additionally, it may be interesting to consider and evaluate the use of different distance metrics for TP-R and PHOC-R. Since both representations have many dimensions, the Manhattan distance or fractional distance metrics may make the represented word images more distinguishable, as pointed out by Aggarwal et al. [1], which could result in a more suitable sample selection.

References

1. Aggarwal, C.C., Hinneburg, A., Keim, D.A.: On the surprising behavior of distance metrics in high dimensional space. In: Van den Bussche, J., Vianu, V. (eds.) ICDT 2001. LNCS, vol. 1973, pp. 420–434. Springer, Heidelberg (2001). https://doi.org/10.1007/3-540-44503-X_27

2. Almazán, J., Gordo, A., Fornés, A., Valveny, E.: Word spotting and recognition with embedded attributes. IEEE Trans. Pattern Anal. Mach. Intell. **36**(12), 2552–2566 (2014)

3. Avidan, S., Shamir, A.: Seam carving for content-aware image resizing. ACM Trans. Graph. (TOG) **26**(3), 10-1–10-9 (2007)

4. Blumenthal, D.B., Boria, N., Gamper, J., Bougleux, S., Brun, L.: Comparing heuristics for graph edit distance computation. VLDB J. **29**(1), 419–458 (2019). https://doi.org/10.1007/s00778-019-00544-1

5. Blumenthal, D.B., Bougleux, S., Gamper, J., Brun, L.: GEDLIB: a C++ library for graph edit distance computation. In: Conte, D., Ramel, J.-Y., Foggia, P. (eds.) GbRPR 2019. LNCS, vol. 11510, pp. 14–24. Springer, Cham (2019). https://doi.org/10.1007/978-3-030-20081-7_2

6. Conte, D., Foggia, P., Sansone, C., Vento, M.: Thirty years of graph matching in pattern recognition. Int. J. Pattern Recogn. Artif. Intell. **18**(03), 265–298 (2004)

7. Doermann, D.: The indexing and retrieval of document images: a survey. Comput. Vis. Image Underst. **70**(3), 287–298 (1998)

8. Elhamifar, E., Sapiro, G., Sastry, S.S.: Dissimilarity-based sparse subset selection. IEEE Trans. Pattern Anal. Mach. Intell. **38**(11), 2182–2197 (2015)

9. Fischer, A., Suen, C.Y., Frinken, V., Riesen, K., Bunke, H.: Approximation of graph edit distance based on Hausdorff matching. Pattern Recogn. **48**(2), 331–343 (2015)

10. Foggia, P., Percannella, G., Vento, M.: Graph matching and learning in pattern recognition in the last 10 years. Int. J. Pattern Recogn. Artif. Intell. **28**(01), 1450001 (2014)

11. Gurjar, N., Sudholt, S., Fink, G.A.: Learning deep representations for word spotting under weak supervision. In: 13th IAPR International Workshop on Document Analysis Systems, pp. 7–12. IEEE (2018)

12. Howe, N.R.: A Laplacian energy for document binarization. In: 2011 International Conference on Document Analysis and Recognition, pp. 6–10. IEEE (2011)

13. Krantz, A., Westphal, F.: Cluster-based sample selection for document image binarization. In: International Conference on Document Analysis and Recognition Workshops, vol. 5, pp. 47–52. IEEE (2019)

14. Krishnan, P., Dutta, K., Jawahar, C.: Word spotting and recognition using deep embedding. In: 13th IAPR International Workshop on Document Analysis Systems, pp. 1–6. IEEE (2018)
15. Krishnan, P., Jawahar, C.V.: Matching handwritten document images. In: Leibe, B., Matas, J., Sebe, N., Welling, M. (eds.) ECCV 2016. LNCS, vol. 9905, pp. 766–782. Springer, Cham (2016). https://doi.org/10.1007/978-3-319-46448-0_46
16. Lavrenko, V., Rath, T.M., Manmatha, R.: Holistic word recognition for handwritten historical documents. In: 2004 Proceedings of the First International Workshop on Document Image Analysis for Libraries, pp. 278–287. IEEE (2004)
17. Pratikakis, I., Gatos, B., Ntirogiannis, K.: ICDAR 2013 document image binarization contest (DIBCO 2013). In: 2013 12th International Conference on Document Analysis and Recognition, pp. 1471–1476. IEEE (2013)
18. Pratikakis, I., Zagoris, K., Gatos, B., Puigcerver, J., Toselli, A.H., Vidal, E.: ICFHR 2016 handwritten keyword spotting competition (H-KWS 2016). In: 15th International Conference on Frontiers in Handwriting Recognition (ICFHR), pp. 613–618. IEEE (2016)
19. Rayar, F., Goto, M., Uchida, S.: CNN training with graph-based sample preselection: application to handwritten character recognition. In: 13th IAPR International Workshop on Document Analysis Systems, pp. 19–24. IEEE (2018)
20. Settles, B.: Active learning. Synth. Lect. Artif. Intell. Mach. Learn. 6(1), 1–114 (2012)
21. Simonyan, K., Zisserman, A.: Very deep convolutional networks for large-scale image recognition. In: Bengio, Y., LeCun, Y. (eds.) 3rd International Conference on Learning Representations (2015). http://arxiv.org/abs/1409.1556
22. Stauffer, M., Fischer, A., Riesen, K.: A novel graph database for handwritten word images. In: Robles-Kelly, A., Loog, M., Biggio, B., Escolano, F., Wilson, R. (eds.) S+SSPR 2016. LNCS, vol. 10029, pp. 553–563. Springer, Cham (2016). https://doi.org/10.1007/978-3-319-49055-7_49
23. Stauffer, M., Fischer, A., Riesen, K.: Keyword spotting in historical handwritten documents based on graph matching. Pattern Recogn. 81, 240–253 (2018)
24. Sudholt, S., Fink, G.A.: Evaluating word string embeddings and loss functions for CNN-based word spotting. In: 14th IAPR International Conference on Document Analysis and Recognition, vol. 1, pp. 493–498. IEEE (2017)
25. Sudholt, S., Fink, G.A.: Attribute CNNS for word spotting in handwritten documents. Int. J. Doc. Anal. Recogn. 21(3), 199–218 (2018). https://doi.org/10.1007/s10032-018-0295-0
26. Wang, H., Kawahara, Y., Weng, C., Yuan, J.: Representative selection with structured sparsity. Pattern Recogn. 63, 268–278 (2017)
27. Wang, H., Yuan, J.: Representative selection on a hypersphere. IEEE Sig. Process. Lett. 25(11), 1660–1664 (2018)
28. Westphal, F., Grahn, H., Lavesson, N.: Efficient document image binarization using heterogeneous computing and parameter tuning. Int. J. Doc. Anal. Recogn. (IJDAR) 21(1–2), 41–58 (2018)
29. Zaeemzadeh, A., Joneidi, M., Rahnavard, N., Shah, M.: Iterative projection and matching: finding structure-preserving representatives and its application to computer vision. In: Proceedings of the IEEE Conference on Computer Vision and Pattern Recognition, pp. 5414–5423 (2019)

Named Entity Recognition in Semi Structured Documents Using Neural Tensor Networks

Khurram Shehzad[1], Adnan Ul-Hasan[2(✉)], Muhammad Imran Malik[1,2], and Faisal Shafait[1,2]

[1] School of Electrical Engineering and Computer Science,
National University of Sciences and Technology (NUST), Islamabad, Pakistan
{kshehzad.mscs15seecs,malik.imran,faisal.shafait}@seecs.edu.pk
[2] Deep Learning Laboratory, National Center of Artificial Intelligence,
Islamabad, Pakistan
adnan.ulhassan@seecs.edu.pk

Abstract. Information Extraction and Named Entity Recognition algorithms derive major applications related to many practical document analysis system. Semi structured documents pose several challenges when it comes to extract relevant information from these documents. The state-of-the-art methods heavily rely on feature engineering to perform layout-specific extraction of information and therefore do not generalize well. Extracting information without taking the document layout into consideration is required as a first step to develop a general solution to this problem. To address this challenge, we propose a deep learning based pipeline to extract information from documents. For this purpose, we define 'information' to be a set of entities that have a label and a corresponding value, e.g., application_number: ADNF8932NF and submission_date: 15FEB19. We form relational triplets by connecting one entity to another via a relationship, such as (max_temperature, is, 100 degrees) and train a neural tensor network that is well-suited for this kind of data to predict high confidence scores for true triplets. Up to 96% test accuracy on real world documents from publicly available GHEGA dataset demonstrate the effectiveness of our approach.

Keywords: Named Entity Recognition · Neural Tensor Networks · Semi structured documents

1 Introduction

Information extraction from documents is crucial for many applications like invoice automation, knowledge management and preservation, and information retrieval. A key components in such systems is the entity extraction for relevant information. These entities could be personal information (like name, date of birth, address, emails, etc.), invoice data (like date, price of certain items,

© Springer Nature Switzerland AG 2020
X. Bai et al. (Eds.): DAS 2020, LNCS 12116, pp. 398–409, 2020.
https://doi.org/10.1007/978-3-030-57058-3_28

total amount, SKUs, etc.), technical data of components (like operating temperatures, power dissipation values, potential hazards, etc.), or legal data (like date of contract, expiration date, specific laws and regulations, etc.).

Entity Recognition can be applied to both speech and textual data. In the current scope of work, we are referring only to textual data and that too contained in paper based documents such as technical specifications, forms, legal contracts. One can divide documents into three broad categories: Unstructured (where information is contained in paragraphs and no explicit labels are present), Structured (where information is presented with specific labels), and Semi structured (where information is partially (un) structured). With respect to the complexity of information extraction from these three types of documents, one can imagine that the structured documents are the easiest form of documents, because each information pair is explicitly available in the document. On the other hand, it is very challenging to recognize entities in an unstructured documents due to the absence of specific tags or labels.

The most commonly found methodology in literature to deal with documents is to first identify the layout and then to apply pattern recognition and machine learning algorithms for entity recognition. Layout based machine learning and rule based approaches mostly utilize the layout and format specific geometric features to extract information from documents [4,6,10,12,13]. These approaches work well when the test image and the training documents have comparable layouts. A challenge in information extraction is that information can be located anywhere on the document page and in any form of data structure, such as tables, figures, logos or the regular text paragraphs. This is a problem with layout based approaches. They are not general solutions to extract information from unseen layouts and document formats. They require human intervention to correct the extraction mistakes in such cases [6,10]. Other challenges in information extraction include variation in fonts of the text, variation in field sizes associated with a label and presence of rare labels of interest that are not found in all documents [10]. Two examples of such documents are shown in Fig. 1 and 2. Figure 1 shows a datasheet document where the labels and their corresponding values are arranged from left to right, horizontally. Figure 2 shows a patent document where the labels and their corresponding values are placed vertically and the field sizes are larger than one sentence.

Both text and geometric features can help in information extraction. For example, one of the ways dates are usually represented is MM/DD/YYYY and the knowledge that where the piece of text appears on the page further helps in associating it with its correct label. Geometric features are mostly layout dependent and if an information extraction system uses them alone, or even in conjunction with textual features, it makes the solution layout dependent as well. Varying layouts makes this problem difficult to solve.

Our work is an effort to develop a general system for information extraction that can work on a broad spectrum of document classes. In this regard, we have chosen to work with the textual features only because they are more general

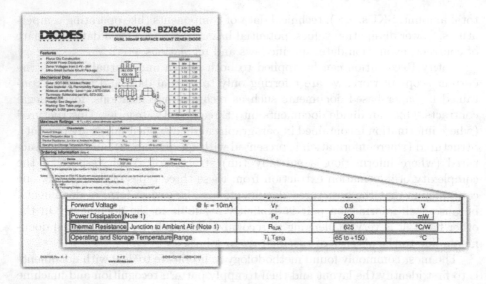

Fig. 1. A sample of datasheet document from GHEGA dataset. The red boxes contain the labels and the blue boxes have their corresponding values. (Color figure online)

compared to geometric features, since information in text is mostly represented in the same way even if the document layouts change.

Natural language processing (NLP) is one of the domains in which machine learning and deep learning in particular have made great progress. One of the most prominent applications of NLP is search via speech where we can talk to our smart phone and our speech gets recognized as a query and the results are returned as speech again. The most popular commercial applications of this technology are found in Google's home, Amazon's Alexa, Window's Cortana and Apple's Siri assistant systems [9]. Other common applications of NLP include language modelling, language translation, speech recognition and image captioning.

Our hypothesis is that if the problem of information extraction can be modeled in a way that it can incorporate the kind of understanding that language models in modern machine learning have, we could get state-of-the-art results without using the knowledge of the document types. Since text is general and common to all documents, such a method could generalize to a variety of document classes. For example, a publication date can be present in a scientific paper, as well as a tender document and even a patent, and it is likely that all three are presented in the same way in text, but different geometrically.

2   Related Work

Documents are broadly classified into three classes based on their layouts [5]; highly structured documents that have a static well-defined layout, semi-structured documents that can be described in terms of templates partially but

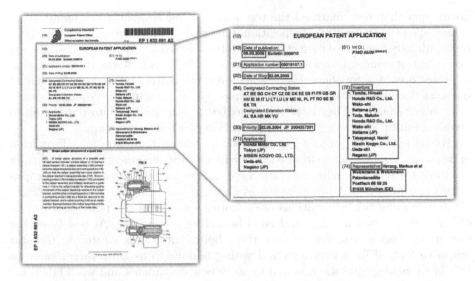

Fig. 2. A sample of patent document from GHEGA dataset. The labels are in the red boxes while the blue ones contain their values. (Color figure online)

are not as well-defined as the former and loosely structured documents that cannot be associated with geometric layouts. Our solution mainly targets the first two document types; the highly structured and semi-structured documents.

OCR error correction can be performed before it is used for information extraction [16]. The OCR text is corrected by string cleaning, substitution, syntactic and semantic checks based on the domain knowledge learned about the labels and their values during training. OCR error has not been taken into account for our work and we chose to work directly with raw OCR output.

Cesarini et al. [4] proposed a method for information extraction from invoices that utilized a three-step pipeline for information extraction. The total dataset they used contained 83 classes of documents with respect to the issuing firm and extracted only two labels, invoice total and invoice number. The labels are called tags and associated values are logical objects. Match is found between them on the basis of the text contained in the two entities or their positions. Document pages are divided into nine regions and distances between the entities on the document are expressed as vectors. A number of thresholds based on hit-and-trial are utilized for this recognition including thresholds for character recognition, average height and average width of bounding boxes. Overall the method is layout based and the class-dependent and even independent knowledge for inference comes from document layouts seen during training.

Esser et al. [6] proposed another layout dependent semi-automatic approach for indexing of required labels in documents. The system begins with an annotated set of training documents which can be augmented with user feedback. The system first performs template matching to find a subset of training documents that best matched the test image using K-nearest neighbour. This subset allows for extracting the positional index data information from the documents already annotated correctly and are transferred for extracting information from the test image. Template matching rate of 98% and information extraction accuracy over 90% for some classes is achieved attributing to the fact the class of business documents they used were assumed to be adhering to the same layouts with minor changes. They introduced feedback from the user into the system to enable addition of new layouts and correction of wrong extractions.

Rusinol et al. [12] proposed incremental structural templates for information extraction. A single image from each provider of the documents is used to train the model. Users manually annotate these images to build weighted star graphs connecting targets to labels where heavy weights are assigned to the physically closer bounding boxes. Relations are encoded by storing the polar coordinates of the vector between target and label bounding boxes (r, θ). At test time, the document class is determined, then the label of interest is located in the star graph to find all its occurrences and voting is done to assign a target value to this label. Star graphs are updated to store new documents and word labels for continuous learning. This approach performed well on a set of known document formats, but the method of connecting labels and targets relies on the document layouts and correct extractions depend on train and test layout similarity.

In Intellix system [13], the extraction pipeline begins by identifying the document class using a variant of kNN on a bag of words of the test document. Next, like other techniques mentioned before, template matching is done using another kNN getting the set of closest resembling layouts of documents in the training set that match the test document layout. Then three indexers are used in combination, namely a fixed-field indexer, a position-based indexer and a context-based indexer followed by a corrector that weighs and combines the scores and assigns a value with the highest score to the label of interest. The extraction steps are different compared to Rusinol et al. [12], but the pipeline seen as a black box as well as the dependence of the approach on seen document layouts is similar.

Medvet et al. [10] divided documents into a set of blocks, and defined rules for each label and a corresponding extraction function for assigning target values to labels. OCR output was used for finding the bounding boxes and a graphical user interface is provided to the users to label the training images of their choice. After that, they system would use the labeled examples to find out the extraction function parameters based on the text and positional relations between the bounding boxes of the labels and its values in the test image. This system, like the rest mentioned before, uses features that are typical to documents coming from the same provider or at least generated using the same software because it relies on layout for extraction.

RelNet, proposed by Bansal et al. [1] is an end-to-end model for entities and relations. RelNet models entities as abstract memory slots and relations between all memory pairs as an additional relational memory. This model is suitable for questioning answering tasks.

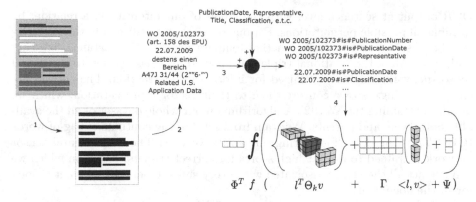

Fig. 3. Our proposed pipeline for information extraction. First, a test image is binarized and deskewed at Step-**1**. Then, OCRopus 2.0 is used for the OCR to get text boxes and text lines at Step-**2**. The obtained text boxes are combined with the predefined set of all the labels to extract, by forming all the possible permutations of labels and text boxes at Step-**3** and relational triplets are formed by combining the labels and the text boxes via the 'is' relation. Finally at Step-**4**, the triplets are given as input to the neural tensor network as the embedding vectors of the two entities, i.e., the labels and the possible corresponding value in the text box to get a prediction score. The last step shows a visualization of a neural tensor network [11], with $k = d = 3$. The 3d cube in the curly braces is the representation of the Θ_k tensor, with size $d \times d \times k$, a 3rd order tensor.

Strubell et al. [17] introduced iterated diluated convolutional neural networks as an alternative to Bidirectional LSTM networks for NER tasks. These networks exploits the power of GPUs massive speed advantages for several NLP tasks like sequence labeling, entity recognition, etc.

Socher et al. [15] proposed Neural Tensor Networks for relational classification. We surmise that this might be a very effective way of modeling the connection between a label and its corresponding value, that we call information. Our work mainly investigates this hypothesis. For knowledge base completion, they compared a variety of network architectures including distance model, single layer fully-connected model, hadamard model and bilinear model with their own proposed neural tensor network to classify whether two entities are connected by a relationship 'R' by forming relational triplets. Our approach for information extraction is inspired by their work and their network is explained in more detail in the next section where we present our solution pipeline.

3   Proposed Solution

We propose the pipeline shown in Fig. 3 for extracting information from documents. As the first step, we preprocess the input document image, which includes adaptive binarization [14] and de-skewing [19] followed by OCR using OCRopus 2.0 [2]. The relational triplets are formed by combining the text boxes in the

OCR output at sentence level with the labels of our interest via a relation, by forming all possible permutations. For our research, we utilize only the "is" relation because we want to verify whether temperature "is" 45°, distance "is" 9 km and so on. All of the relational triplets are passed for classification to the neural tensor network which is optimized for information extraction. The entities are character strings, so we convert them to their vector representation, which we obtain by training the Word2Vec algorithm on our whole dataset of all the available datasheets and patents. We train from scratch the continuous bag-of-words (CBOW) architecture for getting our entity vectors. Then, the neural tensor network is trained to output high scores for correct triplets, based on which we can extract all the values associated with every single label that we want to look for in the document.

A neural tensor network differs from a fully-connected neural network in the sense that instead of having two dimensional weight matrices (which are 2nd order tensors), the weights are tensors of 3rd order which interact with the input entity vectors [15]. The network function is written as

$$p(l, R, v) = \Phi^T f(l^T \Theta_k v + \Gamma <l,v> + \Psi) \qquad (1)$$

where f is a non-linear activation function and l and v are both 100-dimensional entity vectors, label and value vectors respectively. $<l,v>$ is the row concatenation of the two entity vectors. The parameter 'k' in the weight tensor, Θ, is called the *slice size*, which gives a third dimension to the weights, making the weight tensor $\Theta_k \in \mathbb{R}^{d \times d \times k}$, where 'd' is the dimension of entity vectors, and converts an ordinary fully-connected linear multiplication to a bilinear tensor product $l^T \Theta_k v$. Also $\Gamma \in \mathbb{R}^{k \times 2d}$ and $\Phi, \Psi \in \mathbb{R}^k$. More recently, new work has been done on the same problem [3,7,8,18] but we want to demonstrate the effectiveness of modelling information extraction problem in this manner and not the problem of reasoning itself. The problem of knowledge base completion is beyond the scope of our research and we directly use the neural tensor model proposed by [15] and propose a different dimension to look at information extraction from a knowledge modelling point of view that makes the whole solution pipeline independent of the document layout.

For this work, we considered all of the OCR output from the available set of data-sheets and patents for training and testing. Many of the text blocks in the OCR are not associated with any label from our set of labels, so we assign them a sentinel label "unknown". The network was trained by passing the true triplets in the dataset as well as *corrupt triplets* that were constructed by replacing the associated value of a label with any random entity from the dataset. Against each true triplet, we tested different numbers of corrupt triplets for best performance. The ratio of number of corrupt triplets to the number of true triplets is called the *corrupt size*. No GPU was needed for this training because the model is small in size and fits in main memory, which means that inference on a document with a huge number relational triplet permutations can also run on the CPU. The slice size 'k' of the weight tensor is 3 and the words embedding were taken to be 100-dimensional. The following Hinge cost function is used for training the network

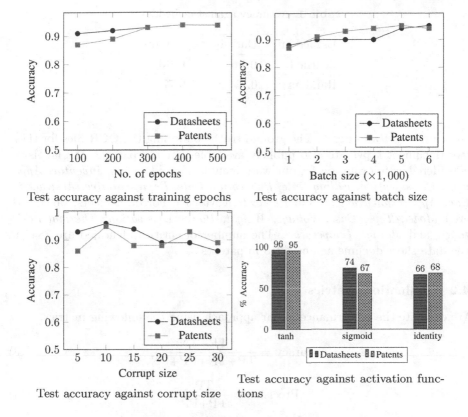

Test accuracy against training epochs

Test accuracy against batch size

Test accuracy against corrupt size

Test accuracy against activation functions

Fig. 4. The results of hyper parameter tuning experiments.

$$C(Y) = \sum_{i=1}^{N_E} \sum_{c=1}^{N_C} \max(0, 1 - p(T^{(i)}) + p(T_c^{(i)})) + \lambda ||Y||_2^2 \qquad (2)$$

where N_E is the number of training examples, N_C is the number of corrupt examples taken against one true triplet (corrupt size), $Y = (\Phi, \Theta_k, \Gamma, \Psi)$ is the set of all the weights that are to be optimized, and $T^{(i)} = (l^{(i)}, R^{(i)}, v^{(i)})$ is a correct training relational triplet while $T_c^{(i)} = (l^{(i)}, R^{(i)}, v_c^{(i)})$ is a corresponding corrupt triplet. Similar to Socher et al. L-BFGS is used for optimizing the weights of the network [15].

4 Experiments

4.1 Data Set

We used Ghega dataset for demonstrating our results, which is the same dataset used by [10], except their invoices have not been made publicly available. It consists of 136 patents and 110 data-sheets documents. One sample from each is

Table 1. Accuracy against embedding

Embeddings	Datasheets	Patents
Learned	**0.96**	**0.96**
Random	0.75	0.77

shown in Fig. 1 and Fig. 2. The ground truth containing the OCR files for these two document types are also publicly available. For the patents, the labels we considered as information for our work include *Title, Priority, Inventor, Applicant, Classification, Filling Date, Publication Date, Representative, Abstract 1st line, Application Number*, and *Publication Number*. For data-sheets, we considered *Model, Type, Case, Voltage, Weight, Power Dissipation, Thermal Resistance*, and *Storage Temperature*. The number of entities identified are 7847 in the data-sheet documents and 6218 in patents.

4.2   Evaluation Metrics

We evaluate the performance of our approach using the following metrics.

$$\text{Accuracy} = \frac{\text{TP+TN}}{\text{TP+TN+FP+FN}} \tag{3}$$

$$\text{Precision} = \frac{\text{TP}}{\text{TP+FP}} \tag{4}$$

$$\text{Recall} = \frac{\text{TP}}{\text{TP+FN}} \tag{5}$$

$$\text{F1} = \frac{2 \times \text{Precision} \times \text{Recall}}{\text{Precision+Recall}} \tag{6}$$

where TP, TN, FP and FN are True Positives, True Negatives, False Positives and False Negatives.

4.3   Results and Discussion

We have tested our approach with different hyper parameters for the neural tensor network. Please refer to Fig. 4 for the results of hyper parameter tuning. A 80/20 train and test split was used for our experiments. We performed cross-validation by constructing five different train and test splits for both document types. The lowest accuracy for any train and test split for our experiments was 93% for data-sheets and 92% for patents. We found that consistent results are obtained for both types of documents by having a tensor network with slice size of 3 and Γ and Ψ parameters set to 0, thus reducing the network to the bilinear product only. If all three components are used with different slice sizes, then a slice size of 1 yields up to 96% accuracy for data-sheets, but the patents perform

Table 2. Per-class results for patents. The sum of values in each column is divided by the number of values to get their average values.

Label	Precision	Recall	F1
Title	0.87	0.91	0.89
Applicant	0.90	0.95	0.92
Inventor	0.97	0.92	0.94
Representative	0.92	0.99	0.95
Filling date	0.91	0.95	0.93
Publication date	0.89	0.96	0.92
Application number	0.93	0.80	0.86
Publication number	0.89	0.99	0.94
Priority	0.96	0.89	0.92
Classification	0.97	0.91	0.94
Abstract 1st line	0.92	0.86	0.89
Unweighted average	0.92	0.92	0.92

slightly worse at 93% even with 3 slices. Our best results with slice size 3 and Γ and Ψ set to 0 are summarized in Tables 1, 2 and 3.

Our accuracy peaks at a corrupt size of 10. For this result, tanh activation was used with learned embeddings. We also tested our method for learned and random embedding vectors for the labels and target values and found that the learned embeddings work better. For this dataset, a tanh activation outperforms sigmoid and identity non-linearities by a big margin. For all the settings we tested, we found that power dissipation in datasheets and abstract 1st line in patents performed the worst. Per-class results are provided in Table 2 and Table 3.

We have also compared our results with those reported in Medvet et al. [10]. They report a "success rate" or precision of 91% for a fixed training set of 15 documents with no human supervision while we get 94% precision for data-sheets and 92% for patents. Since [10] uses layout information, fewer examples lead to better results because the layouts seen at test time are similar to the training inputs. For our approach, since we are learning from text, a larger training set is required to perform at the same level. Medvet et al. also experimented with feedback via human intervention by letting the user correct the results of incorrect extractions. They found out that supervision increases precision to 96%, and our system is capable of producing competitive results using just textual information without any human intervention.

Table 3. Per-class results for datasheets. The sum of values in each column is divided by the number of values to get their average values.

Label	Precision	Recall	F1
Model	0.96	0.98	0.97
Type	0.98	0.97	0.97
Case	0.92	0.98	0.95
Power dissipation	0.94	0.89	0.92
Storage temperature	0.90	0.93	0.91
Voltage	0.91	0.95	0.93
Weight	0.91	0.99	0.95
Thermal resistance	0.99	0.96	0.98
Unweighted average	0.94	0.96	0.95

5   Conclusion

We have presented a layout independent pipeline for information extraction from document images. We are able to obtain highly accurate results on patents and data-sheets for relational classification. Our problem formulation and feature selection are general and layout independent. Results were obtained by forming relational triplets of entities found in the OCR output provided with the dataset with a set of labels defined beforehand. In future, we would develop an end-to-end system, that would incorporate this method of information extraction and work holistically on a document by obtaining OCR, getting the bounding boxes at different levels and testing all the permutations to obtain a fully extracted information result. Newer methods of modelling information other than the neural tensor network of [15] should also be tested for finding more optimal solutions in terms of processing time for faster performance. Since the input features are textual, we would want to test our approach on a variety of other document types such as invoices and tender documents to test its effectiveness.

References

1. Bansal, T., Neelakantan, A., McCallum, A.: RelNet: end-to-end modeling of entities & relations. CoRR abs/1706.07179 (2017). http://arxiv.org/abs/1706.07179
2. Breuel, T.M.: The OCRopus open source OCR system. In: Document Recognition and Retrieval XV, vol. 6815, p. 68150F. International Society for Optics and Photonics (2008)
3. Cai, C.H., Ke, D., Xu, Y., Su, K.: Symbolic manipulation based on deep neural networks and its application to axiom discovery. In: 2017 International Joint Conference on Neural Networks (IJCNN), pp. 2136–2143. IEEE (2017)
4. Cesarini, F., Francesconi, E., Gori, M., Soda, G.: Analysis and understanding of multi-class invoices. Doc. Anal. Recogn. **6**(2), 102–114 (2003). https://doi.org/10.1007/s10032-002-0084-6

5. Dengel, A.R.: Making documents work: challenges for document understanding. In: 7th International Conference on Document Analysis and Recognition, p. 1026. IEEE (2003)
6. Esser, D., Schuster, D., Muthmann, K., Berger, M., Schill, A.: Automatic indexing of scanned documents: a layout-based approach. In: Document Recognition and Retrieval XIX, vol. 8297, p. 82970H. International Society for Optics and Photonics (2012)
7. Liu, Q., et al.: Probabilistic reasoning via deep learning: neural association models. arXiv preprint arXiv:1603.07704 (2016)
8. Liu, Q., Jiang, H., Ling, Z.H., Zhu, X., Wei, S., Hu, Y.: Combing context and commonsense knowledge through neural networks for solving Winograd schema problems. In: AAAI Spring Symposium Series (2017)
9. López, G., Quesada, L., Guerrero, L.A.: Alexa vs. Siri vs. Cortana vs. Google Assistant: a comparison of speech-based natural user interfaces. In: Nunes, I. (ed.) AHFE 2017. Advances in Intelligent Systems and Computing, vol. 592, pp. 241–250. Springer, Cham (2017). https://doi.org/10.1007/978-3-319-60366-7_23
10. Medvet, E., Bartoli, A., Davanzo, G.: A probabilistic approach to printed document understanding. Int. J. Doc. Anal. Recogn. (IJDAR) **14**(4), 335–347 (2011). https://doi.org/10.1007/s10032-010-0137-1
11. Nieze, A.: How to draw a Rubik's cube in Inkscape, September 2014. http://goinkscape.com/how-to-draw-a-rubiks-cube-in-inkscape/
12. Rusinol, M., Benkhelfallah, T., Poulain d'Andecy, V.: Field extraction from administrative documents by incremental structural templates. In: 12th International Conference on Document Analysis and Recognition, pp. 1100–1104. IEEE (2013)
13. Schuster, D., et al.: Intellix-end-user trained information extraction for document archiving. In: 12th International Conference on Document Analysis and Recognition, pp. 101–105. IEEE (2013)
14. Shafait, F., Keysers, D., Breuel, T.M.: Efficient implementation of local adaptive thresholding techniques using integral images. In: Document recognition and retrieval XV, vol. 6815, p. 681510. International Society for Optics and Photonics (2008)
15. Socher, R., Chen, D., Manning, C.D., Ng, A.: Reasoning with neural tensor networks for knowledge base completion. In: Advances in Neural Information Processing Systems, pp. 926–934 (2013)
16. Sorio, E., Bartoli, A., Davanzo, G., Medvet, E.: A domain knowledge-based approach for automatic correction of printed invoices. In: International Conference on Information Society (i-Society 2012), pp. 151–155. IEEE (2012)
17. Strubell, E., Verga, P., Belanger, D., McCallum, A.: Fast and accurate sequence labeling with iterated dilated convolutions. CoRR abs/1702.02098 (2017). http://arxiv.org/abs/1702.02098
18. Trivedi, R., Dai, H., Wang, Y., Song, L.: Know-evolve: deep temporal reasoning for dynamic knowledge graphs. In: Proceedings of the 34th International Conference on Machine Learning, vol. 70, pp. 3462–3471. JMLR. org (2017)
19. Van Beusekom, J., Shafait, F., Breuel, T.M.: Combined orientation and skew detection using geometric text-line modeling. Int. J. Doc. Anal. Recogn. (IJDAR) **13**(2), 79–92 (2010)

Text Detection

SickZil-Machine: A Deep Learning Based Script Text Isolation System for Comics Translation

U-Ram Ko and Hwan-Gue Cho[✉]

Department of Electrical and Computer Engineering, Pusan National University,
Pusan, South Korea
{rhdnfka94,hgcho}@pusan.ac.kr

Abstract. The translation of comics (and Manga) involves removing text from a foreign comic images and typesetting translated letters into it. The text in comics contain a variety of deformed letters drawn in arbitrary positions, in complex images or patterns. These letters have to be removed by experts, as computationally erasing these letters is very challenging. Although several classical image processing algorithms and tools have been developed, a completely automated method that could erase the text is still lacking. Therefore, we propose an image processing framework called 'SickZil-Machine' (SZMC) that automates the removal of text from comics. SZMC works through a two-step process. In the first step, the text areas are segmented at the pixel level. In the second step, the letters in the segmented areas are erased and inpainted naturally to match their surroundings. SZMC exhibited a notable performance, employing deep learning based image segmentation and image inpainting models. To train these models, we constructed 285 pairs of original comic pages, a text area-mask dataset, and a dataset of 31,497 comic pages. We identified the characteristics of the dataset that could improve SZMC performance. SZMC is available at: https://github.com/KUR-creative/SickZil-Machine.

Keywords: Comics translation · Deep learning · Image manipulation system

1 Introduction

Comic literature (or Manga) is globally appreciated. Economically, its market has been growing, especially for the digital form comics [3]. However, the automatic translation of comics is difficult, due to their inherent characteristics. Currently, most comics are translated manually.

Translation of comics could be divided into two phases. First, erasing the foreign language text (Fig. 1-a), and filling the erased areas with a picture to match with their surrounding regions. Second, The translated text is either typeset (font letters) or drawn (calligraphic letters) on the image (Fig. 1-b). Because

© Springer Nature Switzerland AG 2020
X. Bai et al. (Eds.): DAS 2020, LNCS 12116, pp. 413–425, 2020.
https://doi.org/10.1007/978-3-030-57058-3_29

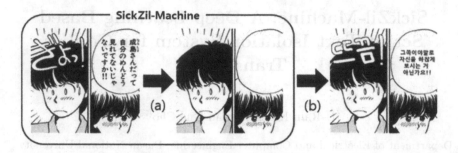

Fig. 1. Example of Japanese to Korean translation of comics. (a) Text removal from the foreign comics (b) Drawing the translated text. SickZil-Machine (SZMC) automates text removal for the comics translation. Original image was extracted from Manga109 dataset [8], ©Yoshi Masako.

the text in comics is drawn at arbitrary places, and often on complex images or patterns, erasing the text from the original comic image is time-consuming, and labor-intensive. The text found in comics include printed font letters and artist-specific handwritten calligraphies, which could be of various styles and sizes. Therefore, detection and segmentation of such text is challenging. Consequently, removing such text, using the classical image processing algorithms, is difficult. To solve this, we propose a deep learning based framework "SickZil-Machine (SZMC)" for effective translation. "SickZil" in Korean means comics editing task.

2 Proposed Approach

Adobe Photoshop (Adobe Inc., San Jose, USA) has been used by professional editors to remove the text in comics. The editors commonly use several classical image processing algorithms that are provided in Photoshop. Although the macros system in Photoshop is useful in removing simple text, it requires the editors' manual intervention for the text in more complex backgrounds.

We approach the removal of comics text as a problem that is combined image segmentation and image inpainting (Fig. 2). SZMC segments the text areas in the comics image (Fig. 2-a) using the image segmentation model, erases the segmented text area, and fills the erased area (Fig. 2-b) using the image inpainting model.

An alternative approach was proposed, which detected and erased the text in an image using a single end-to-end neural network model, known as EnsNet [17]. Although EnsNet had faster processing speed and required lower memory, it had several limitations. Obtaining sufficient data was not feasible, as having comic image pairs, one with the text and the other without, was rare. Further, the majority of them are proprietary data and not public. Second, as the removable text area was implicitly determined by the end-to-end model, determining the erasable area by the user was not possible.

Fig. 2. Proposed approach for automatic text removal during comics translation. (a) image segmentation (b) image inpainting. SZMC segmented the text areas and erased them. The erased area was drawn using image inpainting model, to match with the surrounding. Original image was extracted from Manga109 dataset [8], ©Yoshi Masako.

3 Related Work

3.1 Previous Researches on Comics Image Analysis

Although segmentation of natural or medical images has been extensively studied, the automatic segmentation of text areas from comics has not yet been investigated. Segmentation of common objects, such as lines, speech bubbles, and screen tones was studied [5,7]. Further, detection and analysis of text, in bounding boxes, has been reported [2,11]. [16] was proposed as an open source project for segmenting the text area at the pixel level (Table 1).

Table 1. Summary of the related work on comic content analysis

No	Input	Output	Method	Ref.
1	Manga page	Region	Graph-cut with user stroke	[7]
2	Manga frame	Lines	LoG and Gaussian filter	[5]
3	Manga page	Text region	Connected-Component analysis	[2]
4	Comics text	Plain text	Segmentation-free learning	[11]
5	Comics page	Text mask	Learning-based (MobileNet)	[16]
6	Comics page	Text mask	Learning-based (U-net)	[14]

3.2 Previous Researches on Generic Text Eraser

To the best of our knowledge, we cannot find end-to-end method for removing text from comics. STEraser [10], EnsNet [17], and MTRNet [13] try to remove text from natural images in order to erase personal private information such as telephone numbers, home addresses, etc. Unlike the proposed two-step method, STEraser and EnsNet consider text removal as an image transformation task.

STEraser, a first scene text eraser using a neural net, applies U-net-like [12] structure with residual connection instead of concatenation. It trains and inferences with sliding-window-based 64 × 64 sized crops from the input image. Since STEraser cannot grasp context of the whole image, it cannot erase large text properly.

Unlike STEraser, EnsNet uses the entire image as an input. It follows cGAN [9] framework with novel lateral connection in the generator, and applies multiple losses. Though EnsNet produces a plausible output compare to STEraser, but it is not suitable for the comics translation process. In comics translation, the erasing text depends on the comics editor's decision. Also, those image transformation approach requires input image and the corresponding image with text removed. One serious problem is that such data is hard to collect and very difficult to create.

MTRNet is an inpainting model using modified cGAN. Applying two-step method to remove text, it receives an input image concatenated with a mask that annotates the text location when training and inference. Therefore, a comics editor can select the text to be erased. However, MTRNet is still not applicable to comics translation because it does not consider to erase very large text. In comics, not only small regular letters, but also very large-sized calligraphic characters can exist on a complex background.

We propose two-step approach in order to allow the editor to select the text to erase. To erase large text in comics, we apply more general inpainting model [15] that is not limited to text.

3.3   Previous Researches on Image Inpainting

Adobe Photoshop's content aware fill function has been used by the editors, to erase smaller-sized text in comics. Although this function was effective in erasing small text that was drawn over simple patterns, it failed to naturally erase the text on either the non-stationary images or complex patterns. Further, this function failed to erase large handwritten calligraphic letters, as the classic inpainting techniques disregard the image semantics.

[4] was the first deep learning based end-to-end model for image inpainting. But this model required a longer training period, could be trained only with the square masks, and therefore, overfitted to the square masks. [6] was proposed to reduce the training period and was able to prevent the square mask overfitting. This model was able to acquire the masks as channels in the input image and could inpaint the image by updating the mask for each partial convolution layer, in the feed-forward path. [15] improved [6] by substituting the gated convolution layers for the partial convolution layers. While the partial convolution layers updated the mask channels in the binary values of 0 and 1 (hard gating), the gated convolution layers added the training parameters, for soft gating the mask, to the real values of 0.0 to 1.0.

3.4   Available Comics Datasets

Danbooru2018 [1], and Manga109 [8] are the publicly available comic literature (and Manga) datasets. While Danbooru2018 dataset tags illustrations and comic pages with image-feature metadata, the Manga109 dataset tags the pages from 109 titles of Manga, with the metadata of the script text, the area of the character's face, and the position and contents of the text. However, the mask data indicating a text area at the pixel level, was lacking.

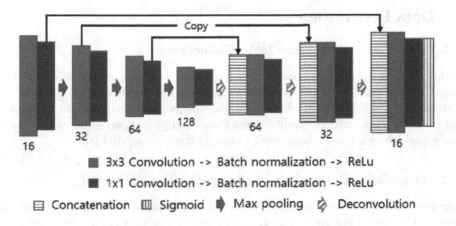

Fig. 3. Modified U-net model for text area segmentation.

4   The Proposed Model

The SZMC segments the text areas to be erased, using a modified U-net [12] and removes the text naturally using [15].

4.1   Step 1 - Text Area Segmentation

U-net is composed of layers that repeat certain basic unit blocks [12]. We set up the U-net basic block with 3×3 and 1×1 filtered convolution layers, followed by a batch normalization layer and a ReLU layer. In our four-layer U-net model, the first, second, third and fourth convolution layers had 16, 32, 64, and 128 filters. We employed maxpooling and deconvolution at the encoder and decoder, respectively (Fig. 3). As the comics text component masks are very unbalanced in classes (background pixels are much more than texts), we used weighted binary cross entropy and Jaccard loss, to train the model.

4.2 Step 2 - Comics Image Inpainting

Comics have not only font-based text, but also calligraphic text drawn by the artist. They can be very difficult to remove naturally due to its large area and irregular shape. Therefore [15], the state-of-the-art image inpainting model, was applied. This model required variously shaped masks in training for best performance. So we used the text components, which were extracted from our comic literature datasets, as mask data.

5 Data Preparation

5.1 Raw Comic Literature Data Collection

Danbooru is a website with anonymously uploaded comics, manga and illustrations. Danbooru2018 dataset is organized by crawling Danbooru [1]. We created a database of selected monochrome-tagged images from Danbooru2018, resulting in a collection of high-quality comics and manga images. Moreover, typical black and white manga images were obtained from Manga109 [8].

5.2 Organizing Dataset for the Image Inpainting Model

The text that could be easily erased with the existing image editing tools, such as that in the speech bubble in Fig. 2, was designated as easyT. However, the text on complex backgrounds that are difficult to erase, such as the calligraphic letters drawn over the man's head in Fig. 2, were designated as hardT. The images in the original comic literature databases were classified into four categories, according to the location of easyT and hardT text: data-1: Images without text, data-2: Images with only easyT text, data-3: Images with easyT and hardT text, data-4: Images with only hardT text.

Two datasets were constructed to train the image inpainting model. The "NoText" dataset was constructed from 20,033 images from data-1, among which 16,033 images were used for training, and 4,000 images were used for testing. The "HasText" dataset was constructed with 27,497 images from datasets-3,4. The NoText test dataset was used for testing the entire system. Since the applied image inpainting model [15] was trained in an unsupervised manner, a validation dataset was not required.

5.3 Dataset Creation for Text Area Segmentation

The dataset for image segmentation consisted of pairs of original comic images and answer masks that could segment the text of the original comic images. "Split" dataset was created to separate the text into two classes: calligraphic and font texts. However, the "All" dataset grouped all the text in one class. The number of images used for training, validation, and testing from both datasets was 200, 57, and 28, respectively.

We used a part of the dataset for image inpainting, to generate the answer image-mask pair data. After clearing easyT from data-2, we generated the answer masks by computing the difference between the original and text-removed images. However, this method could not be applied to create a segmentation mask that had hardT. Thus, the hardT areas in data-3,4 were segmented manually to generate the HardT dataset (Table 2). The first 50 masks were created manually, using GNU Image Manipulation Program. The remaining 235 masks were created by modifying the trained segmentation model output. To prevent the contamination of the validation data, the experimental models were trained separately.

Table 2. Prepared datasets for the system

Model	Name	Train/valid/test	Remarks
Segmentation	Split	200/57/28	Calligraphy, font text separated
Segmentation	All	200/57/28	No separation of text
Inpainting	NoText	16,033/NA/4000	Danbooru2018, Manga109
Inpainting	HasText	27,497/NA/4000	Danbooru2018, Manga109

6   Experiments Results

6.1   Evaluation of Image Segmentation Model

We applied mean intersection over union (IoU) to quantitatively evaluate the image segmentation model. IoU similarity was defined as

$$IoU(G, S) = \frac{G \cap S}{G \cup S},$$

wherein G and S denoted the answer and result sets, respectively.

Table 3. Evaluation of the image segmentation model

Model	Loss	mIoU	Time (sec)
TSII[16]	wbce	0.279	**0.442**
U-net (Split)	wbce	0.424	0.510
U-net (All)	wbce	0.570	0.511
U-net (Split)	Jaccard	0.479	0.511
U-net (All)	Jaccard	**0.602**	0.512

The SZMC was evaluated using an Intel i7 CPU with gtx1070ti GPU. SZMC exhibited slightly slower performance compared to TSII [16]. Notably, SZMC exhibited more than double IoU similarity. Moreover, IoU similarity, for text, of the All dataset trained Unet-All model was 125% higher than the Split dataset trained Unet-Split model. Note that the All and Split datasets had one (all text) and two (calligraphy and font) text classes, respectively. Further, Jaccard loss-trained models exhibited higher IoU metric (Table 3).

6.2   Evaluation of Image Inpainting Model

To evaluate the results of the comics image inpainting, test images I of the NoText dataset and answer masks M of the All dataset were employed. We used M to mask the part of I, to create the input images I'. The restoration I' by the image inpainting model was compared with the original images I. L1 loss, L2 loss, and PSNR were used as performance metrics. The NoText dataset trained model exhibited 2.91%p higher L1 loss compared to the HasText dataset-trained model. However, HasText-trained model exhibited 0.32%p higher L2 loss compared to the NoText-trained model (Table 4). This indicated that the image inpainting model could be trained with the text-containing comics image dataset. Since text-lacking comic images, such as the images in NoText dataset, were difficult to obtain, notably, the SZMC performance was not significantly hampered, even when the training dataset included some text-containing images.

Table 4. Evaluation of image inpainting model

Model	L1 loss	L2 loss	PSNR	Time (sec)
NoText	**8.367%**	5.506%	**74.134**	2.795
HasText	11.286%	**5.190%**	73.939	2.792

6.3   Evaluation of Whole System

To quantitatively evaluate the whole system that is combined image segmentation and inpainting, pairs of text-containing (I) and text-lacking (I') images were required. Public availability of such data is rare. Further, since multiple I' could exist due to removal of text from I, quantitative evaluation of SZMC was unattainable.

Figure 4 depicts the qualitative evaluation of SZMC. Odd and even columns depict the inputs and outputs, respectively. (a,b,c)-1 exhibited better performance, while (a,b,c)-2 recorded somewhat unstable results. Figure 4-a depicts the images with easyT that could be easily processed with the existing tools. SZMC could successfully remove easyT. Figure 4-b depicts the removal of smaller hardT (less than 75×75 pixels in a text-containing bounding box). The hardT is defined as text drawn on a complex picture and that could not be automatically processed with the existing tools. SZMC could successfully remove hardTs of

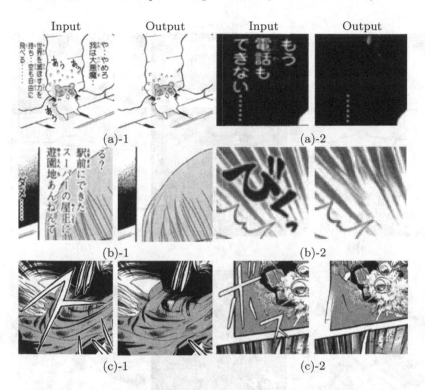

Fig. 4. Experimental results of the whole system. (a) Images that could be processed with existing tools (b) Complex images that could not be edited with existing tools (c) Images with very large calligraphic text. Image source [8], ©Arai Satoshi, ©Miyauchi Saya, ©Shimazaki Yuzuru, ©Miyone Shi

up to 75×5. Figure 4-c depicts removal of larger hardT (bounding box is larger than 125×125 pixels). SZMC exhibited inadequate image segmentation with very large hardTs (greater than 256×256 pixels). Upon adequate segmentation, larger text could be successfully removed from the image.

Figure 5 and 6 depict the experimental evaluation of the whole system. Figure 5 depicts that SZMC could process the entire comic page. Figure 6 depicts the relatively lower performance. Figure 6-a erased the non-text area due to incorrect text area segmentation. Figure 6-b depicts an effective segmentation of text areas. However, erasing was ineffective and failed to match with either the background or screen tone. This could be addressed by increasing the scale of the training dataset and optimizing the model.

7 SZMC GUI Structure

The principal users of SZMC would be professional comics editors. To aid them, we configured the publicly available SZMC in the graphic user interface (GUI) (Fig. 7). It can be downloaded from https://github.com/KUR-creative/SickZil-Machine.

422   U.-R. Ko and H.-G. Cho

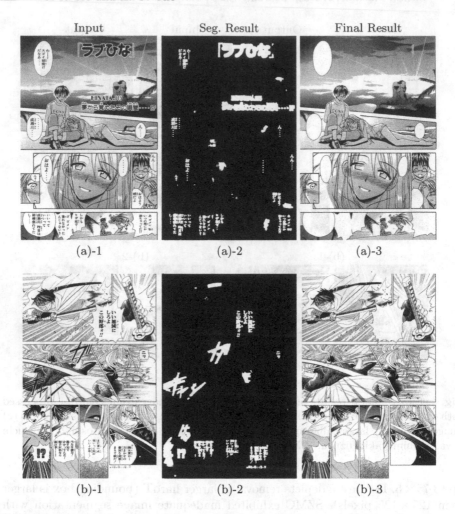

Fig. 5. Experimental results of the whole system when a complete page of comic was input. (a,b)-1 Input images, (a,b)-2 text segmentation, (a,b)-3 image inpainting. Image source: [8], ©Akamatsu Ken, ©Kobayashi Yuki

User can access the comic images directory through:[Open] – [Open Manga Project]. Further, clicking the second button on the toolbar creates the masks for all the images and according to the generated masks, automatically clears the text. Just one click of the RmTxtAll button removes texts from all the images, without any further action by the user. The segmentation masks can be selectively edited, without removing the text, by clicking the first button in the toolbar.

The third and fourth buttons create the mask and remove the text from the currently displayed image, respectively. The fifth to eighth buttons are for editing the generated masks. User can draw or erase masks, using Pen and Rectangle

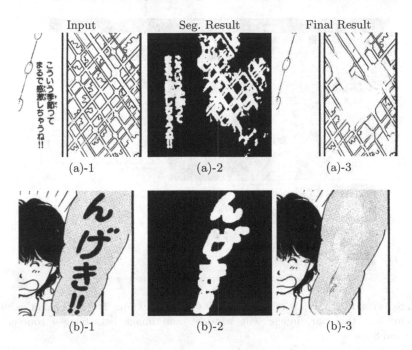

Fig. 6. Substandard cases. (a,b)-1 Input images, (a,b)-2 text segmentation, (a,b)-3 image inpainting. Image source: [8], ©Yoshi Masako

tools. The mask can be edited to either enhance text area segmentation, or to leave some text on purpose.

8 Conclusion

In this paper, we propose a two-step approach to automate the removal of the text from comic images. The SZMC segments the erasable text area from the comic images and inpainted the erased area, naturally. Therefore, the well-studied image segmentation and inpainting techniques could be applied for effective comics text removal.

We created the datasets for effective image segmentation and inpainting, using deep learning based framework, and experimentally verified the effective features. In the text segmentation datasets, grouping the calligraphic letters and font letters as one class, improved the performance. The datasets for image inpainting were not greatly affected even if the dataset had some text-containing images. Moreover, both models were quantitatively evaluated. Compared with the reference model [16], SZMC exhibited accurate image segmentation, with a slightly slower execution.

– We confirm that removing the text from comic images, which has been done manually, could be automated.

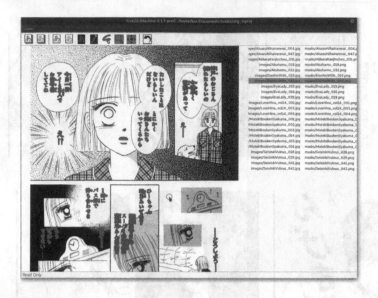

Fig. 7. Screenshot of publicly released SZMC GUI application. The screenshot exhibits the menu bar, tool bar, image edit window, and image list. Images source: [8], ©Miyauchi Saya.

- A deep learning based framework for automating comic image text removal, is presented.
- The proposed framework, SZMC, resulted in improved text area segmentation over the reference model, TSII.
- We created a dataset for removing text from comic images and identified the effective features.

In future work, we plan to release the dataset for the segmentation of the text areas in the comic images. We will create masks for all the images in Manga109 and release only the mask data. Additionally, we will explore more optimized models for better performance.

References

1. Anonymous, The Danbooru Community, Branwen, G., Gokaslan, A.: Danbooru 2018: a large-scale crowdsourced and tagged anime illustration dataset, January 2019. Accessed 1 Jan 2020
2. Aramaki, Y., Matsui, Y., Yamasaki, T., Aizawa, K.: Text detection in manga by combining connected-component-based and region-based classifications. In: 2016 IEEE International Conference on Image Processing (ICIP). IEEE, September 2016
3. Augereau, O., Iwata, M., Kise, K.: A survey of comics research in computer science. J. Imaging **4**(7), 87 (2018)
4. Iizuka, S., Simo-Serra, E., Ishikawa, H.: Globally and locally consistent image completion. ACM Trans. Graph. **36**(4), 1–14 (2017)

5. Ito, K., Matsui, Y., Yamasaki, T., Aizawa, K.: Separation of Manga Line Drawings and Screentones, May 2015
6. Liu, G., et al.: Image inpainting for irregular holes using partial convolutions. In: Ferrari, V., Hebert, M., Sminchisescu, C., Weiss, Y. (eds.) ECCV 2018. LNCS, vol. 11215, pp. 89–105. Springer, Cham (2018). https://doi.org/10.1007/978-3-030-01252-6_6
7. Liu, X., Li, C., Wong, T.-T.: Boundary-aware texture region segmentation from manga. Comput. Vis. Med. **3**(1), 61–71 (2016). https://doi.org/10.1007/s41095-016-0069-x
8. Matsui, Y., et al.: Sketch-based manga retrieval using manga109 dataset. Multimedia Tools Appl. **76**(20), 21811–21838 (2016). https://doi.org/10.1007/s11042-016-4020-z
9. Mirza, M., Osindero, S.: Conditional Generative Adversarial Nets. arXiv:1411.1784 [cs, stat], November 2014
10. Nakamura, T., Zhu, A., Yanai, K., Uchida, S.: Scene Text Eraser. arXiv:1705.02772 [cs], May 2017
11. Rigaud, C., Burie, J., Ogier, J.: Segmentation-free speech text recognition for comic books. In: 2017 14th IAPR International Conference on Document Analysis and Recognition (ICDAR), vol. 03, pp. 29–34, November 2017
12. Ronneberger, O., Fischer, P., Brox, T.: U-Net: convolutional networks for biomedical image segmentation. In: Navab, N., Hornegger, J., Wells, W.M., Frangi, A.F. (eds.) MICCAI 2015. LNCS, vol. 9351, pp. 234–241. Springer, Cham (2015). https://doi.org/10.1007/978-3-319-24574-4_28
13. Tursun, O., et al.: MTRNet: a generic scene text eraser. In: 2019 International Conference on Document Analysis and Recognition (ICDAR). IEEE, September 2019. https://doi.org/10.1109/icdar.2019.00016
14. U-ram, K., Hwan-Gue, C.: A text script removal system for comics using deep learning (Korean). In: Proceedings of Korea Computer Congress, June 2019
15. Yu, J., et al.: Free-form image inpainting with gated convolution. In: The IEEE International Conference on Computer Vision (ICCV), October 2019
16. yu45020: yu45020/Text_segmentation_image_inpainting, October 2019. https://github.com/yu45020/Text_Segmentation_Image_Inpainting. original-date: 2018-06-25T02:48:51Z
17. Zhang, S., Liu, Y., Jin, L., Huang, Y., Lai, S.: EnsNet: Ensconce Text in the Wild. arXiv:1812.00723 [cs], December 2018

Lyric Video Analysis Using Text Detection and Tracking

Shota Sakaguchi[1]([✉]), Jun Kato[2], Masataka Goto[2], and Seiichi Uchida[1]

[1] Kyushu University, Fukuoka, Japan
{shota.sakaguchi,uchida}@human.ait.kyushu-u.ac.jp
[2] National Institute of Advanced Industrial Science and Technology (AIST),
Tsukuba, Japan
{jun.kato,m.goto}@aist.go.jp

Abstract. We attempt to recognize and track lyric words in lyric videos. Lyric video is a music video showing the lyric words of a song. The main characteristic of lyric videos is that the lyric words are shown at frames synchronously with the music. The difficulty of recognizing and tracking the lyric words is that (1) the words are often decorated and geometrically distorted and (2) the words move arbitrarily and drastically in the video frame. The purpose of this paper is to analyze the motion of the lyric words in lyric videos, as the first step of automatic lyric video generation. In order to analyze the motion of lyric words, we first apply a state-of-the-art scene text detector and recognizer to each video frame. Then, lyric-frame matching is performed to establish the optimal correspondence between lyric words and the frames. After fixing the motion trajectories of individual lyric words from correspondence, we analyze the trajectories of the lyric words by k-medoids clustering and dynamic time warping (DTW).

Keywords: Lyric video · Lyric word tracking · Text motion analysis · Video design analysis

1 Introduction

The targets of document analysis systems are expanding because of the diversity of recent document modalities. Scanned paper documents only with texts printed in ordinary fonts were traditional targets of the systems. However, recent advanced camera-based OCR technologies allow us to analyze arbitrary images and extract text information from them. In other words, we can now treat arbitrary images with text information as documents.

In fact, we can consider videos as a promising target of document analysis systems. There have already been many attempts to analyze videos as documents [1]. The most typical attempt is caption detection and recognition in video frames. Another attempt is the analysis of the video from the in-vehicle camera. By recognizing the texts in the in-vehicle videos, it is possible to collect store

© Springer Nature Switzerland AG 2020
X. Bai et al. (Eds.): DAS 2020, LNCS 12116, pp. 426–440, 2020.
https://doi.org/10.1007/978-3-030-57058-3_30

(a) "YOU KNOW WHERE"   (b) "AND THEN SHE"

(c) "I LOVE YOU"   (d) "OOH LOOK WHAT YOU MADE ME DO"

Fig. 1. Frame image examples from lyric video.

and building names and signboard information around the road automatically. There was also an attempt (e.g., [2] as a classical trial) to recognize the sport player's jersey number for identifying individual players and then analyzing their performance.

In this paper, we use *lyric videos* as a new target of document analysis. Lyric videos are music videos published at internet video services, such as YouTube, for promoting a song. The main characteristic of lyric videos is that they show the lyric words of the song (almost) synchronously to the music. Figure 1 shows several frame examples from lyric videos. Lyric words are often printed in various decorated fonts and distorted by elaborated visual designs; they, therefore, are sometimes hard to read even for humans. Background images of the frames are photographic images or illustrations or their mixtures and often too complex to read the lyrics. This means that lyric word detection and recognition for lyric videos is a difficult task even for state-of-the-art scene text detectors and recognizers.

In addition, lyric words in lyric videos move very differently from words in ordinary videos. For example, lyric words often move along with arbitrarily-shaped trajectories while rotating and scaling. This property is very different from video captions since they do not move or just scroll in video frames. It is also different from scene texts in videos from the in-vehicle camera; scene texts move passively according to camera motion, whereas lyric words move actively and arbitrarily. In fact, lyric words and their motion are designed carefully by the creator of the lyric video so that the motion makes the video more impressive. To give a strong impression, lyric words might move like an explosion.

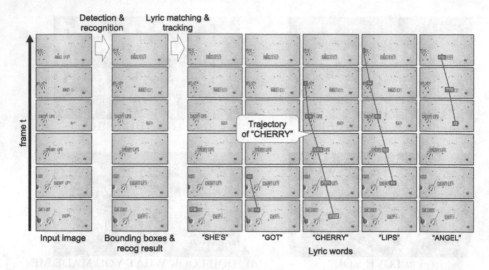

Fig. 2. Our lyric video analysis task. After detecting and recognizing words in video frames, the words in the lyric are tracked over frames. Finally, the word motion trajectories are classified into several clusters to understand the trends of lyric word motions.

The purpose of this paper is to analyze the motion of the lyric words in lyric videos. Specifically, we try to extract and track words in the lyric words automatically and then group (i.e., clustering) their motions into several types to understand their trends. Nowadays, anyone can post her/his music with video. This research will contribute to those non-professional creators to design their own lyric videos. In the future, it will also be possible to realize automatic lyric video generation, by extending this work to understand the relationship between lyric word motion and music beat or music style.

For the above purpose, we develop a lyric word detection and tracking method, where lyric information is fully utilized. As noted before, lyric words are heavily decorated and shown on various background images. Even state-of-the-art scene text detectors and recognizers will have miss-detections (false-negatives), false-detections (false-positives), and erroneous word recognition results. We, therefore, utilize the lyric information as metadata for improving detection and tracking performance. As shown in Fig. 2, we first apply state-of-the-art scene text detectors and recognizers to each video frame. Then, dynamic programming-based optimization is applied to determine the optimal matching between the video frame sequence and the lyric word sequence. We call this novel technique *lyric-frame matching*. Then, using the lyric-frame matching result as a reliable correspondence, spatio-temporal search (tracking and interpolation) is performed for fixing the motion trajectory of each lyric word.

The main contributions of this paper are summarized as follows:

- First, to the authors' best knowledge, this is the first trial of detecting and then tracking lyric words in lyric videos.

- Second, we propose a lyric-frame matching technique where lyric information of the target music is fully utilized as metadata for accurately determining the most confident frame where a lyric word appears. Dynamic programming and state-of-the-art text detection and recognition techniques are used as the modules of the proposed technique.
- Third, we prepared a dataset comprised of 100 lyric videos and attached ground-truth (bounding boxes for lyric words) manually at 10 frames for each of 100 videos. This effort enables us to make a quantitative evaluation of our detection and tracking results.
- Fourth, as an analysis of lyric word motions, we grouped them into several representative word motions by k-medoid clustering. From the clustering result, we could represent each lyric video by a histogram of representative motions.

2   Related Work

To the authors' best knowledge, this paper is the first trial of analyzing lyric videos as a new document modality. We, however, still can find several related tasks, that is, video caption detection and recognition, and text tracking. In the following, we review the past attempts on those tasks, although the readers also refer to a comprehensive survey [1,3]. It should be noted that the performance on all those tasks has drastically been improved by the recent progress of scene text detection and recognition technologies. Since the attempts on scene text detection and recognition are so huge (even if we limit them only to the recent ones), we will not make any review on the topic. A survey [3] will give its good overview.

Caption detection and recognition is a classical but still hot topic of document image analysis research. Captions are defined as the texts superimposed on video frames. Captions, therefore, have different characteristics from scene texts. We can find many attempts on this task, such as [4–13]. Most of them deal with the static captions (i.e., captions without motions), while Zedan [10] deals with moving captions; they assume the vertical or horizontal scrolling of caption text, in addition to the static captions.

Rather recently, scene text tracking and video text tracking [14–22] have also been tried, as reviewed in [1]. Each method introduces its own word tracker. For example, Yang [19] uses a dynamic programming-based tracker to have an optimal spatio-temporal trajectory for each word. A common assumption underlying those attempts is that text motions in video frames are caused mainly by camera motion. Therefore, for example, neighboring words will move to similar directions. We also find "moving MNIST" [23] for a video prediction task, but it just deals with synthetic videos capturing two digits are moving around on a uniform background.

Our trial is very different from those past attempts at the following three points at least. First, our target texts move far more dynamically than the texts that the past attempts assume. In fact, we cannot make any assumption on the

location, appearance, and motion of the words. Second, our trial can utilize the
lyric information as the reliable guide of text tracking and therefore we newly
propose a lyric-frame matching technique for our task. Third, our main focus is
to analyze the text motion trajectory from a viewpoint of video design analysis,
which is totally ignored in the past text tracking attempts.

3 Lyric Video Dataset

We collected 100 lyric videos according to the following steps. First, a list of
lyric videos is generated by searching YouTube with the keyword "official lyric
video"[1]. The keyword "official" was added not only for finding videos with long-
term availability and but also for finding videos created by professional creators.
Videos on the list were then checked visually. A video only with static lyric words
(i.e., a video whose lyric words do not move) is removed from the list. Finally,
the top-100 videos on the list are selected as our targets[2].

The collected 100 lyric videos have the following statistics. Each video is
comprised of 5,471 frames (3 min 38 s) on average, 8,629 frames (5 min 44 s) at
maximum, and 2,280 frames (2 min 33 s) at minimum. The frame image size is
1,920 × 1,080 pixels. The frame rate is 12, 24, 25, and 30 fps on 1, 58, 25, and 16
videos, respectively. The lyrics are basically in English. Each song is comprised
of 338 lyric words on average, 690 words at maximum, and 113 at minimum.

Figure 3 shows four examples showing the variations of the number of lyric
words in a frame. Figure 3 (a) shows a typical frame with lyric words. It is quite
often in lyric videos that a phrase (i.e., consecutive lyric words) is shown over
several frames. It is rather rare that only a single lyric word is shown in a one-
by-one manner synchronously with the music. We also have frames without any
lyric word, like (b). Those frames are often found in introduction parts, interlude
parts, and ending parts of lyric videos. Figures 3 (c) and (d) contain many words
in a frame. In (c), the same word is repeatedly shown in a frame like a refrain of a
song. The words in (d) are just decorations (i.e., they belong to the background)
and not relating to the lyrics.

For a quantitative evaluation in our experiment, we attach the bounding
boxes for lyric words manually at 10 frames for each lyric video. The 10 frames
were selected as follows; for each video, we picked up frames every three seconds
and then select 10 frames with the most word candidates detected automatically
by the procedure of Sect. 4.1. The three-second interval is necessary to avoid the
selection of the consecutive frames. If a lyric word appears multiple times in a
frame (like Fig. 3 (c)), the bounding box is attached to each of them. For the
rotated lyric words, we attached a non-horizontal bounding box[3]. Consequently,
we have 10 × 100 = 1,000 ground-truthed frames with 7,770 word bounding
boxes for the dataset.

---

[1] The search was performed on 18th July 2019.
[2] For the URL list of all the videos and their annotation data, please refer to https://github.com/uchidalab/Lyric-Video.
[3] To attach non-horizontal bounding boxes, we used the labeling tool roLabelImg at https://github.com/cgvict/roLabelImg.

(a) Showing lyric words.

(b) No lyric word (interlude).

(c) Duplicated lyric words (like refrains).

(d) Words unrelated to the lyrics.

Fig. 3. Variations of the spatial distribution of lyric words in a video frame.

4   Lyric Word Detection and Tracking by Using Lyric Information

In this section, we will introduce the methodology to detect and track lyric words in a lyric video. The technical highlight of the methodology is to fully utilize the lyric information (i.e., the lyric word sequence of the song) for accurate tracking results. Note that it is assumed that the lyric word sequence is provided for each lyric video as metadata.

4.1   Lyric Word Candidate Detection

As the first step, lyric word candidates are detected by scene text detectors, as shown in the left side of Fig. 4 (a). Specifically, we use two pretrained state-of-the-art detectors, PSENet [24] and CRAFT [25]. Then, each detected bounding box is fed to a state-of-the-art scene text recognizer; we used TPS-Resnet-BiLSTM-Attn, which achieved the best performance in [26]. If two bounding boxes from those two detectors have 50% overlap and the same recognition result, they are treated as the duplicated bounding boxes and thus one box is removed from the later process.

4.2   Lyric-Frame Matching

After the detection, lyric-frame matching is performed to find the correspondence between the given lyric word sequence and the frame sequence. The red path on the right side of Fig. 4 (a) shows the matching path showing the optimal correspondence. Assume the video and its lyrics are comprised of T frames and K words, respectively, and $t \in [1, T]$ and $k \in [1, K]$ are their indices, respectively.

If the path goes through the node (k,t), the frame t is determined as the most confident frame for the kth lyric word. Note that the other frames where the same kth lyric word appears will be determined later by the tracking process of Sect. 4.3.

The path is determined by evaluating the distance $D(k,t)$ between the kth word and the frame t. The circled number in Fig. 4 (a) represents $D(k,t)$. A

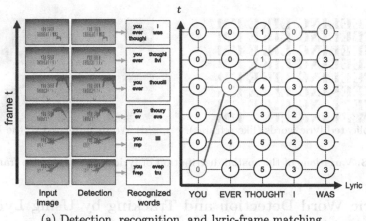

(a) Detection, recognition, and lyric-frame matching.

(b) Tracking by neighbor search.   (c) Interpolation.

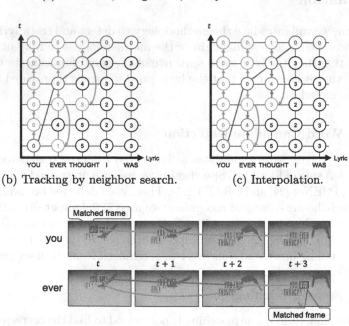

(d) Tracking result of "you" and "ever".

Fig. 4. The detail of lyric word detection and tracking. (Color figure online)

smaller value of $D(k,t)$ means that the kth lyric word will be found at t with a high probability. The distance is calculated as $D(k,t) = \min_{b \in B_t} d(k,b)$, where B_t is the set of bounding boxes (i.e., words) detected at t and $d(k,b)$ is the edit distance function between the kth lyric word and the bth detected word at t. If $D(k,t) = 0$, the kth lyric word is correctly detected at t without misrecognition.

With the distance $\{D(k,t)|\forall k, \forall t\}$, the globally optimal lyric-frame matching is obtained efficiently by dynamic programming (DP), as the red path of Fig. 4 (a). In the algorithm (so-called the DTW algorithm), the following DP recursion is calculated at each (k,t) from $(k,t) = (1,1)$ to (K,T);

$$g(k,t) = D(k,t) + \min_{t-\Delta \leq t' < t} g(k-1,t'),$$

where $g(k,t)$ becomes the minimum accumulated distance from $(1,1)$ to (k,t). The parameter Δ specifies the maximum skipped frames on the path. In the later experiment, we set $\Delta = 1,000$ and it means that we allow about 40-s skip for videos with 24 fps. The computational complexity of the DTW algorithm is $O(\Delta T K)$.

It should be emphasized that this lyric-frame matching process with the lyric information is mandatory for lyric videos. For example, the word "the" will appear in the lyric text many times; this means that there is a large ambiguity of the spatio-temporal location of a certain "the". We, therefore, need to fully utilize the sequential nature of not only video frames but also lyric words by the lyric-frame matching process for determining the most confident frame for each lyric word.

4.3   Tracking of Individual Lyric Words

Although the kth lyric word is matched only to a single frame t by the above lyric-frame matching step, the word will also appear in the neighboring frames of t. We, therefore, search those frames around the tth frame, as shown in Fig. 4 (b). This search is simply done by evaluating not only spatio-temporal similarity but also word similarity to the kth word, at the neighboring frames of t. If both similarities are larger than thresholds at t', we assume the same kth word is also found at t'.

Finally, an interpolation process, shown in Fig. 4 (c), is performed for completing the spatio-temporal tracking process of each lyric word. By the above simple searching process, the lyric word will be missed at a certain frame when a severe misrecognition and/or occlusion occurs on the word at the frame. In the example of Fig. 4, the third lyric word "THOUGHT" is determined at the fifth frame by the lyric-frame matching process and then found at the third and sixth frames by the search. However, it was not found at the fourth frame due to a severe misrecognition of the word. If such a missed frame is found, the polynomial interpolation process determines the location of the lyric word at the frame. Figure 4 (d) shows the final result of the tracking process for two lyric words "you" and "ever."

434   S. Sakaguchi et al.

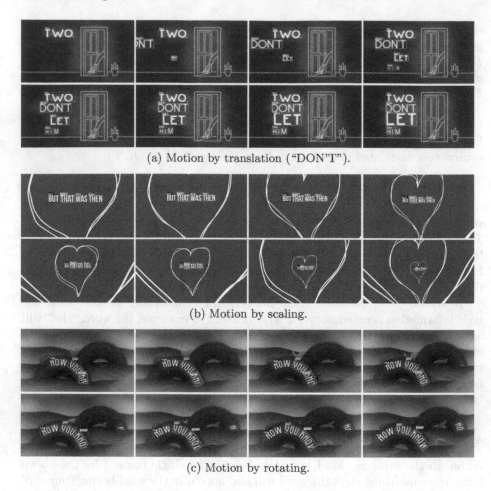

(a) Motion by translation ("DON'T").

(b) Motion by scaling.

(c) Motion by rotating.

Fig. 5. Successful results of lyric word detection and tracking under variable motion types.

5 Experimental Results

By applying the proposed method to all frames of the collected 100 lyric videos (about 547,100 frames in total), we had the tracking results for all lyric words (about 33,800 words in total).

5.1 Qualitative Evaluation

Figure 5 shows several successful results of lyric word detection and tracking. In (a), the word "DON'T", which moves horizontally, is tracked successfully. The words "LET" and "HIM" in (a) and all the words in (b) are tracked correctly, although their size varies. The words in (c) are also tracked successfully even under frame-by-frame rotation.

Fig. 6. The effect of lyric information. Top: The initial word detection results shown as green boxes. Bottom: The final tracking result shown as blue boxes. The lyric words in those frames are "well be alright this time". (Color figure online)

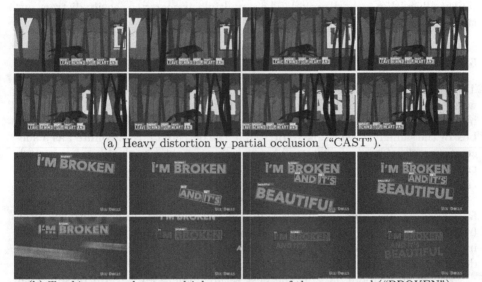

(a) Heavy distortion by partial occlusion ("CAST").

(b) Tracking error due to multiple appearances of the same word ("BROKEN").

Fig. 7. Failure results of lyric word detection and tracking.

Figure 6 confirms the effect of using lyric information at the lyric-frame matching process and the following tracking process for accuracy improvement. Since these frames have a complex background (with character-like patterns), unnecessary bounding boxes are found by the initial word detection step. In contrast, only the correct lyric words remain after lyric-frame matching and tracking.

Figure 7 shows typical failure cases. The failure of (a) was caused by a heavy distortion on a word by an elaborated visual design of the video. Specifically, the word "CAST" is always partially occluded and never detected and recognized correctly over frames. Lyric videos frequently show words which are difficult to read even by the state-of-the-art word detectors and recognizers (and also

Table 1. Performance of the lyric word detection and tracking. MA: Lyric-frame matching. TR: Tracking. IN: Interpolation. TP: #True-positive. FP: #False-positive. FN: #False-negative. P: Precision (%). R: Recall (%). F: F-measure.

MA	TR	IN	TP	FP	FN	$P = \frac{TP}{TP+FP}$	$R= \frac{TP}{TP+FN}$	$F= \frac{2PR}{P+R}$
✓			72	12	7,698	85.71	0.93	0.0183
✓	✓		5,513	462	2,257	92.27	70.95	0.8022
✓	✓	✓	5,547	550	2,223	90.98	71.39	0.8000

by a human). The failure of (b) is caused by a refrain of the same word in the lyrics. Especially in musical lyrics, it is very frequent that the same word appears repeatedly within a short period. These multiple appearances easily distract our tracking process.

5.2   Quantitative Evaluation

Table 1 shows the quantitative evaluation result of the performance of the lyric word detection and tracking, by using 1,000 ground-truthed frames. If a bounding box of a lyric word by the proposed method and the ground-truth bounding box of the same lyric word have IoU > 0.5, the detected box is counted as a successful result. If multiple detected boxes have an overlap with the same ground-truth box with IoU > 0.5, the detected box with the highest IoU is counted as a true-positive and the remaining are false-positives.

The evaluation result shows that the precision is 90.98% and therefore the false positives are suppressed successfully by the proposed method with the lyric-frame matching step and the later tracking step. The interpolation step could increase TPs as expected, but, unfortunately, also increases FPs and slightly degrades the precision value.

The recall is about 71% at the best setup[4]. There are several reasons that degrade recall. First, as shown in Fig. 7 (a), the elaborated visual design often disturbs the word detection and recognition process and degrades the recall. In other words, we need to develop new text detectors and recognizers that are robust to the visual disturbances to the words in lyric videos as future work.

The second reason for the low recall is the errors at the lyric-frame matching step. Since our lyric-frame matching allows the skip of $\Delta = 1,000$ frames (about 40 s) at maximum, to deal with frames without lyrics (interlude). This sometimes causes a wrong skip of the frames where lyric words actually appear and induce errors in the matching step. Multiple appearances of the same word (such as "I") also induce matching errors.

The third reason is the deviation from the official lyrics; even in the official lyric videos, the lyrics shown in video frames are sometimes different from the official lyrics due to, for example, the improvisation of the singer. Since we believe

---

[4] As explained before, only a single frame is determined for each lyric word by the lyric-frame matching. This results in very low recall (0.93%).

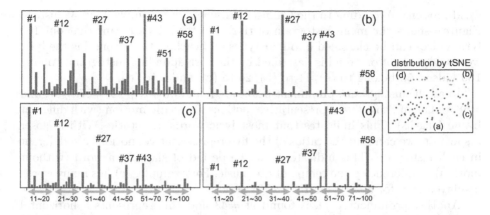

Fig. 8. The mean word motion histograms of four video clusters. Each of 60 bins corresponds to a representative word motion. "11~20" means the 10 representative motions for the trajectories with 11~20 frame length. At the right side, the distribution of the 100 histograms of the individual videos is visualized by tSNE.

the official lyrics for lyric-frame matching, such deviations disturb the matching process and cause the degradation of recall.

5.3 Text Motion Analysis by Clustering

Finally, we analyze the motion trajectory of lyric words. As the result of the proposed lyric word detection and tracking process, we already have the motion trajectory of individual lyric words. Trend analysis of those trajectories will give a hint about how to design lyric videos. Note that we consider that the trajectories subjected to this analysis are reliable enough because the quantitative evaluation in Sect. 5.2 shows a high precision.

As the motion trajectory analysis, we first determine representative motion trajectories via simple clustering. Before the clustering, we divide them into six subsets by their length (11~20 frames (4336 trajectories), 21~30 (4293), 31~40 (3797), 41~50 (3159), 51~70 (4057), 71~100 (2660)). Note that 6123 extremely short ($\leq$10) and 2469 extremely long (>100) trajectories are excluded from the analysis. Then for each subset, we performed k-medoid clustering with DTW distances between trajectories. We set $k = 10$ and thus we have 60 representative motions.

Using those 60 representative motions, we create a histogram with 60 bins for each lyric video. Specifically, like a bag-of-words, all word trajectories in the video are classified into one of 60 representative motions and then we have a histogram whose qth bin represents the number of the lyric words whose motion is the most similar to the qth representative motion. All histograms are then normalized to have the same number of votes.

After this histogram-based representation of all videos, we perform another clustering (k-means) in order to classify 100 videos into clusters with similar word motions. According to the Calinski-Harabasz criterion, we have 4 clusters. Figure 8 shows the mean histogram of those clusters. This means that our 100 lyric videos can be classified into four types and each of them contains the lyric words whose motion trend is described by the corresponding histogram. Among 100 videos, 55 belong to (a), 11 to (b), 25 to (c), and 9 to (d).

In all the histograms, prominent peaks are found at #1, #12, #27, #37, and #58 and all of those five representative motions are static motion (with different frame lengths). This indicates that most lyric words are static. With a closer inspection, we can find the ratios of the five representative motions are different in each histogram. This indicates that the period of showing a word (without motion) is different by the tempo of the music. For example, (c) has many words displayed only for a very short (11~20) period.

Another prominent peak is found at #43 and this represents a horizontal motion. It is interesting to note that a peak at horizontal motions is only found at 51~70; this means that speed of the frequent horizontal motions is almost constant regardless of lyric videos (and also regardless of their tempo). Another observation is that a video that belongs to (a) will contain wide motion varieties. For example, the #51 representative motion is a circular motion.

6   Conclusion and Future Work

To the authors' best knowledge, this paper is the first trial of analyzing lyric videos as novel and dynamic documents. For this difficult analysis task, we developed a novel technique, called lyric-frame matching, where the temporal location, i.e., the frame, of each word in the lyrics is determined automatically by dynamic programming-based optimization. We experimentally showed that the combination of the lyric-frame matching technique and several state-of-the-art word detectors and recognizers could detect lyric words with more than 90% precision and 70% recall on our original 100 official lyric video dataset. Although the recall is lower than the precision, the current tracking performance is already reliable for analyzing the motion trajectories of lyric words for understanding the text motion design in lyric videos. In fact, we could determine four typical motion patterns of lyric videos for our 100 videos.

Since this is the first trial of lyric video analysis, we have multiple future works. First, word detection and recognition performance should be improved for lyric videos. Since the distortions (including elaborated visual designs) on lyric texts are often very different from those of scene texts. This means it is still insufficient to apply state-of-the-art scene text detectors and recognizers to this task. The second and more important future work is to analyze word motion trajectories in lyric videos. Since the moving words in lyric videos are a new target of document analysis, we need to develop many analysis schemes. Word motion analysis for each music genre is straightforward. A more important analysis scheme is the correlation analysis between word motion and music signals, i.e., sounds. We often observe that word motions are strongly affected by the beat and mood of the music. Our final goal is to generate lyric videos automatically or semi-automatically from lyric information and music signals, based on those technologies and analyses. For example, our motion trajectory analysis could contribute to enabling data-driven motion styles on a tool for creating lyric videos [27].

Acknowledgment. This work was supported by JSPS KAKENHI Grant Number JP17H06100.

Videos Shown in the Figures

In the figures of this paper, the following video frames are shown. For URL, the common prefix part "https://www.youtube.com/watch?v=" is omitted in the list. Note that the URL list of all 100 videos and their annotation data can be found at https://github.com/uchidalab/Lyric-Video.

- Fig. 1: (a) MK, 17, `NoBAfjvhj7o`; (b) Demi Lovato, Really Don't Care, `EOEeN9NmyU8`; (c) Alok, Felix Jaehn & The Vamps, All The Lies, `oc218bqEbAA`; (d) Taylor Swift, Look What You Made Me Do, `3KORzZGpyds`;
- Fig. 2: Robin Schulz, Sugar, `jPW5A_JyXCY`.
- Fig. 3: (a) Marshmello x Kane Brown, One Thing Right, `O6RyKbcpBfw`; (b) Harris J, I Promise, `PxN6X6FevFw`; (c) Rita Ora, Your Song, `i95N1b7kiPo`; (d) Green Day, Too Dumb to Die, `qh7QJ_jLam0`;
- Fig. 4: Freya Ridings, Castles, `pL32uHAiHgU`.
- Fig. 5: (a) Dua Lipa, New Rules, `AyWsHs5QdiY`; (b) Anne-Marie, Then, `x9OJpU7O_cU`; (c) Imagine Dragons, Bad Liar, `uEDhGX-UTeI`;
- Fig. 6: Ed Sheeran, Perfect, `iKzRIweSBLA`;
- Fig. 7: (a) Imagine Dragons, Natural, `V5M2WZiAy6k`; (b) Kelly Clarkson, Broken & Beautiful, `618gyacUq4w`;

References

1. Yin, X.C., Zuo, Z.Y., Tian, S., Liu, C.L.: Text detection, tracking and recognition in video: a comprehensive survey. In: IEEE TIP (2016)
2. Bertini, M., Del Bimbo, A., Nunziati, W.: Automatic detection of player's identity in soccer videos using faces and text cues. In: ACM Multimedia (2006)
3. Liu, X., Meng, G., Pan, C.: Scene text detection and recognition with advances in deep learning: a survey. Int. J. Doc. Anal. Recogn. (IJDAR) **22**(2), 143–162 (2019). https://doi.org/10.1007/s10032-019-00320-5
4. Zhao, X., Lin, K.H., Fu, Y., et al.: Text from corners: a novel approach to detect text and caption in videos. In: IEEE TIP (2010)
5. Zhong, Y., Zhang, H., Jain, A.K.: Automatic caption localization in compressed video. In: IEEE TPAMI (2000)
6. Yang, Z., Shi, P.: Caption detection and text recognition in news video. In: International Congress on Image Signal Processing (2012)
7. Lu, T., Palaiahnakote, S., Tan, C.L., Liu, W.: Video caption detection. Video Text Detection. ACVPR, pp. 49–80. Springer, London (2014). https://doi.org/10.1007/978-1-4471-6515-6_3

8. Yang, H., Quehl, B., Sack, H.: A framework for improved video text detection and recognition. Multimedia Tools Appl. **69**(1), 217–245 (2012). https://doi.org/10.1007/s11042-012-1250-6

9. Zhong, D., Shi, P., Pan, D., Sha, Y.: The recognition of Chinese caption text in news video using convolutional neural network. In: IEEE IMCEC (2016)

10. Zedan, I.A., Elsayed, K.M., Emary, E.: Caption detection, localization and type recognition in Arabic news video. In: INFOS (2016)

11. Chen, L.H., Su, C.W.: Video caption extraction using spatio-temporal slices. Int. J. Image Graph. (2018)

12. Xu, Y., et al.: End-to-end subtitle detection and recognition for videos in East Asian languages via CNN ensemble. Sig. Process. Image Commun. **60**, 131–143 (2018)

13. Lu, W., Sun, H., Chu, J., Huang, X., Yu, J.: A novel approach for video text detection and recognition based on a corner response feature map and transferred deep convolutional neural network. IEEE Access **6**, 40198–40211 (2018)

14. Qian, X., Liu, G., Wang, H., Su, R.: Text detection, localization, and tracking in compressed video. Sig. Process. Image Commun. **22**, 752–768 (2007)

15. Nguyen, P.X., Wang, K., Belongie, S.: Video text detection and recognition: dataset and benchmark. In: WACV (2014)

16. Gómez, L., Karatzas, D.: MSER-based real-time text detection and tracking. In: ICPR (2014)

17. Wu, L., Shivakumara, P., Lu, T., Tan, C.L.: A new technique for multi-oriented scene text line detection and tracking in video. IEEE Trans. Multimedia **17**, 1137–1152 (2015)

18. Tian, S., Yin, X.C., Su, Y., Hao, H.W.: A unified framework for tracking based text detection and recognition from web videos. In: IEEE TPAMI (2017)

19. Yang, C., Yin, X.C., Pei, W.Y., et al.: Tracking based multi-orientation scene text detection: a unified framework with dynamic programming. In: IEEE TIP (2017)

20. Pei, W.Y., et al.: Scene video text tracking with graph matching. IEEE Access **6**, 19419–19426 (2018)

21. Wang, Y., Wang, L., Su, F.: A robust approach for scene text detection and tracking in video. In: Hong, R., Cheng, W.-H., Yamasaki, T., Wang, M., Ngo, C.-W. (eds.) PCM 2018. LNCS, vol. 11166, pp. 303–314. Springer, Cham (2018). https://doi.org/10.1007/978-3-030-00764-5_28

22. Wang, Y., Wang, L., Su, F., Shi, J.: Video text detection with fully convolutional network and tracking. In: ICME (2019)

23. Srivastava, N., Mansimov, E., Salakhutdinov, R.: Unsupervised learning of video representations using LSTMs. In: ICML (2015)

24. Wang, W., Xie, E., Li, X., et al.: Shape robust text detection with progressive scale expansion network. In: CVPR (2019)

25. Baek, Y., Lee, B., Han, D., Yun, S., Lee, H.: Character region awareness for text detection. In: CVPR (2019)

26. Baek, J., Kim, G., Lee, J., et al.: What is wrong with scene text recognition model comparisons? dataset and model analysis. In: ICCV (2019)

27. Kato, J., Nakano, T., Goto, M.: TextAlive: integrated design environment for kinetic typography. In: ACM CHI (2015)

Fast and Lightweight Text Line Detection on Historical Documents

Aleksei Melnikov[1] and Ivan Zagaynov[1,2]

[1] R&D Department, ABBYY Production LLC, Moscow, Russia
{aleksei.melnikov,ivan.zagaynov}@abbyy.com
[2] Moscow Institute of Physics and Technology (National Research University),
Moscow, Russia

Abstract. We introduce a novel method for handwriting text line detection, which provides a balance between accuracy and computational efficiency and is suitable for mobile and embedded devices. We propose a lightweight convolutional neural network with only 22K trainable parameters which consists of two parts: downsampling module and context aggregation module (CAM). CAM uses stacked dilated convolutions to provide large receptive field for output features. The downsampling module allows a further reduction of latency. Our network has two output channels, one for baseline prediction and the other for detection of text lines without ascenders/descenders. This modification has a positive impact on accuracy. The proposed method achieves an F-measure of 0.873 on the ICDAR 2017 cBAD complex track with a network inference time of 8 ms on NVIDIA GeForce GTX 1080Ti card and another 12 ms for subsequent postprocessing operations on CPU, which is significantly faster than other existing methods with comparable accuracy.

Keywords: Text detection · Baseline detection · Historical document analysis · CNN · Lightweight network · Low-complexity network

1 Introduction

Text line detection is an important step of the document analysis pipeline. Handwritten document analysis is still a challenging task due to the irregular structure and high variability of text patterns. Working with historical documents makes this task even more difficult as input images may be physically corrupted. Moreover, input documents often contain some geometrical distortions and color artifacts.

In the last decade, deep-learning-based approaches became a general trend for solving document analysis problems. Apart from historical document analysis, where processing time is not very important, there are many real-world scenarios that require handwritten documents to be processed on the fly. State-of-the-art approaches often have high computational and memory requirements, which limits their use on hardware-constrained mobile devices.

X. Bai et al. (Eds.): DAS 2020, LNCS 12116, pp. 441–450, 2020.
https://doi.org/10.1007/978-3-030-57058-3_31

In this paper, we propose a novel method for text line detection that utilizes a deep convolutional network, aiming to achieve decent performance for handwritten text line segmentation task at low computational costs and with a minimal number of trainable parameters. Our network was trained for text baseline detection on handwritten documents and achieved a competitive accuracy, proving itself faster than other methods.

2 Related Work

In recent years, Convolutional Neural Networks (CNNs) have made noticeable improvements in document analysis and currently outperform most of the traditional methods. CNNs achieved best performance on the ICDAR2017 Competition on Baseline Detection in Archival Documents (cBAD). Previous methods of text line detection mainly used handcrafted heuristics, such as projection histograms, smearing based on morphology operations or Gaussian blur, and feature grouping methods. A survey of line detection methods on historical documents can be found in [7].

The approach proposed by Moysser et al. [9] utilizes recurrent neural networks and connectionist temporal classification. This approach is limited to vertical text columns but does not require marked up line boundaries for training: only the number of lines is required for each paragraph.

Pixel-level classification approaches have become very popular over the past few years. Approaches [1,3,4,6,10] utilize U-Net [12] as a segmentation network. These methods use similar network architectures, the main difference among them being postprocessing and additional outputs, such as page zone, vertical table separators, and the end and the beginning of text lines.

Dilated (or atrous) convolutions expand the receptive field of the network without reducing the resolution of the feature map. Since the first version of DeepLab [2], many state-of-the-art neural networks for semantic segmentation have utilized dilated convolutions. Renton et al. [11] proposed a baseline segmentation method based on a fully convolutional network, which used dilated convolutions to detect the center of a text line without ascenders and descenders. The network architecture replicated the first layers of VGG16 [13] network, with dilated convolutions used instead of pooling layers, maintaining the same receptive fields.

Zharkov and Zagaynov [14] proposed a neural network for real-time barcode detection suitable for low-cost mobile devices. This CNN has 3 convolutional layers in the downscale module followed by a context aggregation module (CAM). CAM uses stacked dilated convolutions with an increasing dilation ratio to provide a large receptive field for output features. Because of its low computational cost and applicability to detect non-overlapping objects, we have based our network on this model.

3   Algorithm

The architecture of the proposed network follows [14] with several modifications. We replaced the separable convolutions in the first layers with their vanilla versions, substituting the first two of them with a single layer of 5×5 kernel size, and removing the layer with the dilation rate of 16. The number of filters in layers 1, 5, 6, and 7 was adjusted to 16, as our ablation study (Sect. 4.3) shows that these changes do not decrease performance of the network for text line detection (see Table 2). Our small model is very compact and has only 22,458 trainable parameters.

We also propose an expanded model with residual connections and additional layers.

The ReLU activation function was used for all the convolutional layers except the last one, where the sigmoid function was utilized. Our network has two filters in the output layer, one for baseline prediction and the other one for detection of lines without ascenders/descenders. To get annotations for text lines, we rasterized and filled polygons containing text lines (coords points), then we downscaled the rasterized polygons to the resolution of the segmentation maps. Our experiments demonstrate that this modification results in greater accuracy (see Table 2).

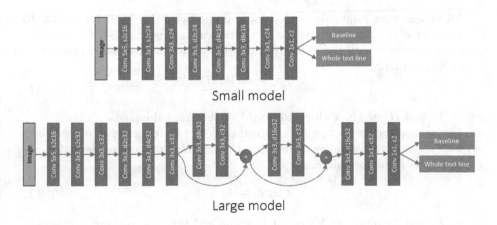

Fig. 1. Architecture of our proposed networks. c – number of filters, d – dilation rate, s – stride.

As overlapping or touching of text baselines is uncommon, one can view baseline detection as a semantic segmentation problem. To speed up inference, low-resolution segmentation maps are used. Network outputs are binarized using a constant threshold value to denote pixels as either a baseline or the background.

Postprocessing is done by extracting all 8-connected components, followed by fitting them into polylines and discarding those whose length is less than a predefined threshold (T_{len}) multiplied by the length of the largest image side. Baselines are smoothed using moving average filter.

Downscaling of the groundtruth segmentation map was performed via nearest-neighbor interpolation, which leads to a systematic error in baseline alignment. To cope with this problem, after upscaling the output segmentation map to the original image size, it is shifted by 6 pixels downward the vertical axis for both Track A and Track B. The choice for this constant was based only on the results for the training parts of the dataset.

4   Experiments

4.1   Training

The model was trained on an Intel i5-8600 with GTX 1080Ti card using Keras with the Tensorflow backend. Training parameters are listed below.

1. Adam optimizer with an initial learning rate of 10^{-3} and zero decay.
2. *ReduceLROnPlateau* callback with $factor = 0.1$, $patience = 20$, $min\_lr = 10^{-5}$ for learning rate scheduling.
3. Mini-batch size 1 to simplify the preprocessing of the training data and to avoid image cropping or border padding effects.

All images were randomly resized for training by setting their larger side to the range of [896, 1152] pixels, the aspect ratio was kept untouched. The training data baselines were assigned to the training segmentation mask as polylines with a width of w pixels.

$$w = 1.5\frac{max(W, H)}{256},\tag{1}$$

where W and H are the width and height of the original image.

The proposed network was also trained to detect the core region of text lines without ascenders/descenders to improve the accuracy of baseline detection.

The loss can be formulated as

$$L = L_b + L_t,\tag{2}$$

where L_b and L_t represent losses for baseline and text line without ascenders/descenders respectively. We define both L_b and L_t as the dice loss [8] defined as

$$L_{dice} = 1 - \frac{2\sum r_{x,y}p_{x,y}}{\sum r_{x,y} + \sum p_{x,y} + \epsilon}\tag{3}$$

where $r_{x,y}$ is 0 or 1 groundtruth label for the pixel with coordinates x and y, and $p_{x,y}$ is the predicted pixel probability, $\epsilon = 10^{-5}$ in our experiments.

4.2 Dataset

The proposed method was tested on the ICDAR 2017 cBAD dataset [5]. The dataset contains only historical documents, but, in our opinion, it is also relevant for scenarios of capturing and processing any types of handwritten documents and text notes on the fly using mobile devices. Unfortunately, due to a lack of publicly available datasets, we have not been able to prove this assumption so far. The dataset is organized into two different tracks. Track A (simple documents) contains only documents with simple layouts and annotated text areas (216 documents for training and 539 for testing). Track B (complex documents) consists of documents with complex layouts, containing tables, images, marginalia, and rotated text lines (270 documents for training and 1010 for testing).

The largest side of the images was limited by 1024 pixels for experiments with the proposed network. T_{len} was adjusted to maximize F1-measure on the training dataset ($T_{len} = 0.03$ for Track A, $T_{len} = 0.0175$ for Track B).

Table 1. Detection performance on the ICDAR 2017 cBAD dataset.

	Track A			Track B		
	Precision	Recall	F-measure	Precision	Recall	F-measure
Renton et al. [11]	0.88	0.88	0.88	0.69	0.77	0.73
P2PaLA [10]	–	–	–	0.848	0.854	0.851
Fink et al. [3]	0.973	0.970	0.971	0.854	0.863	0.859
dhSegment [1]	0.88	0.97	0.92	0.79	**0.95**	0.86
Guerry et al. [6]	–	–	–	0.858	0.935	0.895
ARU-Net [4]	**0.98**	**0.98**	**0.978**	**0.93**	0.92	**0.922**
Ours (small)	0.910	0.956	0.932	0.845	0.903	0.873
Ours (large)	0.943	0.956	0.949	0.888	0.923	0.905

Results of comparison with other methods are shown in Table 1. The proposed method shows performance that is competitive with U-Net based methods. Examples of baseline detection are shown in Fig. 2. Our method has some known issues, such as detection of bleed-through text, missing short lines, missing lines with an unusual font style, joining of lines located too close to each other, and detection of remote ascenders/descenders as separate lines.

4.3 Ablation Study

We conducted an ablation study on the ICDAR 2017 cBAD dataset, as summarized in Table 2. The initial network structure is similar to the one defined in Fig. 1 (small model), but it has 24 filters in layers 1–7, 3 × 3 convolutions in layer 1, and one output class baseline/background. An additional output for the

Fig. 2. Samples of baseline detection on ICDAR 2017 cBAD test set for small model. The ground-truth and predicted baselines are shown in red and green respectively. (Color figure online)

detection of text lines without ascenders/descenders yielded an 1.8% improvement for F-measure on Track B. Training on randomly resized images further improves the F-measure by 0.7% on Track A, and by 1.1% on Track B. Reducing the number of filters in layers 1 and 5–7 to 16 gave us a 16 ms gain in speed with a minor drop in detection performance. Replacing 3×3 convolutions of the first layer with 5×5 convolutions increased the F-measure by 1% on Track A and by 0.6% on Track B.

Experiments were also conducted with an increased input resolution, an additional convolutional layer, and an increased number of filters to demonstrate that a further increase in the size of the model has only negligible impact on accuracy. Further increase in input resolution (i.e., resizing the largest side to 1280 pixels) resulted in a significant decrease in precision. We presume that our network cannot aggregate enough context on increased distances. The addition of a layer with a dilation rate of 16 after layer 7 did not improve performance. Setting the number of filters in layers 1–7 to 32 yielded a 0.7% improvement for F-measure on Track A and Track B.

Our large model (see Fig. 1) gives 1.7%/2% improvement for F-measure over the small model on Tracks A and B when the same input resolution is used. Further increasing input resolution to 1536 pixels boosts F-measure by 0.7% on Track B.

Table 2. Ablation study on the ICDAR 2017 cBAD dataset. **P**, **R** and **F** represent the precision, recall, and F-measure respectively. The latency time was measured on Intel i5-8600 CPU. Selected models (small and large) are shown in bold.

Experiment	Track A			Track B			Latency
	P	R	F	P	R	F	ms
24 filters in layers 1–7 (I)	0.885	0.961	0.921	0.790	0.917	0.848	90
I+Additional output (II)	0.898	0.949	0.922	0.842	0.892	0.866	90
II+Random resize training (III)	0.906	0.953	0.929	0.850	0.906	0.877	90
III+16 filters in layers 1, 5–7 (IV)	0.897	0.948	0.922	0.831	0.905	0.867	74
IV+5 × 5 conv. in layer 1 (V)	**0.910**	**0.956**	**0.932**	**0.845**	**0.903**	**0.873**	**83**
V+1280 px input on largest side	0.884	0.955	0.918	0.834	0.908	0.870	125
V+Layer with dilation rate 16	0.913	0.956	0.934	0.847	0.900	0.873	90
V+32 filters in layers 1–7	0.926	0.957	0.941	0.858	0.904	0.880	127
Large model	0.903	0.950	0.947	0.887	0.899	0.893	199
Large model+1536 px input	**0.943**	**0.956**	**0.949**	**0.888**	**0.923**	**0.905**	446

4.4 Execution Times

Computational complexity is an important factor when it comes to the practical use of any algorithm. We compared the inference time of the proposed method with that of the other methods using their provided implementations. The experiment was conducted on 100 images randomly selected from the ICDAR 2017 cBAD Track B training subset.

448   A. Melnikov and I. Zagaynov

For a fair comparison with the other methods, the image sizes were set based on the respective papers:

1. 1024 px on the largest side, preserving the aspect ratio for the proposed method
2. 10^6 pixels for P2PaLA
3. 1024×768 resolution for dhSegment
4. for ARU-Net, images were downscaled by:
   (a) factor 2 for maximum side length $l_{max} < 2000$
   (b) factor 3 for $2000 \leq l_{max} < 4800$
   (c) factor 4 for $l_{max} \geq 4800$

Table 3 lists the execution times for the proposed method and the other approaches. Execution time was measured separately for network inference and for the processing of the network output.

Table 3. Speed of the proposed method compared against the other methods with published source code. The ARU-Net implementation contains no postprocessing code.

	Intel i5-8600 + GTX 1080Ti		Intel i5-8600	
	Network, ms	Postprocessing, ms	Network, ms	Postprocessing, ms
P2PaLA[a] [10]	47	3750	1180	3755
dhSegment[b] [1]	75	12	2937	12
ARU-Net[c] [4]	57	–	1875	–
Ours (small)	8	12	83	12
Ours (large)	20	18	446	18

[a] https://github.com/lquirosd/P2PaLA
[b] https://github.com/dhlab-epfl/dhSegment
[c] https://github.com/TobiasGruening/ARU-Net

5 Conclusion

In this paper, we propose a new method of baseline detection for handwritten text lines on historical documents. We have demonstrated that the proposed approach has competitive accuracy with some state-of-the-art methods while being several times faster. The proposed architecture for the neural network is very compact (with only 22,458 trainable parameters) and thus may be used in similar tasks on mobile and embedded devices. The proposed method is also suitable for the text line detection on printed documents images captured by mobile devices. Preliminary tests have shown promising results for this scenario.

Further research may study the application of more sophisticated postprocessing techniques, multi-task learning with auxillary data, and use of neural architecture search (NAS) approaches to achieve even better results for text line detection.

References

1. Ares Oliveira, S., Seguin, B., Kaplan, F.: dhSegment: a generic deep-learning approach for document segmentation. In: 2018 16th International Conference on Frontiers in Handwriting Recognition (ICFHR), pp. 7–12, August 2018. https://doi.org/10.1109/ICFHR-2018.2018.00011

2. Chen, L.C., Papandreou, G., Kokkinos, I., Murphy, K., Yuille, A.L.: DeepLab: semantic image segmentation with deep convolutional nets, atrous convolution, and fully connected CRFs. IEEE Trans. Pattern Anal. Mach. Intell. **40**(4), 834–848 (2017)

3. Fink, M., Layer, T., Mackenbrock, G., Sprinzl, M.: Baseline detection in historical documents using convolutional U-nets. In: 2018 13th IAPR International Workshop on Document Analysis Systems (DAS), pp. 37–42, April 2018. https://doi.org/10.1109/DAS.2018.34

4. Grüning, T., Leifert, G., Strauß, T., Michael, J., Labahn, R.: A two-stage method for text line detection in historical documents. Int. J. Doc. Anal. Recogn. (IJDAR) **22**(3), 285–302 (2019). https://doi.org/10.1007/s10032-019-00332-1

5. Grüning, T., Labahn, R., Diem, M., Kleber, F., Fiel, S.: Read-bad: a new dataset and evaluation scheme for baseline detection in archival documents. In: 2018 13th IAPR International Workshop on Document Analysis Systems (DAS), pp. 351–356, April 2018. https://doi.org/10.1109/DAS.2018.38

6. Guerry, C., Coüasnon, B.B., Lemaitre, A.: Combination of deep learning and syntactical approaches for the interpretation of interactions between text-lines and tabular structures in handwritten documents. In: 15th International Conference on Document Analysis and Recognition (ICDAR), Sydney, Australia, September 2019. https://hal.archives-ouvertes.fr/hal-02303293

7. Likforman-Sulem, L., Zahour, A., Taconet, B.: Text line segmentation of historical documents: a survey. Int. J. Doc. Anal. Recogn. (IJDAR) **9**(2–4), 123–138 (2007)

8. Milletari, F., Navab, N., Ahmadi, S.: V-Net: fully convolutional neural networks for volumetric medical image segmentation. CoRR abs/1606.04797 (2016). http://arxiv.org/abs/1606.04797

9. Moysset, B., Kermorvant, C., Wolf, C., Louradour, J.: Paragraph text segmentation into lines with recurrent neural networks. In: 2015 13th International Conference on Document Analysis and Recognition (ICDAR), pp. 456–460, August 2015. https://doi.org/10.1109/ICDAR.2015.7333803

10. Quirós, L.: Multi-task handwritten document layout analysis. CoRR abs/1806.08852 (2018). http://arxiv.org/abs/1806.08852

11. Renton, G., Soullard, Y., Chatelain, C., Adam, S., Kermorvant, C., Paquet, T.: Fully convolutional network with dilated convolutions for handwritten text line segmentation. Int. J. Doc. Anal. Recogn. (IJDAR) **21**(3), 177–186 (2018). https://doi.org/10.1007/s10032-018-0304-3

12. Ronneberger, O., Fischer, P., Brox, T.: U-Net: convolutional networks for biomedical image segmentation. In: Navab, N., Hornegger, J., Wells, W.M., Frangi, A.F. (eds.) MICCAI 2015. LNCS, vol. 9351, pp. 234–241. Springer, Cham (2015). https://doi.org/10.1007/978-3-319-24574-4_28

13. Simonyan, K., Zisserman, A.: Very deep convolutional networks for large-scale image recognition. In: Bengio, Y., LeCun, Y. (eds.) 3rd International Conference on Learning Representations (ICLR 2015), San Diego, CA, USA, 7–9 May 2015, Conference Track Proceedings (2015). http://arxiv.org/abs/1409.1556
14. Zharkov, A., Zagaynov, I.: Universal barcode detector via semantic segmentation. In: 15th International Conference on Document Analysis and Recognition (ICDAR), pp. 837–843, Sydney, Australia, September 2019. https://doi.org/10.1109/ICDAR.2019.00139

From Automatic Keyword Detection
to Ontology-Based Topic Modeling

Marc Beck[1], Syed Tahseen Raza Rizvi[1,2]($\boxtimes$) (iD), Andreas Dengel[1,2] (iD),
and Sheraz Ahmed[2] (iD)

[1] Technische Universität Kaiserslautern, Kaiserslautern, Germany
{m_beck12,s_rizvi14}@cs.uni-kl.de
[2] German Research Center for Artificial Intelligence (DFKI), Kaiserslautern,
Germany
{Syed_Tahseen_Raza.Rizvi,Andreas.Dengel,Sheraz.Ahmed}@dfki.de

Abstract. In this paper, we propose a novel, two-staged system, for
keyword detection and ontology-driven topic modeling. The first stage
specializes in keyword detection in which we introduce a novel graph-
based unsupervised approach called Collective Connectivity-Aware Node
Weight *(CoCoNoW)* for detecting keywords from the scientific literature. *CoCoNoW* builds a connectivity aware graph from a given publication text and eventually assigns weight to the extracted keywords to
sort them in order of relevance. The second stage specializes in topic
modeling, where a domain ontology serves as an attention-map/context
for topic modeling based on the detected keywords. The use of an
ontology makes this approach independent of domain and language.
CoCoNoW is extensively evaluated on three publicly available datasets
Hulth2003, NLM500 and **SemEval2010**. Analysis of results reveals
that *CoCoNoW* consistently outperforms the state-of-the-art approaches
on the respective datasets.

Keywords: Keyword detection · Ontology · Topics · Topic modeling

1 Introduction

Keywords are of significant importance as they carry and represent the essence
of a text collection. Due to the sheer volume of the available textual data, there
has been an increase in demand for reliable keyword detection systems which
can automatically, effectively and efficiently detect the best representative words
from a given text. Automatic keyword detection is a crucial task for various
applications. Some of its renowned applications include information retrieval,
text summarization, and topic detection. In a library environment with thousands or millions of literature artifacts, e.g. books, journals or conference proceedings, automatic keyword detection from each scientific artifact [28] can assist

---

M. Beck, S.T.R. Rizvi—Equal contribution.

X. Bai et al. (Eds.): DAS 2020, LNCS 12116, pp. 451–465, 2020.
https://doi.org/10.1007/978-3-030-57058-3_32

in automatic indexing of scientific literature for the purpose of compiling library catalogs.

In 2014, about 2.5 million scientific articles were published in journals across the world [39]. This increased to more than 3 million articles published in 2018 [15]. It is certainly impractical to manually link huge volumes of scientific publications with appropriate representative keywords. Therefore, a system is imminent which can automatically analyze and index scientific articles. There has been quite a lot of research on the topic of automated keyword detection, however most of the approaches deal with social media like tweets [4,6,7,9,10,12,13,22,26,28,32,33].

A popular approach for keyword detection is representing text as an undirected graph $G = (N, E)$, where the nodes N in graph G correspond to the individual terms in the text and the edges E correspond to the relation between these terms. The most popular relation is term co-occurrence, i.e. an edge is added to the graph between nodes n_1 and n_2 if both corresponding terms co-occur within a given sliding window. The recommended window size depends on the selected approach and often lies in the range between 2 and 10 [19,24,34]. Duari and Bhatnagar [12] note that the window size w has a strong influence on the properties of the resulting graph. With the increase in w, the density also increases while the average path length between any two nodes decreases.

The assumption behind this sliding window is that the words appearing closer together have some potential relationship [34]. There are several variations of the sliding window, e.g. letting the window slide over individual sentences rather than the entire text and stopping at certain punctuation marks [19]. Duari and Bhatnagar [12] proposed a new concept named Connectivity Aware Graph (CAG): Instead of using a fixed window size, they use a dynamic window size that always spans two consecutive sentences. They argue that consecutive sentences are related to one another. This is closely related to the concept of *pragmatics* i.e. transmission of meaning depending on the context, which is extensively studied in linguistics [11,16,23]. In their experiments, they showed that the performance of approaches generally increases when they use CAGs instead of graphs built using traditional window sizes.

The first stage consists of a novel unsupervised keyword detection approach called Collective Connectivity-Aware Node Weight (CoCoNoW). Our proposed approach essentially combines the concepts of Collective Node Weight [6], Connectivity Aware Graphs (CAGs) [12] and Positional Weight [13] to identify, estimate and sort keywords based on their respective weights. We evaluated our approach on three different publicly available datasets containing scientific publications on various topics and with different lengths. The results show that CoCoNoW outperforms other state-of-the-art keyword detection approaches consistently across all three data sets. In the second stage, detected keywords are used in combination with the Computer Science Ontology CSO 3.1[1] [35] to identify topics for individual publications.

The contributions of this paper are as follows:

---

[1] https://cso.kmi.open.ac.uk, accessed Dec-2019.

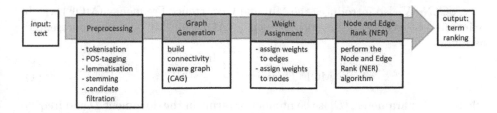

Fig. 1. An overview of Stage 1 (CoCoNoW) for automatic keyword detection

- We present a novel graph-based keyword detection approach that identifies representative words from a given text and assigns weights to rank them in the order of relevance.
- We also evaluated our proposed approach on three different publicly available datasets and consistently outperformed all other existing approaches.
- In this paper, we also complement our keyword detection system with ontology-based topic modeling to identify topics from a given publication.

The rest of the paper is structured as follows: Sect. 2 describes the methodology of the CoCoNoW approach and topic modeling. The performance of the keyword detection and topic modeling is evaluated in Sect. 3. Finally, the presented work is concluded in Sect. 4.

2   Methodology

This paper proposes a two-staged novel approach in which the first stage deals with automatic keyword detection called Collective Connectivity-Aware Node Weight (CoCoNoW) and in the second stage, the detected keywords are consolidated with the Computer Science Ontology [35] to identify topics for a given scientific publication. In CoCoNoW (Fig. 1), we present a unique fusion of Collective Node Weight [6], Connectivity Aware Graphs (CAGs) [12] and Positional Weight [13] to identify keywords from a given document in order to cluster publications with common topics together. Details of the proposed approach are as follows:

2.1   Stage 1: Automatic Keyword Detection Using CoCoNoW

Preprocessing. CoCoNoW uses the standard preprocessing steps like tokenization, part of speech tagging, lemmatization, stemming and candidate filtration. A predefined list of stop words is used to identify stop words. There are several stop word lists available for the English language. For the sake of a fair evaluation and comparison, we selected the stopword list[2] used by the most recent approach by Duari and Bhatnagar [12]. Additionally, any words with less than three characters are considered stop words and are removed from the text.

---

[2] http://www.lextek.com/manuals/onix/stopwords2.html, accessed Dec 2019.

CoCoNoW also introduces the Minimal Occurrence Frequency (MOF) which is inspired by average occurrence frequency (AOF) [6]. MOF can be represented as follows:

$$\text{MOF}(D,\beta) = \beta \frac{\sum_{t \in D} \text{freq}(t)}{|D|} \tag{1}$$

where β is a parameter, $|D|$ is the number of terms in the document D and freq(t) is the frequency of term t. The MOF supports some variation with the parameter β; a higher β means more words get removed, whereas a lower β means fewer words get removed. This allows customizing the CoCoNoW to the document length: Longer Documents contain more words, therefore, having a higher frequency of terms. Parameter optimization techniques on various datasets suggest that the best values for β are about 0.5 for short documents e.g. only analyzing abstracts of papers rather than the entire text; and 0.8 for longer documents such as entire papers.

Graph Building. CoCoNoW is a graph-based approach, it represents the text as a graph. We performed experiments with various window sizes for CoCoNoW, including different numbers of consecutive sentences for the dynamic window size employed by CAGs [12]. The performance dropped when more than two consecutive sentences were considered in one window. Therefore, a dynamic window size of two consecutive sentences was adopted for CoCoNoW. This means that an edge is added between any two terms if they occur within two consecutive sentences.

Weight Assignments. CoCoNoW is based on the Keyword Extraction using Collective Node Weight (KECNW) model developed by Biswas et al. [6]. The general idea is to assign weights to the nodes and edges that incorporate many different features, such as frequency, centrality, position, and weight of the neighborhood.

Edge Weights. The weight of an edge typically depends on the relationship it represents, in our case this relationship is term co-occurrence. Hence, the weight assigned to the edges is the normalized term co-occurrence $w(e)$, which is computed as follows:

$$w(e) = \frac{\text{coocc}(t_u, t_v)}{\text{maxCoocc}} \tag{2}$$

where the weight $w(e)$ of an edge $e = \{u, v\}$ is obtained by dividing the number of times the corresponding terms t_u and t_v co-occur in a sentence $(coocc(t_u, t_v))$ by the maximum number of times any two terms co-occur in a sentence $(maxCoocc)$. This is essentially a normalization of the term co-occurrence.

Node Weights. The final node weight is a summation of four different features. Two of these features, namely *distance to most central node* and *term frequency* are also used by [6]. In addition, we employed *positional weight* [13] and the newly introduced *summary bonus*. All of these features are explained as follows:

Distance to Most Central Node: Let c be the node with the highest degree. This node is considered the most central node in the graph. Then assign the inverse distance $D_C(v)$ to this node as the weight for all nodes:

$$D_C(v) = \frac{1}{d(c,v)+1} \tag{3}$$

where $d(c,v)$ is the distance between node v and the most central node c.

Term Frequency: The number of times a term occurs in the document divided by the total number of terms in the document:

$$\mathrm{TF}(t) = \frac{\mathrm{freq}(t)}{|D|} \tag{4}$$

where freq(t) is the frequency of term t and $|D|$ is the total number of terms in the document D.

Summary Bonus: Words occurring in summaries of documents, e.g. abstracts of scientific articles, are likely to have a higher importance than words that only occur in rest of the document:

$$\mathrm{SB}(t) = \begin{cases} 0 & \text{if t does not occur in the summary} \\ 1 & \text{if t occurs in the summary} \end{cases} \tag{5}$$

where $\mathrm{SB}(t)$ is the summary bonus for term t. If there is no such summary, the summary bonus is set to 0.

Positional Weight: As proposed by Florescu and Caragea [13], words appearing in the beginning of the document have a higher chance of being important. The positional weight $PW(t)$ is based on this idea and is computed as follows:

$$PW(t) = \sum_j^{\mathrm{freq}(t)} \frac{1}{p_j} \tag{6}$$

where freq(t) is the number of times term t occurs in the document and p_j is the position of the jth occurrence in the text.

Final Weight Computation for CoCoNoW: The final node weight W uses all these features described above and combines them as follows:

$$W(v) = \mathrm{SB}(t_v) + D_C(v) + \mathrm{PW}(t_v) + \mathrm{TF}(t_v) \tag{7}$$

where t_v is the term corresponding to node v, SB is the summary bonus, D_C is the distance to the most central node, PW is the positional weight and TF is the term frequency. All individual summands have been normalized in the following way:

$$\mathrm{norm}(x) = \frac{x - \mathrm{minVal}}{\mathrm{maxVal} - \mathrm{minVal}} \tag{8}$$

where x is a feature for an individual node, minVal is the smallest value of this feature and maxVal is the highest value of this feature. With this normalization, each summand in Eq. 7 lies in the interval $[0, 1]$. Thus, all summands are considered to be equally important.

Node and Edge Rank (NER). Both the assigned node and edge weights are then used to perform Node and Edge Rank (NER) [5]. This is a variation of the famous PageRank [31] and is recursively computed as given below:

$$\text{NER}(v) = (1 - d)W(v) + dW(v) \sum_{e=(u,v)} \frac{w(e)}{\sum_{e'=(u,w)} w(e')} \text{NER}(u) \qquad (9)$$

where d is the damping factor, which regulates the probability of jumping from one node to the next one [6]. The value for d is typically set to 0.85. $W(v)$ is the weight of node v as computed in Eq. 7. $w(e)$ denotes the edge weight of edge e, $\sum_{e'=(u,w)} w(e')$ denotes the summation over all weights of incident edges of an adjacent node u of v and $\text{NER}(u)$ is the Node and Edge Rank of node u.

This recursion stops as soon as the absolute change in the NER value is less than the given threshold of 0.0001. Alternatively, the execution ends as soon as a total of 100 iterations are performed. However, it is just a precaution, as the approach usually converges in about 8 iterations. Mihalcea et al. [24] report that the approach needed about 20 to 30 steps to converge for their dataset. All nodes are then ranked according to their NER. Nodes with high values are more likely to be keywords. Each node corresponds to exactly one term in the document, so the result is a priority list of terms that are considered keywords.

2.2   Stage 2: Topic Modeling

In this section, we will discuss the second stage of our approach. The topic modeling task is increasingly popular on social web data [1,3,27,37], where the topics of interest are unknown beforehand. However, this is not the case for the task in hand, i.e. clustering publications based on their topics. All publications share a common topic, for example, all ICDAR papers have *Document Analysis* as a common topic. Our proposed approach takes advantage of the common topic by incorporating an ontology. In this work, an ontology is used to define the possible topics where the detected keywords of each publication are subsequently mapped onto the topics defined by the ontology. For this task, we processed ICDAR publications from 1993 to 2017. The reason for selecting ICDAR publications for this task is that we already had the citation data available for these publications which will eventually be helpful during the evaluation of this task.

Topic Hierarchy Generation. All ICDAR publications fall under the category of *Document Analysis*. The first step is to find a suitable ontology for the ICDAR publications. For this purpose, the CSO 3.1[3] [35] was employed. This

---

[3] https://cso.kmi.open.ac.uk, accessed Dec-2019.

ontology was built using the Klink-2 approach [29] on the Rexplore dataset [30] which contains about 16 million publications from different research areas in the field of computer science. These research areas are represented as the entities in the ontology. The reason for using this ontology rather than other manually crafted taxonomies is that it was extracted from publications with the latest topics that occur in publications. Furthermore, Salatino et al. [36] used this ontology already for the same task. They proposed an approach for the classification of research topics and used the CSO as a set of available classes. Their approach was based on bi-grams and tri-grams and computes the similarity of these to the nodes in the ontology by leveraging word embeddings from word2vec [25].

Computer Science Ontology. The CSO 3.1 contains $23,800$ nodes and $162,121$ edges. The different relations between these nodes are based on the Simple Knowledge Organization System[4] and include eight different types of relations.

Hierarchy Generation. For this task, we processed ICDAR publications from 1993 to 2017. Therefore, in line with the work of Breaux and Reed [8], the node *Document Analysis* is considered the root node for the ICDAR conference. This will be the root of the resulting hierarchy. Next, nodes are added to this hierarchy depending on their relations in the ontology. All nodes with the relation *superTopicOf* are added as children to the root. This continues recursively until there are no more nodes to add. Afterwards, three relation types *sameAs*, *relatedEquivalent* and *preferentialEquivalent* are used to merge nodes. The edges with these relations between terms describe the same concept, e.g. *optical character recognition* and *ocr*. One topic is selected as the main topic while all merged topics are added in the synonym attribute of that node. Note that all of these phrases are synonyms of essentially the same concept. The extracted keywords are later on matched against these sets of synonyms. Additionally, very abstract topics such as *information retrieval* were removed as they are very abstract and could potentially be a super-topic of most of the topics in the hierarchy thus making the hierarchy unnecessarily large and complicated. Lastly, to mitigate the missing topics of specialized topics like *Japanese Character Recognition*, we explored the official topics of interest for the ICDAR community[5]. An examination of the hierarchy revealed missing specialized topics like the only script dependent topic available in the hierarchy was *Chinese Characters*, so other scripts such as Greek, Japanese and Arabic were added as siblings of this node. We also created a default node labeled *miscellaneous* for all those specialized papers which can not be assigned to any of the available topics.

Eventually, the final topic hierarchy consists of 123 nodes and has 5 levels. The topics closer to the root are more abstract topics while the topics further away from the root node represent more specialized topics.

---

[4] https://www.w3.org/2004/02/skos, accessed Dec-2019.
[5] https://icdar2019.org/call-for-papers, accessed Dec-2019.

Table 1. Distribution details of datasets used

Dataset	|D|	L	Avg/SD	Dataset description
Hulth2003 [14]	1500	129	19.5/9.98	Abstracts
NLM500 [2]	500	4854	23.8/8.19	Full papers
SemEval2010 [17]	244	8085	25.5/6.96	Full papers

Topic Assignment. Topics are assigned to papers by using two features of each paper: The title of the publication and the top 15 extracted keywords. The value of $k = 15$ was chosen after manually inspecting the returned keywords; fewer keywords mean that some essential keywords are ignored, whereas a higher value means there are more unnecessary keywords that might lead to a wrong classification.

In order to assign a paper to a topic, we initialize the matching score with 0. The topics are represented as a set of synonyms, these are compared with the titles and keywords of the paper. If a synonym is a substring of the title, a constant of 200 is added to the score. The assumption is that if the title of a paper contains the name of a topic, then it is more likely to be a good candidate for that topic. Next, if all unigrams from a synonym are returned as keywords, the term frequency of all these unigrams is added to the matching score. By using a matching like this, different synonyms will have a different impact on the overall score depending on how often the individual words occurred in the text. To perform matching we used the Levenshtein distance [18] with a threshold of 1. This is the case to accommodate the potential plural terms. The constant bonus of 200 for a matching title comes from assessing the average document frequency of the terms. Most of the synonyms consist of two unigrams, so the document frequencies of two words are added to the score in case of a match. This is usually less than 200 - so the matching of the title is deemed more important.

Publications are assigned to 2.65 topics on average with a standard deviation of 1.74. However, the values of the assignments differ greatly between publications. The assignment score depends on the term frequency, which itself depends on the individual writing style of the authors. For this reason, the different matching scores are normalized: For each paper, we find the highest matching score, then we divide all matching scores by this highest value. This normalization means that every paper will have one topic that has a matching score of 1 - and the scores of other assigned topics will lie in the interval $(0, 1]$. This accounts for different term frequencies and thus, also the different writing styles.

3 Evaluation

In this section, we will discuss the experimental setup and the evaluation of our system where we firstly discuss the evaluation of our first stage CoCoNoW for keyword detection. The results from CoCoNoW are compared with various state-of-the-art approaches on three different datasets: Hulth2003 [14], SemEval2010 [17] and NLM500 [2]. Afterwards, we will discuss the evaluation results of the second stage for topic modeling as well.

3.1 Experimental Setup

Keyword detection approaches usually return a ranking of individual keywords. Hence, the evaluation is based on individual keywords. For the evaluation of these rankings, a parameter k is introduced where only the top k keywords of the rankings are considered. This is a standard procedure to evaluate performance [12,17,20,22,38].

However, as the gold-standard keywords lists contain key phrases, these lists undergo a few preprocessing steps. Firstly, the words are lemmatized and stemmed, then a set of strings called the evaluation set is created. It contains all unigrams. All keywords occur only once in the set, and the preprocessing steps allow the matching of similar words with different inflections. The top k returned keywords are compared with this evaluation set.

Note that the evaluation set can still contain words that do not occur in the original document, which is why an F-Measure of 100% is infeasible. For example, the highest possible F-Measure for the SemEval2010 dataset is only 81% because 19% of the gold standard keywords do not appear in the corresponding text [17].

Table 1 gives an overview of performance on different datasets. These datasets were chosen because they cover different document lengths, ranging from about

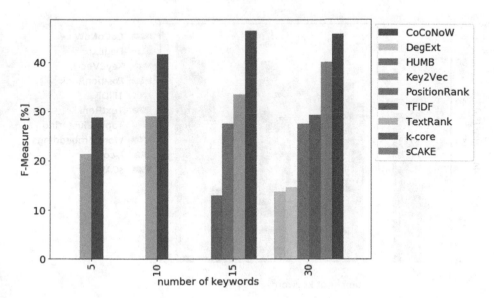

Fig. 2. Evaluation of CoCoNoW on the SemEval2010 [17] dataset.

130 to over 8000 words and belong to different domains: biomedicine, information technology, and engineering.

3.2 Performance Evaluation Stage 1: CoCoNoW

The performance of keyword detection approaches is assessed by matching the top k returned keywords with the set of gold standard keywords. The choice of k influences the performance of all keyword detection approaches as the returned ranking of keywords differs between these approaches. Table 2 shows the performance in terms of Precision, Recall and F-Measure of several approaches for different values of k. Fig. 2 and Fig. 3 compare the performance of our approach with other approaches on the **SemEval2010** [17] dataset and the **Hulth2003** [14] dataset respectively.

By looking at the results of all trials, we can make the following observations:

- CoCoNoW always has the highest F-measure
- CoCoNoW always has the highest Precision
- In the majority of the cases, CoCoNoW also has the highest Recall

For the **Hulth2003** [14] dataset, CoCoNoW achieved the highest F-measure of 57.2% which is about 6.8% more than the previous state-of-the-art. On the **SemEval2010** [17] dataset, CoCoNoW achieved the highest F-measure of 46.8% which is about 6.2% more than the previous state-of-the-art. Lastly, on the **NLM500** [2] dataset, CoCoNoW achieved the highest F-measure of 29.5% which is about 1.2% better than previous best performing approach. For

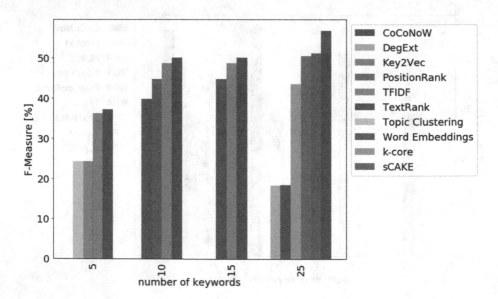

Fig. 3. Evaluation of CoCoNoW on the Hulth2003 [14] dataset.

$k = 5$ CoCoNoW achieved the same Recall as the Key2Vec approach by Mahata et al. [22], however, the Precision was 15.2% higher. For $k = 10$, the approach by Wang et al. [38] had a Recall of 52.8%, whereas CoCoNoW only achieved 41.9%; this results in a difference of 10.9%. However, the Precision of CoCoNoW is almost twice as high, i.e. 73.3% as compared to 38.7%. Lastly, for $k = 25$, the sCAKE algorithm by Duari and Bhatnagar [12] has a higher Recall (0.6% more), but also a lower Precision (9.4% less). All in all, CoCoNoW extracted the most keywords successfully and outperformed other state-of-the-art approaches. Note the consistently high Precision values of CoCoNoW: There is a low number of false positives (i.e. words wrongly marked as keywords), which is crucial in the next stage of clustering publications with respect to their respective topics.

Table 2. Performance comparison of CoCoNoW with several other approaches on the different datasets.

Approach	k	Hulth2003 [14]			SemEval2010 [17]			NLM500 [2]		
		P[%]	R[%]	F1[%]	P[%]	R[%]	F1[%]	P[%]	RD	F1[%]
TF-IDF [20]	5	33.3	17.3	24.2	–	–	–	–	–	–
Topic Clustering [20]		35.4	18.3	24.3	–	–	–	–	–	–
Key2Vec [22]		68.8	**25.7**	36.2	41.0	14.4	21.3	–	–	–
CoCoNoW		**84.0**	**25.7**	**37.3**	**84.1**	**17.5**	**28.7**	48.8	11.4	17.9
TextRank [24]	10	45.4	47.1	39.8	–	–	–	–	–	–
Word Embeddings [38]		38.7	**52.8**	44.7	–	–	–	–	–	–
Key2Vec [22]		57.6	42.0	48.6	35.3	24.7	29.0	–	–	–
CoCoNoW		**73.3**	41.9	**50.0**	**72.3**	**29.8**	**41.6**	43.3	19.8	26.3
Supervised approach [14]	16	25.2	51.7	33.9	–	–	–	–	–	–
TextRank [24]	14	31.2	43.1	36.2	–	–	–	–	–	–
TF-IDF [17]	15	–	–	–	11.6	14.5	12.9	–	–	–
HUMB [21]	15	–	–	–	27.2	27.8	27.5	–	–	–
Key2Vec [22]	15	55.9	50.0	52.9	34.4	32.5	33.4	–	–	–
CoCoNoW	15	**63.5**	**52.9**	**54.2**	**62.2**	**39.2**	**46.5**	37.11	25.2	29.0
TextRank [24]	25	–	–	18.4*	–	–	–	–	–	–
DegExt [19]		–	–	18.2*	–	–	–	–	–	–
k-core [34]		–	–	43.4*	–	–	–	–	–	–
PositionRank [13]		45.7*	64.5*	50.4*	–	–	–	–	–	–
sCAKE [12]		45.4	**66.8**	51.1	–	–	–	–	–	–
CoCoNoW		**54.8**	66.2	**56.8**	47.3	47.8	46.8	29.3	32.6	29.9
TextRank [24]	30	–	–	–	–	–	13.7*	–	–	10.7*
DegExt [19]		–	–	–	–	–	14.6*	–	–	10.9*
k-core [34]		–	–	–	–	–	29.3*	–	–	20.2*
PositionRank [13]		–	–	–	25.3*	31.3*	27.5*	19.7*	26.6*	21.9*
sCAKE [12]		–	–	–	35.8	47.4	40.1	24.5	35.0	28.3
CoCoNoW		52.5	70.1	57.2	**42.6**	**51.5**	**45.8**	**26.7**	**35.3**	**29.5**

* Results reported by Duari and Bhatnagar [12], not by the original authors.

3.3   Performance Evaluation Stage 2: Topic Modeling

This section discusses the evaluations of ontology-based topic modeling. There was no ground truth available for the ICDAR publications, consequently making the evaluation of topic modeling a challenging task. Nevertheless, we employed two different approaches for evaluation: manual inspection and citation count. Details of both evaluations are as follows:

Manual Inspection. The proposed method for topic assignment comes with labels for the topics, so manually inspecting the papers assigned to a topic is rather convenient. This is done by going through the titles of all papers assigned to a topic and judging whether the assignment makes sense.

For specialized topics i.e. the ones far from the root, the method worked very well, as it is easy to identify papers that do not belong to a topic. Manual inspection showed that there are very few false positives, i.e. publications assigned to an irrelevant topic. This is because of the high Precision of the CoCoNoW algorithm: The low number of false positives in the extracted keywords increases the quality of the topic assignment. The closer a topic is to the root i.e. a more generic topic, the more difficult it is to assess whether a paper should be assigned to it: Often, it is not possible to decide whether a paper can be assigned to a general node such as *neural networks* by just reading the title. Hence, this method does not give meaningful results for more general topics. Furthermore, this method was only able to identify false positives. It is difficult to identify false negatives with this method, i.e. publications not assigned to relevant topics.

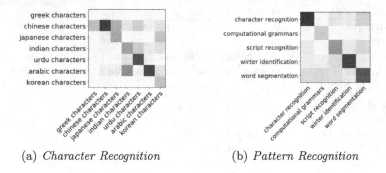

(a) *Character Recognition*   (b) *Pattern Recognition*

Fig. 4. Citation count for different super topics. (Color figure online)

Evaluation by Citation Count. The manual inspection indicated that the topic assignment works well, but that was just a qualitative evaluation. So a second evaluation is performed. It is based on the following assumption: *Papers dealing with a topic cite other papers from the same topic more often than papers dealing with different topics*. We believe that this is a sensible assumption, as papers often compare their results with previous approaches that tackled similar problems. So instead of evaluating the topic assignment directly, it is indirectly evaluated be counting the number of citations between the papers assigned to the topics. However, a paper can be assigned to multiple overlapping topics (e.g. *machine learning* and *neural networks*). This makes it infeasible to compare topics at two different hierarchy levels using this method.

Nevertheless, this method is suited for evaluating the topic assignment of siblings, i.e. topics that have the same super topic. Figure 4 shows heatmaps for citations between child elements of a common supertopic in different levels in the hierarchy. The rows represent the number of citing papers, the columns the number of cited papers. A darker color in a cell represents more citations. Figure 4a shows the script-dependent topics, which are rather specialized and all have the common supertopic *character recognition*. The dark diagonal values clearly indicate that the number of intra-topic citations are higher than inter-topic citations.

Figure 4b shows the subtopics of the node *pattern recognition*, which has a very high level of abstraction and is close to the root of the hierarchy. The cells in the diagonal are clearly the darkest ones again, i.e. there are more citations within the same topic than between different topics. This is a recurring pattern across the entire hierarchy. So in general, results from these evaluations indicate that the topic assignment works reliably if our assumption is correct.

4   Conclusion

This paper presents a two-staged novel approach where the first stage called CoCoNoW deals with automatic keyword detection while the second stage uses the detected keywords and performs ontology-based topic modeling on a given scientific publication. Evaluations of CoCoNoW clearly depict its supremacy over several existing keyword detection approaches by consistently outperforming every single approach in terms of Precision and F-Measure. On the other hand, the evaluation of the topic modeling approach suggests that it is an effective technique as it uses an ontology to accurately define context and domain. Hence, the publications with the same topics can be clustered more reliably.

References

1. Aiello, L.M., et al.: Sensing trending topics in Twitter. IEEE Trans. Multimedia **15**(6), 1268–1282 (2013)
2. Aronson, A.R., et al.: The NLM indexing initiative. In: Proceedings of the AMIA Symposium, p. 17. American Medical Informatics Association (2000)
3. Becker, H., Naaman, M., Gravano, L.: Beyond trending topics: real world event identification on Twitter. In: AAAI (2011)
4. Beliga, S.: Keyword extraction: a review of methods and approaches. University of Rijeka, Department of Informatics, pp. 1–9 (2014)

5. Bellaachia, A., Al-Dhelaan, M.: NE-Rank: a novel graph-based keyphrase extraction in Twitter. In: 2012 IEEE/WIC/ACM International Conferences on Web Intelligence and Intelligent Agent Technology, vol. 1, pp. 372–379. IEEE (2012)
6. Biswas, S.K., Bordoloi, M., Shreya, J.: A graph based keyword extraction model using collective node weight. Expert Syst. Appl. **97**, 51–59 (2018)
7. Boudin, F.: Unsupervised keyphrase extraction with multipartite graphs. arXiv preprint arXiv:1803.08721 (2018)
8. Breaux, T.D., Reed, J.W.: Using ontology in hierarchical information clustering. In: Proceedings of the 38th Annual Hawaii International Conference on System Sciences, p. 111b. IEEE (2005)
9. Carpena, P., Bernaola-Galván, P., Hackenberg, M., Coronado, A., Oliver, J.: Level statistics of words: finding keywords in literary texts and symbolic sequences. Phys. Rev. E **79**(3), 035102 (2009)
10. Carretero-Campos, C., Bernaola-Galván, P., Coronado, A., Carpena, P.: Improving statistical keyword detection in short texts: entropic and clustering approaches. Phys. A: Stat. Mech. Appl. **392**(6), 1481–1492 (2013)
11. Carston, R.: Linguistic communication and the semantics/pragmatics distinction. Synthese **165**(3), 321–345 (2008). https://doi.org/10.1007/s11229-007-9191-8
12. Duari, S., Bhatnagar, V.: sCAKE: semantic connectivity aware keyword extraction. Inf. Sci. **477**, 100–117 (2019)
13. Florescu, C., Caragea, C.: A position-biased PageRank algorithm for keyphrase extraction. In: Thirty-First AAAI Conference on Artificial Intelligence (2017)
14. Hulth, A.: Improved automatic keyword extraction given more linguistic knowledge. In: Proceedings of the 2003 Conference on Empirical Methods in Natural Language Processing, pp. 216–223. Association for Computational Linguistics (2003)
15. Johnson, R., Watkinson, A., Mabe, M.: The STM report. Technical report, International Association of Scientific, Technical, and Medical Publishers (2018)
16. Kecskés, I., Horn, L.R.: Explorations in Pragmatics: Linguistic, Cognitive and Intercultural Aspects, vol. 1. Walter de Gruyter (2008)
17. Kim, S.N., Medelyan, O., Kan, M.Y., Baldwin, T.: SemEval-2010 task 5: automatic keyphrase extraction from scientific articles. In: Proceedings of the 5th International Workshop on Semantic Evaluation, pp. 21–26 (2010)
18. Levenshtein, V.I.: Binary codes capable of correcting deletions, insertions and reversals. Soviet Phys. Doklady **10**, 707 (1966)
19. Litvak, M., Last, M., Aizenman, H., Gobits, I., Kandel, A.: DegExt - a language-independent graph-based keyphrase extractor. In: Mugellini, E., Szczepaniak, P.S., Pettenati, M.C., Sokhn, M. (eds.) Advances in Intelligent Web Mastering - 3. AINSC, vol. 86, pp. 121–130. Springer, Heidelberg (2011). https://doi.org/10.1007/978-3-642-18029-3_13
20. Liu, Z., Li, P., Zheng, Y., Sun, M.: Clustering to find exemplar terms for keyphrase extraction. In: Proceedings of the 2009 Conference on Empirical Methods in Natural Language Processing, vol. 1, pp. 257–266 (2009)
21. Lopez, P., Romary, L.: HUMB: automatic key term extraction from scientific articles in GROBID. In: Proceedings of the 5th International Workshop on Semantic Evaluation, pp. 248–251. Association for Computational Linguistics (2010)
22. Mahata, D., Shah, R.R., Kuriakose, J., Zimmermann, R., Talburt, J.R.: Theme-weighted ranking of keywords from text documents using phrase embeddings. In: 2018 IEEE Conference on Multimedia Information Processing and Retrieval (MIPR), pp. 184–189. IEEE (2018). https://doi.org/10.31219/osf.io/tkvap
23. Mey, J.L.: Whose Language?: A Study in Linguistic Pragmatics, vol. 3. John Benjamins Publishing (1985)

24. Mihalcea, R., Tarau, P.: TextRank: bringing order into text. In: Proceedings of the 2004 Conference on Empirical Methods in Natural Language Processing (2004)
25. Mikolov, T., Chen, K., Corrado, G.S., Dean, J.: Efficient estimation of word representations in vector space. CoRR abs/1301.3781 (2013)
26. Nikolentzos, G., Meladianos, P., Stavrakas, Y., Vazirgiannis, M.: K-clique-graphs for dense subgraph discovery. In: Ceci, M., Hollmén, J., Todorovski, L., Vens, C., Džeroski, S. (eds.) ECML PKDD 2017. LNCS (LNAI), vol. 10534, pp. 617–633. Springer, Cham (2017). https://doi.org/10.1007/978-3-319-71249-9_37
27. O'Connor, B., Krieger, M., Ahn, D.: TweetMotif: exploratory search and topic summarization for Twitter. In: AAAI (2010)
28. Ohsawa, Y., Benson, N.E., Yachida, M.: KeyGraph: automatic indexing by co-occurrence graph based on building construction metaphor. In: Proceedings IEEE International Forum on Research and Technology Advances in Digital Libraries (ADL 1998), pp. 12–18. IEEE (1998). https://doi.org/10.1109/adl.1998.670375
29. Osborne, F., Motta, E.: Klink-2: integrating multiple web sources to generate semantic topic networks. In: Arenas, M., et al. (eds.) ISWC 2015. LNCS, vol. 9366, pp. 408–424. Springer, Cham (2015). https://doi.org/10.1007/978-3-319-25007-6_24
30. Osborne, F., Motta, E., Mulholland, P.: Exploring scholarly data with rexplore. In: Alani, H., et al. (eds.) ISWC 2013. LNCS, vol. 8218, pp. 460–477. Springer, Heidelberg (2013). https://doi.org/10.1007/978-3-642-41335-3_29
31. Page, L., Brin, S., Motwani, R., Winograd, T.: The PageRank citation ranking: bringing order to the web. Technical report, Stanford InfoLab (1999)
32. Pay, T., Lucci, S.: Automatic keyword extraction: an ensemble method. In: 2017 IEEE Conference on Big Data, Boston, December 2017
33. Rabby, G., Azad, S., Mahmud, M., Zamli, K.Z., Rahman, M.M.: A flexible keyphrase extraction technique for academic literature. Procedia Comput. Sci. **135**, 553–563 (2018)
34. Rousseau, F., Vazirgiannis, M.: Main core retention on graph-of-words for single-document keyword extraction. In: Hanbury, A., Kazai, G., Rauber, A., Fuhr, N. (eds.) ECIR 2015. LNCS, vol. 9022, pp. 382–393. Springer, Cham (2015). https://doi.org/10.1007/978-3-319-16354-3_42
35. Salatino, A.A., Thanapalasingam, T., Mannocci, A., Osborne, F., Motta, E.: The computer science ontology: a large-scale taxonomy of research areas. In: Vrandečić, D., et al. (eds.) ISWC 2018. LNCS, vol. 11137, pp. 187–205. Springer, Cham (2018). https://doi.org/10.1007/978-3-030-00668-6_12
36. Salatino, A.A., Osborne, F., Thanapalasingam, T., Motta, E.: The CSO classifier: ontology-driven detection of research topics in scholarly articles. In: Doucet, A., Isaac, A., Golub, K., Aalberg, T., Jatowt, A. (eds.) TPDL 2019. LNCS, vol. 11799, pp. 296–311. Springer, Cham (2019). https://doi.org/10.1007/978-3-030-30760-8_26
37. Slabbekoorn, K., Noro, T., Tokuda, T.: Ontology-assisted discovery of hierarchical topic clusters on the social web. J. Web Eng. **15**(5&6), 361–396 (2016)
38. Wang, R., Liu, W., McDonald, C.: Using word embeddings to enhance keyword identification for scientific publications. In: Sharaf, M.A., Cheema, M.A., Qi, J. (eds.) ADC 2015. LNCS, vol. 9093, pp. 257–268. Springer, Cham (2015). https://doi.org/10.1007/978-3-319-19548-3_21
39. Ware, M., Mabe, M.: The STM report: an overview of scientific and scholarly journal publishing. Technical report, International Association of Scientific, Technical, and Medical Publishers (2015)

A New Context-Based Method for Restoring Occluded Text in Natural Scene Images

Ayush Mittal[1], Palaiahnakote Shivakumara[2(✉)], Umapada Pal[1],
Tong Lu[3], Michael Blumenstein[4], and Daniel Lopresti[5]

[1] Computer Vision and Pattern Recognition Unit, Indian Statistical Institute,
Kolkata, India
mittalayush939@gmail.com, umapada@isical.ac.in
[2] Faculty of Computer Science and Information Technology,
University of Malaya, Kuala Lumpur, Malaysia
shiva@um.edu.my
[3] National Key Lab for Novel Software Technology, Nanjing University,
Nanjing, China
lutong@nju.edu.cn
[4] Faculty of Engineering and Information Technology,
University of Technology Sydney, Ultimo, Australia
Michael.Blumenstein@uts.edu.au
[5] Computer Science and Engineering, Lehigh University, Bethlehem, PA, USA
lopresti@cse.lehigh.edu

Abstract. Text recognition from natural scene images is an active research area because of its important real world applications, including multimedia search and retrieval, and scene understanding through computer vision. It is often the case that portions of text in images are missed due to occlusion with objects in the background. Therefore, this paper presents a method for restoring occluded text to improve text recognition performance. The proposed method uses the GOOGLE Vision API for obtaining labels for input images. We propose to use PixelLink-E2E methods for detecting text and obtaining recognition results. Using these results, the proposed method generates candidate words based on distance measures employing lexicons created through natural scene text recognition. We extract the semantic similarity between labels and recognition results, which results in a Global Context Score (GCS). Next, we use the Natural Language Processing (NLP) system known as BERT for extracting semantics between candidate words, which results in a Local Context Score (LCS). Global and local context scores are then fused for estimating the ranking for each candidate word. The word that gets the highest ranking is taken as the correction for text which is occluded in the image. Experimental results on a dataset assembled from standard natural scene datasets and our resources show that our approach helps to improve the text recognition performance significantly.

Keywords: Text detection · Occluded image · Annotating natural scene images · Natural language processing · Text recognition

© Springer Nature Switzerland AG 2020
X. Bai et al. (Eds.): DAS 2020, LNCS 12116, pp. 466–480, 2020.
https://doi.org/10.1007/978-3-030-57058-3_33

1   Introduction

Over the past decade, the scope and importance of text recognition in natural scene images have been expanding rapidly, driven by new applications, such as robotics, multimedia, as well as surveillance and monitoring. For example, Roy et al. [1] studied forensic applications, where it is necessary to recognize text from multiple views of the same target captured by CCTV cameras. The method uses text detection and recognition for identifying crime location. Shivakumara et al. [2] addressed the issues of text detection and recognition in marathon images. This method detects text on torso images of marathon runners for tracing and studying a person's behavior. Xue et al. [3] proposed curved text detection in blurred and non-blurred video and natural scene images. The method focuses on addressing challenges caused by blur and arbitrary orientation of text in images. For all the above-mentioned methods, it is noted that the main aim is to achieve a better recognition rate irrespective of the applications. However, none of the methods proposed in the literature focuses on images where a part of the text is missing due to occlusion. In such cases, if we run existing methods on such images, the recognition performance degrades severely. Also, recognition results obtained for broken words may give an incorrect interpretation of the image content because the recognition results will miss the actual meaning of the text. Therefore, restoring the missing text is important to determine the semantics of the image.

It is evident from the illustration shown in Fig. 1, where for the input image with occluded text shown in Fig. 1(a) and text detection results shown in Fig. 1(b), the recognition method does not recognize the text correctly as reported in Fig. 1(c). The recognition results of broken words do not exhibit the desired meaning with respect to the content of the image. This is the limitation of state-of-the-art text recognition for the images where a part of the text is missing. Note that for text detection, we use the method called PixelLink [4] as it is robust to complex backgrounds, orientation, and the E2E method [5] for recognition, as it is robust to distortion and different fonts. These limitations of the state-of-the-art approaches motivated us to develop a method for restoring the missing portions of broken words in natural scene images. The content of an image and the text in it indeed have a high degree of similarity at the semantic level. Therefore, the proposed method explores using the GOOGLE Vision API [6] to obtain labels for the input image as shown in Fig. 1(d). This makes sense because the GOOGLE Vision API works based on the relationship between the objects in images. Through the use of a powerful dictionary and lexicon of candidates, we can expect labels which are often close to the broken words because of powerful language models. This shows that one can predict the possible words for incomplete and misspelled words. Based on these observations, the proposed method generates candidate words for each recognized broken word as shown in Fig. 1(e). Then it extracts the context between labels and candidate words based on the distance measure, which results in the global context score. Similarly, the proposed approach uses natural language processing [7] for extracting context between candidate words, which results in a local context score. Furthermore, we combine the global and local context scores to estimate the ranking for each word. The word that gets the highest ranking is considered the most likely replacement for the broken word as shown in Fig. 1(f).

(a) Input natural scene image with occluded text   (b) The results of text detection

(c) Recognition results before prediction for the detected words: "oxfrd" and "ooksre"

(d) Labels given by GOOGLE Vision API: "Building, Advertisement, Outlet store, Display window"

(e) Candidate words for "oxfrd" and "ooksre" using distance measures with word dictionary: "Oxford" and "Books, Bookstore, Boostrap, Booking", respectively.

(f) The proposed prediction results: "oxford book store".

Fig. 1. Illustrating the steps for predicting missing text in a natural scene image

2 Related Work

There are several methods proposed in the past several years for recognizing text in natural scene images. Most recent methods have explored deep learning models. Cheng et al. [8] proposed arbitrarily-oriented text recognition in natural scene images based on a deep learning approach. They use an Arbitrary Orientation Network (AON) to capture deep features of irregular text directly, which generates character sequences using an attention-based decoder. Tian et al. [9] proposed a framework for text recognition in web videos based on tracking text. Their approach combines information from text detection and tracking for recognizing text in images. Luo et al. [10] proposed a multi-object rectified attention network for scene text recognition. They explore a deep learning model, which is invariant to geometric transformation. The approach works well for images affected by rotation, scaling and to some extent – distortion. Raghunandan et al. [11] proposed multi-script-oriented text detection and recognition in video, natural scene and born-digital images. The work extracts features based on the wavelet transform for detecting characters with the help of an SVM classifier. Next, it applies a Hidden Markov Model (HMM) for recognizing characters and words in images. Qi et al. [12] proposed a novel joint character categorization and localization approach for character level scene text recognition. The idea of the method is to categorize characters by a joint learning strategy such that recognition performance improves. Shi et al. [13] proposed an attentional scene text recognizer with flexible rectification. The work uses a thin-plate spline transformation to handle a variety of text irregularities. The idea behind the method is to avoid pre-processing before recognition such that errors can be reduced to improve the recognition rate. Rong et al. [14] employs unambiguous scene text segmentation with referring expression comprehension. The study proposes a unified deep network to jointly model visual and linguistic information at both the region and pixel levels for understanding text in images. Villamizar et al. [15] proposed a multi-scale sequential network for semantic text segmentation and localization. The work explores fully convolutional neural networks that apply to a particular case of slide analysis to understand the text in images. Feng et al. [16] proposed an end-to-end framework for arbitrarily shaped text spotting. The method proposes a new differentiable operator named RoISlide, which detects text and recognizes it in the images.

It is noted from the above discussion that existing methods have addressed challenges such as arbitrary orientation, distortion caused by geometrical transformation, irregularly shaped characters, complex backgrounds, low resolution and low contrast for text recognition. Most of the methods explored deep learning in different ways for achieving their results. However, none of these methods addressed the issue of text occlusion in natural scene images.

There are methods related to restoring missing information in natural scene images. For instance, Lee et al. [17] proposed automatic text detection and removal in video sequences. The method uses Spatio-temporal information for achieving its results. The method identifies locations where the text is missing due to occlusion and then removes that obstacle. However, the scope of the method is limited to text removal but not restoration of missing text. Ye et al. [18] proposed text detection and restoration in natural scene images. The method restores text, which is degraded due to perspective distortion, low resolution and other causes but not missing parts of text information due to occlusion. Tsai et al. [19] proposed text-video completion using structure repair and texture propagation. The method considers text as an obstacle, which occludes object information in images. Therefore, the method detects text information and uses an inpainting approach to restore the missing parts of the object information. Mosleh et al. [20] proposed an automatic inpainting scheme for video text detection and removal. The method is also the same as the text removal approach mentioned above, but it does not restore the missing part of text information in images. Zhang et al. and Wu et al. [21, 22] proposed methods for erasing text in natural scene images based on deep learning models. The scope of the methods is limited to the location of the text information, and to erase it such that it does not alter background information in the images. Hence, one can conclude that restoring the missing part of the text in natural scene images is not addressed for improving recognition performance. Thus, we propose a new method, which combines labels generated from the content of images and recognition results for predicting missing words to replace broken words with the help of a Natural Language Processing (NLP) approach. The main contribution of the proposed work is the way in which our approach combines the labels given by the GOOGLE Vision APIs with candidate words provided by Natural Language Processing to predict the likely words, which are to be replaced as broken words in naturals scene images.

3 Proposed Method

In this work, since the GOOGLE Vision API is available publicly for generating annotations for natural scene images, we use this for generating labels for our input images which contain text within natural scenes [6]. In this work, we propose to use the method called PixelLink [4] because the method is state-of-the-art and works well for images affected by the above-mentioned challenges. For recognizing detected text, we propose to use the E2E method [5] as it is robust to degradations, poor quality, orientation, and irregularly-shaped characters. Also, the method requires fewer training iterations and less training data. The combination of text detection and recognition produces the recognition results for text in images.

For each word that is recognized, the proposed method generates the sequence of candidates by estimating the distance between the recognized word and the words in a predefined large-sized lexicon formed for scene text recognition. This results in a candidate list of the probable words for each of the non-real words. Then the annotated labels are compared with the list of probable candidates to estimate Global Context Score (GCS) using the distance measure. In the same way, the proposed method estimates the Local Context Score (LCS) for the candidates of each of the word recognition results using the NLP approach known as BERT [7]. The proposed method fuses GCS and LCS scores to generate a ranking for each word. The word that gets the highest rank is considered for a predicted word to replace broken words. The steps and flow of the proposed method can be seen in Fig. 2.

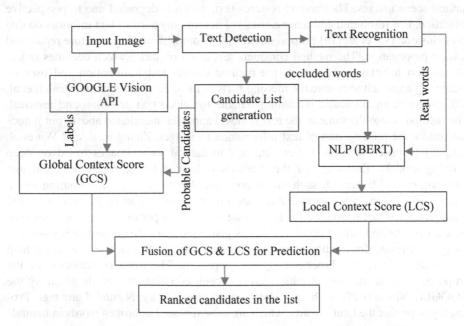

Fig. 2. Block diagram of the proposed method for predicting missing text.

3.1   Generating Labels Using the Google Vision API

According to the website [6], the GOOGLE Vision API was developed based on a large number of features and deep learning. The system generates a list of labels for each input image with a ranking score. According to our experiments on our dataset, it is noted that if a rank score is greater than 85%, the generated keywords are relevant to the content of the images. Therefore, we set the same threshold empirically for all the experiments in this work. For the input image shown in Fig. 3, the system generates labels as listed in the same figure, where it can be seen that all the labels are relevant to the content of the input image. However, sometimes, for images of an unknown place, the system produces irrelevant keywords. In such cases, there are chances of predicting incorrect words by the proposed method. However, since the proposed method considers candidate words of all the words in images with the help of natural language processing, it may not have much of an effect on the overall performance of the proposed method. This step gives a list of labels for the input images.

Labels

- Building
- Advertisement
- Outlet Store
- Display window

Fig. 3. Labels for the natural scene image using GOOGLE Vision API

3.2   Candidate List Generation

Though a portion of the text is missing due to occlusion, still the remaining text can provide some clues for restoring the missing text with the help of lexicons and language processing. Therefore, the proposed method generates possible candidate words for each recognized word including broken words in images. For this, the proposed method uses a deep learning model [4] for text detection in natural scene images to extract the text. Sample results of text detection are shown in Fig. 4, where the method detects almost all the text that appeared in images. For the detected words, we use the recognition method proposed in [5] which also employs deep learning models to recognize the detected words as reported in Fig. 4, where we can see recognition results for the two words present in Fig. 4 as "oxfrd" and "ooksre". The proposed method uses the following steps for producing candidate words for each of the recognized words.

Let $S = \{s_1, s_2, \ldots, s_m\}$ be the sequence of strings separated by spaces obtained from the recognition method which includes good and broken words. It is noted that the broken word creates two new words due to a split in the word. For each word in the list,

A. Mittal et al.

Candidate List for "oxfrd"
Oxford.
Candidate List for "ooksre"
Books, Bookstore, Bootstrap
Booking

Fig. 4. Text detection by the PixelLink method (bounding boxes in the images), recognition results (oxfrd and ooksre) by the E2E method and list of candidate words for the broken words.

the proposed method obtains possible candidate words which can replace the broken words. To achieve this, we propose an iterative process to generate possible words using the Levenshtein distance [23] and the subsequence distance. The subsequence distance between two strings is the length of the longest common subsequence (LCS). This process involves the lexicon of 90 k generic words available publicly[1]. The effectiveness of this process can be seen in Fig. 4, where the list of candidate words is generated for the words "oxfrd" and "ooksre". It is observed from the list of candidate words that the list contains the most likely to be the correct word of the broken words.

3.3 Global and Local Context Score Estimation

The step discussed in Sect. 3.1 outputs the list of labels for the input image, and the steps presented in Sect. 3.2 generates the list of candidate words for the words in the images. To extract the semantic similarity between the labels and candidate words, we calculate the distance between them and this gives a global score for all the words in the list. Furthermore, the proposed method considers the maximum word similarity value calculated against all the image labels as the Global Context Score (GCS). It is illustrated in Fig. 5, where we can see the distance is estimated for all the combinations of the words between labels and the candidate words using the Euclidean distance measure. Therefore, this process results in a vector, which contains GCS in the range of [0, 1].

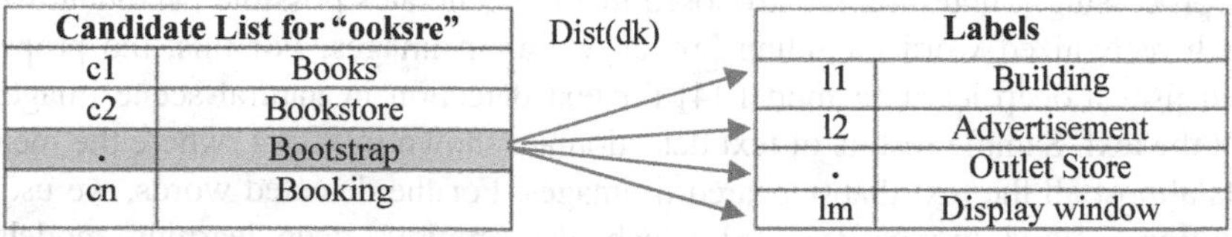

Candidate List for "ooksre"		Dist(dk)		Labels	
c1	Books		11	Building	
c2	Bookstore		12	Advertisement	
.	Bootstrap		.	Outlet Store	
cn	Booking		lm	Display window	

Fig. 5. Estimating the global context score between the label and the candidate words using the Euclidean distance measure. n indicates the list of candidate words, m indicates the list of labels and k indicates the distances.

---

[1] The vocabulary can be downloaded from https://rrc.cvc.uab.es/downloads/GenericVocabulary.txt.

To extract the local semantic similarity between candidate words, we propose to explore the natural language processing approach called BERT's Masked Language Model [7] for predicting missing words. BERT is bidirectional in nature and we use a pre-defined and trained model in this work. The advantage of this model is that it helps in extracting the context with a limited number of words. For n words in a sentence, the proposed method obtains n number of lists. Let $L = \{L_1, L_2, L_3, .., L_n\}$ be the sequence of all those n lists, which contains different possible sequences obtained by the combinations of words, say m number of sequences. For each candidate word in the list, logit values are estimated as follows. For example, for list L_1, the logit values for each candidate word are obtained with all the sequences formed using the remaining lists. These logit values are converted to the range [0, 1] using a sigmoid function over every list, which results in the Local Context Score (LCS). In this way, for each candidate word in every list, LCSs are calculated. The values of GCSs and LCSs are fused as defined in Eq. (1), where the values of a and b are determined empirically based on pre-defined samples chosen randomly from the datasets. This process outputs fused values for each word in every sequence. The fused values are further multiplied with a scalar value, z for every sequence as defined in Eq. (2) and Eq. (3).

$$F_i = a(SC_i) + b(GC_i), \forall i \in l_k \tag{1}$$

$$z_i = \prod_{j=1}^{n} F_j, \forall i \in [1, m] \tag{2}$$

The best sequence z^* is determined based on the maximum z value as defined in equation

$$z^* = \underset{i}{\mathbf{argmax}}\, z_i, \forall i \in [1, m] \tag{3}$$

Sample values of GCS and LCS are reported for candidate words given the word "oxford" in Table 1, where we can see the word "bookstore" has received the highest fused values compared to other words. It is also noted that the next relevant word for "oxford" is "books", which scores the next highest value. For the word "bootstrap", the fused score is zero. This indicates that the word is not relevant to the word "oxford" In this way, the proposed method uses labels generated by the GOOGLE Vision API and NLP for predicting missing words in natural scene images.

Table 1. The fused z scores for predicting the correct word using GCS and LCS

Possible combinations with "oxford"	GCS score	LCS score	Fused (z) score
"books"	0.57	0.090	0.051
"bookstore"	**0.65**	**0.120**	**0.078**
"bootstrap"	0.00	0.001	0.000
"booking"	0.53	0.005	0.003

4 Experimental Results

There is no standard dataset available for restoring missing words to replace broken ones in natural scene images, so we have created our own dataset, which include images captured by ourselves and images collected from FLICKER and SUN datasets, which gives 200 images with occluded text on a different background. As shown in the sample images in Fig. 6, we can see that each image has a different background complexity, and includes occluded text. It is also noted from Fig. 6 that the occlusions are due to trees, poles and other objects. The percentage of occluded words is around 25.9% out of 478 text instances in our dataset. Also, our dataset includes text in multiple scripts, text affected by blur, and text in different resolutions as reported in Table 2. We also collected a few images from standard datasets of natural scene images, namely, ICDAR 2015. This dataset contains very few images with occluded text (2%) as they are created for text detection and recognition but not for the application we are targeting. The images of this dataset have almost the same resolution as reported in Table 1. For the GCS calculation, we use the ground truth of the ICDAR 2015 dataset where we found 3909 instances of text line information.

Fig. 6. Sample images from our dataset with occluded scene text

For measuring the performance of the proposed method, we consider the standard measures, namely, Precision (P) which is defined as the number of words recognized correctly (real word) divided by the number of words recognized correctly (real words) + false positives, and Recall which is defined as the number of words recognized correctly (real words) divided by the total number of ground truth words. The F-measure is a harmonic mean of Recall and Precision.

In this work, we use the text detection method [4] for extracting text from natural scene images. As mentioned in the proposed methodology section, occluded text does not affect text detection performance. This is justified because text detection methods

Table 2. Statistical analysis of both ours and the benchmark dataset

Dataset	Dataset size	Resolution	Blur	MLT script	Uneven illumination	Percentage of occluded words	Number of text instances
Our dataset	200 test	Min: 428 × 476 Max: 4608 × 2304	✓	✗	✓	25.94%	478
ICDAR 2015	500 test	720 × 1280	✗	✗	✗	2%	3909

are capable of detecting a single character as text in natural scene images. Also, for text detection, semantic or context of words in the text line is not necessary unlike recognition which requires a language model for accurate recognition. The method in [4] is considered as the state-of-the-art method for text detection in this work. To validate the above statements, we conducted text detection experiments using different existing methods on our dataset. We implemented the following existing methods including PixelLink and E2E methods. Zhou et al. [24] proposed an efficient and accurate scene text detector, which explores deep learning models for addressing arbitrary orientation of text detection in natural scene images. Shi et al. [25] proposed the detection of oriented text in natural scene images by linking segments that focus on the use of fewer training samples for text detection in natural scene images. Liu et al. [26] proposed a method for detecting curved text in natural scene images, which focuses on the challenge of curved text detection. Busta et al. [5] proposed an end-to-end text detection and recognition method, which focuses on images of multi-lingual scene text. Deng et al. [4] proposed a method for detecting scene text via instance segmentation which addresses the challenges of arbitrary orientation, multi-script and different types of texts. The results of the above methods are reported in Table 3 for our dataset, where it can be noted that the PixelLink method is the best for all three measures. This is the advantage of the PixelLink method compared to the other existing methods. It is also noted from Table 3 that almost all the methods score more than 80% accuracy for our dataset. This demonstrates that the presence of occlusion does not have much of an effect on text detection performance.

Table 3. Text detection performance of the different methods on our dataset.

Methods	P	R	F
Zhou et al. EAST [24]	0.81	0.76	0.78
Shi et al. SegLink [25]	0.83	0.79	0.81
Liu et al. CTD [26]	0.80	0.81	0.80
Busta et al. E2E MLT [5]	0.82	0.83	0.82
Deng et al. PixelLink [4]	**0.84**	**0.86**	**0.84**

In the same way, we use labels given by the GOOGLE Vision API system for input images to restore missing parts. To show that the labels generated by the GOOGLE Vision API are relevant to the content of images, we calculate the Global Context Score (GCS) for the labels with the ground truth provided by the ICDAR 2015 dataset. The process involves 500 images and around 3909 text instances. The GCS values are plotted in Fig. 7, where one can see that most of the GCS values are greater than zero as shown in the scattered graph and pie-chart. This shows that the GOOGLE Vision API generates relevant labels and that they share a high degree of semantic similarity with the corresponding scene text.

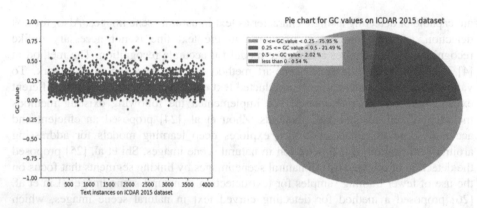

Fig. 7. GCS for labels of the images with ground text instances of the ICDAR 2015 ground truth.

4.1   Evaluating the Proposed Prediction

To show that the proposed method is effective and useful, we calculate the measure before prediction, which provides the recognition rate for the output of text detection without the proposed global and local context scores. The results are compared with those of the proposed method after prediction. It is expected that the recognition rate of the proposed method after prediction should be higher due to the restoration of the missing information. To show the proposed combination (BERT and GOOGLE Vision API system) is better than the individual steps and other Natural Language Processing concepts, namely, edit distance, bigrams, we calculated the recognition rate for each step and the NLP steps, as reported in Table 4. For calculating the recognition rate for only BERT without the results of the GOOGLE Vision API, the proposed method only estimates the Local Context Score for predicting the missing words. Similarly, for calculating the recognition rate for the edit distance and bigram, the proposed method replaces the BERT approach with the edit distance and bigrams for missing word prediction. In Table 4, it is noted that the proposed method achieves the best precision, recall and F-measure compared to the individual steps before prediction. This shows that the GOOGLE Vision API system and the BERT contribute equally to achieve the best results. The reason is that the individual steps miss either local or global context information for accurately predicting the missing words.

Table 4. Analyzing the contribution of key steps of the proposed method for predicting missing text.

Recognition rate before prediction			After prediction											
			Only BERT without Google API			Edit distance			Bigram			Proposed method (BERT + Google API)		
P	R	F	P	R	F	P	R	F	P	R	F	P	R	F
0.43	0.55	0.48	0.64	0.62	0.63	0.59	0.61	0.60	0.48	0.57	0.52	**0.67**	**0.64**	**0.65**

However, when we compare the results of the proposed method before and after prediction, the results of after prediction are better than before, as reported in Table 4. The poor results of the proposed method with edit distance compared to the proposed combination is due to the failure in extracting semantic information between the candidate words. Similarly, the poor results of the proposed method with bigrams are because the bigram model extracts the context in one direction with the preceding word of the missing words. Since the occlusion can be at any position in the sentence, single direction information is not sufficient for complex situations. On the other hand, the proposed combination (BERT + GOOGLE Vision API) has the capability to extract context from both directions (left to right and right to left) in a sentence. Note that in the case of the bigram model, the ground truth of ICDAR 2015 is used for training the model.

To illustrate that text detection does not have a substantial effect on recognizing broken words, we calculate the recognition rate for the output of each detection method listed in Table 5, including the PixelLink and E2E methods for before- and after-prediction. It is observed from Table 5 that the before-prediction results for all the text detection methods is are almost the same. In the same way, though there is a significant improvement in recognition results for after-prediction, the results of before prediction the respective text detection methods are almost the same. This shows that missing characters in the words do not affect text detection and recognition performance of the different methods. It is also noted from Table 5 that the recognition rate of the PixelLink method is better than other existing methods for both before- and after-prediction. Therefore, one can conclude that the proposed combination of the GOOGLE Vision API labels, Text detection-recognition and NLP is the best for improving recognition performance for the images with occluded text.

The advantage of the proposed work is that if a part of the text in an image is missing due to other causes, but it is not necessarily due to occlusion, the proposed method can restore the missing part of the text as shown one example in Fig. 8. In Fig. 8, a few characters of the word "boundary" are missing. As mentioned earlier, for the proposed method it does not matter whether a part of the text is missing due to occlusion or some other causes. As a result, the proposed method predicts the missing part of the word "boundary" as shown in Fig. 8. However, it is hard to generalize that the proposed method works for all the situations and restores the missing text correctly because the success depends on the lexicon size and labels given by the GOOGLE Vision API.

Table 5. Recognition performance before- and after-prediction for the output of different text detection methods on our dataset

Different text detection methods	Before prediction (OCR)			After prediction-proposed method		
	P	R	F	P	R	F
Zhou et al. EAST [22]	0.40	0.47	0.43	0.61	0.62	0.61
Shi et al. SegLink [23]	0.41	0.49	0.44	0.63	0.61	0.62
Liu et al. CTD [24]	0.39	0.44	0.41	0.62	0.60	0.61
Busta et la. E2E-Text Detection [5]	0.42	0.51	0.46	0.65	0.63	0.64
Deng et al. PixelLink [4]	0.43	0.55	0.48	**0.67**	**0.64**	**0.65**

- The recognition results: "bo", "ary and "Road"
- Labels using GOOGLE Vision API: "Nature", "Signage", "Tree", "Grass", "recreation"
- Candidate List for "bo" + "ary": 'bigotry', 'boringly', 'boundary', 'bovary'
- The proposed method results: "boundary road"

Fig. 8. Restoring the missing text without occlusion

- Recognition results before prediction "Raymo"+"vanheusen"
- The proposed method results: " - "

Fig. 9. Limitation of the proposed method

In this work, the labels generated by the GOOGLE Vision API and generating candidate words are critical to the results. If the GOOGLE Vision API mislabels the input image, there are chances of predicting nothing as shown in Fig. 9. For this image, the word is a noun which is not present in the lexicon and hence the proposed method predicts nothing "–". Similarly, if the candidate generation step lists irrelevant choices for a recognized word, the proposed method fails as well. To overcome these limitations, a new method should be developed for labeling images and generating candidates. One possible solution is after getting labels from the GOOGLE Vision API, we can use geographical information about the location to verify or modify the label according to the situational context. Similarly, the same information can be used for verifying generated candidates. However, this is beyond the scope of the proposed work.

5 Conclusions and Future Work

In this work, we have proposed a new method for restoring missing text for replacing broken words due to occluded text in natural scene images. We explore the GOOGLE Vision API system for obtaining labels of the input images. For text in images including broken words, the proposed approach obtains possible candidate words with the help of the lexicons provided by the ICDAR 2015 ground truth. The proposed method finds semantic similarity between labels and candidate words, which results in a global context score. Our method also employs the natural language processing technique called BERT for finding semantic similarity between the candidates, which results in a local context score. Furthermore, the proposed approach fuses global and local context scores to estimate ranking for each word in the list. The word that gets the highest rank is considered as a correct word or a predicted word for replacing broken words. Experimental results on our dataset show that the proposed method is effective and predicts missing parts well for different situations. As shown, the performance of the proposed approach depends on the results of the GOOGLE Vision API and the list of candidate words. Sometimes, for the images with large variations, the GOOGLE Vision API system may not generate correct labels. In this case, the performance of the proposed method degrades. Therefore, there is scope for future work and we can also extend the proposed work to other languages.

References

1. Roy, S., Shivakumara, P., Pal, U., Lu, T., Kumar, G.H.: Delaunay triangulation based text detection from multi-view images of natural scene. Pattern Recogn. Lett. **129**, 92–100 (2020)
2. Shivakumara, P., Raghavendra, R., Qin, L., Raja, K.B., Lu, T., Pal, U.: A new multi-modal approach to bib number/text detection and recognition in Marathon images. Pattern Recogn. **61**, 479–491 (2017)
3. Xue, M., Shivakumara, P., Zhang, C., Lu, T., Pal, U.: Curved text detection in blurred/non-blurred video/scene images. Multimedia Tools Appl. **78**(18), 25629–25653 (2019). https://doi.org/10.1007/s11042-019-7721-2
4. Deng, D., Liu, H., Li, X., Cai, D.: PixelLink: detecting scene text via instance segmentation. In: Proceedings of the AAAI (2018)
5. Patel, Y., Bušta, M., Matas, J.: E2E-MLT-an unconstrained end-to-end method for multi-language scene text. arXiv preprint arXiv:1801.09919 (2018)
6. Google Cloud Vision API

7. Devlin, J., Chang, M.W., Lee, K., Toutanova, K.: BERT: pre-training of deep bidirectional transformers for language understanding. arXiv preprint arXiv:1810.04805 (2018)
8. Cheng, Z., Xu, Y., Bai, F., Niu, Y.: AON: towards arbitrarily-oriented text recognition. In: Proceedings of the CVPR, pp. 5571–5579 (2018)
9. Tian, S., Yin, X.C., Su, Y., Hao, H.W.: A unified framework for tracking based text detection and recognition from web videos. IEEE Trans. PAMI **40**, 542–554 (2018)
10. Luo, C., Jin, L., Sun, Z.: MORAN: a multi-object rectified attention network for scene text recognition. Pattern Recogn. **90**, 109–118 (2019)
11. Raghunandan, K.S., Shivakumara, P., Roy, S., Kumar, G.H., Pal, U., Lu, T.: Multi-script-oriented text detection and recognition in video/scene/born digital images. IEEE Trans. CSVT **29**, 1145–1162 (2019)
12. Qi, X., Chen, Y., Xiao, R., Li, C.G., Zou, Q., Cui, S.: A novel joint character categorization and localization approach for character level scene text recognition. In: Proceedings of the ICDARW, pp. 83–90 (2019)
13. Shi, B., Yang, M., Wang, X., Luy, P., Yao, C., Bai, X.: ASTER: an attentional scene text recognizer with flexible rectification. IEEE Trans. PAMI **41**, 2035–2048 (2019)
14. Rong, X., Yi, C., Tian, Y.: Unambiguous scene text segmentation with referring expression comprehension. IEEE Trans. IP **29**, 591–601 (2020)
15. Villamizar, M., Canevert, O., Odobez, J.M.: Multi-scale sequential network for semantic text segmentation and localization. Pattern Recogn. Lett. **129**, 63–69 (2020)
16. Feng, W., He, W., Yin, F., Zhang, X.Y., Liu, C.L.: TextDragon: an end-to-end framework for arbitrary shaped text spotting. In: Proceedings of the ICCV, pp. 9076–9085 (2019)
17. Lee, C.W., Jung, K., Kim, H.J.: Automatic text detection and removal in video sequences. Pattern Recogn. Lett. **24**, 2607–2623 (2003)
18. Ye, Q., Jiao, J., Huang, J., Yu, H.: Text detection and restoration in natural scene images. J. Vis. Commun. Image Represent. **18**, 504–513 (2007)
19. Tsai, T.H., Fang, C.L.: Text-video completion using structure repair and texture propagation. IEEE Trans. MM **13**, 29–39 (2011)
20. Mosleh, A., Bouguila, N., Hamaza, A.B.: Automatic inpainting scheme for video text detection and removal. IEEE Trans. IP **22**, 4460–4472 (2013)
21. Zhang, S., Liu, Y., Jin, L., Huang, Y., Lai, S.: EnsNet: ensconce text in the wild. In: Proceedings of the AAAI (2019)
22. Wu, L., et al.: Editing text in the wild. In: Proceedings of the ACM MM, pp. 1500–1508 (2019)
23. Tong, X., Evans, D.A.: A statistical approach to automatic OCR error correction in context. In: Proceedings of the WVLC, pp. 88–100 (1996)
24. Zhou, X., et al.: East: an efficient and accurate scene text detector. In: Proceedings of the CVPR, pp. 2642–2651 (2017)
25. Shi, B., Bai, X., Belongie, S.: Detecting oriented text in natural images by linking segments. In: Proceedings of the CVPR, pp. 3482–3490 (2017)
26. Liu, Y., Jin, L., Zhang, S., Zhang, S.: Detecting curve text in the wild: new dataset and new solution. arXiv:1712.02170 (2017)

New Benchmarks for Barcode Detection Using Both Synthetic and Real Data

Andrey Zharkov[1,2]([✉]) [iD], Andrey Vavilin[1] [iD], and Ivan Zagaynov[1,2] [iD]

[1] R&D Department, ABBYY Production LLC, Moscow, Russia
{andrew.zharkov,andrey.vavilin,ivan.zagaynov}@abbyy.com
[2] Moscow Institute of Physics and Technology (National Research University), Moscow, Russia

Abstract. Given the wide use of barcodes, there is a growing demand for their efficient detection and recognition. However, the existing publicly available datasets are insufficient and of poor quality. Moreover, recently proposed approaches were trained on different private datasets, which makes the comparison of proposed methods even more unfair. In this paper, we propose a simple yet efficient technique to generate realistic datasets for barcode detection problem. Using the proposed method, we synthesized a dataset of ~30,000 barcodes that closely resembles real barcode data distribution in terms of size, location, and number of barcodes on an image. The dataset contains a large number of different barcode types (Code128, EAN13, DataMatrix, Aztec, QR, and many more). We also provide a new real test dataset of 921 images, containing both document scans and in-the-wild photos, which is much more challenging and diverse compared to existing benchmarks. These new datasets allow a fairer comparison of existing barcode detection approaches. We benchmarked several deep learning techniques on our datasets and discuss the results. Our code and datasets are available at https://github.com/abbyy/barcode_detection_benchmark.

Keywords: Barcode detection · Barcode · Barcode dataset · Object detection · Semantic segmentation

1 Introduction

The domain of machine-readable codes or barcodes has a long history. Initially, barcode recognition required specific hardware, such as laser scanners, but due to the rapid progress in mobile camera technology, it is now possible to detect and read barcodes on images taken with end users' mobile devices.

Currently, neural networks significantly outperform traditional computer vision approaches in the majority of tasks where large datasets are available. However, this is not true for barcode detection, where the best-performing deep learning methods [8, 15, 17] are comparable in quality with the best of traditional approaches [9].

© Springer Nature Switzerland AG 2020
X. Bai et al. (Eds.): DAS 2020, LNCS 12116, pp. 481–493, 2020.
https://doi.org/10.1007/978-3-030-57058-3_34

We believe there may be two reasons for such a result. First, there is insufficient training data and second, and more important, too simple evaluation benchmarks have been used. At the time of writing, there are only two small datasets publicly available. On these datasets, the state-of-the-art approaches have already achieved nearly 100% quality and, as shown in our prior work [17], the only errors that remain may in fact be caused by incorrect ground-truth labels rather than by detection algorithm flaws.

Another issue with existing public datasets is that they are not diverse nor difficult enough, having only one large EAN13 barcode in the center of each image in most cases. In reality, however, there exist many different types of barcodes, such as QR, DataMatrix, Postnet, Aztec, PDF417, and more than 30 others. Also, a single image may contain multiple different barcodes at different scales, ranging from very small and almost invisible barcodes to very large objects that may fill the entire image.

In this paper, we address the problem of insufficiency of publicly available data for barcode detection purposes by providing new datasets to facilitate further research and establish fairer comparison. Our main contributions are as follows:

- We have collected a new diverse and challenging real dataset that contains different barcode types at different scales, both on the document images and on in-the-wild photos.
- We propose a simple strategy to generate arbitrary large synthetic datasets of *documents* with barcodes. Using this strategy, we created a dataset of 30,000 images.
- We have evaluated some of the existing techniques on the proposed datasets, obtaining baseline results for future research.

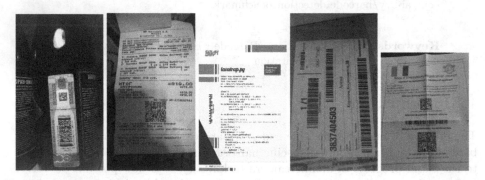

Fig. 1. Examples of images from the real dataset (some were rotated to fit the width)

2 Related Work

2.1 Barcode Detection

First camera-based approaches for barcode detection used traditional computer vision techniques, such as morphology operations [7], corner detection [10], or gradient images [3]. More recent ideas from traditional computer vision are mainly based on various dark bar detection methods [1,2,9].

A pioneering deep learning approach in barcode detection was [14], where authors utilized deep learning to detect dark bars in Hough space. The authors of [8] used YOLO to detect entire barcodes on raw images. However, there are certain barcode types that may be very long and narrow, thus having a very high aspect ratio. Such objects would require the YOLO detector to have a very large context and it would still be hard for the detector to efficiently predict objects with very high aspect ratios. For this reason, a simpler approach may be to solve semantic segmentation problem instead, as done in [15]. But the target task we are interested in is still detection, not segmentation, so after predicting output segmentation map we should apply some postprocessing, i.e. finding connected components and treating these found components as detected objects. In our previous work [17], we introduced such simple postprocessing method and also proposed lighter and faster segmentation backbone network architecture.

2.2 Barcode Datasets

There are two publicly available barcode detection datasets. One is the WWU Muenster Barcode Database (Muenster), introduced in Sörös et al. [13]. It contains 595 images with several 1D barcode types, with 1–3 barcodes per image. The other is the ArTe-Lab Medium Barcode Dataset (Artelab), proposed by Zamberletti et al. [14]. This dataset contains 365 images, with exactly one barcode (EAN13) on each. Barcodes in both of these datasets are typically relatively large compared to the image size. Unfortunately, these datasets have several incorrectly labeled images, and since current state-of-the-art methods are approaching 100% quality on these datasets, we must question their further use for detection algorithms comparison.

One notable example of a synthetic barcode dataset can be found in [15], whose authors synthesized a dataset of 30,000 barcodes by simply putting QR and 1D codes into image backgrounds. However, the authors have not published their dataset and offer only a qualitative evaluation of their method on real data. Our synthetic generation strategy is similar to that proposed in [15], the main difference being that we not just replace background pixels with a barcode, but "adapt" the inserted barcode image to make the entire image look more realistic (see Sect. 3.1 for details).

Some other works, e.g. [15,17], use private training (or even testing) datasets, which does not allow a fair comparison of the proposed methods.

To sum up, currently available real datasets are insufficient in terms of diversity and size, are too simple, and contain only a few 1D barcode types. They also contain some markup errors, which prevents fair comparison, as state-of-the-art methods are in fact competing to approximate these markup errors. There are no publicly available synthetic benchmarks either. Finally, there has been an increasing trend to use private datasets for training, which makes it even more difficult to reproduce and compare proposed methods.

3 Creating Synthetic Datasets

3.1 Synthesis Methodology

In this section, we explain our approach to synthesizing datasets of document images with barcodes.

The image generation process can be divided into four stages: analyzing the background of an image, generating barcodes, finding the appropriate locations for the barcodes, and placing the barcode onto the image.

During background analysis, we apply adaptive binarization to detect foreground and background pixels. Based on the resulting binarized image, we compute an integral image to speed up the subsequent steps. We also estimate the average intensity for white (background) I_{avg}^{b} and black (foreground) I_{avg}^{f} pixels, using the binarized image as a mask for the black pixels.

The generation process starts with a random selection of some barcodes to be added onto the document images. Each barcode is either read from a predefined set of clean barcode images or generated with parameters (size, type, bar width, etc.) randomly selected from a predefined range. Additionally, a barcode image can be rotated by a random angle. As a result of this step, we obtain a set of barcode bitmaps that can be placed on the document images.

For each generated barcode bitmap, we must select an appropriate location where the barcode will not overlap with the surrounding data (because barcodes do not normally overlap with surrounding text). The following algorithm is applied:

1. Generate a binary image in which 1 indicates pixels where the barcode bitmap does not intersect with surrounding data. For each pixel in this image, check if the rectangle enclosing the bitmap has any intersections with any surrounding data. This can be easily checked using the integral image.
2. Apply connected components labeling to split up regions where a barcode could be placed.
3. Randomly select one of the regions, then randomly select a pixel inside the selected region.

Finally, we put a barcode bitmap in the selected location, preserving the texture of the background image. The most straightforward and bold approach for drawing a clean barcode in a selected location would be to simply replace background image pixels with the pixels of the clean barcode. However, this approach will produce unrealistic images with high-contrast barcodes inside. We propose a simple "adaptation" technique that produces more realistic results:

- Select the intensity of the topmost left background pixel in the selected location as reference intensity I_{ref}.
- For all the white pixels of the barcode, keep the intensity of the background image intact.
- For the black pixels, intensity is calculated as average black intensity I_{avg}^f (computed during image analysis step) plus the difference between the background intensity at the pixel's location I_{ij} and the reference intensity I_{ref}. So the formula for the black pixels is $I_{avg}^f + (I_{ij} - I_{ref})$.

3.2 Synthetic Dataset Structure

Using the proposed method, we generated a dataset of ~30,000 images with barcodes. Ten different strategies were used, with ~3,000 images per strategy.

The best way to collect backgrounds would probably be to use a lot of real documents. Since we were unable to find a sufficient number of documents, we used a subset of the synthetic document dataset from DDI-100 [16]. For clean barcodes, we generated several (10–25 per type) barcode samples of the following types: AustralianPost, Aztec, DataMatrix, EAN8, EAN128, Kix, MaxiCode, PDF417, Postnet, and QRCode.

Table 1 lists the hyperparameters used for the generation process. The values in the "Background" column specify the respective portion of the background dataset.

The "Side Relations" column contains the range of allowed side relations (minimal barcode side/minimal document side) in percent, so using the values from [5, 30] we can prevent too small or too large barcodes from appearing in the generated data, as they are uncommon in real documents. However, very small and very large barcodes sometimes do occur in real data, so we used different values for some portions of our dataset.

In the "Generator" column, we specify if a barcode generator was used and, if yes, for which barcode type. If this value is specified, the barcode was selected from *both* a predefined directory with clean barcodes (50% chance) and a generator for the specified type (50% chance). If no value in this column is specified, only clean barcode images were used.

The "Max Barcodes" column specifies the maximum number of barcodes to be placed on each image (if, after placing $0 < m < max\_barcodes$, we could not find a place for additional barcodes, an image with only m barcodes was generated). The barcodes were rotated by an angle uniformly sampled from $[-max\_rotation, max\_rotation]$. Some examples from our synthetic dataset are shown in Fig. 2.

3.3 Real Dataset

We also collected a real dataset, which is larger than its publicly available predecessors. The collected dataset has 971 images with 1214 barcodes in total, with a maximum of 7 barcodes per image. One image may contain barcodes of different types. There are 18 different barcode types. The exact distribution of the types is

Table 1. Synthetic dataset

Strategy	Images	Background	Max rotation	Side relations	Max barcodes	Generator
01	3,000	part01	10	[5, 30]	3	–
02	3,000	part10	5	[5, 30]	5	Code128
03	3,009	part11	0	[5, 30]	3	Code128
04	3,000	part03	45	[5, 30]	5	–
05	3,000	part03	5	[5, 30]	20	Code128
06	3,000	part04	45	[5, 30]	3	Code128
07	2,999	part06	0	[1, 30]	10	Standard2of5
08	3,000	part01	0	[1, 15]	5	EAN13
09	3,000	part01	0	[1, 30]	20	EAN13
10	2,999	part01	45	[1, 7]	3	EAN13

Fig. 2. Examples of images from the synthetic dataset

shown in Table 2. There are some challenging images, e.g. containing very small barcodes or barcodes of extreme aspect ratios (i.e. very wide but small in hight). Some examples are shown in Fig. 1.

Table 2. Barcode types in the real dataset

Barcode type	EAN13	PDF417	RoyalMail	QR	Code128	DataMatrix
Total count	308	194	192	146	101	92
Barcode type	Code39	Aztec	JapanPost	Interleaved25	UPCA	Kix
Total Count	51	42	25	22	10	8
Barcode type	Postnet	UCC128	2-digit	IATA25	IntelligentMail	EAN8
Total count	7	6	3	3	2	2

4   Baseline Approach

In all of our experiments, we decided to use semantic segmentation for barcode detection as described in [17], as this approach is suitable for detecting objects of arbitrary shapes, including very wide barcodes with small hights (which are quite common in real documents). In the next subsection, we briefly explain the original idea from [17]. After that, we describe the changes that we made (mostly for simplicity of implementation).

4.1   Object Detection via Semantic Segmentation

The main idea is to use semantic segmentation, then binarize the heatmap of the output "objectness" channel, and finally, find connected components in the binarized output. Next, the connected components may be appropriately convexified (in the case of barcodes, we find minimal enclosing rectangle) and regarded as detected objects.

If several object classes are detected (i.e. we need to differentiate between 1D and 2D barcodes), additional network output channels may be used to determine the class of the found objects, which is achieved by (1) averaging these additional channel values in the heatmap area inside the found object and then (2) determining the class of the found object as the class that corresponds to the output channel with the maximum value computed in (1). The whole approach is visualized in Fig. 3.

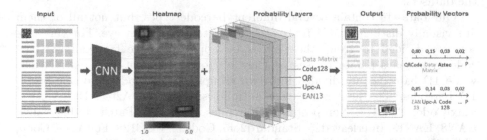

Fig. 3. Object detection via semantic segmentation. Input image is fed into CNN. The output channels of the CNN are related to 0) objectness, 1) EAN13, 2) UPC-A, 3) QR, 4) Code128, and 5) DataMatrix (six channels in total). To locate the output objects, we binarize channel 0, find connected components on the resulting image, and then find minimal enclosing rectangles for them. For each of the found objects, the probabilities for the class channels are averaged of the current object area. Finally, the class of the object is determined with respect to its related channel with maximum average probability within the found object area.

The first output channel of the segmentation network is used to detect arbitrary barcodes, and the rest of the channels are used for further classifying the already detected objects. In [17], it was shown that a barcode classification subtask along with detection improved detection results, so we used this additional classification subtask in all of our experiments.

4.2   Differences Between the Original Approach and Our Baseline

We have made two slight modifications to the original approach. The first is that we classify barcodes not by their exact type but by visually distinguishable groups (which we formulate in Sect. 5.1).

The second modification is related to the training process. Instead of using softmax loss for classification, we compute a sigmoid for each classification channel separately. This approach is used, for instance, in the newer YOLO versions [11], resulting in more accurate classification.

5   Experiments

5.1   Datasets

In our experiments, we used our proposed synthetic dataset (ZVZ-synth) and a real dataset (ZVZ-real). For ZVZ-synth we used 10% of images for validation and the rest for training. ZVZ-real was split into 512 training images, 102 validation images, and 306 test images (our earlier experiments had shown that training on synthetic data alone did not produce good results, so we had to use a portion of the real dataset for training; however, we also provide numbers for the entire real dataset for comparison in future research, where ZVZ-real may be used for evaluation only). To allow a fair comparison, we provide the splits along with the datasets.

Our datasets contain several different barcode types, but not all of them are visually distinguishable from each other to the human eye. For this reason, we decided to group these types by appearance: we treat all 2D barcode types as separate classes as they are quite different from each other, while 1D barcodes are divided into two groups - postal codes and non-postal 1D barcodes. The final classes are as follows: QRCode, Aztec, DataMatrix, MaxiCode, PDF417, Non-postal-1D-Barcodes (Code128, Patch, Industrial25, EAN8, EAN13, Interleaved25, Standard2of5, Code32, UCC128, FullASCIICode, MATRIX25, Code39, IATA25, UPCA, UPCE, CODABAR, Code93, 2-Digit), and Postal-1D-Barcodes (Postnet, AustraliaPost, Kix, IntelligentMail, RoyalMailCode, JapanPost).

5.2   Metrics

For comparison, we used object-based *precision, recall, F-measure*, and image-based *detection rate* as defined in [17]. We used *detection rate* mainly to compare our results with prior research, even though this measure has two big disadvantages. First, it is not very sensitive to missed small objects (this type of errors will only be visible at very high thresholds). Second, it is typically very high compared to F-measure (see our experiments in this section, for example), which makes it more difficult to compare two models, as good models typically have very high detection rates. Since our target task is *detection*, we propose to use object-based precision, recall, and F-measure instead of image-based detection rate in further studies.

Table 3. Results obtained on the synthetic validation subset

Arch	Pretrained on	Precision	Recall	F1-score	Detection rate
ResNet18U-Net	ImageNet	**0.971**	**0.963**	**0.967**	**0.993**
ResNet18U-Net	–	0.970	0.956	0.964	0.989
DilatedModel	–	0.886	0.947	0.915	0.988

5.3 Baseline Models

We used two semantic segmentation architectures. The first is U-Net [12] with a ResNet18 [5] backbone, while the second is a small dilated model from [17], where we replaced all the separable convolutions with simple vanilla convolutions.

Since we have 7 barcode type groups, the total number of output channels for each network is 8 (1 detection + 7 classification). The number of input channels is 3, representing RGB image (we chose RGB over grayscale input to be able to use ImageNet pretrained weights).

5.4 Training Setup

For training, we resized all of the images to the maximum side of 512 pixels. To simplify the training procedure, we zero-padded all of the images to 512 × 512 and grouped them into batches. For inference prediction, we always used one image with the maximum side of 512 pixels *without paddings*, all sides evenly divisible by 32 (which is the maximum scale for ResNet18).

For all the models, we used batch size 8 and Adam optimizer with a learning rate 0.001 for the first 50 epochs and 0.0003 after that. For data augmentation, we applied small rotations, perspective transformations, and color and intensity changes.

The loss function is defined as

$$L = \frac{1}{1 + w_{classification}} L_{detection} + \frac{w_{classification}}{1 + w_{classification}} L_{classification},$$

where $w_{classification} = 0.05$. $L_{detection}$ and $L_{classification}$ are computed as the average base loss function for the respective detection or classification channels. We use the unweighted average of three losses (binary cross entropy, dice-loss, and iou-loss) as the base loss function: $L_{base} = (L_{bce} + L_{dice} + L_{iou})/3$. The reason for this choice is to make the loss function smoother, which is a known technique to improve generalization [6].

5.5 Experiments on Synthetic Data

For ZVZ-synth, we trained all the models for 10 epochs, which was quite enough to converge. The results are provided in Table 3.

We observed two kinds of errors (see Fig. 4). The most frequent error is that very close barcodes are joined into a single large connected component, which leads to missed detection. The second frequent error is a failure to detect extremely small barcodes.

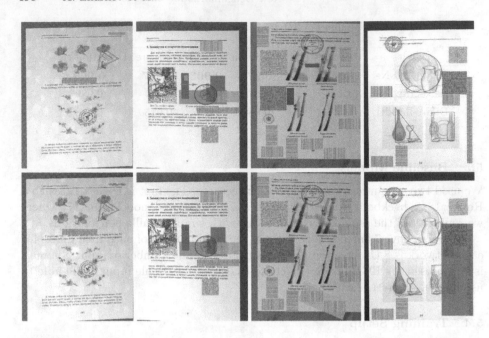

Fig. 4. Example of errors on the synthetic data (top - ground truth, bottom - predictions)

5.6   Experiments on Real Data

As ZVZ-real contains a lot of in-the-wild photos, it was to be expected that training on ZVZ-synth (which contains only document images) alone did not result in good performance on real data (approximately 40% precision and 80% recall). Therefore, we had to fine-tune all the models on real data. We compared ImageNet pretraining (we used weights of ResNet18 from torchvision) with pre-training on ZVZ-synth.

For ZVZ-real, we trained all the models for 500 epochs. It is worth mentioning that for the models which were pretrained on other datasets we used *the same learning rate* as for the models trained from scratch. Surprisingly, it turned out that lowering the learning rate resulted into *worse performance* compared to the models trained on real data alone.

Figure 5 shows typical recall errors on the test dataset for the best-performing model (ImageNet pretrained). Again, as well as with the synthetic dataset, it was hard for the model to predict extremely small objects. Interestingly, another type of errors which we observed seemed to be due some confusion of the model when there are several objects located close to one another, in that case model can completely miss some of the objects. We did not observe any problems with barcode separation, as the test dataset was pretty clear and there were no overlapping barcodes.

Table 4. Results obtained on the real dataset (P = precision, R = recall, F = F1-score, D = detection rate). FULL states for entire real dataset. (train + valid + test)

Arch	Pretrained on	TEST				FULL			
		P	R	F	D	P	R	F	D
ResNet18U-Net	ImageNet	0.830	**0.959**	0.889	**0.967**	0.905	**0.972**	0.937	**0.978**
ResNet18U-Net	ZVZ-synth	**0.865**	0.937	**0.900**	0.944	**0.924**	0.962	**0.943**	0.969
ResNet18U-Net	–	0.809	0.915	0.858	0.925	0.878	0.948	0.912	0.953
DilatedModel	ZVZ-synth	0.730	0.951	0.827	**0.967**	0.812	0.964	0.881	0.974
DilatedModel	–	0.731	0.932	0.820	0.944	0.799	0.944	0.866	0.962

6   Discussion

6.1   Strength of Our Baseline

Before discussing results and making conclusions we want to warn the reader about the strength of our baselines.

As our primary goal was to provide good public datasets we have not spent much time for developing the best training pipeline (e.g. we did not try different optimizers, learning rate schedules, augmentation strength, loss functions, etc.). So we believe the baseline results may be strengthened by the improved training protocol, but we left this for further studies.

6.2   Effects of Pretraining

When comparing the results with and without pretraining on our simple learning pipeline, it appears as if pretraining (either on synthetic data or on ImageNet) helps. However, we believe that this particular difference may be minimized (see [4] for evidence).

But at least for our simple training pipeline pretraining seems to provide useful features, so if you do not want to spend too much time on parameter tuning it may be a good idea to pretrain your network on a larger dataset.

6.3   Baseline Models Compared

In our experiments, we used two segmentation models - U-Net with a ResNet18 backbone and DilatedModel as described in Sect. 5.3. The number of parameters and inference speeds are listed in Table 5.

Table 5. Baseline backbones compared. GPU time for Nvidia GTX 1080, CPU time for Intel Core i5-8600 3.10 GHz. Average over 100 runs. Input image size 512 × 512. Run in PyTorch 1.3.1 with cudatoolkit 10.1.

Arch	Parameters	Inference GPU	Inference CPU
ResNet18U-Net	26M	27.4 ± 0.5 ms	9,570 ± 50 ms
DilatedModel	42K	1.8 ± 0.3 ms	203 ± 2 ms

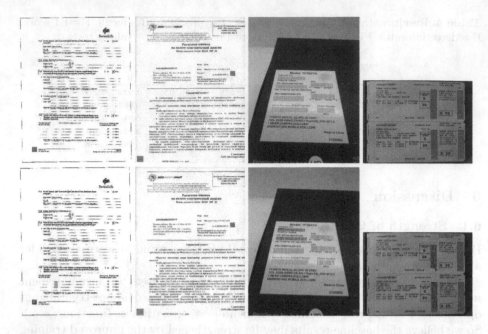

Fig. 5. Examples of errors on the real data (top - ground truth, bottom - predictions)

When we compare precision and recall of the two models (see Table 3 and Table 4), we see that they have very similar levels of recall (and sometimes the smaller DilatedModel has better recall than the much larger ResNet18U-Net) and the main difference is that ResNet18U-Net is much more precise (2–5 times better precision error).

To sum up, DilatedModel is ∼600 times lighter, ∼15 times faster when using a GPU, almost as good (or even slightly better) in terms of recall, but ∼2–5 times worse in terms of precision compared to ResNet18U-Net. Since the two models have approximately the same recall, one may want to choose the faster model and additional garbage hypothesis post-filtering which will probably give better efficiency.

7   Conclusion

In this paper, we present new publicly available synthetic and real datasets for barcode detection. The provided datasets are larger, much more challenging and diverse than the existing ones. On these new datasets we have built baseline deep learning methods and analyse their results quantitatively and qualitatively. We also proposed synthetic data generation technique on documents which may be used to create arbitrary large datasets which can potentially contain not only barcodes but any other objects.

References

1. Creusot, C., Munawar, A.: Real-time barcode detection in the wild. In: 2015 IEEE Winter Conference on Applications of Computer Vision, pp. 239–245, January 2015. https://doi.org/10.1109/WACV.2015.39
2. Creusot, C., Munawar, A.: Low-computation egocentric barcode detector for the blind. In: IEEE International Conference on Image Processing (ICIP), pp. 2856–2860, September 2016. https://doi.org/10.1109/ICIP.2016.7532881
3. Gallo, O., Manduchi, R.: Reading 1D barcodes with mobile phones using deformable templates. IEEE Trans. Pattern Anal. Mach. Intell. **33**, 1834–1843 (2011)
4. He, K., Girshick, R., Dollar, P.: Rethinking imagenet pre-training. In: The IEEE International Conference on Computer Vision (ICCV), October 2019
5. He, K., Zhang, X., Ren, S., Sun, J.: Deep residual learning for image recognition. In: The IEEE Conference on Computer Vision and Pattern Recognition (CVPR), June 2016
6. Iglovikov, V.I., Seferbekov, S., Buslaev, A.V., Shvets, A.: TernausNetV2: fully convolutional network for instance segmentation (2018)
7. Katona, M., Nyúl, L.G.: Efficient 1D and 2D Barcode detection using mathematical morphology. In: Hendriks, C.L.L., Borgefors, G., Strand, R. (eds.) ISMM 2013. LNCS, vol. 7883, pp. 464–475. Springer, Heidelberg (2013). https://doi.org/10.1007/978-3-642-38294-9_39
8. Hansen, D.K., Nasrollahi, K., Rasmusen, C.B., Moeslund, T.: Real-time barcode detection and classification using deep learning, pp. 321–327, January 2017. https://doi.org/10.5220/0006508203210327
9. Namane, A., Arezki, M.: Fast real time 1D barcode detection from webcam images using the bars detection method, July 2017
10. Ohbuchi, E., Hanaizumi, H., Hock, L.A.: Barcode readers using the camera device in mobile phones. In: International Conference on Cyberworlds, pp. 260–265, November 2004. https://doi.org/10.1109/CW.2004.23
11. Redmon, J., Farhadi, A.: YOLO9000: better, faster, stronger (2016)
12. Ronneberger, O., Fischer, P., Brox, T.: U-net: convolutional networks for biomedical image segmentation. ArXiv abs/1505.04597 (2015)
13. Sörös, G., Flörkemeier, C.: Blur-resistant joint 1D and 2D barcode localization for smartphones. In: Proceedings of the 12th International Conference on Mobile and Ubiquitous Multimedia, MUM 2013, pp. 11:1–11:8. ACM, New York (2013). https://doi.org/10.1145/2541831.2541844
14. Zamberletti, A., Gallo, I., Albertini, S.: Robust angle invariant 1d barcode detection. In: 2013 2nd IAPR Asian Conference on Pattern Recognition, pp. 160–164, November 2013. https://doi.org/10.1109/ACPR.2013.17
15. Zhao, Q., Ni, F., Song, Y., Wang, Y., Tang, Z.: Deep dual pyramid network for barcode segmentation using barcode-30k database (2018)
16. Zharikov, I., Nikitin, F., Vasiliev, I., Dokholyan, V.: DDI-100: dataset for text detection and recognition (2019)
17. Zharkov, A., Zagaynov, I.: Universal barcode detector via semantic segmentation. In: 15th International Conference on Document Analysis and Recognition (ICDAR), Sydney, Australia, pp. 837–843, September 2019. https://doi.org/10.1109/ICDAR.2019.00139

Font Design and Classification

Character-Independent Font Identification

Daichi Haraguchi$^{(\boxtimes)}$, Shota Harada, Brian Kenji Iwana, Yuto Shinahara, and Seiichi Uchida

Kyushu University, Fukuoka, Japan
{daichi.haraguchi,shota.harada,brian,uchida}@human.ait.kyushu-u.ac.jp

Abstract. There are a countless number of fonts with various shapes and styles. In addition, there are many fonts that only have subtle differences in features. Due to this, font identification is a difficult task. In this paper, we propose a method of determining if any two characters are from the same font or not. This is difficult due to the difference between fonts typically being smaller than the difference between alphabet classes. Additionally, the proposed method can be used with fonts regardless of whether they exist in the training or not. In order to accomplish this, we use a Convolutional Neural Network (CNN) trained with various font image pairs. In the experiment, the network is trained on image pairs of various fonts. We then evaluate the model on a different set of fonts that are unseen by the network. The evaluation is performed with an accuracy of 92.27%. Moreover, we analyzed the relationship between character classes and font identification accuracy.

Keywords: Font identification · Representation learning · Convolutional Neural Networks

1 Introduction

In this paper, we tackle font identification from different character classes. Specifically, given a pair of character images from different character classes[1], we try to discriminate whether the images come from the same font or not. Figure 1(a) shows example input pairs of the task; we need to discriminate the same or different font pairs. It is easy to identify that Fig. 1(b) is the same font pair. It is also easy to identify that Fig. 1(c) is a different font pair. In contrast, the examples in Fig. 1(d) and (e) are more difficult. Figure 1(d) shows the same font pairs, whereas (e) shows different font pairs.

This task is very different from the traditional font identification task, such as [9,20,34]. In the traditional task, given a character image (a single letter

---

[1] Throughout this paper, we assume that the pairs come from different character classes. This is simply because our font identification becomes a trivial task for the pairs of the same character class (e.g., 'A'); if the two images are exactly the same, they are the same font; otherwise, they are different.

© Springer Nature Switzerland AG 2020
X. Bai et al. (Eds.): DAS 2020, LNCS 12116, pp. 497–511, 2020.
https://doi.org/10.1007/978-3-030-57058-3_35

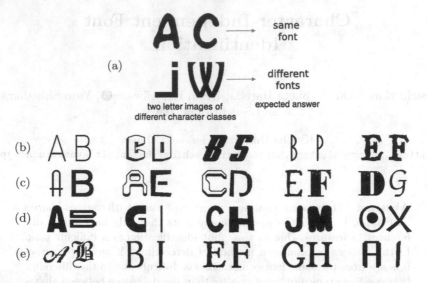

Fig. 1. Explanation of our task and examples of character image pairs from different classes. (a) is an explanation of our task. The pairs in, (b) shows the same font pairs, whereas (c) shows different font pairs. In contrast, (d) and (e) are difficult cases; (d) shows the same pairs, whereas (e) shows different pairs.

image or a single word image or a sentence image), we need to answer the font name (e.g., Helvetica) and the type name of font (e.g. Blackletter). In a sense, it is rather a recognition task than an identification task. This is because, in the traditional task, we only can identify the fonts which are registered in the system in advance. In other words, it is a multi-class font recognition task and each class corresponds to a known font name. In contrast, our font identification task is a two-class task to decide whether a pair of character images come from the same font or different fonts, without knowing those font names beforehand.

We propose a system for our font identification task for a pair of character images from different character classes. The system is practically useful because the system will have more flexibility than the systems for the traditional font identification task. As noted above, the traditional systems only "recognize" the input image as one of the fonts that are known to the system. However, it is impossible to register all fonts to the system because new fonts are generated every day in the world. (In the future, the variations of fonts will become almost infinite since many automatic font generation systems have been proposed, such as [11, 16, 21, 31]). Accordingly, traditional systems will have a limitation with dealing with those fonts that are "unknown" to them. Since the proposed system does not assume specific font classes, it can deal with arbitrary font images.

Moreover, since the proposed system assumes single character images as its input, we can perform font identification even if a document contains a small number of characters. For example, analysis of incunabula or other printed historical documents often needs to identify whether two pieces of documents are printed in the same font or not. A similar font identification task from a limited number of characters can be found in forensic research. For example, forensic experts need to determine whether two pieces of documents are printed by the same printer or not.

In addition to the above practical merits, our font identification is a challenging scientific task. In fact, our task is very difficult even though it is formulated as just a binary classification problem. Figure 2 illustrates the distribution of image samples in a feature space. As a success of multi-font optical character recognition (OCR) [33] proves, the samples from the same character class form a cluster and the clusters of different character classes are distant in the feature space. This is because inter-class variance is much larger than intra-class variance; that is, the difference by the character classes is larger than the difference by the fonts. This fact can be confirmed by imagining the template matching-based identification. Although we can judge the class identity of two images (in different fonts) even by template matching, we totally cannot judge the font identity of two images (in different character classes). Consequently, our system needs to disregard large differences between character classes and emphasize tiny differences (such as the presence or absence of serif) in fonts. We find a similar requirement in the text-independent writer identification task, such as [23].

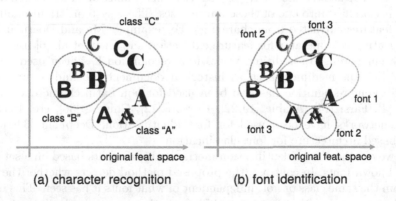

(a) character recognition   (b) font identification

Fig. 2. Comparing multi-font character recognition (a), our font identification (b) is a more difficult classification task.

In this paper, we experimentally show that even a simple two-stream convolutional neural network (CNN) can achieve high accuracy for our font identification task, in spite of the above anticipated difficulty. Our CNN is not very modern (like a CNN with a feature disentanglement function [17,18,27]) but simply accepts two-character image inputs and makes a decision for the binary classification (i.e., the same font or not). In addition, we show - detailed analysis of the identification results. For example, we will observe which alphabet pairs (e.g., 'A'-'K') are easier or more difficult for the identification. In fact, there is a large difference in the identification performance among alphabet pairs.

The main contributions of our paper are summarized as follows:

- To the authors' best knowledge, this is the first attempt of font identification for different character classes.
- Through a large-scale experiment with more than 6,000 different fonts, we prove that even a simple two-stream CNN can judge whether two-letter images come from the same font or not with high accuracy ($>90\%$), in spite of the essential difficulty of the task. It is also experimentally shown that trained CNN has a generalization ability. This means that the representation learning by the simple CNN is enough to extract font style features while disregarding the shape of the character class.
- By analyzing the experimental results, we prove the identification accuracy is dependent on the character class pairs. For example, 'R' and 'U' are a class pair with high accuracy, whereas 'I' and 'Z' have a lower accuracy.

2   Related Work

2.1   Font Identification and Recognition

To the authors' best knowledge, this is the first trial of font identification in our difficult task setting. Most past research on font identification is visual font recognition, where a set of fonts are registered with their names and an input character image is classified into one of those font classes. These systems traditionally use visual features extracted from characters. For example, Ma and Doermann [20] use a grating cell operator for feature extraction and Chen et al. [6] use a local feature embedding. In addition, visual font recognition has been used for text across different mediums, such as historical documents [9] and natural scene text [6]. Font recognition has also been used for non-Latin characters, such as Hindi [4], Farsi [14], Arabic [19,22], Korean [12], Chinese [36], etc. Recently, neural networks have been used for font identification. DeepFont [34] uses a CNN-based architecture for font classification.

However, these font identification methods classify fonts based on a set number of known fonts. In contrast, the proposed method detects whether the fonts are from the same class or not, independent of what fonts it has seen. This means that the proposed method can be used for fonts that are not in the dataset, which can be an issue given the growing popularity of font generation [2,11,21].

In order to overcome this, an alternative approach would be to only detect particular typographical features or groups of fonts. Many classical font recognition models use this approach and detect typographical features such as typeface, weight, slope, and size [5,13,30]. In addition, clustering has been used to recognize groups of fonts [3,24].

2.2   Other Related Identification Systems

The task of font identification can be considered as a subset of script identification. Script identification is a well-established field that aims to recognize the script of text, namely, the set of characters used. In general, these methods are designed to recognize the language for individual writing-system OCR modules [8]. Similar to font identification, traditional script identification use visual features such as Gabor filters [26,32] and text features [7,25].

Furthermore, font identification is related to the field of signature verification and writer identification. In particular, the task of the proposed method is similar to writer-independent signature verification in that both determine if the text is of the same source or different sources. Notably, there are methods in recent times which use CNNs [10,37] and Siamese networks [28,35] that resemble the proposed method.

3   Font Identification by Convolutional Neural Networks

Given a pair of character images $\mathbf{x}_c$ and $\mathbf{x}_d$ of font class c and d respectively, our task is to determine if the pair of characters are of the same font $(c = d)$ or different fonts $(c \neq d)$. In this way, the classifier assigns a binary label indicating a positive match and a negative match. The binary label is irrespective of the character or actual font of the character used as an input pair.

In order to perform font identification, we propose a two-stream CNN-based model. As shown in Fig. 3, a pair of input characters are fed to separate streams of convolutional layers which are followed by fully-connected layers and then the binary classifier. In addition, the two streams of convolutional layers have the same structure and shared weights. This is similar to a Siamese network [15], typically used for metric learning, due to the shared weights. However, it differs in that we combine the streams before the fully connected layers and have a binary classifier with cross-entropy loss.

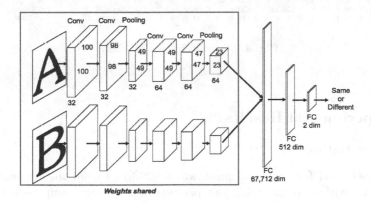

Fig. 3. Structure of the neural networks for font identification.

Each stream is comprised of four convolutional layers and two max-pooling layers. The kernel size of the convolutions is 3×3 with stride 1 and the kernel size of the pooling layers is 2×2 with stride 2. The features from the convolutional layers are concatenated and fed into three fully-connected layers. Rectified Linear Unit (ReLU) activations are used for the hidden layers and softmax is used for the output layer. During training, dropout with a keep probability of 0.5 is used after the first pooling layer and between the fully-connected layers.

Table 1. Confusion matrix of the test set

GT/predicted	Same	Different
Same	196,868 ± 1,758	7,232 ± 1,758
Different	24,331 ± 2,014	179,769 ± 2,014

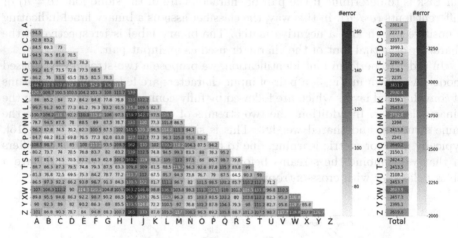

Fig. 4. Misidentification by class

4   Experimental Results

4.1   Font Dataset

The dataset used for the experiment was 6,628 fonts from the Ultimate Font Download[2]. Although the total font package is originally comprised of about 11,000, we removed "dingbat" fonts (i.e., icon-like illustrations and illegible fonts) for the experiments and the 6,628 fonts remain. This font dataset still contains main fancy fonts; in an appendix, we will discuss another dataset with more formal fonts. To construct the dataset, we rasterize the 26 uppercase alphabet characters into 100 × 100 binary images. We only use uppercase characters in

---

[2] http://www.ultimatefontdownload.com/.

this paper for experimental simplicity. Although, it should be noted that there are some fonts that contain lowercase character shapes as uppercase characters.

The 6,628 fonts were divided into three font-independent sets, 5,000 for training, 1,000 for validation, and 628 for testing. Within each set, we generated uppercase alphabet pairs from the same font (positive pairs) and different fonts (negative pairs). Each of the pairs contains different alphabetical characters. Furthermore, each combination of characters is only used one time, i.e. either 'A'-'B' or 'B'-'A' is used, not both. Therefore, we have $_{26}C_2 = 325$ total pairs of each font. Consequently, the training set has $5,000 \times 325 \approx 1.60 \times 10^6$ positive pairs. An equal number of negative pairs are generated by randomly selecting fonts within the training set. Using this scheme, we also generated $2 \times 3.25 \times 10^5$ for validation and approximately $2 \times 2.04 \times 10^5$ for testing. In addition, as outlined in Appendix Appendix A, a second experiment was performed on an external dataset to show the generalization ability of the trained model on other fonts.

The 20 worst
'I' - 'Z', 'I' - 'W', 'I' - 'Q', 'I' - 'S', 'I' - 'M', 'J' - 'W', 'I' - 'X', 'I' - 'O', 'A' - 'I', 'J' - 'M', 'C' - 'I', 'I' - 'J', 'I' - 'R', 'W' - 'Z', 'L' - 'W', 'G' - 'I', 'B' - 'I', 'E' - 'I', 'T' - 'W', 'I' - 'Y'

The 20 best
'R' - 'U', 'D' - 'H', 'K' - 'U', 'B' - 'P', 'B' - 'D', 'D' - 'U', 'D' - 'P', 'D' - 'O', 'D' - 'S', 'D' - 'R', 'D' - 'E', 'U' - 'V', 'K' - 'Y', 'C' - 'U', 'D' - 'K', 'D' - 'G', 'C' - 'D', 'C' - 'G', 'B' - 'R', 'N' - 'U'

Fig. 5. The character pairs with the 20 worst and 20 best accuracies

4.2 Quantitative Evaluation

We conduct 6-fold cross-validation to evaluate the accuracy of the proposed CNN. The identification accuracy for the test set was $92.27 \pm 0.20\%$. The high accuracy demonstrates that it is possible for the proposed method to determine if the characters come from the same font or not, even when they come from different characters. Table 1 shows a confusion matrix of the test results. From this table, it can be seen that different font pairs have more errors than the same font pairs. This means that similar but different font pairs are often misidentified as the same font.

Among the misidentifications, there are some character pairs that are more difficult to classify than others. As shown in Fig. 4, we find that the pairs including 'I' or 'J' are more difficult. This is due to there being few distinct features in 'I' and 'J' due to their simplicity.

Additionally, we found that character pairs with similar features are predictably easier to differentiate and character pairs with different features are difficult. In other words, the amount of information that characters have, such as angles or curves, is important for separating matching fonts and different fonts. For example, in Fig. 4, the number of misidentifications of the 'I'-'T' pair is the lowest of any pair combination which includes an 'I' because 'T' has a similar shape to 'I'. We also find that the number of misidentifications for 'D', 'K', 'R', and 'U' are the least because they have the most representative features of straight lines, curves, or angles.

This is consistent with other characters with similar features. The character pair with the worst classification rate is 'I'-'Z' and the character pair with the highest accuracy is 'R'-'U,' as outlined in Fig. 5. From this figure, we can see that many characters with similar features have high accuracies. For example, 'B'-'P,' 'B'-'D,' and 'O'-'D.' As a whole, 'C'-'G' and 'U'-'V' pairs have fonts that are easy to identify. These pairs are not likely to be affected by the shape of the characters.

Interestingly, the top 5 easiest characters paired with 'B' for font identification are 'P,' 'D,' 'R,' 'H,' and 'E' and the top 5 for 'P' are 'B,' 'D,' 'R,' 'E,' and 'F.' In contrast, the top 5 easiest font identifications with 'R' are 'U,' 'D,' 'B,' 'C,' and 'K.' 'B' and 'P' have the same tendency when identifying fonts. However, font identification with 'R' seems to use different characteristics despite 'B,' 'P,' and 'R' having similar shapes. This is because 'B' and 'P' are composed of the same elements, curves, and a vertical line, whereas 'R' has an additional component.

Fig. 6. Examples of correctly identified pairs (GT: same → prediction: same). The font pair marked by the red box has a nonstandard character. (Color figure online)

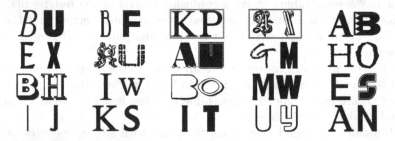

Fig. 7. Examples of correctly identified pairs (GT: different → prediction: different). The font pairs marked by the blue boxes have similar but different fonts. The red box indicates fonts that are almost illegible. (Color figure online)

4.3   Qualitative Evaluation

We show some examples of correctly identified pairs in Fig. 6. In the figure, the proposed method is able to identify fonts despite having dramatically different features such as different character sizes. However, the font weight of the correctly identified fonts tends to be similar. Also notably, in Fig. 6, in the 'A'-'O' pair, the 'O' does not have a serif, yet, the proposed method is able to identify them as a match. Furthermore, the character pair highlighted by a red box in Fig. 6 is identified correctly. This is surprising due to the first character being unidentifiable and not typical of any character. This reinforces that the matching fonts are determined heavily by font weight.

It is also easy for the proposed method to correctly identify different font pairs that have obviously different features to each other. Examples of different font pairs that are correctly identified are shown in Fig. 7. Almost all of the pairs have different features like different line weights or the presence of serif. On the other hand, the proposed method was also able to distinguish fonts that are similar, such as 'K'-'P' highlighted by a blue box.

There are also many examples of fonts that are difficult with drastic intra-font differences. For example, Fig. 8 shows examples of fonts that had the same class but predicted to be from different classes. Some of these pairs are obviously the same fonts, but most of the pairs have major differences between each other including different line weights and different shapes. In particular, the font in Fig. 9 is difficult as there is seemingly no relation between the characters. This font had the lowest accuracy for the proposed method.

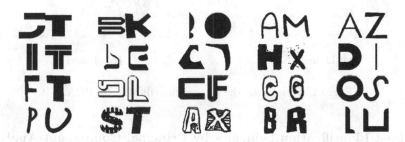

Fig. 8. Examples of misidentified pairs (GT: same → prediction: different).

There are many fonts that look similar visually but are different which makes it difficult to identify with the proposed method. Figure 10 shows examples of font pairs that are misidentified as the same font when they are actually different fonts. These fonts are very similar to each other. It is also difficult even for us to identify as different. Their font pairs have similar features, including line weights, slant lines, and white areas.

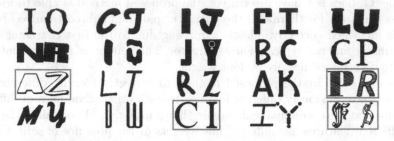

Fig. 9. The font with the most identification errors (GT: same → prediction: different).

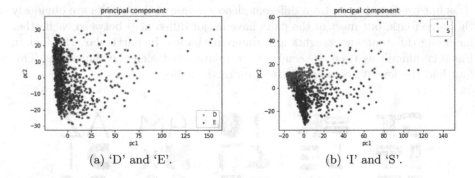

Fig. 10. Examples of misidentified pairs (GT: different→ prediction: same). The green boxes indicate font pairs which are outline fonts and the font pair with the blue box is the font is difficult even for humans. (Color figure online)

(a) 'D' and 'E'.	(b) 'I' and 'S'.

Fig. 11. Visualizing the feature distribution by PCA.

4.4   Font Identification Difficulty by Principal Component Analysis

We analyze the difference in the identification difficulty of font pairs using Principal Component Analysis (PCA). In order to do this, PCA is applied to flattened vectors of the output of each stream. Figure 11 shows two-character comparisons, 'D'-'E' and 'I'-'S,' with the test set fonts mapped in a 2D space. In the figure, the fonts of the first character are mapped in red and the second character blue, which allows us to compare the similarity of the output of each stream. From this figure, we can observe that feature distribution between characters that the proposed method had an easy time identifying, e.g. 'D'-'E,' have significant overlap. Conversely, characters that were difficult, e.g. 'I'-'S,' have very few features that overlap. From these figures, we can expect that font identification between characters that contain very different features is difficult for the proposed method.

4.5 Explanation Using Grad-CAM

We visualize the contribution map of the font pairs toward font identification using Gradient-weighted Class Activation Mapping (Grad-CAM) [29]. Grad-CAM is a neural network visualization method which uses the gradient to weight convolutional layers in order to provide instance-wise explanations. In this case, we use Grad-CAM to visualize the contribution that regions on the pair of inputs have on the decision process. Specifically, the last convolutional layer of each stream is used to visualize the important features of the input.

We first visualize font pairs which are difficult due to having a similar texture fill. Figure 12 shows two fonts that were easily confused by the proposed method. Figure 12(a) is examples where the first font which was correctly identified as the same and (b) is examples of the second font. From these, we can confirm that the presence of the lined fill contributed heavily to the classification. Note that even characters with dramatically different shapes like 'I'-'O' put a large emphasis on contribution on the filling.

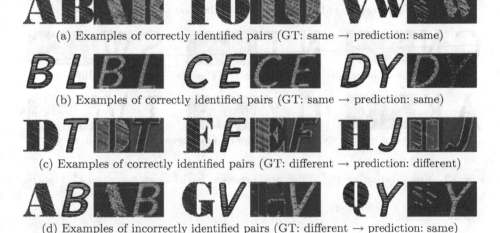

(a) Examples of correctly identified pairs (GT: same → prediction: same)

(b) Examples of correctly identified pairs (GT: same → prediction: same)

(c) Examples of correctly identified pairs (GT: different → prediction: different)

(d) Examples of incorrectly identified pairs (GT: different → prediction: same)

Fig. 12. Visualization of difficult pairs using Grad-CAM

Next, we compared the results of Grad-CAM on correctly identified different pairs. The results are shown in Fig. 12(c). In this case, Grad-CAM revealed that the network focused on the outer regions of the second font. This is due to that font containing a subtle outline. Accordingly, the network focused more on the interior of the first font. We also visualize fonts that are misidentified as the same. In Fig. 12(d), the striped texture of the second font is inappropriately matching the fonts. From these examples, we can infer that the proposed method is able to use features such as character fill and outline to identify the fonts.

In the next example, in Fig. 13, we demonstrate font identification between two fonts which are very similar but one having serifs and the other not having serifs. Compared to Fig. 12, the results of Grad-CAM in Fig. 13 show that specific regions and features are more important than overall textures. For example, in Fig. 13(a) focuses on the curves of 'D', 'R', 'O', 'Q', 'B', and 'C'. In comparison, Fig. 13(b) puts importance on the vertical straight edges.

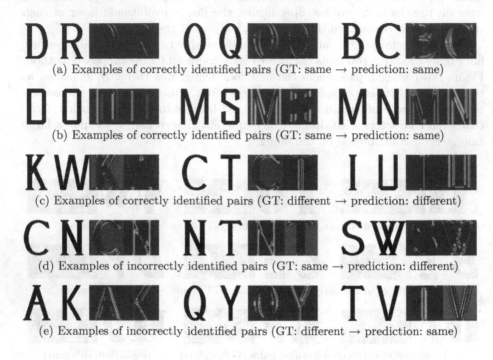

(a) Examples of correctly identified pairs (GT: same → prediction: same)

(b) Examples of correctly identified pairs (GT: same → prediction: same)

(c) Examples of correctly identified pairs (GT: different → prediction: different)

(d) Examples of incorrectly identified pairs (GT: same → prediction: different)

(e) Examples of incorrectly identified pairs (GT: different → prediction: same)

Fig. 13. Visualization of the difference between "serif" and "sans serif" font by Grad-CAM

On the other hand, the font pairs correctly identified as a different focus on different regions of the pairs. Figure 13(c) shows examples of contribution maps which were correctly identified as different. In this case, the top and bottom regions of the characters are highlighted. This is expected as the differences between the fonts should be the presence of serifs.

As for the characters misidentified as different when the fonts were the same and same when the fonts were different, examples of Grad-CAM visualizations are shown in Figs. 13(d) and (e), respectively. In the former case, it seems as those there were not enough common features between the two characters for the system to judge them as the same. For example, 'C' and 'S' are almost entirely composed of curves, while 'N' and 'W' are made of lines and angles. A similar phenomenon happens in Fig. 13 where the similarity of the features outweighs the differences in serif. In these cases, Grad-CAM shows that the serif regions are barely emphasized and the network focuses on the straight edges more.

5   Conclusion

Character-independent font identification is a challenging task due to the differences between characters generally being greater than the differences between fonts. Therefore, we propose the use of a two-stream CNN-based method which determines whether two characters are from the same font or different fonts. As a result, we were able to demonstrate that the proposed method could identify fonts with an accuracy of 92.27 ± 0.20% using 6-fold cross-validation. This is despite using different characters as representatives of their font.

Furthermore, we perform qualitative and quantitative analysis of the results of the proposed method. Through the analysis, we are able to identify that the specific characters involved in the identification contribute to the accuracy. This is due to certain characters containing information about the font within their native features and without common features, it is difficult for the proposed method to realize that they are the same font. To further support this claim, we perform an analysis of the results using PCA and Grad-CAM. PCA is used to show that it is easier to differentiate fonts with similar convolutional features and Grad-CAM is used to pinpoint some of the instance-wise features that led to the classifications.

In the future, we have the plan to analyze the difficulty of font identification between classes of fonts such as serif, sans serif, fancy styles, and so on. In addition, we will try to identify fonts of other languages, including intra- and inter-language comparisons. It also might be possible to use transfer learning to identify fonts between datasets or languages.

Acknowledgment. This work was supported by JSPS KAKENHI Grant Number JP17H06100.

Appendix A   Font Identification Using a Dataset with Less Fancy Fonts

The dataset used in the above experiment contains many fancy fonts and thus there was a possibility that our evaluation might overestimate the font identification performance; this is because fancy fonts are sometimes easy to be identified by their particular appearance. We, therefore, use another font dataset, called the Adobe Font Folio 11.1[3]. From this font set, we selected 1,132 fonts, which are comprised of 511 Serif, 314 Sans Serif, 151 Serif-Sans Hybrid, 74 Script, 61 Historical Script, and (only) 21 Fancy fonts. Note that this font type classification for the 1,132 fonts is given by [1]. We used the same neural network trained by the dataset of Sect. 4.1, i.e., trained with the fancy font dataset and tested on the Adobe dataset. Note that for the evaluation, 367,900 positive pairs and 367,900 negative pairs are prepared using the 1,132 fonts. Using the Adobe fonts as the test, the identification accuracy was 88.33 ± 0.89%. This was lower than 92.27% of the original dataset. However, considering the fact that formal fonts are often

---

[3] https://www.adobe.com/jp/products/fontfolio.html.

very similar to each other, we can still say that the character-independent font identification is possible even for the formal fonts.

References

1. Type Identifier for Beginners. Seibundo Shinkosha Publishing (2013)
2. Abe, K., Iwana, B.K., Holmér, V.G., Uchida, S.: Font creation using class discriminative deep convolutional generative adversarial networks. In: Asian Conference on Pattern Recognition, pp. 232–237 (2017)
3. Avilés-Cruz, C., Villegas, J., Arechiga-Martínez, R., Escarela-Perez, R.: Unsupervised font clustering using stochastic versio of the EM algorithm and global texture analysis. In: Sanfeliu, A., Martínez Trinidad, J.F., Carrasco Ochoa, J.A. (eds.) CIARP 2004. LNCS, vol. 3287, pp. 275–286. Springer, Heidelberg (2004). https://doi.org/10.1007/978-3-540-30463-0_34
4. Bagoriya, Y., Sharma, N.: Font type identification of Hindi printed document. Int. J. Res. Eng. Technol. **03**(03), 513–516 (2014)
5. Chaudhuri, B., Garain, U.: Automatic detection of italic, bold and all-capital words in document images. In: International Conference on Pattern Recognition (1998)
6. Chen, G., et al.: Large-scale visual font recognition. In: Conference on Computer Vision and Pattern Recognition, pp. 3598–3605 (2014)
7. Elgammal, A.M., Ismail, M.A.: Techniques for language identification for hybrid Arabic-English document images. In: International Conference on Document Analysis and Recognition (2001)
8. Ghosh, D., Dube, T., Shivaprasad, A.: Script recognition-a review. IEEE Trans. Pattern Anal. Mach. Intell. **32**(12), 2142–2161 (2010)
9. Gupta, A., Gutierrez-Osuna, R., Christy, M., Furuta, R., Mandell, L.: Font identification in historical documents using active learning. arXiv preprint arXiv:1601.07252 (2016)
10. Hafemann, L.G., Sabourin, R., Oliveira, L.S.: Learning features for offline handwritten signature verification using deep convolutional neural networks. Pattern Recogn. **70**, 163–176 (2017)
11. Hayashi, H., Abe, K., Uchida, S.: GlyphGAN: style-consistent font generation based on generative adversarial networks. Knowl.-Based Syst. **186**, 104927 (2019)
12. Jeong, C.B., Kwag, H.K., Kim, S., Kim, J.S., Park, S.C.: Identification of font styles and typefaces in printed Korean documents. In: International Conference on Asian Digital Libraries, pp. 666–669 (2003)
13. Jung, M.C., Shin, Y.C., Srihari, S.: Multifont classification using typographical attributes. In: International Conference on Document Analysis and Recognition (1999)
14. Khosravi, H., Kabir, E.: Farsi font recognition based on sobel-roberts features. Pattern Recogn. Lett. **31**(1), 75–82 (2010)
15. Koch, G., Zemel, R., Salakhutdinov, R.: Siamese neural networks for one-shot image recognition. In: ICML Deep Learning Workshop (2015)
16. Li, Q., Li, J.P., Chen, L.: A bezier curve-based font generation algorithm for character fonts. In: International Conference on High Performance Computing and Communications, pp. 1156–1159 (2018)
17. Liu, A.H., Liu, Y.C., Yeh, Y.Y., Wang, Y.C.F.: A unified feature disentangler for multi-domain image translation and manipulation. In: Advances in Neural Information Processing Systems, pp. 2590–2599 (2018)

18. Liu, Y., Wei, F., Shao, J., Sheng, L., Yan, J., Wang, X.: Exploring disentangled feature representation beyond face identification. In: Conference on Computer Vision and Pattern Recognition, pp. 2080–2089 (2018)
19. Amer, I.M., ElSayed, S., Mostafa, M.G.: Deep Arabic font family and font size recognition. Int. J. Comput. Appl. **176**(4), 1–6 (2017)
20. Ma, H., Doermann, D.: Font identification using the grating cell texture operator. In: Document Recognition and Retrieval XII, vol. 5676, pp. 148–156. International Society for Optics and Photonics (2005)
21. Miyazaki, T., et al.: Automatic generation of typographic font from small font subset. IEEE Comput. Graph. Appl. **40**, 99–111 (2019)
22. Moussa, S.B., Zahour, A., Benabdelhafid, A., Alimi, A.M.: New features using fractal multi-dimensions for generalized arabic font recognition. Pattern Recogn. Lett. **31**(5), 361–371 (2010)
23. Nguyen, H.T., Nguyen, C.T., Ino, T., Indurkhya, B., Nakagawa, M.: Text-independent writer identification using convolutional neural network. Pattern Recogn. Lett. **121**, 104–112 (2019)
24. Oöztuörk, S.: Font clustering and cluster identification in document images. J. Electron. Imaging **10**(2), 418 (2001)
25. Pal, U., Chaudhuri, B.B.: Identification of different script lines from multi-script documents. Image Vis. Comput. **20**(13–14), 945–954 (2002)
26. Pan, W., Suen, C., Bui, T.D.: Script identification using steerable Gabor filters. In: International Conference on Document Analysis and Recognition (2005)
27. Press, O., Galanti, T., Benaim, S., Wolf, L.: Emerging disentanglement in auto-encoder based unsupervised image content transfer (2018)
28. Ruiz, V., Linares, I., Sanchez, A., Velez, J.F.: Off-line handwritten signature verification using compositional synthetic generation of signatures and siamese neural networks. Neurocomputing **374**, 30–41 (2020)
29. Selvaraju, R.R., Cogswell, M., Das, A., Vedantam, R., Parikh, D., Batra, D.: Grad-CAM: Visual explanations from deep networks via gradient-based localization. Int. J. Comput. Vis. (2019)
30. Shinahara, Y., Karamatsu, T., Harada, D., Yamaguchi, K., Uchida, S.: Serif or sans: visual font analytics on book covers and online advertisements. In: International Conference on Document Analysis and Recognition, pp. 1041–1046 (2019)
31. Suveeranont, R., Igarashi, T.: Example-based automatic font generation. In: International Symposium on Smart Graphics, pp. 127–138 (2010)
32. Tan, T.: Rotation invariant texture features and their use in automatic script identification. IEEE Trans. Pattern Anal. Mach. Intell. **20**(7), 751–756 (1998)
33. Uchida, S., Ide, S., Iwana, B.K., Zhu, A.: A further step to perfect accuracy by training CNN with larger data. In: International Conference on Frontiers in Handwriting Recognition, pp. 405–410 (2016)
34. Wang, Z., et al.: DeepFont: identify your font from an image. In: ACM International Conference on Multimedia, pp. 451–459 (2015)
35. Xing, Z.J., yi-chao wu, Liu, C.L., Yin, F.: Offline signature verification using convolution siamese network. In: Yu, H., Dong, J. (eds.) International Conference on Graphic and Image Processing (2018)
36. Yang, Z., Yang, L., Qi, D., Suen, C.Y.: An FMD-based recognition method for chinese fonts and styles. Pattern Recogn. Lett. **27**(14), 1692–1701 (2006)
37. Zheng, Y., Ohyama, W., Iwana, B.K., Uchida, S.: Capturing micro deformations from pooling layers for offline signature verification. In: International Conference on Document Analysis and Recognition, pp. 1111–1116 (2019)

A New Common Points Detection Method for Classification of 2D and 3D Texts in Video/Scene Images

Lokesh Nandanwar[1], Palaiahnakote Shivakumara[1(✉)], Ahlad Kumar[2],
Tong Lu[3], Umapada Pal[4], and Daniel Lopresti[5]

[1] Faculty of Computer Science and Information Technology,
University of Malaya, Kuala Lumpur, Malaysia
lokeshnandanwar150@gmail.com, shiva@um.edu.my

[2] Dhirubhai Ambani Institute of Information and Communication Technology,
Gandhinagar, Gujarat, India
ahlad_kumar@daiict.ac.in

[3] National Key Lab for Novel Software Technology, Nanjing University,
Nanjing, China
lutong@nju.edu.cn

[4] Computer Vision and Pattern Recognition Unit, Indian Statistical Institute,
Kolkata, India
umapada@isical.ac.in

[5] Computer Science and Engineering, Lehigh University, Bethlehem, PA, USA
lopresti@cse.lehigh.edu

Abstract. Achieving high quality recognition result for video and natural scene images that contain both standard 2D text as well as decorative 3D text is challenging. Methods developed for 2D text may fail for 3D text due to the presence of pixels representing shadow and depth in the 3D text. This work aims at classification of 2D and 3D texts in video or scene images such that one can choose an appropriate method in the classified text for achieving better results. The proposed method explores Generalized Gradient Vector Flow (GGVF) for finding dominant points for input 2D and 3D text line images based on opposite direction symmetry. For each dominant point, our approach finds distance between neighbor points and plots a histogram to choose points which contribute to the highest peak as candidates. Distance symmetry between a candidate point and its neighbor points is checked and if a candidate point is visited twice, a common point is created. Statistical features such as the mean and standard deviation of the common points and candidate points are extracted to feed to Neural Network (NN) for classification. Experimental results on dataset of 2D-3D text line images and the dataset collected from standard natural scene images show that the proposed method outperforms exiting methods. Furthermore, recognition experiments before and after classification show recognition performance improves significantly as a result of applying our method.

Keywords: Gradient Vector Flow · Edge points · Candidate points · 2D text · 3D text · Text recognition · Video/scene images

© Springer Nature Switzerland AG 2020
X. Bai et al. (Eds.): DAS 2020, LNCS 12116, pp. 512–528, 2020.
https://doi.org/10.1007/978-3-030-57058-3_36

1   Introduction

When we look at the literature on text detection and recognition in video and natural scene images, new applications are emerging, such as extracting exciting events from sports videos, finding semantic labels for natural scene images [1] etc. At the same time, the challenges of text detection and recognition are also rising. There are methods for addressing some of these in the literature. For instance, Roy et al. [2] proposed a method for text detection from multiple views of natural scenes by targeting forensic application, where it is expected different views captured by CCTV cameras for the same location. Shivakumara et al. [3] proposed a method for Bib number detection and recognition in marathon images to trace the runner. Xue et al. [4] proposed a method for addressing challenges posed by blur and arbitrary orientation of text in images. Tian et al. [5] proposed a method for tracking the text in the web videos. Shi et al.'s [6] method is developed for solving the issues such as rectification caused by perspective

2D text

3D text

(a) Image with 2D and 3D texts.

(b) Text lines are extracted

"adventures", "harkboy"

"adventures" ,   "sharkbox "

(c) The recognition results of the ASTER and MORAN methods before classification.

"adventures", "sharkboy"

"adventures",   "sharkboy"

(d) The recognition results of the ASTER and MORAN methods after classification

Fig. 1. Example of recognition results for 2D and 3D text before and after classification. The recognition results are displayed over the images for respective methods.

distortion and different camera angles to improve recognition rate. Luo et al. [7] proposed a method for text of arbitrary shaped characters in natural scene images.

It is noted from the above discussion that the main focus of the methods is to find a solution to several new challenges of 2D text recognition but not images that contain both 2D and 3D texts. The 3D text usually provides depth information representing 3D plane instead of standard 2D plane. This makes difference between 2D and 3D text in the images. However, the presence of 3D text in the images does not affect much for text detection performance in contrast to recognition performance. It is evident from the results shown in Fig. 1(a), where the method called CRAFT (Character region awareness for text detection) [8] that employs deep learning for arbitrary oriented text detection in natural scene images detects both 2D and 3D texts well as shown in Fig. 1 (b). At the same time, it is observed from the recognition results shown in Fig. 1(b) that the methods called ASTER [6] and MORAN [7], which uses deep learning models for achieving better recognition results, detect 2D text correctly but not 3D text. This is understandable because the methods are developed for 2D text recognition but not for 3D, where one can expect the following challenge. In case of natural images, 3D effect can be due to real depth of letters, or can be rendered by the artist to draw attention of the viewer in the case of synthetic images. Furthermore, the 3D text can contain shadow information because of capturing images at different angles. The effect can arise due to carving in stone or wood, or through embossing on paper. As a result, the extracted features may not be effective for differentiating text and non-text pixels in the images. One such example can be seen in Fig. 1(a) and (b).

Due to this effect, the above methods are not adequate for recognizing 3D text. This limitation motivated us to propose a new method for the classification of 2D and 3D texts in both video and natural scene images so that we can choose an appropriate method or modify the existing methods for achieving better results for 2D and 3D text recognition. It can be verified from the results shown in Fig. 1(c), where the same methods report correct recognition results for both 2D and 3D text after classification. Therefore, it is expected the recognition methods should score better results after classification compared to before classification.

2 Related Work

As mentioned in the previous section, several methods are proposed for recognizing text in video and natural scene images in the literature. Cheng et al. [9] proposed a method for arbitrarily oriented text recognition in natural scene images based on deep learning. The method proposes Arbitrary Orientation Network (AON) to capture deep features of irregular texts directly, which generate character sequences using an attention based decoder. Tian et al. [5] proposed a framework for text recognition in web videos based on text tracking. The method combines the information of text detection and tracking for recognizing texts. Luo et al. [7] proposed a multi-object rectified attention network for scene text recognition. The method explores a deep learning model, which is invariant to geometric transformation. The method works well for images affected by rotation, scaling and some extent to distortion. Raghunandan et al. [1] proposed multi-script-oriented text detection and recognition in video, natural scene and born digital images. The method extracts features based on wavelet trans-form for detecting characters with the help of an SVM classifier. Next, the method explores Hidden Markov Model (HMM) for recognizing characters and words in images.

Qi et al. [10] proposed a novel joint character categorization and localization approach for character level scene text recognition. The idea of the method is to categorize characters by a joint learning strategy such that recognition performance improves. Shi et al. [6] proposed an attentional scene text recognizer with flexible rectification. The method uses thin plate spline transformation to handle a variety of text irregularities. The idea behind the method is to avoid pre-processing before recognition such that errors can be reduced to improve recognition rate. Rong et al. [11] proposed unambiguous scene text segmentation with referring expression compre-hension. The method proposed a unified deep network to jointly model visual and linguistic information on both region and pixels levels for understanding texts. Villamizar et al. [12] proposed a multi-scale sequential network for semantic text segmentation and localization. The method explores fully convolutional neural net-works that apply to particular cases of slide analysis to understand texts. Feng et al. [13] proposed an end-to-end framework for arbitrary shaped text spotting. The method proposes a new differentiable operator named RoISlide, which detects and recognize texts in images.

In the light of above review on text recognition methods, it is noticed that the methods find solutions to several challenges. However, the main aim of the methods is to recognize 2D text but not 3D text. As shown in Fig. 1, when we run the existing methods with a pre-defined network on 3D text, the recognition performance degrades. This is due to the effect of 3D, where we can expect depth and shadow information, which makes the problem more complex compared to 2D. Therefore, the scope of the above methods is limited to 2D images. Hence, this work aims at proposing a new method for classification of 2D and 3D texts in video and natural scene images such that the same methods can be modified to obtain better results for 3D text. However, there is an attempt to solve this classification problem due to, Xu et al. [14] where a method for multi-oriented graphics-scene 3D text classification in video is proposed. The method explores medial axis points of a character for classifying 2D and 3D texts in video. This method is sensitive to images with complex background and shadow because gradient directional symmetry depends on edge image. Zhong et al. [15] proposed to use shadow detection for 3D text classification. However, the method is not robust to complex background images because the threshold used in the method does not work well for different images. In addition, if an image does not contain enough shadow information, the method may not work well. Therefore, there is a need for developing a new method, which can overcome the above problems to improve recognition performance.

3   Proposed Method

The work described here takes as its input text line images that are produced by existing text detection methods for classification 2D and 3D text. We use the method called CRAFT [8] for text detection as it uses a powerful deep learning model and is robust to several challenges, which is evident from the results shown in Fig. 1(a) and (b).

Motivated by the method [14] where it is stated that stroke width distance, which is thickness of the stroke, is almost the same for each whole character, the proposed method exploits the same property for extracting features that can discriminate 2D and 3D texts in this work. As a result, it is expected that stroke width distance for 2D text exhibits regular patterns while does not for 3D text. It is illustrated in Fig. 2 for 2D and 3D text line images in Fig. 1(b), where it can be seen that the histogram which is drawn for stroke width distance *vs* frequencies appear like a normal distribution but from the histogram of 3D text, we cannot predict the behavior. This is true because for 2D text, the number of pixels which satisfy stroke width is larger than the other distances, while for 3D, it does not. The proposed method finds the pixels that have opposite directions with a certain degree in 3×3 window of every edge pixel in the image as dominant points. The special property of Gradient Vector Flow (GVF) is that the GVF arrow pointing towards edges [16] due to force at edges. Therefore, for the pixels that represent text edges, one can expect opposite GVF direction symmetry with a certain degree. This works well for text with and without shadows. This step results in dominant points, which generally represent edge pixels including edges of shadow. However, we explore Generalized Gradient Vector Flow (GGVF) [17] for obtaining GGVF direction for pair pixels unlike traditional GVF, which is not robust to the pixels at corners and low resolution [16].

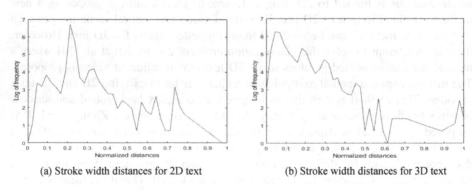

(a) Stroke width distances for 2D text   (b) Stroke width distances for 3D text

Fig. 2. Histogram for the stroke width distances of 2D and 3D texts.

The proposed method finds the distance between each dominant point and its eight neighbors. We perform histogram operation on distance value for choosing the pixel which contributes to the highest peak as candidate points. In other words, the proposed method chooses a pair of pixels which satisfy the distance symmetry. This step eliminates false dominant points. If it is 3D text, one can expect a shadow. If there is a shadow, we can expect candidate points which share the stroke width of shadow information, resulting in common points which are visited by twice while checking distance symmetry. This is the cue for extracting statistical features for using common point as well candidate points. The extracted features are passed to a Neural Network (NN) classifier for the classification of 2D and 3D texts.

3.1 Dominant Point Detection

Inspired by the work [17] where GGVF is proposed for medical image segmentation, we explore GGVF for finding common points detection in this work. According to the method [17], a GGVF arrow usually points towards edges because of high force at edges. As a result, every edge pixel can have opposite GGVF arrow directions in this work. If a pixel represents text, it is expected that the two opposite arrows have almost the same angle, else the two opposite arrows have different angles. The general equation for GGVF is defined in Eq. (1), which is an energy function of the GGVF snake model. We use the same equation to obtain GGVF for the input image. This GGVF accepts edge images as the input for finding GGVF, and the proposed method obtains Canny edge image for the input image to obtain GGVF arrows as shown in Fig. 4(a), where we can see GGVF arrows for 2D and 3D text images. The energy function of GGVF field $z(x, y) = (u(x, y), v(x, y))$ is defined as,

$$E = \iint g(|\nabla f|)(\emptyset(|\nabla u|) + \emptyset|\nabla v|) + h(|\nabla f|)(z - \nabla f)dx\,dy \qquad (1)$$

where $\emptyset(|\nabla v|) = \sqrt{1 + |\nabla v|^2}$ and $\emptyset(|\nabla u|) = \sqrt{1 + |\nabla u|^2}$ with

$$g(|\nabla f|) = e^{\frac{-\nabla f}{k}} \quad \text{and} \quad h(|\nabla f|) = 1 - g(|\nabla f|)$$

In Eq. (1), the first term denotes smoothing, which produce a vector field. The second term is the data fidelity that drives the vector field z close to the gradient of the image i.e. ∇f. Also parameter k acts as a weighing parameter that balance the smoothing and data fidelity term. Here, higher the value of noise indicates a larger value of k.

For each edge pixel in the image, the proposed method defines a window of 3×3 dimension over the input image. The proposed method checks GGVF arrow directions of vertical, horizontal and diagonal pixels pairs in 3×3 window as shown in Fig. 3(b). If any pair satisfies the opposite arrow direction symmetry as defined in Eq. (2), the pair of pixels are considered as dominant points as shown in Fig. 3(c), where one can see almost all the edge pixels of text are detected irrespective of 2D and 3D texts. It is noted that the results in Fig. 3(c) contain still edges of background information. This is due to complex background and shadow of images.

$$GGVF(p1) - GGVF(p2) = \pi \qquad (2)$$

where $p1$ and $p2$ are the pair of pixels in the 3×3 window, which can represent vertical, horizontal and diagonal pixels.

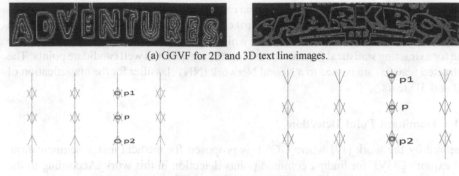

(a) GGVF for 2D and 3D text line images.

(b) Opposite direction symmetry using GGVF for 2D and 3D texts.

(c) Detecting dominant points for 2D and 3D text line images.

Fig. 3. Dominant points detection for 2D and 3D using GGVF opposite direction symmetry.

3.2 Candidate Points Detection

In order to eliminate false dominant points, inspired by the statement that the pixels which represent edges of characters have almost the same stroke width distance [16], the proposed method finds the distance between the eight neighbor points of each dominant point as defined in Eq. (3). It is illustrated in Fig. 4(a) and (b), where (a) shows the results of dominant points detection for 2D and 3D texts, and Fig. 4(b) gives directions of eight neighbor points to find the distances between the center and its eight neighbors. This process continues for all the dominant pints in the image in Fig. 4(a).

To extract the point which represents stroke width of characters, the proposed method performs histogram operation on distance values, and considers the pixels that contributes to the highest peak in the histogram as candidate points as shown in Fig. 4(c) and (d), respectively. Note: before performing histogram operation, we normalize the distance values using a log function to balance distance values, and ignore the first highest peak in the histogram because it considers adjacent dominant point of the center point, which does not give stroke width of the text. Usually the distance between adjacent candidate points is one or two and hence this value corresponds to the highest peak in the histogram. Therefore, the proposed method considers the second highest peak for detecting candidate points, which is considered as distance symmetry. It is observed from the results in Fig. 4(d) that some of the dominant points which represent background as well as text are removed, especially for 2D text. Removing a few text pixels does not affect for the classification of 2D and 3D texts in this work. It is also noted from the results of 3D text in Fig. 4(d) that the detected candidates represent

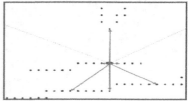

(a) Dominant pints of 2D and 3D texts

(b) Calculating distance between the dominant points and its eight neighbors marked in rectangle in respective 2D and 3D text in (a).

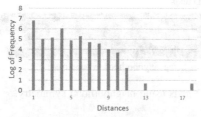
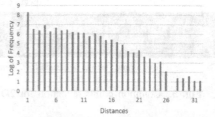

(c) Histogram of 2D and 3D texts for the distances calculated in (b)

(d) Candidate points detection for 2D and 3D texts.

Fig. 4. Candidate points detection for 2D and 3D text using distance symmetry.

both text and shadow pixels. This shows that stroke width of the pixel that represents shadow have almost the same distance as text pixels. However, still we can see some points which represent background information.

$$CN(x,y) = \{a_i(x,y), b_i(x,y)\} \ \forall \ \{a(x,y), b(x,y)\} \in DH \leftrightarrow H_i = h2 \qquad (3)$$

where $h2$ denotes the value of distance at second highest peak, H be the set of all distance between the eight neighbor points of each dominant point and set DH contains set of pair points $\{a_i(x,y), b_i(x,y)\}$ corresponding to distance $H_i, i \in length(H)$ and $CN(x,y)$ denote candidate points.

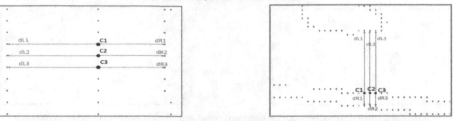

(a) Candidate point of 2D and 3D text

(b) Neighborhood symmetry for candidate point and its two adjacent pixels marked rectangle in (a) for respective 2D and 3D texts.

(c) Common points detection for 2D and 3D texts using neighborhood symmetry.

Fig. 5. Common points detection for 2D and 3D texts using neighborhood symmetry.

3.3 Common Points Detection

It is noted from the results of candidate point detection shown in Fig. 5(a) for 2D and 3D text that candidate points are detected irrespective of 2D and 3D. It is also noted that due to shadow, one can expect more candidate points which share the stroke width of the shadow as well as characters in the case of 3D text. But for 2D, we cannot expect as many as candidate points because the points share the stroke width only at character boundaries. To extract the above observation, the proposed method finds stroke widths for each candidate points and its two nearest neighbor candidate points based on gradient direction and opposite gradient direction of the points as shown in Fig. 5(b), where we can see stroke width for three candidate points for both 2D and 3D texts. When the proposed method finds stroke width distances, it is noted that some of the points are hit twice; these are common points to the text and shadow in the case of 3D, and character boundary points in the case of 2D as shown in Fig. 5(c). The common point detection is defined in Eq. (4). In Fig. 5(c), one can notice a few pixels for 2D and as many as for 3D. If the points share the stroke width of character and shadow in case of 3D and character boundaries in case of 2D, the points considered as common points and the condition is called neighborhood symmetry.

$$CP(x, y) = (x_i, y_i) \leftrightarrow (dR1_i = dR2_i = dR3_i) \ \forall \ (x_i, y_i) \in CN(x_i, y_i) \quad (4)$$

where $\{dL1, dL2, dL3\}$ denote the nearest pixel distance in gradient direction, $\{dR1, dR2, dR3\}$ denote the nearest pixel distance in the opposite of gradient direction, and $CP(x, y)$ denote the common points.

The results in Fig. 5(c) show that spatial distribution of candidate and common points indicates clear difference between 2D and 3D texts. In addition, it is also true that the points which represent shadow have low intensity values compared to character pixels. Based on these observations, the proposed method extracts the following statistical features for candidate and common points. Since every common points associated with two neighbor points, the proposed method consider the point detected using gradient direction to the common point as Neighbor point-1 and the point detected using opposite gradient direction to the common point as Neighbor point-2 as shown in Fig. 5(b). The proposed method considers pixels between the common point to Neighbor point-1 and the common point to Neighbor point-2 as two separate groups for feature extraction, say, Group-1 and Group-2. Group-1 and Group-2 include the pixels of all the common points in the image. The proposed method calculates the mean of intensity values of the pixels in the respective groups, which gives two features as defined in Eq. (5). The proposed method also calculates standard deviation for the intensity values of the pixels in the respective groups, which gives two more features as defined in Eq. (6). In the same way, instead of intensity values, the proposed method considers distance between the common point to Neighbour points to calculate mean and standard deviation for the respective groups, which gives four features as defined in Eq. (7) and Eq. (8).

In total, 8 features are extracted from common points. In addition, the proposed method calculates the mean and standard deviation for the distances of candidate points of respective two groups, which gives four features. Overall, the proposed method extracts 12 features using common points and candidate points, which represent spatial distribution and difference between intensity values of points for classification of 2D and 3D texts. The distribution of 12 features are shown for the input 2D and 3D texts in Fig. 7, where one can see smooth variation for 2D and large variations for 3D. This is expected because the space between the characters does not vary much in the case of 2D text, while we can expect large variations in the case of 3D text due to presence of shadows. This shows that the proposed feature extracts distinct property of 2D and 3D texts for classification. Note that the values are normalized to the range of 0 and 1 before plotting the graphs in Fig. 6.

$$MI = \frac{\sum m}{n} \tag{5}$$

$$SI = \sqrt{\frac{\sum (m - \bar{m})^2}{n}} \tag{6}$$

$$MD = \frac{\sum d}{n} \tag{7}$$

$$SD = \sqrt{\frac{\sum (d - \bar{d})^2}{n}} \tag{8}$$

where MI denotes the mean, m represents the intensity values, n is the total number of pixels, SI denotes the standard deviation, MD, SD denote the mean and standard deviation of distance values and d is the distance between the common point and its neighbor points.

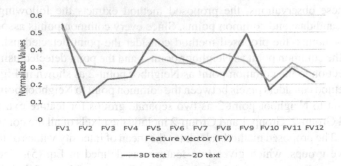

Fig. 6. Statistical Feature Vector for 2D and 3D text classification

Motivated by the ability of neural networks [18] for classification, we propose such a network for classifying 2D and 3D text in this work by feeding 12 features into the network. The structure of proposed network is as follows. It has 5 intermediate dense layers with 50, 100, 50, 25 and 10 units/features each. Input layer has 12 features and final output layer has 1 feature {0 for 2D and 1 for 3D}. We use Rectified linear activation function (ReLU) for all intermediate layers and Sigmoid [19] activation function for the final layer. Dropout with drop rate of 20% between the intermediate layers is applied to reduce the overfitting problem. Binary cross entropy loss is estimated [20] as defined in Eq. (9), where y is the label (Ground Truth), and p(y) is the predicted probability for the total number of N samples. Adam [21] optimizer with learning rate of 0.01 is used during the training process and the batch size used is 8 during training. Each model is trained for 200 epochs with model checkpoint to store the best trained model using Keras framework [22]. For training and testing, we use 80% samples for training and 20% for testing in this work. The details of proposed deep neural network are listed in Table 1.

$$BCE(q) = -\frac{1}{N}\sum_{i=1}^{N}(y_i.\log(p(y_i)) + (1 - y_i).\log(1 - p(y_i)))\tag{9}$$

Table 1. The details of the deep neural network classifier

Layer number	Out size
Input layer	12 × 1, ReLU
Dense layer 1	50 × 1, Dropout = 20%, ReLU
Dense layer 2	100 × 1, Dropout = 20%, ReLU
Dense layer 3	50 × 1, Dropout = 20%, ReLU
Dense layer 4	30 × 1, Dropout = 20%, ReLU
Dense layer 5	10 × 1, ReLU
Output layer	1 × 1, Sigmoid

4   Experimental Results

For 2D and 3D text classification, we create our own dataset assembled from different sources, including YouTube, scene images and movie posters that are fond online. Our dataset includes text line images with complex backgrounds, low resolution, low contrast, font font-size variations, arbitrary shaped characters and some extent to distortion as shown sample images in Fig. 7(a), which gives 513 2D and 505 3D text line images. For objective evaluation of the proposed method, we also collect 2D and 3D text line images from the benchmark datasets of natural scene images, namely, IIIT5K, COCO-Text, ICDAR 2013 and ICDAR 2015, which have a few 3D text line images as shown sample images in Fig. 7(b). We collect 317 2D, 305 3D from IIIT5K dataset, 472 2D, 530 3D from COCO-Text, 123 2D, 74 3D from ICDAR 2013 and 111 2D, 90 3D from ICDAR 2015 datasets, which gives total 1023 2D text line images and 999 3D text line images. This dataset is considered as the standard dataset for experimentation in this work.

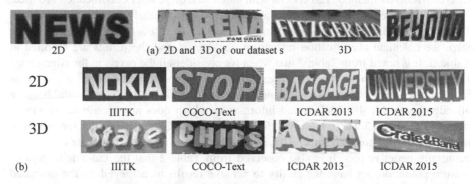

Fig. 7. Sample images of 2D and 3D text from our dataset and standard dataset

In total, 3040 text line images, which include 1536 2D images and 1504 3D images are considered for evaluating the proposed and existing methods.

To show effectiveness of the proposed method, we implement two existing methods which have the same objective of 2D and 3D text classification. Xu et al. [14] method employs gradient inward and outward directions for 2D and 3D text image classification. Zhong et al.'s [15] method proposes shadow detection for 2D and 3D text image classification. The motivation to choose these two existing methods is to show that stroke width features used in Xu et al. for classifying 2D and 3D text image is not enough to handle the images considered in this work. Similarly, the features used for shadow detection in Zhong et al. are not adequate. Similarly, to validate the proposed classification, we implement two recognition methods, namely, ASTER [6] and MORAN [7], which explore powerful deep learning models for recognizing text of different complexities in natural scene images. The codes of these methods are available to the public.

For measuring performance, we use the standard metrics, namely, a confusion matrix and average classification rates. The classification rate is defined as the number of images classified correctly by the proposed method divided by the actual number of images. The average classification rate is defined as the mean of diagonal elements of the confusion matrix. For recognition experiments, we use recognition rate, which is defined as the number of characters recognized correctly divided by the actual number of characters. To test the utility of the proposed classification, we calculate the recognition rate of the methods before after classification. Before classification includes text lines of both 2D and 3D for calculating recognition rate, while after classification includes text lines of 2D and 3D separately. It is expected that the recognition rate will be higher after classification compared to before classification. This is because after classification the complexity of the problem is reduced and the method can be modified/trained according to the complexity of individual classes. In this work, we use pre-trained model with the same parameters and values, while for after classification, we tune the parameters, namely, "Epochs" and "Batch size" according to complexity of individual classes. In case of ASTER method, 200 epochs with batch size of 16 and in case of MORAN method, 120 epochs with batch size of 32 after classification are used.

The proposed method involves a few key steps, namely, Canny edge pixels, Dominant, Candidate and Common points detection. To assess the contribution of each step, we calculate classification rate using our dataset and the results are reported in Table 2. It is noted from Table 2 that when we consider all the pixels in the edge image for the classification of 2D and 3D text lines, the proposed method does not score as high as other steps and the proposed method with common points. This is valid because all edge pixels involve background information which does not contribute for classification of 2D and 3D text. However, for Dominant and Candidate points, the proposed method scores almost the same. This shows that both the steps contribute equally for achieving better results. It is also observed from Table 2 that the individual steps or feature points do not have the ability to achieve results as achieved by the proposed method. This is due to the inclusion of many background pixels, while common points does not.

Table 2. Confusion matrix and average classification rate of the different steps for 2D and 3D texts classification on our dataset.

Features	All edge pixels		Dominant points		Candidate points		Proposed (common points)	
Classes	2D	3D	2D	3D	2D	3D	2D	3D
2D	66.86	33.14	69.78	30.22	62.76	37.24	92.62	7.38
3D	28.32	71.68	18.02	81.98	8.52	91.48	6.93	93.07
Average	69.27		75.88		72.37		**92.8**	

4.1   Evaluating the 2D and 3D Text Classification

For testing the proposed classification on each dataset and the full dataset, we calculate average classification rate as reported in Table 3, where one can see as sample size increases, the results also increase. This shows that each dataset has its own complexity. For the full dataset, which is the collection from all the dataset, the proposed method scores promising results. Quantitative results of the proposed and existing methods for our dataset and the standard full-dataset are reported in Table 3, where it is noted that the proposed method is the best at average classification rate for both our and the standard full dataset compared to the existing methods. This is justifiable because the features used in the methods are not adequate to cope with the challenges of our dataset and the standard datasets. In addition, the features are sensitive to complex background. When we compare the results of two existing methods, Zhong et al. achieve poor results compared to Xu et al. This is due to constant thresholds used for shadow detection of 3D text in the method. Since the dataset contains a large variation, constant thresholds do not work, while Xu et al. use dynamic rules for classification, which is better than fixing constant threshold values. On the other hand, the proposed method explores Generalized Gradient Vector Flow (GGVF) for finding directional and distance symmetry, which is effective compare to the features used in the existing methods. It is observed from Table 4 that the proposed method scores low results for the standard full-dataset compared to our dataset. This shows that the standard dataset is much more complex than our dataset.

Table 3. Confusion matrix and average classification rate of the proposed method on different standard natural scene datasets and full dataset.

Datasets	IIIT5K		COCO-Text		ICDAR 2013		ICDAR 2015		Standard full-dataset	
Classes	2D	3D	2D	3D	2D	3D	2D	3D	2D	3D
Size	317	305	472	530	123	74	111	90	1023	999
2D	85.61	14.38	91.79	8.21	73.61	26.38	70.45	29.54	86.33	13.6
3D	13.27	86.72	7.92	92.07	31.42	68.58	20	80	12.6	87.3
Average	86.15		91.93		71.1		75.27		**86.83**	

Table 4. Confusion matrix and average classification rate of the proposed and existing methods on both our and standard datasets (in %)

Dataset	Our dataset						Standard full-dataset					
Methods	Zhong et al.		Xu et al.		Proposed		Zhong et al.		Xu et al.		Proposed	
Classes	2D	3D	2D	3D	2D	3D	2D	3D	2D	3D	2D	3D
2D	32.74	67.25	66.28	33.72	92.62	7.38	53.86	46.13	46.6	53.4	86.33	13.67
3D	46.33	53.67	28.31	71.69	6.93	93.07	32.8	67.2	31.2	68.8	12.66	87.34
Average	43.2		70.48		**92.8**		60.53		57.7		**86.83**	

4.2 Recognition Experiments for Validating the Proposed Classification

As mentioned earlier, to show the usefulness of the proposed classification, we conduct recognition experiments before and after classification using different recognition methods on both our own and the standard datasets. The recognition methods run on all the images regardless of 2D and 3D for before classification, while the methods run on each class individually after classification. The results of recognition methods on the above mentioned datasets are reported in Table 5, where it is noticed clearly that the recognition rates given by two methods after classification are improved significantly for both the datasets compared to the recognition rates before classification. This shows that the proposed classification is useful for improving recognition performance on text detection in 3D video and natural scene images. It is observed from Table 5 that both the recognition methods achieve low results for 3D and compared to 2D after classification. This indicates that the recognition methods are not capable of handling 3D texts. This is true because in this work, we train recognition methods on 2D and 3D classes individually after classification step. We modify the architecture using transfer learning to achieve better results according to complexity. However, the proposed classification improves the overall performance of the recognition methods compared to before classification.

Table 5. Recognition performance of different methods before and after classification on our and standard datasets (in%)

Methods	Our dataset				Standard dataset-FULL			
	Before classification	After classification			Before classification	After classification		
	2D + 3D	2D	3D	Average	2D + 3D	2D	3D	Average
ASTER [6]	78.15	97.0	85.7	91.35	88.5	96.1	85.4	90.75
MORAN[7]	85.5	94.1	87.6	90.85	89.78	96.09	86.4	91.25

Sometimes, when the images affected by multiple adverse effects as shown in Fig. 8, there are chances of misclassification. When the features overlap with the background information, it is hard to find common points for differentiating 2D and 3D texts, which shows there is a scope for improvement in future work.

    (a) Our dataset   Standard   (b) Our dataset   Standard

Fig. 8. Example of unsuccessful classification of the proposed method on our and standard datasets. (a) 2D misclassified as 3D and vice versa for the images in (b)

5   Conclusion and Future Work

In this paper, we have proposed a new method for the classification of 2D and 3D text in video images which includes natural scene images such that recognition performance can be improved significantly. We explore generalized gradient vector flow for directional and distance symmetry checking, unlike conventional gradient vector flow which is not robust for low contrast and corners. Based on the property of stroke width distance, our approach defines directional and distance symmetry for detecting dominant and candidate points. The proposed method defines neighborhood symmetry for finding common points which share the stroke width of characters as well as shadow information. For the common points, our approach extracts statistical features, and the features are further passed to a deep neural network for the classification of 2D and 3D texts in video and natural scene images. Experimental results on different datasets show that the proposed classification outperforms the existing classification methods. The recognition results before and after classification show that recognition performance improves significantly after classification compared to before classification. However, there are still some limitations as discussed in the Experimental Section, thus we plan to improve the method in an attempt to address such limitations in the future.

References

1. Raghunandan, K.S., Shivakumara, P., Roy, S., Kumar, G.H., Pal, U., Lu, T.: Multi-script-oriented text detection and recognition in video/scene/born digital images. IEEE Trans. CSVT **29**, 1145–1162 (2019)
2. Roy, S., Shivakumara, P., Pal, U., Lu, T., Kumar, G.H.: Delaunay triangulation based text detection from multi-view images of natural scene. PRL **129**, 92–100 (2020)
3. Shivakumara, P., Raghavendra, R., Qin, L., Raja, K.B., Lu, T., Pal, U.: A new multi-modal approach to bib number/text detection and recognition in Marathon images. PR **61**, 479–491 (2017)
4. Xue, M., Shivakumara, P., Zhang, C., Lu, T., Pal, U.: Curved text detection in blurred/non-blurred video/scene images. MTAP **78**(18), 25629–25653 (2019). https://doi.org/10.1007/s11042-019-7721-2
5. Tian, S., Yin, X.C., Su, Y., Hao, H.W.: A unified framework for tracking based text detection and recognition from web videos. IEEE Trans. PAMI **40**, 542–554 (2018)
6. Shi, B., Yang, M., Wang, X., Luy, P., Yao, C., Bai, X.: ASTER: an attentional scene text recognizer with flexible rectification. IEEE Trans. PAMI **41**, 2035–2048 (2019)
7. Luo, C., Jin, L., Sun, Z.: MORAN: a multi-object rectified attention network for scene text recognition. PR **90**, 109–118 (2019)
8. Baek, Y., Lee, B., Han, D., Yun, S., Lee, H.: Character region awareness for text detection. In: Proceedings of CVPR (2019)
9. Cheng, Z., Xu, Y., Bai, F., Niu, Y.: AON: towards arbitrarily-oriented text recognition. In: Proceedings of CVPR, pp. 5571–5579 (2018)
10. Qi, X., Chen, Y., Xiao, R., Li, C.G., Zou, Q., Cui, S.: A novel joint character categorization and localization approach for character level scene text recognition. In: Proceedings of ICDARW, pp. 83–90 (2019)
11. Rong, X., Yi, C., Tian, Y.: Unambiguous scene text segmentation with referring expression comprehension. IEEE Trans. IP **29**, 591–601 (2020)

12. Villamizar, M., Canevert, O., Odobez, J.M.: Multi-scale sequential network for semantic text segmentation and localization. PRL **129**, 63–69 (2020)
13. Feng, W., He, W., Yin, F., Zhang, X.Y., Liu, C.L.: TextDragon: an end-to-end framework for arbitrary shaped text spotting. In: Proceedings of ICCV, pp. 9076–9085 (2019)
14. Xu, J., Shivakumara, P., Lu, T., Tan, C.L., Uchida, S.: A new method for multi-oriented graphics-scene-3D text classification in video. PR **49**, 19–42 (2016)
15. Zhong, W., Raj, A.N.J., Shivakumara, P., Zhuang, Z., Lu, T., Pal, U.: A new shadow detection and depth removal method for 3d text recognition in scene images. In: Proceedings of ICIMT, pp. 277–281 (2018)
16. Khare, V., Shivakumara, P., Chan, C.S., Lu, T., Meng, L.K., Woon, H.H., Blumenstein, M.: A novel character segmentation-reconstruction approach for license plate recognition. ESWA **131**, 219–239 (2019)
17. Zhu, S., Gao, R.: A novel generalized gradient vector flow snake model using minimal surface and component-normalized method for medical image segmentation. BSPC **26**, 1–10 (2016)
18. Silva, I.N.D., Spatti, D.H., Flauzino, R.A., Liboni, L.H.B., Reis Alves, S.F.D.: Artificial Neural Networks, vol. 39. Springer, Heidelberg (2017). https://doi.org/10.1007/978-3-319-43162-8
19. Narayan, S.: The generalized sigmoid activation function: competitive supervised learning. IS **99**, 69–82 (1997)
20. Nasr, G.E., Badr, E.A., Joun, C.: Cross entropy error function in neural networks: forecasting gasoline demand. In: Proceedings of FLAIRS, pp. 381–384 (2002)
21. Kingma, P.D., Bai, J.L.: Adam: a method for stochastic optimization. In: Proceedings of ICLR, pp. 1–15 (2015)
22. Keras: Deep learning library for theano and tensorflow (2015). https://keras.io/

Analysis of Typefaces Designed for Readers with Developmental Dyslexia
Insights from Neural Networks

Xinru Zhu[1]([✉]), Kyo Kageura[1], and Shin'ichi Satoh[2]

[1] Graduate School of Education, The University of Tokyo, Bunkyō, Tokyo, Japan
{shushinjo,kyo}@p.u-tokyo.ac.jp
[2] National Institute of Informatics, Chiyoda, Tokyo, Japan
satoh@nii.ac.jp

Abstract. Developmental dyslexia is a specific learning disability that is characterized by severe difficulties in learning to read. Amongst various supporting technologies, there are typefaces specially designed for readers with dyslexia. Although recent research shows the effectiveness of these typefaces, the visual characteristics of these typefaces that are good for readers with dyslexia are yet to be revealed.

This research aims to explore the possibilities of using neural networks to clarify the visual characteristics of Latin dyslexia typefaces and apply those to typefaces in other languages, in this case, Japanese.

As a first step, we conducted simple classification tasks to see whether a CNN identifies subtle differences between Latin dyslexia typefaces and standard typefaces, and whether it can be applied to classify Japanese characters.

The results show that CNNs are able to learn visual characteristics of Latin dyslexia typefaces and classify Japanese typefaces with those features. This indicates the possibility of further utilizing neural networks for research regarding typefaces for readers with dyslexia across languages.

Keywords: Latin typeface · Japanese typeface · Developmental tyslexia · Convolutional neural network · Classification

1 Introduction

Developmental dyslexia, or simply dyslexia, is "a specific learning disability that is neurobiological in origin. It is characterized by difficulties with accurate and/or fluent word recognition and by poor spelling and decoding abilities" [14]. Evidence shows that 5–17% of the population in English-speaking countries [22] and 3–5% of the population in Japan [15] have developmental dyslexia.

Although dyslexia is generally considered as a phonological deficit [19], some subtypes of dyslexia are often accompanied by visual symptoms such as character reversals, character distortion, blurring, and superimposition [23]. Figure 1 illustrates the typical visual symptoms of dyslexia.

© Springer Nature Switzerland AG 2020
X. Bai et al. (Eds.): DAS 2020, LNCS 12116, pp. 529–543, 2020.
https://doi.org/10.1007/978-3-030-57058-3_37

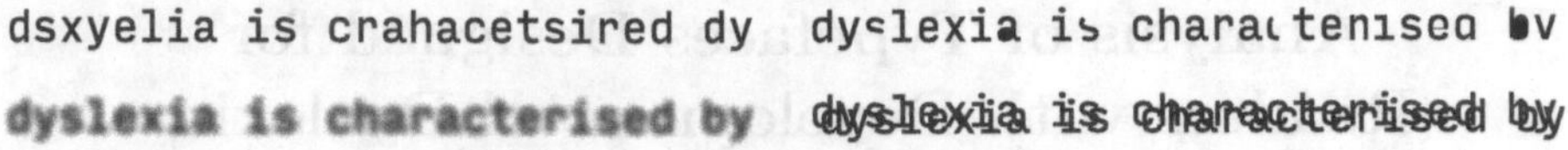

dsxyelia is crahacetsired dy   dy<lexia i> charactenisea bv

dyslexia is characterised by   dyslexia is characterised by

Fig. 1. Visual symptoms of dyslexia

Typefaces play an important role in document design [20] due to their function regarding readability and legibility [25]. Research demonstrates that standard typefaces are not as effective for readers with dyslexia as for other readers [12].

Amongst various supporting strategies, several specially designed Latin typefaces have been created in order to increase the ease of reading for readers with dyslexia [29]. Read Regular [7], Lexie Readable [2], Sylexiad [11], Dyslexie [5], and OpenDyslexic [9] are existing typefaces designed for readers with dyslexia. Empirical research regarding the effectiveness of these typefaces shows that with the specially designed typefaces, readers with dyslexia are able to read more accurately or feel more comfortable with reading [12,17].

Figure 2 shows the letterforms of dyslexia typefaces compared to a standard typeface, Arial. We can see that dyslexia typefaces have different visual characteristics from standard typefaces.

Arial
hamburgefontsiv
Dyslexie
hamburgefontsiv
Lexie Readable
hamburgefontsiv
OpenDyslexic
hamburgefontsiv

Fig. 2. Arial, Dyslexie, Lexie Readable, and OpenDyslexic

In addition to that, some characters that have similar letterforms to others are specifically modified in dyslexia typefaces to be more distinguishable. For instance, 'b' and 'd' or 'p' and 'q' are character pairs where one can be a mirror image of another in standard typefaces, but in dyslexia typefaces, they are modified to be less symmetric to reduce the possibilities of mirroring. Another example is 'a', 'c', 'e', and 's'. These are characters with the same height and can be easily confused with one another in small sizes. In dyslexia typefaces, they are modified to have larger counters (white space inside a character) to be more legible. We show those characters in Fig. 3.

Since the visual characteristics of a typeface decide its readability and legibility [25], clarifying common visual characteristics of dyslexia typefaces is a way to understand what kind of visual characteristics are effective for readers with dyslexia. Once the common visual characteristics of dyslexia typefaces are revealed, dyslexia typefaces in different languages can be created based on these visual characteristics because visual symptoms of dyslexia in different writing

Arial   Dyslexie   Lexie Readable   OpenDyslexic

b d p q   b d p q   b d p q   b d p q

a c e s   a c e s   a c e s   a c e s

Fig. 3. Similar letterforms in Arial, Dyslexie, Lexie Readable, and OpenDyslexic

systems are very similar [22]. Also, this will contribute to dyslexia research by providing a basis for further investigating the correlation between specific visual characteristics of letters and symptoms of dyslexia.

However, the existing dyslexia typefaces are designed based on designers' own experiences and no research has been done to systematically analyze the common visual characteristics of dyslexia typefaces.

Since visual characteristics of typefaces and their impacts on human brains are so subtle and ambiguous to be described by using a definite set of features, we consider training neural networks and observing how they perceive different typefaces an appropriate starting point. In this research, we explore the possibilities of utilizing neural networks for research of typefaces for readers with dyslexia.

As a first step, we conducted three typeface classification tasks on the character level to see (1) whether a network learns subtle differences between Latin dyslexia typefaces, and (2) whether it can be applied to classify typefaces from different languages, in this case, Japanese.

The first and second experiments address the first question, and the third experiment corresponds to the second question. This paper reports the design and results of the three experiments.

2   Related Work

With the emergence of deep learning techniques, breakthroughs have been made in the field of document analysis, including printed and handwritten character recognition [21,28].

In recent years, CNNs have been applied to typeface-related problems and achieved significant results. We review the literature for two types of tasks regarding typefaces: classification and analysis.

2.1   Typeface Classification

Tensmeyer et al. used CNNs (ResNet-50 in specific) to classify Latin medieval manuscripts at the page level and 40 Arabic computer typefaces at the line level. Their method achieved state-of-the-art performance with 86.6% accuracy in Latin manuscripts and 98.8% accuracy in Arabic typefaces [24].

Zhang et al. classified four standard calligraphy styles of Chinese typefaces at the character level by using a novel CNN structure. They achieved 97.88% accuracy [27].

These results show that CNNs may learn the visual characteristics of typefaces even if the differences are very subtle.

2.2 Typeface Analysis

Ide and Uchida analyzed how CNNs manage different typefaces, specifically, whether and how CNNs fuse or separate the same character of different typefaces when predicting, by classifying digit images in different typefaces at the character level. From this analysis, they concluded that "types are not fully fused in convolutional layers but the distributions of the same class from different types become closer in upper layers" [13].

Although not adopting CNNs, Uchida et al. aimed to reveal the principles in typeface design by adopting engineering approaches, i.e. congealing images based on a nonlinear geometric transformation model and representing font variations as a large-scale relative neighborhood graph, to discover the most standard typeface shape and typeface designs that have not been explored before in design space [26].

In this research, we use deep learning techniques of typeface classification tasks to analyze and discover common visual features of dyslexia typefaces. In terms of the purpose, our work inherits the work of Uchida et al. and bridges the field of engineering and design.

2.3 Dyslexia Typefaces

Aside from empirical research on Latin dyslexia typefaces that we have mentioned in Sect. 1, we have conducted research on Japanese typefaces that are readable for readers with dyslexia.

In this research, we defined the visual characteristics of Latin dyslexia typefaces with heuristic methods and mapped those characteristics to Japanese typefaces [29] to create a set of Japanese dyslexia typefaces. We also conducted empirical experiments regarding the readability and legibility of these newly created Japanese typefaces and the results show that they are both objectively and subjectively more readable for readers with dyslexia [30].

3   Dataset and Architecture

3.1   Dataset

We created a dataset consisting of 2795 images of characters from 13 typefaces of two classes: dyslexia and standard. Figure 4 shows three dyslexia typefaces and 10 standard typefaces included in the dataset.

Dyslexie, Lexie Readable, and OpenDyslexic are selected as dyslexia typefaces because, amongst all dyslexia typefaces, they are either evaluated in previous research and/or free to acquire so that they are relatively widely used.

Arial, Calibri, Century Gothic, Comic Sans, Trebuchet, and Verdana are selected as standard typefaces based on the recommendation of British

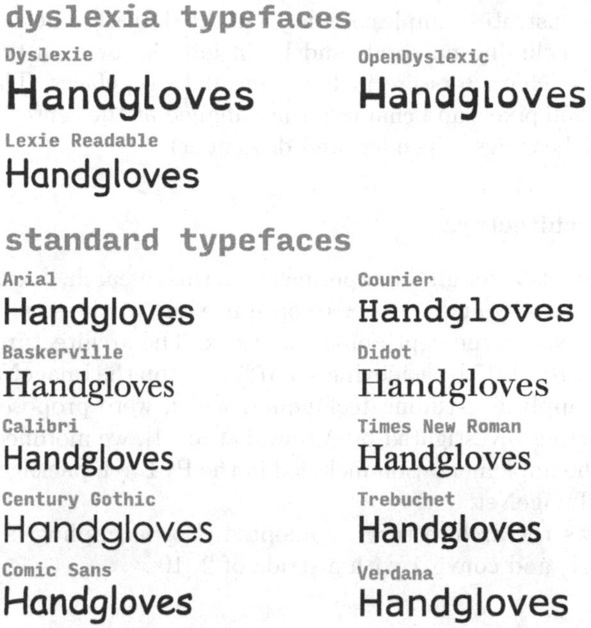

dyslexia typefaces
Dyslexie
Handgloves
OpenDyslexic
Handgloves
Lexie Readable
Handgloves
standard typefaces
Arial
Handgloves
Courier
Handgloves
Baskerville
Handgloves
Didot
Handgloves
Calibri
Handgloves
Times New Roman
Handgloves
Century Gothic
Handgloves
Trebuchet
Handgloves
Comic Sans
Handgloves
Verdana
Handgloves

Fig. 4. Dyslexia typefaces and standard typefaces in the dataset

215 characters
dyslexia typefaces
standard typefaces
0 1 2 3 4 5 6 7 8 9 A B C D E F ... ż Ž ž

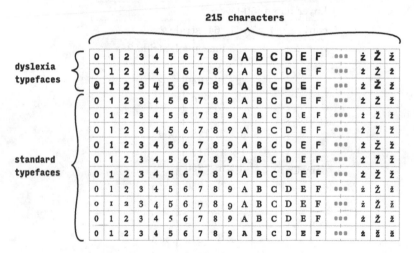

Fig. 5. Samples of the dataset

Dyslexia Association [6], which designates them being more readable for readers with dyslexia than other standard typefaces. In addition, serif typefaces (i.e. Baskerville, Didot, and Times New Roman) and a monospace typeface (i.e. Courier) are added to the standard typefaces for better representing a wide range of styles in real-world typeface collections.

Figure 5 demonstrates samples of images in the dataset. The dataset contains 215 characters, including numerals and Latin letters, for each typeface for the reason that these characters are included in all 13 typefaces. The size of each image is 256 × 256 pixels and characters are aligned at the center based on their metrics (i.e. sidebearings, ascender, and descender).

3.2 CNN Architecture

We adopted ResNet-50 for all the experiments in this research, due to the fact that it has lower complexity and is easier to optimize and at the same time can gain accuracy by increasing the depth of the network. The architecture won the first place in the ILSVRC 2015 by achieving a 3.57% error on the ImageNet test set [10].

In order to apply fine-tuning techniques, which were proposed by Girshick et al. [8] and further investigated by Agrawal et al. [1], we modified the architecture based on the implementation included in the PyTorch package, with weights pre-trained on ImageNet.

Table 1 shows the architecture we adopted. Downsampling is performed by conv3_1, conv4_1, and conv5_1 with a stride of 2 [10].

Table 1. The architecture of ResNet-50 used for fine-tuning [10]

Layer name	50-layer
conv1	7 × 7, 64, stride 2
conv2_x	3 × 3 max pool, stride 2
	$\begin{bmatrix} 1 \times 1, & 64 \\ 3 \times 3, & 64 \\ 1 \times 1, & 256 \end{bmatrix} \times 3$
conv3_x	$\begin{bmatrix} 1 \times 1, & 128 \\ 3 \times 3, & 128 \\ 1 \times 1, & 512 \end{bmatrix} \times 4$
conv4_x	$\begin{bmatrix} 1 \times 1, & 256 \\ 3 \times 3, & 256 \\ 1 \times 1, & 1024 \end{bmatrix} \times 6$
conv5_x	$\begin{bmatrix} 1 \times 1, & 512 \\ 3 \times 3, & 512 \\ 1 \times 1, & 2048 \end{bmatrix} \times 3$
	Average pool, 2-d fc, softmax

3.3 Data Augmentation

In order to reduce overfitting due to the small size of the dataset, we artificially enlarged the dataset by using label-preserving transformations [16].

Amongst various transformation strategies [18], we employed affine transformations limited to scaling and translating in order to preserve the visual characteristics of typeface styles [29]. Thus two kinds of transformations are applied at random on training data and can be formulated as (1). The background of the images is filled white after the transformations.

$$\begin{bmatrix} x' \\ y' \\ 1 \end{bmatrix} = \begin{bmatrix} S & 0 & T \\ 0 & S & T \\ 0 & 0 & 1 \end{bmatrix} \begin{bmatrix} x \\ y \\ 1 \end{bmatrix}, S \in [1, 1.2], T \in [-25.6, 25.6] \tag{1}$$

3.4 Hyperparameter Optimization

We adopted the tree-structured Parzen estimator approach (TPE) of the sequential model-based global optimization [3] to optimize the hyperparameters of the network. This is a reproducible and unbiased optimization approach that yielded state of the art performance over hand-tuning approaches and other automated approaches [4].

We used the implementation of this algorithm in the Optuna package. Table 2 shows the ranges of hyperparameters we have searched.

Table 2. Hyperparameter optimization

Hyperparameter	Range to search
Batch size	$[4, 64]$
Optimization algorithm	{SGD, Adam}
Weight decay of optimizer	$[10^{-10}, 10^{-3}]$
Learning rate of optimizer	$[10^{-5}, 10^{-1}]$
Step size of scheduler	$[5, 20]$
Gamma of scheduler	$[10^{-3}, 10^{-1}]$
Epochs of training	$[50, 100]$

4   Experimental Design

We designed three experiments to address the questions. While Experiment A and B focus on the question of whether neural networks can learn visual characteristics of Latin dyslexia typeface, Experiment C examines that whether the trained network can classify various Latin typefaces and even Japanese typefaces that it has never seen.

In Experiment A, we trained a network to distinguish dyslexia typefaces from standard typefaces. We divided the dataset at random and used 80% of the data for training and 20% for testing without considering typeface and character types. Figure 6(a) demonstrates the split of the data.

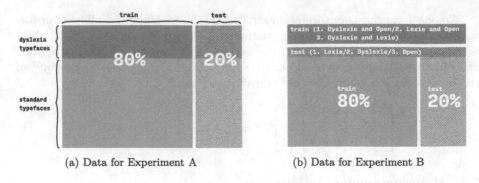

(a) Data for Experiment A   (b) Data for Experiment B

Fig. 6. Data for Experiment A and B

(a) Latin typefaces 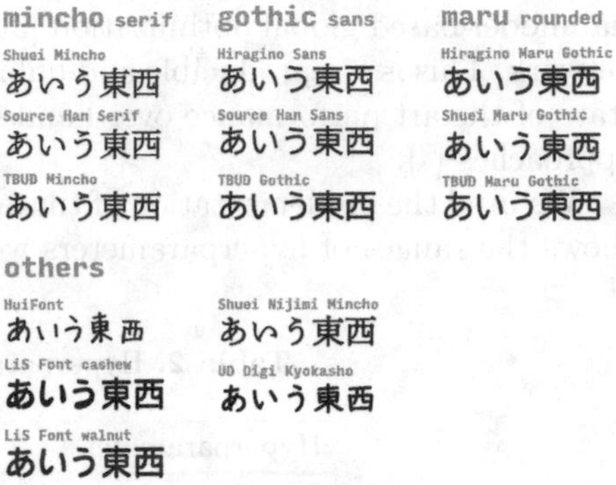   (b) Japanese typefaces

Fig. 7. Latin and Japanese typefaces tested in Experiment C

In Experiment B, we trained a network with the data of two out of three dyslexia typefaces and 80% of the data of standard typefaces and used the network to classify the rest of the data. Since there are three dyslexia typefaces, this experiment includes three sub experiments. Figure 6(b) shows the split of the data for this experiment.

In Experiment C, we used the network trained in Experiment A to classify Latin and Japanese typefaces that are new to the network.

We created two test datasets for Experiment C. The first dataset contains 1134 character images from 18 Latin typefaces (63 characters for each) shown in Fig. 7(a). The second dataset contains 32256 character images (including Latin characters and Japanese kana and kanji characters) from 14 Japanese typefaces (2304 characters for each) shown in Fig. 7(b). We included typefaces designed for readers with dyslexia, namely LiS Font walnut and LiS Font cashew, from our previous research to try to get the idea of how different from or how similar to human brains the neural networks perceive visual characteristics of typefaces across languages.

The size of each image is 256 × 256 pixels and each character is aligned at the center, which is consistent with the train data.

5 Experimental Results

In this section, we report the results of the experiments.

5.1 Experiment A

In Experiment A, the classifier achieved 98.211% accuracy overall (95.333% for dyslexia typefaces and 99.267% for standard typefaces).

From Fig. 8, which illustrates the result of the classification for each typeface, we can see that a portion of characters from Comic Sans, a standard typeface, is classified as dyslexia. This agrees with our prior knowledge that the style of Comic Sans is similar to that of Lexie Readable, a dyslexia typeface. Moreover, it would be intriguing to point out that although Comic Sans has been disliked by designers, it is considered readable among children with dyslexia [6].

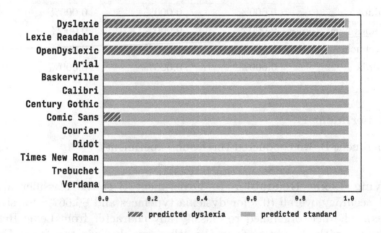

Fig. 8. The result of Experiment A

Figure 9 shows misclassified characters in this experiment. We found that numbers, superscripts, and bar-like characters, which are intuitively hard cases, tend to be misclassified.

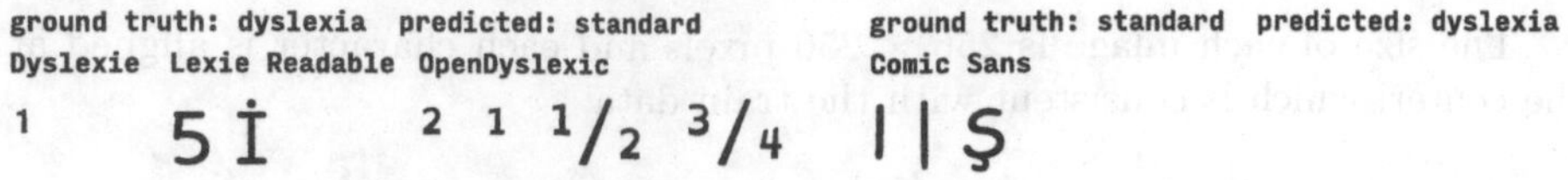

ground truth: dyslexia   predicted: standard   ground truth: standard   predicted: dyslexia
Dyslexie Lexie Readable OpenDyslexic   Comic Sans

1   5 İ   ² 1 ¹/₂ ³/₄   I | Ş

Fig. 9. Misclassified characters in Experiment A

Table 3. Ratio of characters predicted dyslexia in Experiment B

Typeface	Experiment B-1	Experiment B-2	Experiment B-3
Dyslexie	–	78.140%	–
Lexie Readable	0.000%	–	–
OpenDyslexic	–	–	83.721%
Arial	0.000%	0.000%	0.000%
Baskerville	0.000%	0.000%	0.000%
Calibri	0.000%	0.000%	0.000%
Century Gothic	0.000%	0.000%	0.000%
Comic Sans	0.000%	2.564%	10.256%
Courier	0.000%	0.000%	2.778%
Didot	0.000%	0.000%	0.000%
Times New Roman	0.000%	0.000%	0.000%
Trebuchet	0.000%	0.000%	0.000%
Verdana	0.000%	0.000%	0.000%

5.2 Experiment B

For Experiment B, we conducted three sub experiments.

Classifying Lexie Readable. In Experiment B-1, the classifier achieved 66.667% accuracy overall (0% for dyslexia typefaces and 66.667% for standard typefaces). The network failed to classify any character from Lexie Readable after training with characters from the other two dyslexia typefaces, Dyslexie and OpenDyslexic, as shown in the first column of Table 3. This result implies that Lexie Readable has rather different visual characteristics from Dyslexie and OpenDyslexic.

Classifying Dyslexie. In Experiment B-2, the classifier achieved 92.558% accuracy overall (78.140% for dyslexia typefaces and 99.767% for standard typefaces). A small portion of characters from Comic Sans are classified as dyslexia, which again indicates that Lexie Readable shares similar characteristics with Comic Sans.

Classifying OpenDyslexic. In Experiment B-3, the classifier achieved 92.558% accuracy overall (78.140% for dyslexia typefaces and 99.767% for standard typefaces). With a portion of characters from Comic Sans and a smaller portion of those from Courier classified as dyslexia, the result may indicate that the combined visual characteristics of Lexie Readable and Dyslexie have some similarities with Comic Sans and Courier which is a monospace typeface.

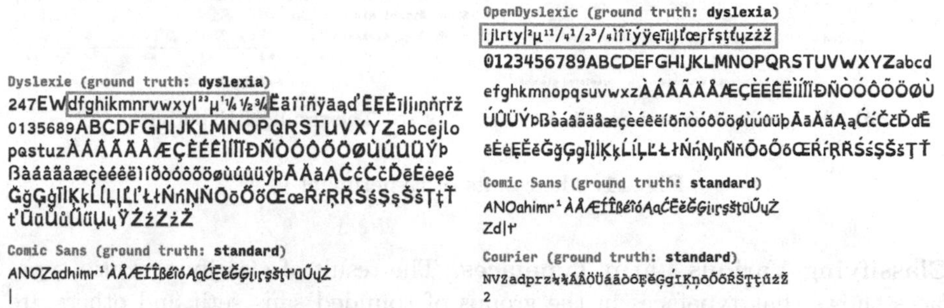

(a) Experiment B-2

(b) Experiment B-3

Fig. 10. Characters in Experiment B-2 and B-3 (blue: classified as 'standard'; red: classified as 'dyslexia') (Color figure online)

We show classified characters in Experiment B-2 and B-3 in Fig. 10. Characters classified as 'standard' are in blue and those classified as 'dyslexia' are in red. It may show that certain types of characters, such as superscripts and lowercases that are highlighted in Fig. 10, are more difficult to classify. However, further qualitative analysis is needed in future research.

The overall results of Experiments A and B show that neural networks do learn subtle differences between dyslexia and standard typefaces and even differences among different dyslexia typefaces. The results also suggest that the networks may perceive typefaces in the same manner as people with and without dyslexia do and could be a useful tool for typeface analysis.

5.3 Experiment C

The network trained in Experiment A was used in this experiment to examine various Latin and Japanese typefaces. The results are shown in Fig. 11.

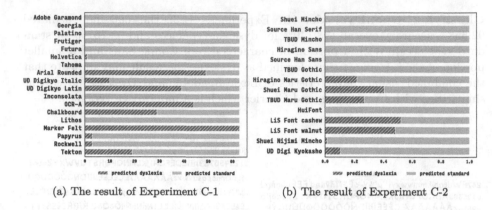

(a) The result of Experiment C-1   (b) The result of Experiment C-2

Fig. 11. The results of Experiment C

Classifying Various Latin Typefaces. The result of classifying Latin typefaces shows that typefaces in the groups of rounded sans serif and others are most likely to be classified as dyslexia. This agrees with the analysis of dyslexia typefaces using the typeface classification system in our previous research [29].

Classifying Various Japanese Typefaces. The result in terms of typeface styles is similar in classifying Japanese typefaces that typefaces in the groups of rounded sans serif and others are most likely to be classified as dyslexia. In both Latin and Japanese, those with rounded ends (rounded sans serif in Latin typefaces and maru gothic in Japanese typefaces), bolder strokes and irregular letterforms tend to be predicted as dyslexia typefaces despite the fact that they have diverging character sets from one another.

We also noticed that LiS Font cashew and LiS Font walnut—typefaces specially designed for readers with dyslexia—have more characters classified as dyslexia than others.

Therefore the results of Experiment C not only show the possibilities of using neural networks for identifying typefaces with similar characteristics across languages, but also indicates that to some extent we have successfully transferred the characteristics of Latin typefaces to Japanese typefaces with the heuristic methods.

Figure 12 shows classification of Japanese kana characters in Experiment C-2. We can observe that aside from characters with weighted bottom included in LiS Font cashew, katakana characters that are highlighted in Fig. 12, are more likely to be classified as 'dyslexia'. But again, further qualitative analysis is to be conducted to uncover the association between typefaces and character shapes.

Figure 13 illustrates the classification results of Japanese kanji (Chinese) characters of LiS Font cashew in Experiment C-2. Kanji characters classified as 'dyslexia' and 'standard' is about 1:1 but we did not find any qualitative differences between those two groups. Since LiS Font cashew shares kanji characters with LiS Font walnut but has a different set of kana characters, we presume that

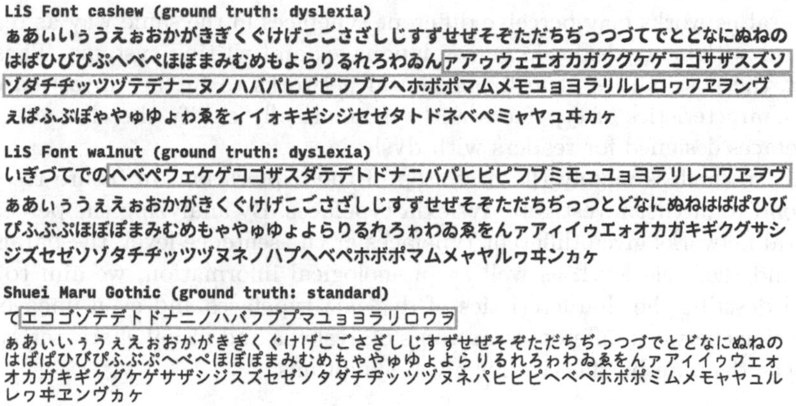

LiS Font cashew (ground truth: dyslexia)
ああいいううえおおかがきぎくぐけげこごさざしじすずせぜそぞただちぢっつづてでとどなにぬねの
はばぱひびぴふぶぷへべぺほぼぽまみむめもよらりるれろわゐんアウウェエオカガクグケゲコゴサザスズソ
ゾダチヂッツヅテデナニヌノハバパヒビピフブプヘホボポマムメモユヨヨラリルレロゥワエヲンヴ
えばふぶぱやゃゆゅょわゑをィイオキギシジセゼタトドネベペミャヤユヰカケ

LiS Font walnut (ground truth: dyslexia)
いぎづてでのヘベペウェケゲコゴザスダテデトドナニバパヘビピフブミモュユヨヨラリレロワエヲゥ
ああいううええおおかがきくぐけげけこごささざしじすずせぜそぞただちぢっつっとどなにぬねはばぱひび
ぴふぶぷほぼぽまみむめもゃやゅゆよょよらりるれろわわゐゑをんァアィイゥエオオカガキギクグサシ
ジズセゼソゾタチヂッツヅヌネノハプヘベペホボポマムメャヤルワヰンカケ

Shuei Maru Gothic (ground truth: standard)
てエココゾテデトドナニノハバフブプマユヨヨラリロワヲ
ああいいううええおおかおかがきぎくぐけげけこごさざしじすずせぜそぞただちぢっつづでとどなにぬねの
はばぱひびぴふぶぷへべぺほぼぽまみむめもゃやゅゆよょよらりるれろわわゐゑをんァアィイゥウエオ
オカガキグクゲゲサザシジスズセゼソタダチヂッツヅヌネパヒビピヘベペホボポミムメモャヤユル
レワヰヱンヴカケ

Fig. 12. Kana characters in Experiment C-2 (blue: classified as 'standard'; red: classified as 'dyslexia') (Color figure online)

weighted bottoms of kana characters of LiS Font cashew are a reason for its performance in Experiment C-2. This also supports our observation that irregular letterforms, along with rounded ends and bolder strokes can be extracted features of dyslexia typefaces.

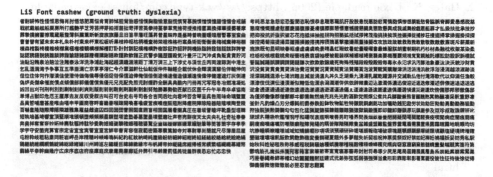

LiS Font cashew (ground truth: dyslexia)

Fig. 13. Kanji characters of LiS Font cashew in Experiment C-2 (blue: classified as 'standard'; red: classified as 'dyslexia') (Color figure online)

6   Conclusions

In this paper, we reported our first attempt at utilizing neural networks for typeface analysis, especially aiming at clarifying the visual characteristics of the Latin typefaces specially designed for readers with dyslexia and transferring the results to typefaces in other languages.

From the three experiments we conducted, the conclusions can be drawn that neural networks do learn subtle differences among different typefaces and different groups of typefaces. Furthermore, the results of the experiments show that the neural networks may perceive different typefaces in the same way as readers with and without dyslexia do across languages and writing systems. These all suggest the possibilities of utilizing neural networks to further investigate the visual characteristics of typefaces and systematically clarify the characteristics of typefaces designed for readers with dyslexia.

This research is merely the first step of the exploration. In future research, we intend to further investigate dyslexia typefaces. By analyzing the perception of neural networks given different typefaces at the sentence level, the paragraph level, and the page level, as well as phonological information, we aim to clarify and describe the characteristics of dyslexia typefaces and even uncover the associations between different symptoms of dyslexia and visual characteristics of typefaces that can be effective for given symptoms.

Acknowledgment. This research is supported by Grant-in-Aid for JSPS Fellows 19J11843.

References

1. Agrawal, P., Girshick, R., Malik, J.: Analyzing the performance of multilayer neural networks for object recognition. In: Fleet, D., Pajdla, T., Schiele, B., Tuytelaars, T. (eds.) ECCV 2014. LNCS, vol. 8695, pp. 329–344. Springer, Cham (2014). https://doi.org/10.1007/978-3-319-10584-0_22
2. Bates, K.: Lexie readable (2006). https://www.k-type.com/fonts/lexie-readable/
3. Bergstra, J., Bardenet, R., Bengio, Y., Kégl, B.: Algorithms for hyper-parameter optimization. In: Proceedings of NIPS, pp. 2546–2554 (2011)
4. Bergstra, J., Yamins, D., Cox, D.D.: Making a science of model search: hyperparameter optimization in hundreds of dimensions for vision architectures. In: Proceedings of ICML, pp. 115–123 (2013)
5. Boer, C.: Dyslexie font (2008). https://www.dyslexiefont.com/en/typeface/
6. British Dyslexia Association: Dyslexia Style Guide (2015). https://www.bdadyslexia.org.uk/advice/employers/creating-a-dyslexia-friendly-workplace/dyslexia-friendly-style-guide
7. Frensch, N.: Read regular (2003). http://www.readregular.com/english/regular.html
8. Girshick, R., Donahue, J., Darrell, T., Malik, J.: Rich feature hierarchies for accurate object detection and semantic segmentation. In: Proceedings of CVPR. IEEE (2014). https://doi.org/10.1109/cvpr.2014.81
9. Gonzalez, A.: OpenDyslexic (2011). https://www.opendyslexic.org/
10. He, K., Zhang, X., Ren, S., Sun, J.: Deep residual learning for image recognition. In: Proceedings of CVPR, pp. 770–778 (2016). https://doi.org/10.1109/CVPR.2016.90
11. Hillier, R.A.: Sylexiad (2006). http://www.robsfonts.com/fonts/sylexiad
12. Hillier, R.A.: A typeface for the adult dyslexic reader. Ph.D. thesis, Anglia Ruskin University (2006)
13. Ide, S., Uchida, S.: How does a CNN manage different printing types? In: Proceedings of ICDAR, pp. 1004–1009 (2018). https://doi.org/10.1109/ICDAR.2017.167
14. International Dyslexia Association: Definition of dyslexia (2002). https://dyslexiaida.org/definition-of-dyslexia/

15. Karita, T., Sakai, S., Hirabayashi, R., Nakamura, K.: Trends in Japanese developmental dyslexia research [in Japanese]. J. Dev. Disord. Speech Lang. Hear. **8**, 31–45 (2010)
16. Krizhevsky, A., Sutskever, I., Hinton, G.E.: ImageNet classification with deep convolutional neural networks. Commun. ACM **60**(6), 84–90 (2017). https://doi.org/10.1145/3065386
17. Marinus, E., Mostard, M., Segers, E., Schubert, T.M., Madelaine, A., Wheldall, K.: A special font for people with dyslexia: does it work and if so, why? Dyslexia **22**(3), 233–244 (2016)
18. Perez, L., Wang, J.: The effectiveness of data augmentation in image classification using deep learning. arXiv:1712.04621 [cs.CV] (2017)
19. Reid, G., Fawcett, A.J., Manis, F., Siegel, L.S. (eds.): The SAGE Handbook of Dyslexia. SAGE Publications, London (2008)
20. Schriver, K.A.: Dynamics in Document Design. Wiley, New York (1997)
21. Simard, P.Y., Steinkraus, D., Platt, J.C.: Best practices for convolutional neural networks applied to visual document analysis. In: Proceedings of ICDAR, pp. 958–963 (2003). https://doi.org/10.1109/ICDAR.2003.1227801
22. Smythe, I., Everatt, J., Salter, R. (eds.): The International Book of Dyslexia: A Guide to Practice and Resources. Wiley, New Jersey (2005)
23. Stein, J.: The current status of the magnocellular theory of developmental dyslexia. Neuropsychologia **130**, 66–77 (2019). https://doi.org/10.1016/j.neuropsychologia.2018.03.022
24. Tensmeyer, C., Saunders, D., Martinez, T.: Convolutional neural networks for font classification. In: Proceedings of ICDAR, pp. 985–990 (2018). https://doi.org/10.1109/ICDAR.2017.164
25. Tracy, W.: Letters of Credit. David R. Godine, Jaffrey (2003)
26. Uchida, S., Egashira, Y., Sato, K.: Exploring the world of fonts for discovering the most standard fonts and the missing fonts. In: Proceedings of ICDAR, pp. 441–445 (2015). https://doi.org/10.1109/ICDAR.2015.7333800
27. Zhang, J., Guo, M., Fan, J.: A novel CNN structure for fine-grained classification of Chinese calligraphy styles. Int. J. Doc. Anal. Recogn. (IJDAR) **22**(2), 177–188 (2019). https://doi.org/10.1007/s10032-019-00324-1
28. Zhong, Z., Jin, L., Feng, Z.: Multi-font printed Chinese character recognition using multi-pooling convolutional neural network. In: Proceedings of ICDAR, pp. 96–100 (2015). https://doi.org/10.1109/ICDAR.2015.7333733
29. Zhu, X.: Characteristics of Latin typefaces for dyslexic readers [in Japanese]. IPSJ SIG Tech. Rep. **2016–CE–13**(4), 1–9 (2016)
30. Zhu, X., Kageura, K.: Research on Japanese typefaces and typeface customisation system designed for readers with developmental dyslexia. In: International Association of Societies of Design Research (2019)

Neural Style Difference Transfer and Its Application to Font Generation

Gantugs Atarsaikhan$^{(\boxtimes)}$, Brian Kenji Iwana, and Seiichi Uchida

Kyushu University, Fukuoka, Japan
{gantugs.atarsaikhan,brian,uchida}@human.ait.kyushu-u.ac.jp

Abstract. Designing fonts requires a great deal of time and effort. It requires professional skills, such as sketching, vectorizing, and image editing. Additionally, each letter has to be designed individually. In this paper, we introduce a method to create fonts automatically. In our proposed method, the difference of font styles between two different fonts is transferred to another font using neural style transfer. Neural style transfer is a method of stylizing the contents of an image with the styles of another image. We proposed a novel neural style difference and content difference loss for the neural style transfer. With these losses, new fonts can be generated by adding or removing font styles from a font. We provided experimental results with various combinations of input fonts and discussed limitations and future development for the proposed method.

Keywords: Convolutional neural network · Style transfer · Style difference

1 Introduction

Digital font designing is a highly time-consuming task. It requires professional skills, such as sketching ideas on paper and drawing with complicated software. Individual characters or letters has many attributes to design, such as line width, angles, stripes, serif, and more. Moreover, a designer has to design all letters character-by-character, in addition to any special characters. For example, the Japanese writing system has thousands of Japanese characters that needs to be designed individually. Therefore, it is beneficial to create a method of designing fonts automatically for people who have no experience in designing fonts. It is also beneficial to create a way to assist font designers by automatically drawing fonts.

On the other hand, there are a large number of fonts that have already designed. Many of them have different font styles, such as **bold**, *italic*, SERIF and SANS SERIF. There are many works done to create new fonts by using already designed fonts [27,36]. In this paper, we chose an approach to find the difference between two fonts and transfer it onto a third font in order to create a new font. For example, using a font with serifs and a font without serifs to transfer the serif difference to a third font that originally lacked serifs, as shown in Fig. 1.

© Springer Nature Switzerland AG 2020
X. Bai et al. (Eds.): DAS 2020, LNCS 12116, pp. 544–558, 2020.
https://doi.org/10.1007/978-3-030-57058-3_38

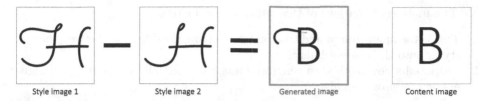

Style image 1   Style image 2   Generated image   Content image

Fig. 1. An example results of the proposed method. Style difference between the style image 1 and 2 is transferred onto the content image by equalling to the style difference between the newly generated image and the content image.

Content image   Style image   Generated image

Fig. 2. An example of the NST. Features of the style image are blended into the structure of the content image in the generated result image.

In recent years, the style transfer field has progressed rapidly with the help of Convolutional Neural Networks (CNN) [20]. Gatys et al. [12] first used a CNN to synthesize an image using the style of an image and the content of another image using Neural Style Transfer (NST). In the NST, the content is regarded as feature maps of a CNN and the style is determined by the correlation of feature maps in a Gram matrix. The Gram matrix calculates how correlated the feature maps are to each other. An example of the NST is shown in Fig. 2. The content image and style image are mixed by using features from the CNN to create a newly generated image that has contents (buildings, sky, trees, etc.) from the content image and style (swirls, paint strokes, etc.) from the style image. There are also other style transfer methods, such as a ConvDeconv network for real-time style transfer [17] and methods that utilize Generative Adversarial Networks (GAN) [15].

The purpose of this paper is to explore and propose a new method to create novel fonts automatically. Using NST, the contents and styles of two different fonts have been found and their difference is transferred to another font to generate a new synthesized font. Figure 1 shows an example results of our proposed method. We provided experimental results and inspected the performance of our method with various combinations of content and style images.

The main contributions of this paper are as follows.

1. This is the first attempt and trial on transferring the difference between neural styles onto the content image.
2. Proposed a new method to generate fonts automatically to assist font designers or non-professionals.

The remaining of this paper is arranged as follows. In Sect. 2, we discuss previous works on font generation, style transfer fields, and font generation using CNN. The proposed method is explained in Sect. 3 in detail. Then, Sect. 4 examines the experiments and the results. Lastly, we conclude in Sect. 5 with concluding remarks and discussions for improvements.

2   Related Work

2.1   Font Generation

Various attempts have been made to create fonts automatically. One approach is to generate font using example fonts. Devroye and McDougall [10] created handwritten fonts from handwritten examples. Tenenbaum and Freeman [34] clustered fonts by its font styles and generated new fonts by mixing font styles with other fonts. Suveeranont and Igarashi [32,33] generated new fonts from user-defined examples. Tsuchiya et al. [35] also used example fonts to determine features for new fonts. Miyazaki et al. [27] extracted strokes from fonts and generated typographic fonts.

Another approach in generating fonts is to use transformations or interpolations of fonts. Wada and Hagiwara [38] created new fonts by modifying some attributes of fonts, such as slope angle, thickness, and corner angle. Wand et al. [39] transformed strokes of Chinese characters to generate more characters. Campbell and Kautz [7] created new fonts by mapping non-linear interpolation between multiple existing fonts. Uchida et al. [36] generated new fonts by finding fonts that are simultaneously similar to existing fonts.

Lake et al. [19] generated handwritten fonts by capturing example patterns with the Bayesian program learning method. Baluja et al. [6] learned styles from four characters to generate other characters using CNN-like architecture. Recently, many studies use machine learning for font design. Atarsaikhan et al. [3] used the NST to synthesize a style font and a content font to generate a new font. Also, GANs have been used to generate new fonts [1,24,31]. Lastly, fonts have been stylized with patch-based statistical methods [40], NST [4] and GAN methods [5,41,43].

2.2   Style Transfer

The first example-based style transfer method was introduced in "Image Analogies" by Hertzmann et al. [14]. Recently, Gatys et al. developed the NST [12] by utilizing a CNN. There are two types of NST methods, i.e. image

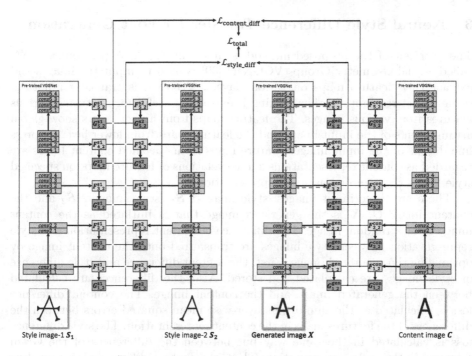

Fig. 3. The overview font generation with neural style difference transfer. Yellow blocks show feature maps from one layer and purple blocks show Gram matrices calculated using the feature maps. (Color figure online)

optimization-based and network optimization-based. NST is the most popular image optimization-based method. However, the original NST [12] was introduced for artistic style transfer and works have been done for photorealistic style transfer [23,26]. New losses are introduced for stable results and preserving low-level features [21,28]. Additionally, improvements, such as semantic aware style transfer [8,25], controlled content features [11,13] were proposed. Also, Mechrez et al. [25] and Yin et al. [42], achieved semantic aware style transfer.

In network optimization-based style transfer, first, neural networks are trained for a specific style image and then this trained network is used to stylize a content image. A ConvDeconv neural network [17] and a generative network [37] are trained for style transfer. They improved for photorealistic style transfer [22] and semantic style transfer [8,9].

There are many applications of the NST due to its non-heuristic style transfer qualities. It has been used for video style transfer [2,18], portrait [29], fashion [16], and creating doodles [8]. Also, character stylizing techniques have been proposed by improving the NST [4] or using NST as part of the bigger network [5].

3 Neural Style Difference Transfer for Font Generation

The overview of the proposed method is shown in Fig. 3. A pre-trained CNN called Visual Geometry Group (VGGNet) [30] is used to input the images and extract their feature maps on various layers. VGGNet is trained for natural scene object recognition, thus making it extremely useful for capturing various features from various images. The feature maps from higher layers show global arrangements of the input image and the feature maps from lower layers express fine details of the input image. Feature maps from specified content layers are regarded as content representations and correlations of feature maps on specified style layers are regarded as style representations.

There are three input images: style image-1 S_1, style image-2 S_2 and the content image C. X is the generated image that is initiated as the content image or as a random image. The difference of content representation and style representation between style images are transferred onto the content image by optimizing the generated image. First, the content difference and style difference of style images are calculated and stored. Also, the differences are calculated between the generated image and the content image. The content difference loss is calculated as the sum of the layer-wise mean squared errors between the differences of the features maps in the content representation. The style difference loss is calculated in the same way but between the differences of the Gram matrices in the style representation. Then, the content difference loss and style difference loss are accumulated into the total loss. Lastly, the generated image is optimized through back-propagation to minimize the total loss. By repeatedly optimizing the generated image with this method, the style difference between the style images is transferred to the content image.

3.1 Neural Style Transfer

Before explaining the proposed method in detail, we will briefly discuss NST. In the NST, there are two inputs: a content image C and a style image S. The image to be optimized is the generated image X. It also uses a CNN, i.e. VGGNet to capture features of the input images and create a Gram matrix for the style representation from the captured feature maps. A Gram matrix is shown in Eq. 1, where D_l is a matrix that consists of flattened feature maps of layer l as shown in Eq. 2. The Gram matrix calculates the correlation value of feature maps from one layer to each other and stores it into a matrix.

$$G_l = D_l(D_l)^{\top}, \tag{1}$$

where,

$$D_l = \{F_1, ..., F_{n_l}, ..., F_{N_l}\}. \tag{2}$$

First, the style image S is input to the VGGNet. Its feature maps $\mathbb{F}^{\text{style}}$ on given style layers are extracted and their Gram matrices $\mathbb{G}^{\text{style}}$ are calculated. Next, the content image C input to the VGGNet and its feature maps $\mathbb{F}^{\text{content}}$ on given content layers are extracted and stored. Lastly, the generated image X

is input to the network. Its Gram matrices $\mathbb{G}^{\text{generated}}$ on style layers and feature maps $\mathbb{F}^{\text{generated}}$ from content layers are found.

Then, by using the feature maps and Gram matrices, the content loss and style loss are calculated as,

$$\mathcal{L}_{\text{content}} = \sum_{l}^{L_c} \frac{w_l^{\text{content}}}{2N_l M_l} \sum_{n_l}^{N_l} \sum_{m_l}^{M_l} (F_{n_l,m_l}^{\text{generated}} - F_{n_l,m_l}^{\text{content}})^2, \tag{3}$$

and

$$\mathcal{L}_{\text{style}} = \sum_{l}^{L_s} \frac{w_l^{\text{style}}}{4N_l^2 M_l^2} \sum_{i}^{N_l} \sum_{j}^{N_l} (G_{l,i,j}^{\text{generated}} - G_{l,i,j}^{\text{style}})^2, \tag{4}$$

where L_c and L_s are the number of layers, N_l is the number of feature maps, M_l is the number of elements in one feature map, w_l^{content} and w_l^{style} are weights for layer l. Lastly, the content loss $\mathcal{L}_{\text{content}}$ and the style loss $\mathcal{L}_{\text{style}}$ are accumulated into the total loss $\mathcal{L}_{\text{total}}$ with weighting factors α and β:

$$\mathcal{L}_{\text{total}} = \alpha \mathcal{L}_{\text{content}} + \beta \mathcal{L}_{\text{style}}. \tag{5}$$

Once the total loss $\mathcal{L}_{\text{total}}$ is calculated, the gradients of content layers, style layers, and generated image $\boldsymbol{X}$ are determined by back-propagation. Then, to minimize the total loss $\mathcal{L}_{\text{total}}$, only the generated image $\boldsymbol{X}$ is optimized. By repeating these steps multiple times, the style from the style image are transferred to the content image in the form of a generated image.

In the NST, the goal of the optimization process is to match the styles of the generated image to those of the style image, and feature maps of the generated image to those of the content image. However, in the proposed method, the goal of the optimization process is to match the style differences between the generated image and the content image to those of style images, as well as, differences of content difference between the generated image and the content image to those of style images.

3.2 Style Difference Loss

Let $\boldsymbol{G}_l^{\text{style1}}$ and $\boldsymbol{G}_l^{\text{style2}}$ be the Gram matrices of feature maps on layer l, when style image-1 $\boldsymbol{S}_1$ and style image-2 $\boldsymbol{S}_2$ are input respectively. Then, the style difference between the style images on layer l is defined as,

$$\Delta \boldsymbol{G}_l^{\text{style}} = \boldsymbol{G}_l^{\text{style1}} - \boldsymbol{G}_l^{\text{style2}}, \tag{6}$$

Similarly, the style difference between the generated image $\boldsymbol{X}$ and the content image $\boldsymbol{C}$ is defined as,

$$\Delta \boldsymbol{G}_l^{\text{generated}} = \boldsymbol{G}_l^{\text{generated}} - \boldsymbol{G}_l^{\text{content}}. \tag{7}$$

Consequently, the style loss is the mean squared error between the style differences:

$$\mathcal{L}_{\text{style\_diff}} = \sum_{l}^{L} \frac{w_l^{\text{style}}}{4N_l^2 M_l^2} \sum_{i}^{N_l} \sum_{j}^{N_l} (\Delta G_{l,i,j}^{\text{generated}} - \Delta G_{l,i,j}^{\text{style}})^2, \tag{8}$$

where w_l^{style} is a weighting factor for an individual layer l. Note, it can be set to zero to ignore a specific layer. The style difference loss in the proposed method means that the difference of correlations of feature maps (Gram matrix) of the generated and content images (X and C) are forced to match that of style images (S_1 and S_2) through optimization of the generated image, so that the style difference (e.g. bold and light fonts styles, italic or regular fonts styles) are transferred.

3.3   Content Difference Loss

Extracting the feature maps on a layer l as F_l^{style1} and F_l^{style2} for the style images, the content difference between the style images on layer l are defined as follows,

$$\Delta F_l^{\text{style}} = F_l^{\text{style1}} - F_l^{\text{style2}}. \tag{9}$$

Using the same rule, the content difference between the generated and content images on layer l is defined as,

$$\Delta F_l^{\text{generated}} = F_l^{\text{generated}} - F_l^{\text{content}}, \tag{10}$$

where $F_l^{\text{generated}}$ is the feature maps on layer l when the generated image X is input, and F_l^{content} is the feature maps when the content image C is input. By using content differences of the two style images and the generated and content images, the content difference loss is calculated as,

$$\mathcal{L}_{\text{content\_diff}} = \sum_l^L \frac{w_l^{\text{content}}}{2 N_l M_l} \sum_{n_l}^{N_l} \sum_{m_l}^{M_l} (\Delta F_{l,n_l,m_l}^{\text{generated}} - \Delta F_{l,n_l,m_l}^{\text{content}})^2, \tag{11}$$

where w_l^{content} is weighting factor for layer l. Layers also can be ignored by setting w_l^{content} to zero. The content difference loss captures the difference in global feature of the style images, e.g, difference in serifs.

3.4   Style Transfer

For style transfer, a generated image X is optimized to simultaneously match the style difference on style layers and the content difference on content layers. Thus, a loss function is created and minimized: The total loss $\mathcal{L}_{\text{total}}$ which is the accumulation of the content difference loss $\mathcal{L}_{\text{content\_diff}}$ and the style difference loss $\mathcal{L}_{\text{style\_diff}}$ written as,

$$\mathcal{L}_{\text{total}} = \mathcal{L}_{\text{content\_diff}} + \mathcal{L}_{\text{style\_diff}}. \tag{12}$$

With the total loss, the gradients of pixels on the generated image are calculated using back-propagation and used for an optimization method, such as L-BFGS or Adam. We found from experience that L-BFGS requires fewer iterations and produces better results. Also, we use the same image size for each input image

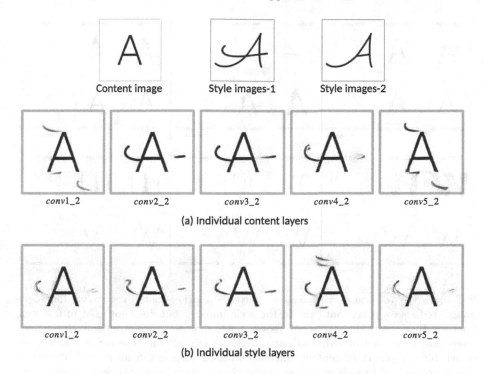

Fig. 4. Various weights for content and style layers.

for comparable feature sizes. Due to plain subtracting operation in Eq. 6, Eq. 7, Eq. 9, and Eq. 10, input images have to be chosen carefully. Font styles of S_2 must be similar to that of C. Because of the subtraction process, the generated image X is most likely optimized to have similar font styles with style image-1 S_1. Moreover, contrary to the NST, we do not use weighting factors for content loss and style loss, instead, individual layers are weighted.

4   Experimental Results

In the experiments below except specified, feature maps for the style difference loss are taken from the style layers, $conv1\_2$, $conv2\_2$, $conv3\_2$, $conv4\_2$, $conv5\_2$ with style weights $\boldsymbol{w}^{\text{style}} = \{\frac{10^3}{64^2}, \frac{10^3}{128^2}, \frac{10^3}{256^2}, \frac{10^3}{512^2}, \frac{10^3}{512^2}\}$, and feature maps for the content difference loss is taken from content layer $conv4\_2$ with content weight $w_{conv4\_2} = 10^4$ on VGGNet. Generated image X is initialized with the content image C. The optimization process is stopped at 1,000th step with more than enough convergence. Also, due to black pixels having a zero value, input images are inverted before inputting to the VGGNet and inverted back for the visualization.

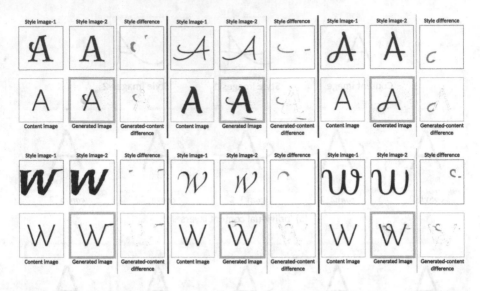

Fig. 5. Results of transferring missing parts to another font. In the style difference image, red shows parts that exist in the style image-1 but does not exist in the style image-2. Blue shows the reverse. Moreover, parts that are transferred onto the content image are visualized in red, and parts that are erased from the content image are shown blue in generated-content difference images. As a clarification, style difference and generated-content difference are mere visualizations from the input images and generated image, they are not input or results images themselves. (Color figure online)

4.1   Content and Style Layers

Figure 4 shows results using various content and style layers individually. We used a sans-serif font for the content image, and tried to transfer the horizontal line style difference between style fonts. In Fig. 4a, we experimented on using each content layers while the weights for the style layers are fixed as $\boldsymbol{w}^{\text{style}} = \{\frac{10^3}{64^2}, \frac{10^3}{128^2}, \frac{10^3}{256^2}, \frac{10^3}{512^2}, \frac{10^3}{512^2}\}$. Using w_{conv1_2} and w_{conv5_2} as content layers resulted the style difference appear in random places. Moreover, results using w_{conv2_2} and w_{conv3_2} has too firm of a style difference. On the other hand, using content layer w_{conv4_2} resulted in not too firm or not random style difference. In Fig. 4, we experimented on individual style layers while fixing the content layer to w_{conv4_2} with weight of $w_{conv4\_2} = 10^4$. Each of the results show not incomplete but not overlapping style difference on the content image. Thus, we used all of the style layers to capture complete style difference of the style images.

4.2   Complementing the Content Font

Transferring Missing Parts. We experimented on transferring missing parts of a font. As visualized in red in style difference images of Fig. 5, style image-2 lacks some parts compared to the style image-1. We will try to transfer this

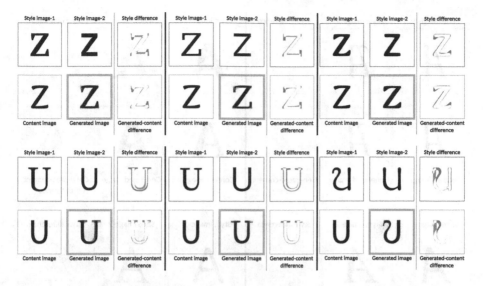

Fig. 6. Examples of generating serifs on the content font using the difference between serif font and sans-serif font from the same font family.

difference of parts to the content image. By using the above parameter settings, the proposed method was able to transfer the missing parts onto the content image as shown in the figure. The transferred parts are visualized in red in the "Generated-content difference" images of the Fig. 5. The proposed method transfers the style difference while trying to match the style of the content font. Thus, the most appended part of the content image is connected to the content font. The best results were achieved when the missing parts are relatively small or narrower than the content fonts. Using wider fonts works most of the time. However, the proposed method struggled to style transfer when the difference part is too large or separated. Moreover, missing parts do not only transfer onto the content image, but the style of the missing part is changed to match the style of the content image.

Generating Serifs. Figure 6 shows the experiments on generating serifs on the content image. Both style images are taken from the same font family. Style image-1 includes serif font, style image-2 includes sans-serif fonts and the content image includes a sans-serif font from different the font family of the style fonts. As shown in the figure, serifs are generated on the content image successfully. Moreover, not only the content font is extended by the serif, parts of it are morphed to include the missing serif as shown in the lower right corner of the figure.

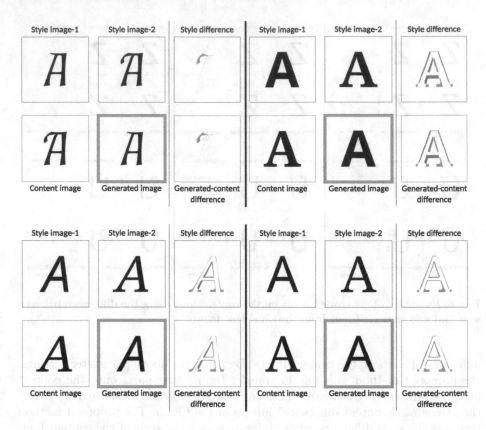

Fig. 7. Experiments for removing serifs from the content image by style difference.

4.3   Removing Serifs

Figure 7 shows the experiments on removing serifs from the content image fonts. Style image-1 includes a font that does not have serif, and style image-2 includes a font that has serifs. By using this difference in serifs, we experimented on removing serifs from the content font. As shown in the figure, serifs have been removed from the content image. However, the font styles of the generated image became different than those of the content image.

4.4   Transferring Line Width

Figure 8 shows transferring difference of font line width between the style images to the content image to change the font line width from narrow to wide or from wide to narrow. The proposed method was able to change the content font with a narrow line to a font that has a wider line in most cases and vice-versa. However, it struggled to change the wide font line to a narrow font line, when the content font line is wider than the font line in the style image-2.

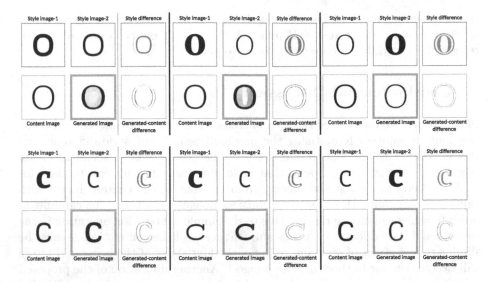

Fig. 8. Transferring font line width difference. The first two columns show experiments for widening the font line width of the content font. The third column shows eroding the content font by style difference.

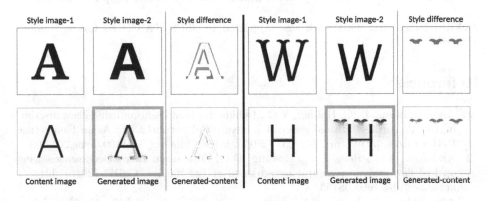

Fig. 9. Failure cases. The first experiment shows a failure when styles of the content image and style image-2 are not similar. The second experiment shows a failure case when the input characters are different.

4.5   Failure Cases

Figure 9 shows failure cases for font generation. On the left side of Fig. 9, fonts from the content image and style image-2 do not have a similar font style. The font in the style image-2 has wide lines, whereas, the font in the content image as narrow lines. The style difference between the style fonts is serif in the foot of the font. Consequently, the proposed method tried to transfer both wideness and serifs, so it resulted in an incomplete font. Moreover, the right side of Fig. 9

shows input images that have different characters. Although the font style match between content image and style image-2, the content image is not suitable to receive the font difference between the style images.

5 Conclusion and Discussion

In this paper, we introduced the idea of transferring the style difference between two fonts to another font. Using the proposed neural font style difference, we showed that it is possible to transfer the differences between styles to create new fonts. Moreover, we showed that style difference transfer can be used for both the complementing and erasing from the font in the content image with experimental results. However, the input font images must be chosen carefully in order to achieve plausible results. Due to the simple subtraction operation in the content and style difference calculation, font styles of the content image must be similar to font styles of style image-2. So, the font styles of the generated image will become similar to those of style image-1. Another limitation of the proposed method is the processing time due to the back-propagation in each stylization step. These issues can be solved by utilizing the encoder-decoder style transfer method [17] or adversarial method [5]. However, these methods will have to be trained for the style transfer process first is contrary to the proposed method.

Acknowledgment. This work was supported by JSPS KAKENHI Grant Number JP17H06100.

References

1. Abe, K., Iwana, B.K., Holmér, V.G., Uchida, S.: Font creation using class discriminative deep convolutional generative adversarial networks. In: Asian Conference Pattern Recognition, pp. 232–237 (2017). https://doi.org/10.1109/acpr.2017.99
2. Anderson, A.G., Berg, C.P., Mossing, D.P., Olshausen, B.A.: Deepmovie: using optical flow and deep neural networks to stylize movies. CoRR (2016). http://arxiv.org/abs/1605.08153
3. Atarsaikhan, G., Iwana, B.K., Narusawa, A., Yanai, K., Uchida, S.: Neural font style transfer. In: International Conference on Document Analysis and Recognition, pp. 51–56 (2017). https://doi.org/10.1109/icdar.2017.328
4. Atarsaikhan, G., Iwana, B.K., Uchida, S.: Contained neural style transfer for decorated logo generation. In: International Workshop Document Analysis Systems, pp. 317–322 (2018). https://doi.org/10.1109/das.2018.78
5. Azadi, S., Fisher, M., Kim, V.G., Wang, Z., Shechtman, E., Darrell, T.: Multicontent GAN for few-shot font style transfer. In: IEEE Conference Computer Vision and Pattern Recognition, pp. 7564–7573, June 2018. https://doi.org/10.1109/cvpr.2018.00789
6. Baluja, S.: Learning typographic style: from discrimination to synthesis. Mach. Vis. Appl. **28**(5–6), 551–568 (2017)
7. Campbell, N.D.F., Kautz, J.: Learning a manifold of fonts. ACM Trans. Graphics **33**(4), 1–11 (2014). https://doi.org/10.1145/2601097.2601212

8. Champandard, A.J.: Semantic style transfer and turning two-bit doodles into fine artworks. CoRR (2016). http://arxiv.org/abs/1603.01768
9. Chen, Y.L., Hsu, C.T.: Towards deep style transfer: A content-aware perspective. In: British Machine Vision Conference, pp. 8.1-8.11 (2016). https://doi.org/10.5244/c.30.8
10. Devroye, L., McDougall, M.: Random fonts for the simulation of handwriting. Electronic Publishing (EP—ODD) **8**, 281–294 (1995)
11. Gatys, L.A., Bethge, M., Hertzmann, A., Shechtman, E.: Preserving color in neural artistic style transfer. CoRR (2016). http://arxiv.org/abs/1606.05897
12. Gatys, L.A., Ecker, A.S., Bethge, M.: Image style transfer using convolutional neural networks. In: IEEE Conference on Computer Vision and Pattern Recognition, pp. 2414–2423 (2016). https://doi.org/10.1109/cvpr.2016.265
13. Gatys, L.A., Ecker, A.S., Bethge, M., Hertzmann, A., Shechtman, E.: Controlling perceptual factors in neural style transfer. In: IEEE Conference on Computer Vision and Pattern Recognition, pp. 3730–3738 (2017). https://doi.org/10.1109/cvpr.2017.397
14. Hertzmann, A., Jacobs, C.E., Oliver, N., Curless, B., Salesin, D.H.: Image analogies. In: Proceedings of the 28th Annual Conference on Computer Graphics and Interactive Techniques, pp. 327–340. ACM (2001)
15. Isola, P., Zhu, J.Y., Zhou, T., Efros, A.A.: Image-to-image translation with conditional adversarial networks. In: IEEE Conference on Computer Vision and Pattern Recognition, pp. 1125–1134 (2016). https://doi.org/10.1109/cvpr.2017.632
16. Jiang, S., Fu, Y.: Fashion style generator. In: International Joint Conference on Artificial Intelligence, pp. 3721–3727, August 2017. https://doi.org/10.24963/ijcai.2017/520
17. Johnson, J., Alahi, A., Fei-Fei, L.: Perceptual losses for real-time style transfer and super-resolution. In: Leibe, B., Matas, J., Sebe, N., Welling, M. (eds.) ECCV 2016. LNCS, vol. 9906, pp. 694–711. Springer, Cham (2016). https://doi.org/10.1007/978-3-319-46475-6_43
18. Joshi, B., Stewart, K., Shapiro, D.: Bringing impressionism to life with neural style transfer in come swim. In: ACM SIGGRAPH Digital Production Symposium, p. 5 (2017). https://doi.org/10.1145/3105692.3105697
19. Lake, B.M., Salakhutdinov, R., Tenenbaum, J.B.: Human-level concept learning through probabilistic program induction. Science **350**(6266), 1332–1338 (2015)
20. LeCun, Y., Bottou, L., Bengio, Y., Haffner, P.: Gradient-based learning applied to document recognition. Proc. IEEE **86**(11), 2278–2324 (1998). https://doi.org/10.1109/5.726791
21. Li, S., Xu, X., Nie, L., Chua, T.S.: Laplacian-steered neural style transfer. In: ACM International Conference Multimedia, pp. 1716–1724 (2017). https://doi.org/10.1145/3123266.3123425
22. Li, Y., Liu, M.Y., Li, X., Yang, M.H., Kautz, J.: A closed-form solution to photorealistic image stylization. In: Proceedings of the European Conference on Computer Vision (ECCV), pp. 453–468 (2018)
23. Luan, F., Paris, S., Shechtman, E., Bala, K.: Deep photo style transfer. In: IEEE Conference on Computer Vision and Pattern Recognition, pp. 4990–4998 (2017). https://doi.org/10.1109/cvpr.2017.740
24. Lyu, P., Bai, X., Yao, C., Zhu, Z., Huang, T., Liu, W.: Auto-encoder guided GAN for Chinese calligraphy synthesis. In: International Conference Document Analysis and Recognition, pp. 1095–1100 (2017). https://doi.org/10.1109/icdar.2017.181
25. Mechrez, R., Talmi, I., Zelnik-Manor, L.: The contextual loss for image transformation with non-aligned data. CoRR (2018). http://arxiv.org/abs/1803.02077

26. Mechrez, Roey, S.E., Zelnik-Manor, L.: Photorealistic style transfer with screened poisson equation. In: British Machine Vision Conference (2017)
27. Miyazaki, T., et al.: Automatic generation of typographic font from a small font subset. CoRR abs/1701.05703 (2017). http://arxiv.org/abs/1701.05703
28. Risser, E., Wilmot, P., Barnes, C.: Stable and controllable neural texture synthesis and style transfer using histogram losses. CoRR (2017). http://arxiv.org/abs/1701.08893
29. Selim, A., Elgharib, M., Doyle, L.: Painting style transfer for head portraits using convolutional neural networks. ACM Trans. Graphics **35**(4), 129 (2016). https://doi.org/10.1145/2897824.2925968
30. Simonyan, K., Zisserman, A.: Very deep convolutional networks for large-scale image recognition. In: International Conference Learning Representations (2015)
31. Sun, D., Zhang, Q., Yang, J.: Pyramid embedded generative adversarial network for automated font generation. CoRR (2018). http://arxiv.org/abs/1811.08106
32. Suveeranont, R., Igarashi, T.: Feature-preserving morphable model for automatic font generation. In: ACM SIGGRAPH ASIA 2009 Sketches, p. 7. ACM (2009)
33. Suveeranont, R., Igarashi, T.: Example-based automatic font generation. In: Taylor, R., Boulanger, P., Krüger, A., Olivier, P. (eds.) SG 2010. LNCS, vol. 6133, pp. 127–138. Springer, Heidelberg (2010). https://doi.org/10.1007/978-3-642-13544-6_12
34. Tenenbaum, J.B., Freeman, W.T.: Separating style and content with bilinear models. Neural Comput. **12**(6), 1247–1283 (2000)
35. Tsuchiya, T., Miyazaki, T., Sugaya, Y., Omachi, S.: Automatic generation of kanji fonts from sample designs. In: Tohoku-Section Joint Convention of Institutes of Electrical and Information Engineers, p. 36 (2014)
36. Uchida, S., Egashira, Y., Sato, K.: Exploring the world of fonts for discovering the most standard fonts and the missing fonts. In: International Conference on Document Analysis and Recognition, pp. 441–445. IEEE (2015). https://doi.org/10.1109/icdar.2015.7333800
37. Ulyanov, D., Lebedev, V., Vedaldi, A., Lempitsky, V.: Texture networks: feedforward synthesis of textures and stylized images. In: International Conference on Machine Learning, pp. 1349–1357 (2016)
38. Wada, A., Hagiwara, M.: Japanese font automatic creating system reflecting user's kansei. In: IEEE International Conference on Systems, Man and Cybernetics. Conference Theme-System Security and Assurance, vol. 4, pp. 3804–3809. IEEE (2003)
39. Wang, Y., Wang, H., Pan, C., Fang, L.: Style preserving Chinese character synthesis based on hierarchical representation of character. In: IEEE International Conference on Acoustics, Speech and Signal Processing, pp. 1097–1100. IEEE (2008)
40. Yang, S., Liu, J., Lian, Z., Guo, Z.: Awesome typography: statistics-based text effects transfer. In: IEEE Conference on Computer Vision and Pattern Recognition, pp. 2886–2895, July 2017. https://doi.org/10.1109/CVPR.2017.308
41. Yang, S., Liu, J., Wang, W., Guo, Z.: TET-GAN: text effects transfer via stylization and destylization. CoRR (2018). http://arxiv.org/abs/1812.06384
42. Yin, R.: Content aware neural style transfer. CoRR (2016). http://arxiv.org/abs/1601.04568
43. Zhao, N., Cao, Y., Lau, R.W.: Modeling fonts in context: font prediction on web designs. Comput. Graphics Forum **37**(7), 385–395 (2018). https://doi.org/10.1111/cgf.13576

A Method for Scene Text Style Transfer

Gaojing Zhou[✉], Lei Wang, Xi Liu, Yongsheng Zhou, Rui Zhang,
and Xiaolin Wei

Meituan-Dianping Group, Beijing, China
{zhougaojing,wanglei102,liuxi12,zhougyongsheng,
zhangrui36,weixiaolin02}@meituan.com

Abstract. Text style transfer is a challenging problem in optical character recognition. Recent advances mainly focus on adopting the desired text style to guide the model to synthesize text images and the scene is always ignored. However, in natural scenes, the scene and text are a whole. There are two key challenges in scene text image translation: i) transfer text and scene into different styles, ii) keep the scene and text consistency. To address these problems, we propose a novel end-to-end scene text style transfer framework that simultaneously translates the text instance and scene background with different styles. We introduce an attention style encoder to extract the style codes for text instances and scene and we perform style transfer training on the cropped text area and scene separately to ensure the generated images are harmonious. We evaluate our method on the ICDAR2015 and MSRA-TD500 scene text datasets. The experimental results demonstrate that the synthetic images generated by our model can benefit the scene text detection task.

Keywords: Style transfer · Scene text style transfer · GANs · Scene text detection

1 Introduction

Text style transfer is regarded as the task of generating the specified text style image based on the guidance image. It has been adapted to text detection and visual creation like advertisement. There are research made on this direction and achieved fascinating results [11–14]. For instance, Yang [13] raises the topic of text effects transfer that turns plain texts into fantastic artworks. They further [11] propose a controllable artistic text style transfer method to create artistic typography by migrating the style from a source image to the target text to create artistic typography. Gomez et al. [12] explore the possibilities of image style transfer applied to text maintaining the original transcriptions.

As shown in Fig. 1, these text style transfer methods focus on the textual area style transfer and ignore the relationship between textual area and background scene. However, scene text images contain not only text instances but also scene background. Scene background is also critical for the text detection task.

It is quite challenging to translate text instance and background separately with different styles. There exist three main problems: (i) how to ensure that the stylized images are harmonious while transferring text instances and scene background with

© Springer Nature Switzerland AG 2020
X. Bai et al. (Eds.): DAS 2020, LNCS 12116, pp. 559–571, 2020.
https://doi.org/10.1007/978-3-030-57058-3_39

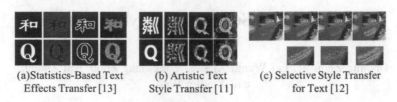

(a)Statistics-Based Text
Effects Transfer [13]

(b) Artistic Text
Style Transfer [11]

(c) Selective Style Transfer
for Text [12]

Fig. 1. The works focus on textual area style transfer.

different styles, (ii) the difference between text instance and scene background makes the transfer task difficult to model uniformly.

In this paper, we propose a novel scene text transfer method to address these challenges. Our key idea is to transfer the text instance and scene background to different styles using an end-to-end framework. In order to guarantee the stylized image are harmonious, inspired by INIT [2], during the training process, we crop the textual area from the entire image, the text and scene are input to our scene text style transfer model to get the stylized image, and we design loss functions for text and scene to optimize the model. For the differences between textual area and scene, we introduce the attention mechanism in the text scene style transfer model to improve the stylized image. The contribution of our work is three-fold: (i) we extend an existing style transfer framework to transfer the text instance and scene background to different styles in an end-to-end framework, (ii) we propose a novel attention encoder to extract the style code of text instance and scene, (iii) we use the proposed text style transfer method as a data augmentation tool to generate the synthetic images for scene text detection task and achieve better detection results compared with the original data.

2   Related Work

In recent years, translating images from one domain to another has received significant attention in the computer vision community, since many vision and graphics problems can be formulated as a domain translation problem like super-resolution, neural style transfer, colorization. The target of Image-to-Image translation is to learn the mapping between two different domains, with the rise of generative models, there are many researchers focus on this topic. Pix2Pix [31] was used to learn the mapping from input images to outputs images using U-Net in an adversarial way. Zhu et al. proposed Bicycle-GAN [40] which can model multimodal distributions and produce both diverse and realistic results.

The text style transfer aims at modifying the style of text image while preserving its content, which is closely related to image-to-image translation. The most of style transfer methods are example-based [19–23]. The image analogy method [19] which takes a pair of images as input tends to determine the relationship between them, and then applies it to stylize other images. As it tries to find dense correspondence, analogy-based approaches [20–23] generally require that the images should compose a pair depicting the same type of scene as much as possible.

Gatys [24] first proposed the idea that the style can be formulated as the Gram matrix of deep features, which can be extracted from the neural networks. Li [25, 26] considered that the neural patches can preserve the structures of image styles. In parallel, Johnson [1] trained a deep network StyleNet for style transfer [7, 27–29]. Besides, some works regarded the problem of style transfer as a translation task of converting an image to another image [31, 32], which exploited Generative Adversarial Network(GAN) to transfer image style, such as cartoongan [33]. GAN yields better results compared to Gram-based and patch-based methods, because GAN can learn the representation of style from the image data directly.

Yang [13] first raised the problem of artistic text style transfer. The authors regarded image patches as text style, which brought in a heavy computational burden because of the patch matching. Li [34] utilized a Markov Random Field over the patches to decide which patch is useful. Zhao [25] used two models for style transfer: the first was to generate masks and the second was to transfer style. However, the problem of the method was that it applied the style transfer to the whole image. Inspired by the progress of CNN, Azadi [34] built a net called MC-GAN for text style transfer, which was only applied in 26 capital letters. Recently, Yang [35] collected a large dataset of text effects to train the network to transfer text effects for any glyph, which showed promise for more application scenarios. Gatys [36] proposed a method to control perceptual factors such as color, luminance, and spatial location and scale, which had reference meaning to text style transfer.

3   Scene Text Style Transfer Model

In order to produce diverse scene text images, our goal is to realize the scene and text translation with different styles. Inspired by MUNIT [3] and INIT, we leveraging these methods to build our scene text transfer model.

In style transfer task, the encoder is usually defined as extracting the content code and the style code of the input image, the decoder was used for rebuilding the image according to the content code and the style code. Based on these concepts, there are two tasks for scene text style transfer: First, the style encoder can automatically extract the style codes and content codes corresponding to scene and text. Second, the encoders can combine the content codes and replaced text/scene style codes to generate the stylized image. The common difficulty in both tasks is how to disentangle style codes on textual area and scene, the content image consists of text and scene, and the encoder needs to extract the style codes of scene and text. Considering the difference between scene and text, it is difficult to express scene and text in one style code. To solve this problem, we build the training process of scene text style transfer model based on the INIT.

As shown in Fig. 2(a), in the training process of our scene text style transfer framework, four images are fed into the framework, which contains the content image and style image for text instance, and the content image and style image for scene background. For simplicity, we define I^{con} as content image and I^{sty} as style image for both scene and text. The attention style encoders E_{att}^{con} and content encoder E^{con} decomposes images I^{con}/I^{sty} into a shared content space and a specific style space to obtain the style codes S_1^{con}/S_1^{sty} and content codes C_1^{con}/C_1^{sty}. Then, we input style codes and content codes into the decoder D^{con}/D^{sty} in different combinations to obtain stylized images $I_1^{con\_sty}/I_1^{sty\_con}$ and reconstructed image $I_1^{con\_rescon}/I_1^{sty\_rescon}$ for scene and text respectively. In addition, we paste the stylized text image on stylized scene image and we generate the $I_1^{sty\_con\_paste}/I_1^{con\_sty\_paste}$ for adversarial loss to help transfer model generate more harmonious stylized image. The stylized image $I_1^{sty\_con}/I_1^{con\_sty}$ is input to encoder again to get the content code C_2^{con} and C_2^{sty}. After that, the decoder reconstructs text image and scene image based on the S_1^{con}/C_2^{con} and S_1^{sty}/C_2^{sty}.

During inference time, the attention style encoder E_{att} extracts the style codes of text and scene by attention block. The decoder D^{sty} generates the stylized image according to the style codes and content codes provided by the encoders. Figure 2(b) summarizes the process of translating scene text image.

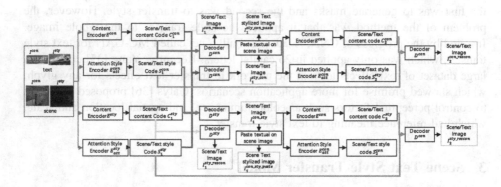

(a)   The training process of scene text style transfer. Text image and scene
       background image do not share the parameters for decoder and encoder.

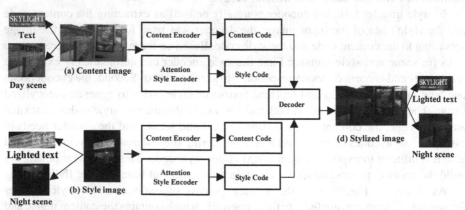

(b) Illustration of scene text style translation.

Fig. 2. Overview of our scene text style transfer framework. In the training process, we input the cropped style text image and style scene image for training, in inference process, we input the entire style image for generate the stylized image.

Below we will first explain the structure of style attention encoder, which is used to distinguish the style of textual area and background in one image. We show the details of the attention block. Then we introduce the loss function of our scene text style transfer.

3.1 Style Attention Encoder

In order to distinguish the style of textual area and background in one image, we proposed the attention encoder E_S^T which takes text instance images and scene images as inputs to generating style codes. As shown in Fig. 3, the style of the lighted textual area is different with night scene. In order to guarantee that the generated images are harmonious, we use one encoder to extracting the different style codes for text instance and night scene.

(a) The samples of lighted text images.

(b) The samples of night scene images.

Fig. 3. The samples of lighted text images and night scene images.

Besides that, we assume that these style code can be extracted from the style attention encoder. As shown in Fig. 4, the style attention encoder consists of style code extraction module and style selection module. Inspired by SENet [5], we forward the style codes through the global average pooling and two fully-connected layers around the non-linearity, to obtain the weights W_s of style codes, these weights are applied to the style codes to generate the output of the attention style encoder which can be fed into subsequent decoder.

$$E_{att}(I) = S_{ori} \cdot W_s \tag{1}$$

$$S_{ori} = E_{ori}(I) \tag{2}$$

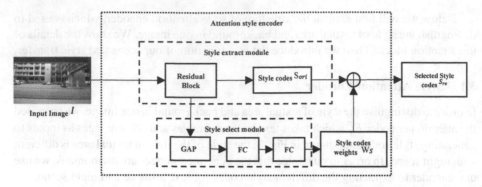

Fig. 4. The structure of our attention style encoder, the attention style encoder is an extension of encoder, the style extract module followed by a global average pooling layer and two fully connected layer, the style select module produce weights from the style features, then the style codes multiply the weights, and finally outputted the selected style codes by attention style encoder.

$$W_S^c = Sigmoid\left(F_2\left(ReLU\left(F_1\left(\frac{1}{H \times W} \sum_{i,j}^{w,h} S_{ori,i,j}^c \right) \right) \right) \right) \qquad (3)$$

where I refers to the input image, E_{ori} denotes the style encoder, S_{re} is the output of the style encoder. H/W is the width/height of channels of S_{ori}, c is the c-th channel of S_{ori}, The content/style image of textual areas and the background shared the attention style encoder levels.

3.2 Loss Function

As shown in Fig. 2(a), we decompose style image I^{sty} and content image I^{con} into specific style spaces and content spaces. The attention style encoder E_{att}^{con} generate style codes S_1^{con} from I^{con}, and E_{att}^{sty} generate S_1^{sty} from I^{sty}. The content encoder extracts the content codes, and the decoder reconstruct the style image and content image with S_1^{sty}/S_1^{con} and C_1^{sty}/C_1^{con}, the reconstruct loss is define as follow:

$$S_1^{con} = E_{att}^{con}(I^{con}), C_1^{con} = E^{con}(I^{con}), S_1^{sty} = E_{att}^{sty}(I^{sty}), C_1^{con} = E^{sty}(I^{sty}) \qquad (4)$$

$$L_{recon}^{con} = \left\| D^{con}\left(S_1^{con}, C_1^{con} \right) - I^{con} \right\|_1, L_{recon}^{sty} = \left\| D^{sty}\left(S_1^{sty}, C_1^{sty} \right) - I^{sty} \right\|_1 \qquad (5)$$

Where E^{con} is the content encoder for content images, and the E^{sty} is the content encoder for style images.

The decoder generated the stylized images $I_1^{sty-con}$ from S_1^{con}/C_1^{sty}, and $I_1^{con-sty}$ from S_1^{sty}/C_1^{con}. Then, the encoder taking the stylized images as inputs generate inputs for the decoder. And we use AdaIN [7] as the decoder to generate the stylized image $I_1^{sty-con}$ (with S_1^{sty} and C_1^{con}) and $I_1^{con-sty}$ (with S_1^{con} and C_1^{sty}), the process is shown in Fig. 5.

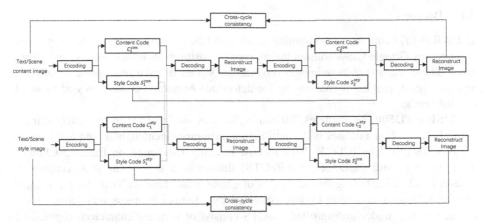

Fig. 5. The cross-cycle consistency of scene text style transfer.

The encoders decomposed the stylized image $I_1^{sty-con}$ and $I_1^{con-sty}$, so we get S_2^{sty}/C_2^{con} and S_2^{con}/C_2^{sty}. We swapping the S_1 and S_2 and the AdaIN reconstruct the image $I_2^{sty-con}$ with the S_1 and C_2 to perform the cycle reconstruction, With encoded S_2^{sty}/C_2^{con} and S_2^{con}/C_2^{sty}, the decoders decode these codes back to original input to perform the cross-cycle consistency [6, 8].

$$L_{cyc-recon}^{con} = \left\| E_{att}\left(D\left(S_1^{con}, C_2^{con}\right)\right) - I^{con}\right\|_1 \tag{6}$$

$$L_{cyc-recon}^{sty} = \left\| E_{att}\left(D^{sty}\left(S_1^{sty}, C_2^{sty}\right)\right) - I^{sty}\right\|_1 \tag{7}$$

To avoid repetition, we refer to textual areas and entire image as I^{con}/I^{sty}. In addition, we adopt GANs [9] to discriminate the $I_2^{sty-con}/I_2^{con-sty}$ and I^{con}/I^{sty} and $I_1^{sty\_con\_paste}/I_1^{con\_sty\_paste}$, which can yield higher quality images. Our full objective function is:

$$L_{text} = L_{cyc-recon}^{text} + L_{recon}^{text} + L_{cyc-consist}^{text} + L_{gan}^{text} \tag{8}$$

$$L_{global} = L_{cyc-recon}^{global} + L_{recon}^{global} + L_{cyc-consist}^{global} + L_{gan}^{global} \tag{9}$$

$$L_{all} = \lambda^{global} L_{global} + \lambda^{text} L_{text} \tag{10}$$

4 Experiments

In our experiment, we first give the qualitative results on two scene text datasets ICDAR15 and MSRA-TD500. Then we conduct our experiment on scene text detection task to verify the effectiveness of data augmentation using our text style transfer method.

4.1   Datasets

ICDAR15 [37] contains 1000 training images and 500 testing images. The images are collected by Google Glass without taking care of positioning, image quality, and viewpoint. Therefore, text in these images is of various scales, orientations, contrast, blurring, and viewpoint, making it challenging for detection. Annotations are provided as word quadrilaterals.

MSRA-TD500 [38] includes 300 training images and 200 test images collected from natural scenes. It is a dataset with multilingual, arbitrary-oriented, and long text lines.

ReCTS [39] The ICDAR 2019 Robust Reading Challenge on Reading Chinese Text on Signboard (ICDAR 2019-ReCTS) dataset is a new scene text recognition dataset that has 25,000 signboard images of Chinese and English Text. All the images are from Meituan-Dianping Group, collected by Meituan business merchants, using phone cameras under uncontrolled. ReCTS consist of Chinese characters, digits, and English characters, with Chinese characters taking the largest portion. The 25,000 images are split training set contain 20,000 images and test set contain 5,000 images.

4.2   Experiment Settings

To make it fair, our scene text style transfer method and MUNIT use the same training data. The training data is divided into two parts: style images and content images, of which there are 120 style images (20 images form ICDAR15, 100 images from ReCTS) and 1485 content images that came from ReCTS. We translate text instance and scene into lighted text instance and night scene. For our method, we crop the text region from the training image as the style and content image for text instance, and then resized the

(a) The samples of content image. (left column is daytime image, right column is cropped textual area image).

(b) The samples of style image(left column is night image, right column is cropped lighted textual area image).

Fig. 6. The training data of style transfer model.

images to 512 * 512. As for MUNIT, the training image is resized to 512 * 512 directly. The samples of training images are shown in Fig. 6.

During the training phase of scene text style transfer, we adopt the Adam optimizer whose $\beta 1 = 0.5$, $\beta 2 = 0.999$, and the initial learning rate is 0.0001 and decreased by half every 100,000 iterations. Besides that, in all experiments, the batch size is 1 and total epoch is 60.

For the generated scene text detection training data, we resize the short side of generated images to 512. All the generated images have the lighted text and night scene style.

4.3 Scene Text Style Transfer

The visualization of the stylized scene text image is shown in Fig. 7, the first column is the original image, the second column is the result of MUNIT, the third column is our result. Thanks to the attention style encoder and the training process of our scene text transfer framework, our scene text style transfer model successfully translates the content image to lighted text night scene stylized image. We observe that our model adept in translating textual style and the entire stylized image are harmonious, and MUNIT sometimes fails to translate the textual area. Besides, compared with MUNIT, the stylized images generated by our scene text style transfer model retain richer image details.

4.4 Data Augmentation

In this section, we perform experiments that demonstrate the effectiveness of the proposed scene text style transfer model as a data enhancement tool to improve the performance of scene text detectors. The reason for using style transfer as a data augmentation technique is that some datasets cannot provide the large number of images needed to effectively train large neural networks. In this case, style transfer can artificially increase the size of the dataset by combining different style and content. We use the MUNIT and our scene text style transfer method to generate lighted text and night scene images. In order to quantitatively verify the effectiveness of our method, we perform experiments using two text detectors, an improved PixelLink [17] and MaskRCNN [18], in which the training data are augmented by our method and MUNI. For ICDAR15, we translate 110 daytime scene images and add them to the dataset. For MSRA-TD500, we translate 46 daytime scene images.

Table 1 shows the results of the improved PixelLink and Table 2 shows the results of Mask-RCNN. Note that the H-mean of the PixelLink and MaskRCNN with MUNIT is even lower than the performance training with the original data. However, as seen in Table 1 and Table 2, adding the images generated by our method, both Pixellink and Mask-RCNN achieve the best results. Specifically, compared with the original data, the H-mean of the improved PixelLink improves from 86.26% to 86.46% in ICDAR15 and improves from 84.73% to 87.36% in MSRA-TD500, while the H-mean of MaskRCNN improves from 80.67% to 80.68% in ICDAR15 and improves from 71.32% to 73.24% in MSRA-TD500. For the improved Pixellink model, the H-mean improved from 86.26% to 86.46%, which is 0.2%, maybe different models have different sensitivity to

568 G. Zhou et al.

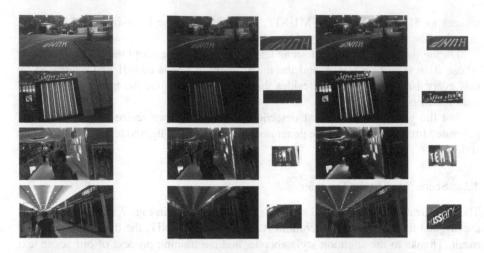

(a) Example results of ICDAR15 images translation. The original image(left). Outputs of MUNIT(mid), Output of ours(right).

(b) Example results of MSRA-TD500 images translation. The original image(left). Outputs of MUNIT(mid), Output of ours(right).

Fig. 7. Qualitative comparison on ICDAR15 and MSRA-TD500.

the data. The ICDAR15 trainset contains 1000 images, and it contains richer lighting conditions, while the MSRA-TD500 trainset contains 300 images and less variation in lighting conditions of the dataset. We think that is why the improvement of MSRA-TD500 is better than ICDAR15.

Table 1. The result of an improved Pixellink on ICDAR15 and MSRA-TD500.

Dataset	Augmentations	H-mean
ICDAR15	None	86.26%
ICDAR15	MUNIT	85.92%
ICDAR15	**Ours**	**86.46%**
MSRA-TD500	None	84.73%
MSRA-TD500	MUNIT	83.01%
MSRA-TD500	**Ours**	**87.36%**

Table 2. The result of mask-RCNN on ICDAR15 and MSRA-TD500.

Dataset	Augmentations	H-mean
ICDAR15	None	80.67%
ICDAR15	MUNIT	80.41%
ICDAR15	**Ours**	**80.68%**
MSRA-TD500	None	71.32%
MSRA-TD500	MUNIT	72.66%
MSRA-TD500	**Ours**	**73.24%**

5 Conclusion

In this work, we propose a scene text transfer model for generating scene text image with different styles. It is an end-to-end framework that can simultaneously transfer the text instance and scene background to different styles. An attention style encoder is introduced to extract the style codes for text instances and scene background. Both the cropped text and scene images are input to the style transfer framework to ensure the generated images are harmonious. Qualitative results demonstrate that our scene text transfer model can produce realistic scene text image in different styles. Furthermore, the scene text style transfer is evaluated as a data augmentation tool to expand scene text detection datasets, resulting in a boost of text detectors performance on scene text datasets ICDAR15 and MSRA-TD500.

References

1. Johnson, J., Alahi, A., Fei-Fei, L.: Perceptual losses for real-time style transfer and super-resolution. In: Leibe, B., Matas, J., Sebe, N., Welling, M. (eds.) ECCV 2016. LNCS, vol. 9906, pp. 694–711. Springer, Cham (2016). https://doi.org/10.1007/978-3-319-46475-6_43
2. Shen, Z., et al.: Towards instance-level image-to-image translation. arXiv preprint arXiv: 1905.01744 (2019)
3. Huang, X., Liu, M.Y., Belongie, S., Kautz, J.: In: Proceedings of the European Conference on Computer Vision (ECCV), pp. 172–189 (2016)

4. Cheung, B., et al.: Discovering hidden factors of variation in deep networks. arXiv preprint arXiv:1412.6583 (2014)

5. Hu, J., Shen, L., Sun, G.: Squeeze-and-excitation networks. In: Proceedings of the IEEE Conference on Computer Vision and Pattern Recognition, pp. 7132–7141 (2018)

6. Yi, Z., Zhang, H., Tan, P., Gong, M.: Dualgan: unsupervised dual learning for image-to-image translation. In: Proceedings of the IEEE International Conference on Computer Vision, pp. 2849–2857 (2017)

7. Huang, X., Belongie, S.J.: Arbitrary style transfer in real-time with adaptive instance normalization. In: Proceedings of the IEEE International Conference on Computer Vision, pp. 1501–1510 (2017)

8. Kim, T., Cha, M., Kim, H., Lee, J.K., Kim, J.: Learning to discover cross-domain relations with generative adversarial networks. In: Proceedings of the 34th International Conference on Machine Learning-Volume 70, pp. 1857–1865 (2017)

9. Goodfellow, I., et al.: Generative adversarial nets. In: Advances in Neural Information Processing Systems, pp. 2672–2680 (2014)

10. Karatzas, D., et al.: ICDAR 2015 competition on robust reading. In: 2015 13th International Conference on Document Analysis and Recognition, pp. 1156–1160 (2015)

11. Yang, S., et al.: Controllable artistic text style transfer via shape-matching GAN. arXiv preprint arXiv:1905.01354 (2019)

12. Gomez, R., Biten, A.F., Gomez, L., et al.: Selective style transfer for text. arXiv preprint arXiv:1906.01466 (2019)

13. Yang, S., Liu, J., Lian, Z., Guo, Z.: Awesome typography: statistics-based text effects transfer. In: Proceedings of the IEEE Conference on Computer Vision and Pattern Recognition, pp. 7464–7473 (2017)

14. Yang, S., Liu, J., Yang, W., et al.: Context-aware text-based binary image stylization and synthesis. IEEE Trans. Image Process. **28**(2), 952–964 (2018)

15. Ulyanov, D., Vedaldi, A., Lempitsky, V.: Improved texture networks: maximizing quality and diversity in feed-forward stylization and texture synthesis. In: Proceedings of the IEEE Conference on Computer Vision and Pattern Recognition, pp. 6924–6932 (2017)

16. He, K., Zhang, X., Ren, S., Sun, J.: Deep residual learning for image recognition. In: Proceedings of the IEEE Conference on Computer Vision and Pattern Recognition, pp. 770–778 (2016)

17. Deng, D., Liu, H., Li, X., Cai, D.: Pixellink: detecting scene text via instance segmentation. In: Thirty-Second AAAI Conference on Artificial Intelligence (2018)

18. He, K., Gkioxari, G., Dollár, P., Girshick, R.: Mask r-cnn. In: Proceedings of the IEEE International Conference on Computer Vision, pp. 2961–2969 (2017)

19. Hertzmann, A., Jacobs, C.E., Oliver, N., Curless, B., Salesin, D.H.: Image analogies. In: Proceedings of the 28th annual Conference on Computer Graphics and Interactive Techniques, pp. 327–340. ACM (2001)

20. Shih, Y., Paris, S., Barnes, C., Freeman, W.T., Durand, F.: Style transfer for headshot portraits. ACM Trans. Graphics (TOG) **33**(4), 148 (2014)

21. Shih, Y., Paris, S., Durand, F., Freeman, W.T.: Data-driven hallucination of different times of day from a single outdoor photo. ACM Trans. on Graphics (TOG) **32**(6), 200 (2013)

22. Frigo, O., Sabater, N., Delon, J., Hellier, P.: Split and match: example-based adaptive patch sampling for unsupervised style transfer. In: Proceedings of the IEEE Conference on Computer Vision and Pattern Recognition, pp. 553–561 (2016)

23. Liao, J., et al.: Visual attribute transfer through deep image analogy. arXiv preprint arXiv: 1705.01088 (2017)

24. Gatys, L.A., Ecker, A.S., Bethge, M.: Image style transfer using convolutional neural networks. In: Proceedings of the IEEE Conference on Computer Vision and Pattern Recognition, pp. 2414–2423 (2016)
25. Li, C., Wand, M.: Combining markov random fields and convolutional neural networks for image synthesis. In: Proceedings of the IEEE Conference on Computer Vision and Pattern Recognition, pp. 2479–2486 (2016)
26. Li, C., Wand, M.: Precomputed real-time texture synthesis with markovian generative adversarial networks. In: Leibe, B., Matas, J., Sebe, N., Welling, M. (eds.) ECCV 2016. LNCS, vol. 9907, pp. 702–716. Springer, Cham (2016). https://doi.org/10.1007/978-3-319-46487-9_43
27. Wang, X., Oxholm, G., Zhang, D., Wang, Y.F.: Multimodal transfer: a hierarchical deep convolutional neural network for fast artistic style transfer. In: Proceedings of the IEEE Conference on Computer Vision and Pattern Recognition, pp. 5239–5247 (2017)
28. Li, Y., Fang, C., Yang, J., Wang, Z., Lu, X., Yang, M.H.: Diversified texture synthesis with feed-forward networks. In: Proceedings of the IEEE Conference on Computer Vision and Pattern Recognition, pp. 3920–3928 (2017)
29. Li, Y., Fang, C., Yang, J., Wang, Z., Lu, X., Yang, M.H. Universal style transfer via feature transforms. In: Advances in Neural Information Processing Systems, pp. 386–396 (2017)
30. Chen, D., Yuan, L., Liao, J., Yu, N., Hua, G.: Stylebank: an explicit representation for neural image style transfer. In: Proceedings of the IEEE Conference on Computer Vision and Pattern Recognition, pp. 1897–1906 (2017)
31. Isola, P., Zhu, J.Y., Zhou, T., Efros, A.A.: Image-to-image translation with conditional adversarial networks. In: Proceedings of the IEEE Conference on Computer Vision and Pattern Recognition, pp. 5967–5976 (2017)
32. Zhu, J.Y., Park, T., Isola, P., Efros, A.A.: Unpaired image-to-image translation using cycle-consistent adversarial networks. In: Proceedings of the IEEE International Conference on Computer Vision, pp. 2242–2251 (2017)
33. Chen, Y., Lai, Y.K., Liu, Y.J.: Cartoongan: generative adversarial networks for photo cartoonization. In: Proceedings of the IEEE Conference on Computer Vision and Pattern Recognition, pp. 9465–9474 (2018)
34. Azadi, S., Fisher, M., Kim, V.G., Wang, Z., Shechtman, E., Darrell, T.: Multi-content gan for few-shot font style transfer. In: Proceedings of the IEEE Conference on Computer Vision and Pattern Recognition, pp. 7564–7573 (2018)
35. Yang, S., Liu, J., Wang, W., Guo, Z.: Tet-gan: text effects transfer via stylization and destylization. In: Proceedings of the AAAI Conference on Artificial Intelligence, pp. 1238–1245 (2019)
36. Gatys, L.A., Ecker, A.S., Bethge, M., Hertzmann, A., Shechtman, E.: Controlling perceptual factors in neural style transfer. In: Proceedings of the IEEE Conference on Computer Vision and Pattern Recognition, pp. 3985–3993 (2017)
37. Karatzas, D.: ICDAR 2015 competition on robust reading. In: ICDAR 2015 (2015)
38. Yao, C., Bai, X., Liu, W.Y., Ma, Y., Tu, Z.W.: Detecting texts of arbitrary orientations in natural images. In Proceedings IEEE Conference Computer Vision and Pattern Recognition (2012)
39. Liu, X.: Icdar 2019 robust reading challenge on reading chinese text on signboard (2019)
40. Zhu, J.Y., Zhang, R., Pathak, D., et al.: Toward multimodal image-to-image translation. In: Advances in Neural Information Processing Systems (2017)

Re-Ranking for Writer Identification and Writer Retrieval

Simon Jordan[1], Mathias Seuret[1], Pavel Král[2], Ladislav Lenc[2], Jiří Martínek[2],
Barbara Wiermann[3], Tobias Schwinger[3], Andreas Maier[1],
and Vincent Christlein[1(✉)]

[1] Pattern Recognitition Lab, FAU Erlangen-Nuremberg, Erlangen, Germany
{simon.simjor.jordan,mathias.seuret,andreas.maier,
vincent.christlein}@fau.de
[2] Department of Computer Science and Engineering, University of West Bohemia,
Plzeň, Czech Republic
{pkral,llenc,jimar}@kiv.zcu.cz
[3] Sächsische Landesbibliothek, Staats- und Universitätsbibliothek Dresden, Dresden,
Germany
barbara.wiermann@slub-dresden.de,ortus_ts@t-online.de

Abstract. Automatic writer identification is a common problem in document analysis. State-of-the-art methods typically focus on the feature extraction step with traditional or deep-learning-based techniques. In retrieval problems, re-ranking is a commonly used technique to improve the results. Re-ranking refines an initial ranking result by using the knowledge contained in the ranked result, e.g., by exploiting nearest neighbor relations. To the best of our knowledge, re-ranking has not been used for writer identification/retrieval. A possible reason might be that publicly available benchmark datasets contain only few samples per writer which makes a re-ranking less promising. We show that a re-ranking step based on k-reciprocal nearest neighbor relationships is advantageous for writer identification, even if only a few samples per writer are available. We use these reciprocal relationships in two ways: encode them into new vectors, as originally proposed, or integrate them in terms of query-expansion. We show that both techniques outperform the baseline results in terms of mAP on three writer identification datasets.

Keywords: Writer identification · Writer retrieval · Re-ranking

1 Introduction

In the past decades, vast amounts of historical documents have been digitized and made publicly available. Prominent providers, among others, are the British Library[1] or the 'Zentral- und Landesbibliothek Berlin'.[2] Searchable catalogues

---

[1] http://www.bl.uk/manuscripts/.
[2] https://digital.zlb.de.

© Springer Nature Switzerland AG 2020
X. Bai et al. (Eds.): DAS 2020, LNCS 12116, pp. 572–586, 2020.
https://doi.org/10.1007/978-3-030-57058-3_40

and archives can ease historical investigations, but considering the tremendous amount of available documents, manual examination by historical experts is no longer feasible. Hence automatic systems, which allow for investigations on a greater scale, are desired. Since the historical documents are provided as digital images, they are well suited for machine learning methods. In this work, we will look at a particular case, namely writer retrieval, which is particularly useful in the field of digital humanities, as the identities of writers of historical documents are frequently unknown.

This task is challenging to most robust classification systems, suitable for identification tasks, since they are only applicable in supervised scenarios. Therefore, those are not well suited for our particular case, where we would like to also identify *new* writers, who are unknown to the system during the training stage, a. k. a. zero-shot classification. Hence, we focus on robust writer retrieval, which does not directly identify the scribe of a document, but provides a ranked list consisting of the most similar, already known writers.

Our main contribution is the introduction of a re-ranking step, which aims to improve the overall result by refining the initial ranking in an unsupervised manner. Therefore, we propose two different re-ranking methods, which we adapt to our case. First, we employ a well-known re-ranking method based on the Jaccard distance and k-reciprocal neighbors. These neighbors are furthermore exploited in a novel re-ranking technique through combination with query expansion. We show that even retrieval tasks with low gallery size can benefit from re-ranking methods. Another contribution consists in the newly created CzByCHRON corpus which will be made freely available for research purposes.

The structure of this work is as follows. After discussing related work in the field of writer identification/retrieval and re-ranking in Sect. 2, we describe our writer-retrieval pipeline in Sect. 3. There, we also propose two different re-ranking methods, which can theoretically be applied to most information retrieval tasks, but which we adapted to better fit our particular problem. Finally, we present a detailed evaluation on these methods on three different diverse datasets (MUSICDOCS, CzByCHRON, ICDAR17) in Sect. 4.

2   Related Work

2.1   Writer Identification

Offline text-independent writer identification and retrieval methods can be grouped into codebook-based methods and codebook-free methods. In codebook-based methods, a codebook is computed that serves as background model. Popular codebooks are based on GMMs [7,8,14] or k-means [9,10]. Such a model is then used to compute statistics that form the global descriptor, e. g., first order statistics are computed in VLAD encoding [15].

Conversely, codebook-free methods compute a global image descriptor directly from the handwriting, such as the width of the ink trace [2] or the so called Hinge descriptor [3] which computes different angle combinations evaluated at the contour.

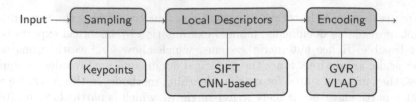

Fig. 1. Processing pipeline to obtain a global descriptor from an input sample.

Recent methods rely on deep learning, which can still fall in both groups depending on their usage. However, apart from some exceptions [22], the most methods use Convolutional Neural Networks (CNN) to compute strong local descriptors by using the activations of a layer as features [8,9,14,16] followed by an encoding method to compute a global descriptor. In this work, we rely on both, traditional SIFT [18] descriptors and VLAD-encoded CNN activation features provided by the work of Christlein et al. [9], while we evaluate different methods to improve the initial rankings.

2.2 Re-Ranking

Since we tackle a writer retrieval problem, research in the field of information retrieval is also of interest. Relevant work in this field was proposed by Arandjelović and Zisserman [1], and Chum et al. [12], both in general looking at specific Query Expansion (QE) related methods. For a detailed survey about automatic QE in information retrieval see Carpineto and Romano [5].

Arandjelović and Zisserman [1] propose discriminative QE, which aims to consider 'negative' data, provided by the bottom of the initial ranking. While positive data can be added to the model by averaging (AQE), incorporating negative data is a more complex process. Beside QE, special nearest neighbor (NN) relationships, provided by an initial retrieval result, can be used to improve performance. In the work of Qin et al. [21], unidirectional NN-graphs are used to treat different parts of the ranked retrieval list with different distance measures. Since in our case however, the top of the list is of special interest, the re-ranking method proposed by Zhong et al. [24] is better suited, which we were able to adapt to our scenario, and will describe in detail in Sect. 3.2.

3   Methodology

3.1   Pipeline

This paper focuses on the re-ranking of given embeddings. These embeddings were created using two different instances of the same pipeline, cf. Fig. 1. In case of the ICDAR17 dataset, we rely on the embeddings of Christlein et al. [6], where the embeddings are computed as follows. First, local descriptors are computed by means of a deep neural network and evaluated at patches located at SIFT keypoints [18]. The local descriptors are aggregated and encoded using GVR encoding [25] and normalized using power normalization [20], i. e., each element of the encoding vector is normalized by applying the square-root to it. Afterwards, a feature transformation based on Exemplar SVMs (ESVM-FE) [7] is applied. Therefore, the embeddings are transformed to a new space using each sample of the test set as single positive sample and all samples of the (disjoint) training set as negatives. The new features are the coefficients (weight vector) of the fitted linear SVM, normalized such that the ℓ^2 norm of this representation is 1.

For the newly created datasets CzByChron and MusicDocs, we are using in essence a similar pipeline, but with several modifications. Instead of binarizing the image, we are using the contour image, acquired with the Canny edge detector [4]. The hysteresis thresholds are determined by using a Gaussian mixture model with two mixtures [23]. Using this image as input, we extract 10 000 SIFT descriptors computed from SIFT keypoints. Afterwards, the SIFT descriptors are Dirichlet normalized [17] and a PCA-whitening is applied to counter overfrequent bins and correlation, respectively. The local descriptors are encoded using VLAD encoding in combination with power normalization and Generalized Max Pooling [19], which showed to be beneficial for writer identification [10]. Afterwards, the global vector representations are once more PCA-whitened and transformed via ESVM-FE.

3.2   Re-Ranking

Once an initial ranking is obtained by the pipeline described above, it is desirable to improve this result by using knowledge included in the ranking itself. This is possible with both Query Expansion and Re-ranking methods.

Query Expansion tries to ease the problem by expanding the original model with information provided by the top-ranked samples. Relevant work in this field was published by Arandjelović and Zisserman [1] and Chum et al. [11,12]. Re-ranking employs reciprocal rank order knowledge, and thus looks at whether or not two samples are ranked within the top-n results of each other.

In this section, we first look into a recent re-ranking method proposed by Zhong et al. [24] using Reciprocal Nearest Neighbours (rNN) and the Jaccard distance. Then, we present our own query expansion approaches, which extend the original ESVM and use rNN to overcome the lack of spatial verification when using our features.

Jaccard Distance Re-Ranking. The problem tackled by Zhong et al. [24] is, like ours, an image based retrieval task. Instead of scanned documents from different writers, Zhong et al. aim to re-identify a person (query) based on images stored in a database. For a given query image, the database is searched in order to return a list of the most similar images (persons). This initial ranking is then improved using the following two techniques.

k-reciprocal Nearest Neighbours. While it is possible to directly work with the initial ranking for computing the re-ranking, false matches are likely to appear and their presence has a negative effect on the re-ranking [24]. Thus Zhong et al. make use of the more constrained reciprocal nearest-neighbour (rNN) relationship instead. This relationship is defined as follows: two samples $\mathbf{q}$ and $\mathbf{t}$ are considered k-rNN, when both appear within the k highest ranked samples of each other. While the rNN relationship itself is symmetric, the resulting sets of k-rNNs are not. Given a query $\mathbf{q}$ with initial ranking result π_q, the set $\mathcal{Q}_k$ of its k-rNNs holds only those of the first k samples in π_q, which have sample $\mathbf{q}$ appear within their highest k ranked samples. Thus the number of samples in the set $\mathcal{Q}_k$ depends on the query $\mathbf{q}$ and the value of k. It is possible, especially for low values of k, that a sample has no k-rNNs at all.

Jaccard Distance. Two such sets, $\mathcal{Q}_k$ and $\mathcal{T}_k$, can then be used to define a similarity between the corresponding samples $\mathbf{q}$ and $\mathbf{t}$. For this purpose, we follow the approach from Zhong et al. [24] and use the Jaccard distance, which gets a lower value, the more samples are shared between both sets, and is maximal if no sample is shared at all. It is given as:

$$d_{\text{Jaccard}}\left(\mathbf{q}, \mathbf{t}\right) = 1 - \frac{|\mathcal{Q}_k \cap \mathcal{T}_k|}{|\mathcal{Q}_k \cup \mathcal{T}_k|}. \tag{1}$$

It is convenient to encode the set information into binary vectors $\boldsymbol{\eta}$, with $\eta_i = 1$ when sample $\mathbf{x}_i$ is part of the corresponding k-rNNs set and $\eta_i = 0$ otherwise. This way it is also possible to incorporate the distance $d\left(\mathbf{q}, \mathbf{x}_i\right)$ between the samples and the query, by changing the set vector accordingly,

$$\eta_i = \begin{cases} e^{-d(\mathbf{q}, \mathbf{x}_i)} & \text{if} \mathbf{x}_i \in \mathcal{Q}_k \\ 0 & \text{otherwise} \end{cases}. \tag{2}$$

Then, the Jaccard distance of the sets $\mathcal{Q}_k$ and $\mathcal{T}_k$ with set-vectors $\boldsymbol{\eta}^q$ and $\boldsymbol{\eta}^t$ can be calculated as

$$d_{\text{Jaccard}}\left(\mathbf{q}, \mathbf{t}\right) = 1 - \frac{\sum_{i=1}^{N} \min\left(\eta_i^q, \eta_i^t\right)}{\sum_{i=1}^{N} \max\left(\eta_i^q, \eta_i^t\right)}, \tag{3}$$

where N is the total number of vector elements, $\min\left(\eta_i^q, \eta_i^t\right)$ the minimum of the respective vector elements at position i and $\max\left(\eta_i^q, \eta_i^t\right)$ the maximum of those elements [24]. The minimum Jaccard distance, for two equal set-vectors $\boldsymbol{\eta}^q = \boldsymbol{\eta}^t$ is $d_{\text{Jaccard}} = 1 - 1 = 0$. With the maximum distance being 1, the range of the Jaccard distance is $[0, 1]$. While the initial distance between the samples is already incorporated in the vector entries, see Eq. (2), and thus does influence the Jaccard distance, Zhong et al. [24] re-rank the samples based on a weighted sum of the original distance and the Jaccard distance, resulting in the following final distance $d_{J\lambda}$,

$$d_{J\lambda}\left(\mathbf{q}, \mathbf{t}\right) = \left(1 - \lambda\right) d_{\text{Jaccard}}\left(\mathbf{q}, \mathbf{t}\right) + \lambda d\left(\mathbf{q}, \mathbf{t}\right), \tag{4}$$

with $d(\mathbf{q}, \mathbf{t})$ being the distance between $\mathbf{q}$ and $\mathbf{t}$ the initial ranking was based on. The weight term λ lies within $[0, 1]$. If λ is set to 1, only the original distance will be considered and the initial ranking will stay unchanged. With $\lambda = 0$, the original distance is ignored and the new ranking is solely based on the Jaccard distance.

Note that our implementation differs slightly from (3), since we add a constant of $\varepsilon = 10^{-8}$ to the denominator in order to enhance numerical stability in case the query $\mathbf{q}$ and test sample $\mathbf{t}$ share no rNN at all. By adding a constant small ε, we ensure to avoid a division by zero. We observe such cases as we use a lower value of k and work on other data than Zhong et al. [24].

Query Expansion. A more common approach to boost the performance in information retrieval systems is automatic query expansion (QE). With an initial ranking given, the original query gets expanded, using information provided by the top-n highest ranked samples. Since those samples can also include false matches, a spatial verification step is often applied to ensure that only true matches are used. Such verification needs to be obtained by employing a different measure than the one used for the initial ranking. For image based tasks, typically it is also important to consider the region of interest (ROI) in the retrieved images, where a single image can contain several query objects. In our case, the features do not encode any global spatial information, so we do not have a meaningful spatial verification procedure at hands. With the lack of spatial information, it is also not possible to make use of meaningful regions of interest.

Chum et al. [12] proposed the following QE approach. After the spatially verified top-n samples $\mathcal{F}$ from the initial ranking have been obtained, a new query sample can be formed using average query expansion (AQE),

$$\mathbf{q}_{\text{avg}} = \frac{1}{n+1} \left(\mathbf{q}_0 + \sum_{i=1}^{n} \mathbf{f}_i \right), \tag{5}$$

with $\mathbf{f}_i$ being the ith sample in $\mathcal{F}$ and n being the total number of samples $|\mathcal{F}|$. To obtain the improved ranking, the database gets re-queried using $\mathbf{q}_{\text{avg}}$ instead of $\mathbf{q}$.

Pair & Triple SVM. When we consider our feature vectors, obtained using ESVMs, as feature encoders, another way of query expansion is more straightforward. Instead of forming a new query feature vector by averaging, we can expand our model, the ESVM, by adding more samples to the positive set. When a set of suitable *friend* samples $\mathcal{F}$ is given, we can extend the original positive set from the ESVM, which only holds the query sample $\mathbf{q}$, and obtain a new SVM. This new SVM, like the original ESVM, provides a weight vector which allows us to use it as a feature encoder like before. Note that the number of positive samples does not affect the dimension of the weight vector. The cosine distance may again be applied as a similarity measure. Beside the information we provide for the SVM to correctly classify the positive set, nothing has changed. Thus,

adding more positive samples will affect the decision boundary and thus the
weight vector, but not how we define a similarity based on it.

While this approach may benefit from larger *friend* sets $\mathcal{F}$, we choose to focus
on pair and triple SVMs, by using the most suited *friend* sample, and two most
suited *friend* samples, respectively. This way, our QE-re-ranking methods are not
limited to scenarios with large gallery sizes n_G. In case of the ICDAR17 dataset
for example, only four true matches per query are present within the whole test
set, and thus using more than two additional positive samples increases the risk
of adding a false match to the positive set tremendously.

Also since we lack any form of spatial verification, we still risk to impair our
model when the selected *friend* samples contain a false match. To minimize this
risk, we use the more constrained k-rNNs of the query $\mathbf{q}$ as basis for our *friend*
set $\mathcal{F}_{rNN}$ which we use for rNN-SVM. This however introduces a new hyper-
parameter k, which needs to be optimized. Since the intention of the *friend* set
$\mathcal{F}$ is to hold only true matches, it follows naturally that optimal values for k
may lie within the range of the gallery size n_G.

In case of very small gallery sizes, small values for k may be chosen, which
then affects the size of $\mathcal{F}_{rNN}$. In some cases, $\mathcal{F}_{rNN}$ may only consist of a single
sample, and thus only a pair SVM is possible. If $\mathcal{F}_{rNN}$ holds more than two
samples, those with the smallest original distance to $\mathbf{q}$ are selected for the triple
SVM. In cases where $\mathcal{F}_{rNN}$ holds no samples at all, we will stick to the original
ESVM-FE method; which will result in the original ranking. That way, our
proposed QE approach is dynamically adapted to the information provided by
the initial ranking and we minimize the risk to end up with a worse result
compared to what was our initial guess.

4   Evaluation

4.1   Error Metrics

Mean Average Precision (mAP) is calculated as the mean over all examined
queries q of set $\mathcal{Q}$:

$$mAP = \frac{\sum_{q \in \mathcal{Q}} \text{AveP}(q)}{|\mathcal{Q}|}, \tag{6}$$

with $\text{AveP}(q)$ being the average precision of query q, given as:

$$\text{AveP}(q) = \frac{\sum_{k=1}^{n} (P(k) \times \text{rel}(k))}{\text{number of relevant documents}}, \tag{7}$$

with n being the total number of retrieved documents, $\text{rel}(k)$ being an indicator
function, which is 1 if the retrieved sample at rank k is relevant (from the same
writer as the query) and 0 otherwise, and $P(k)$ being the precision at position k.

Top-1. Top-1 is equal to precision at 1 ($P@1$) and thus tells us the percentage
of cases when we retrieved a relevant document as highest ranked ($k = 1$) document. Generally speaking, it represents the probability that the highest ranked

(a) ICDAR17 dataset, ID: 1379-IMG_MAX_309740. (b) CZBYCHRON dataset, ID: 30260559/0060 (c) MUSICDOCS datasets, Source: SLUB, digital.slub-dresden.de/id324079575/2 (Public Domain Mark 1.0)

Fig. 2. Examples images/excerpts of the used datasets.

sample is a true match (from the same writer as the query sample q). This makes the Top-1 score very informative for our Pair- and Triple-SVM methods, where we rely on a true match as highest ranked result to expand our model in the re-ranking step.

Hard-k & Soft-k. Hard-k represents the probability of retrieving only true matches within the first k retrieved samples. Thus, with only four true matches possible in total, the Hard-k probability for $k \geq 5$ is 0% in case of the ICDAR17 dataset. Unlike the Hard-k metric, Soft-k is increasing for increasing k, since it represents the probability of retrieving *at least one* relevant sample within the first k retrieved samples. We chose $k = 5$ and $k = 10$ for our Soft-k measurements and $k = 2, 3, 4$ for Hard-k.

4.2 Datasets

For the evaluation of different re-ranking techniques, we use three datasets: (1) ICDAR17, (2) CZBYCHRON, and (3) MUSICDOCS. Example images can be seen in Fig. 2.

Icdar17. The ICDAR17 dataset was proposed by Fiel et al. [13] for the "ICDAR 2017 competition on historical document writer identification". It provides two disjunct sets, one for training and one for testing. The former consists of 1182 document images, written by 394 writers (3 pages per writer). The test set consists of 3600 samples, written by 720 different writers, providing 5 pages each. This results in a gallery size $n_G = 5$ for the test set. For any test sample used as query, only four true matches remain within the test set. With the documents dating from 13th to 20th century, some of the pages had been damaged over the years, resulting in some variation of the dataset.

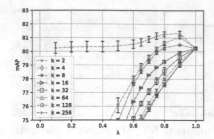 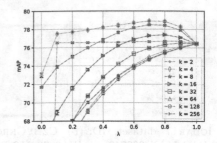

(a) mAP for various combinations of k and λ performed on the ICDAR17 training set (initial ranking achieved using cosine distances).

(b) mAP for various combinations of k and λ performed on the ICDAR17 test set (initial ranking achieved using ESVM-FE distances).

Fig. 3. Parameter influence (k, λ) on the mAP performance of the Jaccard Distance method for the ICDAR17 training and test set. For $\lambda = 1$, the distance used by this method is just the distance used for the initial ranking. Hence for $\lambda = 1$, the performance is equal to the initial ranking (for all values of k). Regarding the test set, best results are achieved for $k = 4$, closely followed by $k = 8$. For varying values of k, best results are achieved within the range $0.6 \leq \lambda \leq 0.8$. Note that the optimal parameter values (k, λ) differ for training and test set, which may be due to the fact that the gallery size n_G is slightly higher in the test set.

CzByChron. The CZBYCHRON dataset[3] consists of city chronicles, documents made between the 15th and 20th century and provided by Porta Fontium[4] portal. We used a set of 5753 images written by 143 scribes. Documents from 105 writers were used for testing while the remaining ones were used for training the codebook and PCA rotation matrices. The training set consists of 1496 samples while the test set contains 4257 images. The number of samples per writer vary and range between 2 and 50 samples (median: 44).

MusicDocs. The MUSICDOCS dataset consists of historical hand-written music sheets of the mid 18th century provided by the SLUB Dresden[5]. They commonly do not contain much text but staves with music notes. It contains 4381 samples, 3851 of which are used for testing coming from 35 individuals. The remainder of 530 samples are used for training and stems from 10 individuals. The median gallery size is $n_G^{median} = 40$, with a minimum gallery size of $n_G^{min} = 3$ and maximum $n_G^{min} = 660$.

---

[3] https://doi.org/10.5281/zenodo.3862591.

[4] http://www.portafontium.eu.

[5] https://www.slub-dresden.de.

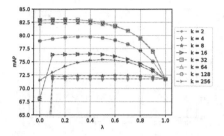 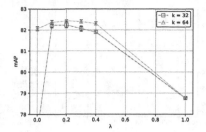

(a) mAP for various combinations of k and λ performed on the CzByChron training set (initial ranking achieved using cosine distances).

(b) mAP for selected combinations of k and λ performed on the CzByChron test set (initial ranking achieved using ESVM-FE distances).

Fig. 4. Parameter influence (k, λ) on the mAP performance of the Jaccard Distance method for the CzByChron training and test set. For $\lambda = 1$, the distance used by this method is just the distance used for the initial ranking. Hence for $\lambda = 1$, the performance is equal to the initial ranking (for all values of k).

4.3   Results

Hyper-parameter Estimation. Re-ranking based on k-reciprocal neighbors adds new hyper-parameters (k, λ) which need to be chosen deliberately. For this purpose, we use the training set and cosine distances. However, estimating optimal hyper-parameters using a training set builds on the assumptions that the test set is in fact represented by the data in the training set. In case of the ICDAR17 set however, this assumption is violated because the gallery size n_G differs between training and test set. In this case, the optimal values for (k, λ) differ between test and training set.

Note that re-ranking via the QE methods Pair- & Triple SVM, only introduces k as new hyper-parameter. However for robust estimation of k, we recommend to evaluate k with respect to λ, even if λ is unused.

In the following sections, we will discuss the results achieved for each dataset, individually.

Jaccard Distance Re-Ranking. Using the initial ranking, along with the initial distances, provided by the baseline model, we are able to receive a new ranking. The new final distances $d_{J\lambda}$, (4), depend on two parameters, k for the k-reciprocal nearest neighbours (krNN), and λ for the weighting between initial distance and Jaccard distance term.

ICDAR17. Because the test of the ICDAR17 dataset is not well represented by the data in the training set (the gallery size n_G differs among the sets), it is of no surprise that the optimal values for (k, λ) differ between test and training set. To give an impression of the potential best parameters on the test set, we evaluated the influence of both parameters k and λ on both datasets, individually. Results are presented in Fig. 3a and Fig. 3b respectively.

CzByChron. In case of the CzByChron dataset, the gallery size n_G is not fixed but varies among the writers within $6 \le n_G \le 50$. Since the test set is

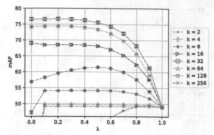 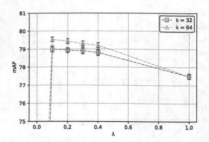

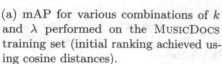

(a) mAP for various combinations of k and λ performed on the MusicDocs training set (initial ranking achieved using cosine distances).

(b) mAP for selected combinations of k and λ performed on the MusicDocs test set (initial ranking achieved using ESVM-FE distances).

Fig. 5. Parameter influence (k, λ) on the mAP performance of the Jaccard Distance re-ranking for the MusicDocs training and test set. For $\lambda = 1$, the distance used by this method is just the distance used for the initial ranking. Hence for $\lambda = 1$, the performance is equal to the initial ranking (for all values of k).

drawn randomly, we can assume that the distribution of n_G is roughly equal for both training and test set. Thus, we can assume that the optimal hyper-parameters (k, λ) for the training set will also work well on the test set. Since applying ESVM would require us to split off a 'negative' set from the training set and hence reduce the training set size, we will again rely on cosine distances for hyper-parameter optimization instead of ESVM-FE.

Various combinations of k and λ using cosine distances and the training set are presented in Fig. 4a. The optimal values for k are $k = 32$, closely followed by $k = 64$. Regarding λ, $0.1 \leq \lambda \leq 0.3$ works best. Hence we recommend using $k = 32$ and $\lambda = 0.2$ for the CzByChron test set. The results are presented in Fig. 4b, along with some other selected combinations of (k, λ). Both $k = 32$ and $k = 64$, the best performing values for k regarding the training set, were able to improve the performance on the test set by more than 3% mAP. A drawback is the loss in Top-1 performance, which drops by about 1%.

MusicDocs. We follow the aforementioned procedure using the training set and cosine distances to estimate the optimal hyper-parameter values (k, λ) for re-ranking. Similar to the optimal values for the CzByChron training set, $k = 32$ and $k = 64$ perform best, again for $\lambda = 0.2$. The results are presented in Fig. 5a and Fig. 5b showing a similar behavior than with the CzByChron dataset.

Pair & Triple-SVM. As described in detail in Sect. 3.2, a query-expansion based re-ranking step can be added in straightforward manner by replacing the original ESVM with Pair- & Triple-SVMs. The additional positive samples are chosen carefully by looking at the krNN, hence we use the optimal values for k already estimated for the Jaccard distance method.

Comparison of Re-Ranking Techniques. Table 1 shows a comparison between the different re-ranking techniques (more results containing Soft and Hard

Table 1. Pair- & Triple SVM methods denoted as 'krNN-P' and 'krNN-T', respectively. Baseline (ESVM-FE) also listed for comparison. For more complete results, see Table 1.

Method	Top-1	mAP
Cosine [6]	83.00	64.03
Baseline [6]	89.32	76.43
krNN-P$_{k=2}$	**89.43**	**78.20**
krNN-T$_{k=2}$	89.38	78.19
Jaccard$_{k=2,\lambda=0.9}$	89.19	76.70

(a)ICDAR17

Method	Top-1	mAP
Cosine	97.72	77.84
Baseline	97.95	78.80
krNN-P$_{k=32}$	98.03	80.06
krNN-T$_{k=32}$	**98.04**	80.10
Jaccard$_{k=32,\lambda=0.2}$	96.67	**82.21**

(b) CzByChron

Method	Top-1	mAP
Cosine	98.28	76.89
Baseline	98.48	77.49
krNN-P$_{k=32}$	**98.62**	78.64
krNN-T$_{k=32}$	98.60	78.67
Jaccard$_{k=32,\lambda=0.2}$	98.28	**78.95**

(c) MusicDocs

values can be found in Table 2, see appendix). All proposed re-ranking methods outperform the cosine distance-based ranking (cosine) and the ranking based on the ESVM feature transform (baseline) in terms of mAP significantly. Jaccard distance re-ranking is mAP-wise the best performing re-ranking technique. While the results show that the overall ranking gets improved, the Top-1 accuracy using the Jaccard distance re-ranking gets affected negatively. In case of Pair/Triplet-SVM this is not the case and can therefore be seen as a good compromise.

5 Conclusion

In this paper, we investigated the benefits of including a re-ranking step to a writer retrieval pipeline. We presented two ways of re-ranking, one based on the Jaccard distance and another one on an improved query expansion. We show that both methods achieve significantly higher mAP than the state-of-the-art baseline. While the Jaccard distance re-ranking returns the highest retrieval rates, the Top-1 accuracy drops. This makes a usage of Pair-SVM more suitable in case of identification.

Our results strongly hint that the gallery size has an impact on the optimal value for the number of reciprocal neighbors k. For larger galleries, larger k values are better, and the re-ranking is more efficient. However, even for smaller amounts of data, such as the ICDAR17 dataset, using one of the proposed re-ranking approaches improves the current state-of-the-art results.

There is however still room for improvement, and further investigations. We believe that the proposed method could benefit from the assignment of weights to the friend samples in the positive set in order to attribute them different importance based on their ranking. The two different techniques, Pair/Triple-SVMs and Jaccard distance re-ranking, could potentially also be combined to improve the results further.

Acknowledgement. This work has been partly supported by the Cross-border Cooperation Program Czech Republic – Free State of Bavaria ETS Objective 2014–2020 (project no. 211).

Appendix

Table 2. Full results. Pair- & Triple SVM methods denoted as 'krNN-P' and 'krNN-T' respectively. Baseline (ESVM-FE) also listed for comparison.

Method	Top-1	Hard-2	Hard-3	Hard-4	Soft-5	Soft-10	mAP	
Cosine	83.00	66.86	50.11	30.59	87.92	89.43	64.03	±0.29
Baseline	89.32	79.09	67.68	48.74	**92.96**	93.83	76.43	±0.06
krNN-P$_{k=2}$	89.43	**81.68**	71.83	54.49	92.06	93.31	78.20	±0.14
krNN-P$_{k=4}$	**89.55**	81.49	71.89	**57.78**	91.92	92.95	78.80	±0.25
krNN-T$_{k=2}$	89.38	81.63	71.79	54.44	92.06	93.30	78.19	±0.16
krNN-T$_{k=4}$	89.52	81.25	71.61	57.51	91.95	93.01	78.75	±0.26
Jaccard$_{k=2,\lambda=0.9}$	89.19	80.33	68.25	49.03	92.95	93.83	76.70	±0.01
Jaccard$_{k=4,\lambda=0.7}$	88.47	81.48	**73.35**	57.49	92.71	**93.84**	**78.91**	±0.10

(a) ICDAR17

Method	Top-1	Hard-2	Hard-3	Hard-4	Soft-5	Soft-10	mAP	
Cosine	97.72	95.46	93.53	91.77	98.64	98.90	77.84	±0.04
Baseline	97.95	95.69	94.00	92.07	**98.80**	**98.99**	78.80	±0.14
krNN-P$_{k=32}$	98.03	**96.47**	**94.73**	**93.20**	98.61	98.80	80.06	±0.20
krNN-T$_{k=32}$	**98.04**	96.44	94.61	93.11	98.60	98.80	80.10	±0.24
Jaccard$_{k=32,\lambda=0.2}$	96.67	94.88	93.37	91.93	97.83	98.30	**82.21**	±0.06

(b) CZBYCHRON

Method	Top-1	Hard-2	Hard-3	Hard-4	Soft-5	Soft-10	mAP	
Cosine	98.28	97.43	96.82	95.94	99.12	99.21	76.89	±0.30
Baseline	98.48	97.69	96.92	96.10	**99.15**	99.31	77.49	±0.15
krNN-P$_{k=32}$	**98.62**	**98.04**	97.67	96.86	99.13	**99.32**	78.64	±0.11
krNN-T$_{k=32}$	98.60	98.03	**97.70**	**96.89**	99.13	99.31	78.67	±0.14
Jaccard$_{k=32,\lambda=0.2}$	98.28	97.45	97.01	96.39	99.00	99.17	**78.95**	±0.03

(c) MUSICDOCS

References

1. Arandjelović, R., Zisserman, A.: Three things everyone should know to improve object retrieval. In: 2012 IEEE Conference on Computer Vision and Pattern Recognition, pp. 2911–2918. IEEE (2012)
2. Brink, A., Smit, J., Bulacu, M., Schomaker, L.: Writer identification using directional ink-trace width measurements. Pattern Recogn. **45**(1), 162–171 (2012)
3. Bulacu, M., Schomaker, L.: Automatic handwriting identification on medieval documents. In: 14th International Conference on Image Analysis and Processing (ICIAP 2007), Modena, pp. 279–284, September 2007
4. Canny, J.: A computational approach to edge detection. IEEE Trans. Pattern Anal. Mach. Intell. **8**(6), 679–698 (1986)
5. Carpineto, C., Romano, G.: A survey of automatic query expansion in information retrieval. ACM Comput. Surv. (CSUR) **44**(1), 1 (2012)
6. Christlein, V.: Handwriting analysis with focus on writer identification and writer retrieval. doctoral thesis, Friedrich-Alexander-Universität Erlangen-Nürnberg (FAU) (2019)
7. Christlein, V., Bernecker, D., Hönig, F., Maier, A., Angelopoulou, E.: Writer identification using GMM supervectors and exemplar-SVMs. Pattern Recogn. **63**, 258–267 (2017)
8. Christlein, V., Bernecker, D., Maier, A., Angelopoulou, E.: Offline writer identification using convolutional neural network activation features. In: Gall, J., Gehler, P., Leibe, B. (eds.) GCPR 2015. LNCS, vol. 9358, pp. 540–552. Springer, Cham (2015). https://doi.org/10.1007/978-3-319-24947-6_45
9. Christlein, V., Gropp, M., Fiel, S., Maier, A.: Unsupervised feature learning for writer identification and writer retrieval. In: 2017 14th IAPR International Conference on Document Analysis and Recognition (ICDAR), vol. 1, pp. 991–997. IEEE (2017)
10. Christlein, V., Maier, A.: Encoding CNN activations for writer recognition. In: 13th IAPR International Workshop on Document Analysis Systems, Vienna, pp. 169–174, April 2018
11. Chum, O., Mikulik, A., Perdoch, M., Matas, J.: Total recall ii: query expansion revisited. In: CVPR 2011, pp. 889–896. IEEE (2011)
12. Chum, O., Philbin, J., Sivic, J., Isard, M., Zisserman, A.: Total recall: automatic query expansion with a generative feature model for object retrieval. In: 2007 IEEE 11th International Conference on Computer Vision, pp. 1–8. IEEE (2007)
13. Fiel, S., et al.: Icdar 2017 competition on historical document writer identification (historical-wi). In: 2017 14th IAPR International Conference on Document Analysis and Recognition (ICDAR), vol. 01, pp. 1377–1382, November 2018
14. Fiel, S., Sablatnig, R.: Writer identification and writer retrieval using the fisher vector on visual vocabularies. In: 2013 12th International Conference on Document Analysis and Recognition (ICDAR), Washington DC, pp. 545–549, August 2013
15. Jégou, H., Perronnin, F., Douze, M., Sánchez, J., Pérez, P., Schmid, C.: Aggregating local image descriptors into compact codes. IEEE Trans. Pattern Anal. Mach. Intell. **34**(9), 1704–1716 (2012)
16. Keglevic, M., Fiel, S., Sablatnig, R.: Learning features for writer retrieval and identification using triplet CNNs. In: 2018 16th International Conference on Frontiers in Handwriting Recognition, Niagara Falls, pp. 211–216, August 2018
17. Kobayashi, T.: Dirichlet-based histogram feature transform for image classification. In: 2014 IEEE Conference on Computer Vision and Pattern Recognition (CVPR), Columbus, pp. 3278–3285, June 2014

18. Lowe, D.G.: Distinctive image features from scale-invariant keypoints. Int. J. Comput. Vis. **60**(2), 91–110 (2004). https://doi.org/10.1023/B:VISI.0000029664.99615.94
19. Murray, N., Jegou, H., Perronnin, F., Zisserman, A.: Interferences in match kernels. IEEE Trans. Pattern Anal. Mach. Intell. **39**(9), 1797–1810 (2016)
20. Perronnin, F., Sánchez, J., Mensink, T.: Improving the fisher kernel for large-scale image classification. In: Daniilidis, K., Maragos, P., Paragios, N. (eds.) ECCV 2010. LNCS, vol. 6314, pp. 143–156. Springer, Heidelberg (2010). https://doi.org/10.1007/978-3-642-15561-1_11
21. Qin, D., Gammeter, S., Bossard, L., Quack, T., Van Gool, L.: Hello neighbor: accurate object retrieval with k-reciprocal nearest neighbors. In: CVPR 2011, pp. 777–784. IEEE (2011)
22. Tang, Y., Wu, X.: Text-independent writer identification via CNN features and joint Bayesian. In: 2016 15th International Conference on Frontiers in Handwriting Recognition (ICFHR), Shenzhen, pp. 566–571, October 2016
23. Wahlberg, F., Mårtensson, L., Brun, A.: Large scale style based dating of medieval manuscripts. In: 3rd International Workshop on Historical Document Imaging and Processing (HIP 2015), Nancy, pp. 107–114. ACM, August 2015
24. Zhong, Z., Zheng, L., Cao, D., Li, S.: Re-ranking person re-identification with k-reciprocal encoding. In: Proceedings of the IEEE Conference on Computer Vision and Pattern Recognition, pp. 1318–1327 (2017)
25. Zhou, X., Zhuang, X., Tang, H., Hasegawa-Johnson, M., Huang, T.S.: Novel gaussianized vector representation for improved natural scene categorization. Pattern Recogn. Lett. **31**(8), 702–708 (2010)

Author Index

Printed in the United States
By Bookmasters